Signal Transduction in the Cardiovascular System in Health and Disease

For other titles published in this series, go to
http://www.springer.com/7064

Ashok K. Srivastava • Madhu B. Anand-Srivastava
Editors

Signal Transduction in the Cardiovascular System in Health and Disease

Springer

Editors:
Ashok K. Srivastava
Department of Medicine
University of Montreal and Research Centre of the
Centre Hospitalier de l' Universite de Montreal
Montreal, PQ, H1W 4A4
Canada

Madhu B. Anand-Srivastava
Department of Physiology
Faculty of Medicine
University of Montreal
Montreal, PQ, H3C3J7
Canada

ISBN 978-1-4899-9856-9   ISBN 978-0-387-09552-3 (eBook)

Cover illustration:

Printed on acid-free paper

9 8 7 6 5 4 3 2 1

springer.com

Preface

In recent years, there has been a surge of interest in studies related to the role of a variety of signaling pathways in the control of cardiovascular physiology. Evidence has also accumulated to suggest that an aberration in the signal transduction pathways contributes to the pathophysiology of cardiovascular disease. Several components of the signaling pathways have been identified as potential targets for the development of new therapies of cardiovascular disease. Therefore, this volume has been compiled to highlight the contributions of different signaling systems in modulating normal cardiovascular functions and how a perturbation in these signaling events leads to abnormal cell functions and cardiovascular disorders.

This volume has been divided into five sections dealing with five key signaling pathways regulating different aspects of cardiovascular physiology. The first section describes the role of G-protein-coupled receptor (GPCR) signaling in cardiovascular functions. In this section, Anand-Srivastava has elegantly summarized studies showing that the expression levels of various G-proteins as well as responsiveness of adenylyl cyclase systems to various stimuli such as β-adrenergic receptor (βAR) agonist and vasoactive peptides are defective in various models of hypertension, congestive heart failure (CHF), cardiac hypertrophy, and other diseases. Dent et al. have highlighted studies showing how the alterations in different components of the βAR signaling system contribute to CHF and suggest that βAR blockade could be used as a strategy to treat CHF. Continuing on the same theme, Vacek et al. have reviewed the pathophysiological mechanisms involved in CHF and sudden cardiac death, with an emphasis on the role of homocysteine-induced cross talk between NMDA receptor and GPCRs, while Moolman et al. have elaborated on the contributions of adenosine, cAMP/PKA system as well as p^{38mapk} in eliciting a cardioprotective response during early preconditioning. This section also has two elegant articles on the role of angiotensin II in cardiovascular pathophysiology: Engberding and Grindling and Schaffer and Mozaffari have provided in-depth accounts of various signaling pathways induced by angiotensin II and how the dysregulation of this pathway contributes to heightened growth, proliferation, hypertrophy, and cell survival death responses associated with various cardiovascular abnormalities.

The second section focuses on the role of redox-induced signaling system in cardiovascular biology and complications of diabetes. In this section, Das and Goswami provide experimental evidence supporting a role of redox-regulating proteins in mitigating ischemia-induced oxidative stress and conferring a cardioprotective response, while Turan has demonstrated how an interplay between βAR signaling and redox pathways can modify mechanical performance and energy homeostasis in heart. Additional articles by Anand-Srivastava and Srivastava and by Wu have examined the role of hyperglycemia and methylglyoxal-related advanced glycation end products-induced activation of MAP kinase, PKB, GPCR, G-proteins, adenylyl cyclases, and inflammatory genes in the cardiovascular complications associated with diabetes.

The third section contains articles focused on the regulatory role of growth factor and their receptors in cardiac hypertrophy, vascular remodeling, and therapeutic angiogenesis. In this section, Bouallegue and Srivastava have reviewed the concept of growth factor receptor transactivation as a triggering mechanism to transduce the downstream effects of vasoactive peptides, whereas Calderone has provided a comprehensive analysis of the contributions of peptide growth factors, G_q proteins, and phosphatidylinositol 3-kinase (PI3K)-dependent signaling events in physiological/pathophysiological cardiac hypertrophy. Further, Dixon et al. have elegantly reviewed the role of TGF-β and R-Smad signaling pathways in remodeling of the extracellular matrix in failing hearts, and Luo et al. have demonstrated that activation of MAPK and PI3K may contribute to the pathogenetic mechanism involved in coxsackievirus-induced myocarditis. Two articles in this section by Maulik and Rajalakshmi et al. have provided evidence supporting the use of growth factors such as basic fibroblast growth factors and vascular endothelial growth factors to enhance angiogenesis and vasculogenesis. Finally, Selvakumar and Sharma summarize studies on the characterization and biological significance of *N*-myristoyltransferase (NMT) and its binding proteins which are involved in myristoylation of several signaling proteins, including protein kinases, thereby altering their functions.

The role of calcium in regulating cardiovascular physiology is presented in the fourth section of this volume. House et al. have provided an excellent review on the structure of calmodulin-dependent protein kinase II and its role in the contractility as well as proliferation and migration of VSMC. Banderali et al. have examined in detail the cellular regulation and pharmacological properties of calcium-activated potassium channels and their role in control of vascular tone by endothelium.

The final section of this volume contains articles by Karmazyn et al. and An et al. who have examined in detail the roles of leptin and lipid-induced signaling pathways in the pathogenesis of cardiometabolic syndrome.

Overall, this volume provides a detailed analysis of a wide range of signal transduction systems that mediate hypertrophy, intimal hyperplasia, oxidative damage, contractility, cardiovascular protection, and remodeling. Many components of these signaling pathways are potential targets to develop new therapeutics to treat cardiovascular disorders.

We wish to take this opportunity to thank the contributing authors for their co-operation and sustained interest and the staff at Springer in putting this volume together.

Montreal, 2008

Madhu B. Anand-Srivastava, Ph.D.
Ashok K. Srivastava, Ph.D.

Contents

Part IV Calcium Signaling

Part V [illegible] Signaling

[illegible]

Contributors

Ahmed Almaqrhi
Institute of Cardiovascular Sciences, St. Boniface General Hospital Research Centre, Winnipeg, MB, Canada R2H 2A6

Ding An
Division of Pharmacology and Toxicology, Faculty of Pharmaceutical Sciences, University of British Columbia, Vancouver, BC, Canada V6T 1Z3

Umberto Banderali
Human Performance Laboratory, University of Calgary, Faculty of Kinesiology, Calgary, AB, Canada, T2N 1N4

Kristen M. Bedosky
Institute of Cardiovascular Sciences, St. Boniface General Hospital Research Centre, Winnipeg, MB, Canada R2H 2A6

Ali Bouallegue
Laboratory of Cell Signaling, Research Center, Centre Hospitalier de l'Université de Montréal (CHUM) – Angus Campus and Department of Medicine, Université de Montréal, Montreal, QC, Canada

Angelino Calderone
Département de Physiologie, Université de Montréal, Montréal, QC, Canada, and the Montreal Heart Institute, Montréal, QC, Canada

Ryan H. Cunnington
Institute of Cardiovascular Sciences, St. Boniface General Hospital Research Centre, Winnipeg, MB, Canada R2H 2A6

Dipak K. Das
Cardiovascular Research Center, University of Connecticut School of Medicine, Farmington, CT, USA

Melissa R. Dent
Department of Human Anatomy and Cell Science, Faculty of Medicine, University of Manitoba, Winnipeg, MB, Canada

Naranjan S. Dhalla
Department of Human Anatomy and Cell Science, Faculty of Medicine, University of Manitoba, Winnipeg, MB, Canada

Ian M. C. Dixon
Institute of Cardiovascular Sciences, St. Boniface General Hospital Research Centre, Winnipeg, MB, Canada R2H 2A6

Niels Engberding
Division of Cardiology, Emory University, Atlanta, GA 30322, USA

Line Garneau
Department of Physiology, Membrane Protein Research Group, Université de Montréal, Montréal, PQ, Canada, H3T 1J4

Sonia Genade
Department of Biomedical Sciences (Division Medical Physiology), Faculty of Health Sciences, Tygerberg, Stellenbosch University, Republic of South Africa

Roman G. Ginnan
Center for Cardiovascular Sciences, Albany Medical College, Albany, NY 12208, USA

Aghila Rani Koippallil Gopalakrishnan
Division of Cellular and Molecular Cardiology, Sree Chitra Tirunal Institute for Medical Sciences and Technology, Thiruvananthapuram 695011, India

Shyamal K. Goswami
School of Life Sciences, Jawaharlal Nehru University, New Delhi, India

Kathy K. Griendling
Division of Cardiology, Emory University, Atlanta, GA 30322, USA

Suzanne J. House
Center for Cardiovascular Sciences, Albany Medical College, Albany, NY 12208, USA

Morris Karmazyn
Department of Physiology and Pharmacology, Schulich School of Medicine and Dentistry, University of Western Ontario, London, ON, Canada N6A 5C1

Chandrasekharan Cheranellore Kartha
Division of Cellular and Molecular Cardiology, Sree Chitra Tirunal Institute for Medical Sciences and Technology, Thiruvananthapuram 695011, India

Min-Suk Kim
Division of Pharmacology and Toxicology, Faculty of Pharmaceutical Sciences, University of British Columbia, Vancouver, BC, Canada V6T 1Z3

Hélène Klein
Department of Physiology, Membrane Protein Research Group, Université de Montréal, Montréal, PQ, Canada, H3T 1J4

Amanda Lochner
Department of Biomedical Sciences (Division Medical Physiology), Faculty of Health Sciences, Tygerberg, Stellenbosch University, Republic of South Africa

Honglin Luo
Department of Pathology and Laboratory Medicine, The James Hogg iCAPTURE Centre for Cardiovascular and Pulmonary Research, University of British Columbia-St. Paul's Hospital, Vancouver, BC, Canada

Syanda Makaula
Department of Physiology, Faculty of Health Sciences, University of the Witwatersrand, Johannesburg, Republic of South Africa

Erna Marais
Department of Biomedical Sciences (Division Medical Physiology), Faculty of Health Sciences, Tygerberg, Stellenbosch University, Republic of South Africa

David Marchant
Department of Pathology and Laboratory Medicine, The James Hogg iCAPTURE Centre for Cardiovascular and Pulmonary Research, University of British Columbia-St. Paul's Hospital, Vancouver, BC, Canada

Nilanjana Maulik
Department of Surgery, Molecular Cardiology and Angiogenesis Laboratory, University of Connecticut Medical Center, Farmington, CT, USA

Bruce M. McManus,
Department of Pathology and Laboratory Medicine, The James Hogg iCAPTURE Centre for Cardiovascular and Pulmonary Research, University of British Columbia-St. Paul's Hospital, Vancouver, BC, Canada

Naira Metreveli
Department of Physiology and Biophysics, University of Louisville School of Medicine, Louisville, KT, USA

Johan Moolman
Department of Biomedical Sciences (Division Medical Physiology), Faculty of Health Sciences, Tygerberg, Stellenbosch University, Republic of South Africa

Karni S. Moshal
Department of Physiology and Biophysics, University of Louisville School of Medicine, Louisville, KT, USA

Mahmood S. Mozaffari
Department of Oral Biology/Maxillofacial Pathology, Medical College of Georgia, Augusta, GA, USA

Daniel M. Purdham
Department of Physiology and Pharmacology, Schulich School of Medicine and Dentistry, University of Western Ontario, London, ON, Canada N6A 5C1

Dhanya Chandrasekharan Rajalakshmi
Division of Cellular and Molecular Cardiology, Sree Chitra Tirunal Institute for Medical Sciences and Technology, Thiruvananthapuram 695011, India

Venkatesh Rajapurohitam
Department of Physiology and Pharmacology, Schulich School of Medicine and Dentistry, University of Western Ontario, London, ON, Canada N6A 5C1

Sunil G. Rattan
Institute of Cardiovascular Sciences, St. Boniface General Hospital Research Centre, Winnipeg, MB, Canada R2H 2A6

Dee Van Riper
Center for Cardiovascular Sciences, Albany Medical College, Albany, NY 12208, USA

Brian Rodrigues
Division of Pharmacology and Toxicology, Faculty of Pharmaceutical Sciences, University of British Columbia, Vancouver, BC, Canada V6T 1Z3

Dorothea Rosenberger
Department of Physiology and Biophysics, University of Louisville School of Medicine, Louisville, KT, USA

Rémy Sauvé
Department of Physiology, Membrane Protein Research Group, Université de Montréal, Montréal, PQ, Canada, H3T 1J4

Stephen W. Schaffer
Department of Pharmacology, College of Medicine, University of South Alabama, Mobile, AL, USA

Ponniah Selvakumar
Department of Pathology and Laboratory Medicine, College of Medicine and Health Research Division, Saskatchewan Cancer Agency, University of Saskatchewan, Saskatoon, SK, Canada S7N 4H4

Utpal Sen
Department of Physiology and Biophysics, University of Louisville School of Medicine, Louisville, KT, USA

Rajat Sethi
Department of Pharmaceutical Sciences, Irma Lerma Rangel College of Pharmacy, Texas A&M University, Kingsville, TX, USA

Rajendra K. Sharma
Department of Pathology and Laboratory Medicine, College of Medicine and Health Research Division, Saskatchewan Cancer Agency, University of Saskatchewan, Saskatoon, SK, Canada S7N 4H4

Xiaoning Si
Department of Pathology and Laboratory Medicine, The James Hogg iCAPTURE Centre for Cardiovascular and Pulmonary Research, University of British Columbia-St. Paul's Hospital, Vancouver, BC, Canada.

Manuel Simoes
Laboratoire MAEM-Faculté des Sciences et Techniques-Entrée 4A-Bld des Aiguillettes-BP239-Vandoeuvreles-Nancy, cedex, France

Tushi Singal
Department of Human Anatomy and Cell Science, Faculty of Medicine, University of Manitoba, Winnipeg, MB, Canada

Harold A. Singer
Center for Cardiovascular Sciences, Albany Medical College, Albany, NY 12208, USA

Madhu B. Anand-Srivastava
Department of Physiology, Research Centre, Centre hospitalier de l'Université de Montréal (CHUM) – Angus Campus, Faculty of Medicine, University of Montreal, Montreal, QC, Canada

Ashok K. Srivastava
Department of Medicine, Research Centre, Centre hospitalier de l'Université de Montréal (CHUM) – Angus Campus, Faculty of Medicine, University of Montreal, Montreal, QC, Canada

Paramjit S. Tappia
Department of Human Anatomy and Cell Science, Faculty of Medicine, University of Manitoba, Winnipeg, MB, Canada

Suresh C. Tyagi
Department of Physiology and Biophysics, University of Louisville School of Medicine, Louisville, KT, USA

Neetu Tyagi
Department of Physiology and Biophysics, University of Louisville School of Medicine, Louisville, KT, USA

Belma Turan
Department of Biophysics, School of Medicine, Ankara University, Ankara, Turkey

Thomas P. Vacek
Department of Physiology and Biophysics, University of Louisville School of Medicine, Louisville, KT, USA

Baiqiu Wang
Brigham and Women's Hospital, Harvard University, Boston, MA, USA

Lingyun Wu
Department of Pharmacology, College of Medicine, University of Saskatchewan, Saskatoon, SK, Canada S7N 5E5

Decheng Yang
Department of Pathology and Laboratory Medicine, The James Hogg iCAPTURE Centre for Cardiovascular and Pulmonary Research, University of British Columbia-St. Paul's Hospital, Vancouver, BC, Canada

Melissa M. Zachar
Center for Cardiovascular Sciences, Albany Medical College, Albany, NY 12208, USA

Asad Zeidan
Department of Physiology and Pharmacology, Schulich School of Medicine and Dentistry, University of Western Ontario, London, ON, Canada N6A 5C1

Part I
G-protein-Coupled Receptor Signaling

Chapter 1
Implication of G-proteins in Cardiovascular Disease

Madhu B. Anand-Srivastava

Abstract Guanine nucleotide regulatory proteins (G-proteins) play a key role in the regulation of various signal transduction systems including adenylyl cyclase/cAMP and phospholipase C (PLC)/phosphatidylinositol turnover (PI) which are implicated in the modulation of a variety of physiological functions such as platelet functions, including platelet aggregation, secretion, and clot formation, and cardiovascular functions, including arterial tone and reactivity. Several abnormalities in adenylyl cyclase activity, cAMP levels, G-proteins, and PLC/PKC have been shown to be responsible for the altered cardiac performance and vascular functions observed in cardiovascular disease states. The enhanced or unaltered levels of inhibitory G-proteins (Giα-2 and Giα-3) and mRNA have been reported in different models of hypertension, whereas Gsα levels were shown to be unaltered. These changes in G-protein were associated with functions. The enhanced levels of Giα proteins precede the development of blood pressure and suggest that overexpression of Gi proteins may be one of the contributing factors for the pathogenesis of hypertension. The augmented levels of Giα proteins and associated adenylyl cyclase signaling in hypertension were shown to be attributed to the enhanced levels of vasoactive peptides. In addition, enhanced oxidative stress in hypertension may also be responsible for the enhanced expression of Giα proteins observed in hypertension. The levels of Gqα/11 and PLCβ have been shown to be upregulated in different models of hypertension. On the other hand, the levels of Gsα and not of Giα proteins were decreased in volume- or pressure-overload hypertrophy. The responsiveness of adenylyl cyclase to β-adrenergic agonists was also attenuated. In addition, the levels of Gqα were augmented in hypertrophy and the β-adrenergic receptor levels were decreased. Furthermore, the role of PKC in the development and progression of cardiac hypertrophy was also shown. Similarly, ischemia was shown to be associated with decreased, increased, or unaltered levels of Gsα, with decreased levels of Giα, and with decreased responsiveness of adenylyl cyclase to various stimuli such as β-adrenergic agonists, guanine nucleotides, forskolin, and others. Thus, the altered levels of G-proteins and associated signaling may be responsible for the impaired cardiovascular functions observed in hypertension, hypertrophy, and cardiac failure.

A. K. Srivastava, M. B. Anand-Srivastava (eds.) *Advances in Biochemistry in Health and Disease.* Signal Transduction in the Cardiovascular System in Health and Disease,

Introduction

Guanine nucleotide regulatory proteins (G-proteins) are a family of guanosine triphosphate (GTP) binding proteins that play a key regulatory role as transducers in a variety of signal transduction systems. These include the adenylyl cyclase/cAMP system (Rodbell et al. 1971), the receptor-mediated activation of phospholipase C and A2 (Cockroft and Gompert 1985; Litosch et al. 1985), and a number of hormone and neurotransmitter-regulated ionic channels (Breiwieser and Szabo 1985; Pfaffinger et al. 1985). G-proteins are heterotrimeric proteins composed of three distinct subunits; α, β, and γ (Gilman 1984). The α-subunits bind and hydrolyze GTP and confer specificity in receptor and effector interactions (Gilman 1984). The GDP-bound form of α binds tightly to βγ and is inactive, whereas the GTP-bound form of α dissociates from βγ and serves as a regulator of effector proteins. All α-subunits possess intrinsic GTPase activity and hydrolyze the terminal phosphate of bound GTP to yield bound GDP and free inorganic phosphate (P_i). Upon hormone binding and receptor activation, the receptor interacts with the heterotrimeric protein to promote a conformational change and dissociation of bound GDP from the guanine nucleotide binding site. GDP is released and replaced by GTP. Binding of GTP to α induces a conformational change and promotes the dissociation of hormone receptor complex (HR) and the holo G-protein into α and βγ. Both α-GDP and βγ-subunits can interact with effectors. This activation cycle is terminated by intrinsic GTPase activity of α-subunit. The GDP-bound form of α-subunit has high affinity for βγ and then reassociates with the βγ dimer to form the heterotrimer in the basal resting state. The family of G-protein α-subunits can be subclassified according to functional or structural relationship. More than 20 mammalian Gα gene products and several alternatively spliced isoforms have been identified. These can be divided into four major subfamilies according to amino acid homology and are represented by Gsα, Giα, Gqα/α11, and α12/α13. The G-proteins Gsα and Giα are implicated in the regulation of adenylyl cyclase/cAMP signal transduction system.

The hormone-sensitive adenylyl cyclase system is composed of three components: the receptor, the catalytic subunit, and G-proteins—stimulatory (Gs) and inhibitory (Gi). Molecular cloning has revealed four different forms of Gsα having molecular weights of 42, 45, 47 and 52 kDa resulting from the different splicing of one gene (Bray et al. 1986; Robishaw et al. 1986; Murakami and Yasuda 1988). Gsα is positively coupled to adenylyl cyclase and mediates the stimulatory responses of hormones on adenylyl cyclase (Stryer and Bourne 1986; Spiegel 1987). The Gs-mediated activation of adenylyl cyclase results in the increased formation of cAMP. cAMP activates cAMP-dependent protein kinase A that induces the phosphorylation of contractile filaments, sarcolemmal and sarcoplasmic proteins, and regulates intracellular calcium homeostasis (Wankerl and Schwartz 1995). In addition, Gsα was also shown to open the Ca^{2+} channels directly by a cAMP-independent mechanism (Yatani and Brown 1989). In contrast, Giα protein is associated with adenylyl cyclase inhibition (Stryer and Bourne 1986; Spiegel 1987). Three distinct forms of Giα, namely, Giα-1, Giα-2, and Giα3, have been cloned and encoded by three distinct genes (Itoh et al. 1986, 1988 Jones and Reed 1987). All three forms of Giα (Giα1–3) have been shown to be implicated in adenylyl cyclase inhibition (Wong et al. 1992) and activation of atrial ACh-K^+ channels (Brown et al. 1988). Both the Gα and Gβγ dimer mediate G-protein signaling. Five different β-subunits of 35–36 kDa and 12 γ-subunits of 8–10 kDa have been identified by molecular cloning. The βγ dimer is tightly associated with GDP-bound chain and facilitates interaction of G-protein with a receptor molecule. The effectors regulated by Gβγ include K^+ channels, phospholipase C-β, and adenylyl cyclase (Simon et al. 1991; Tang and Gilman 1991; Wickman et al. 1994). Like the α-subunit, the γ-subunit is subject to a cascade of posttranscriptional modification including isoprenylation and myristoylation that contributes to βγ membrane association and the interaction of the subunits (Wedegaertner et al. 1995). The Gqα family consists of at least four members (G11, G14, G15, and G16) encoded by individual genes with different expression pattern and is involved in the pertussis toxin-insensitive regulation of PLCβ-isoforms. The fourth family of Gα-subunit is composed of Gα12 and Gα13 (Strathmann and Simon 1991). These G-proteins are pertussis toxin insensitive and are not implicated in the stimulation of phosphatidylinositol turnover. The role of Gα12 and Gα13 in growth regulation and activation of Na^+/H^+ exchange has been shown. Gα13 has been reported to directly bind to and activate PYK2 (Shi et al. 2000). In addition, Gα13 has also been shown to directly associate and activate p^{115} RhoGEF (Hart et al. 1998) which in turn stimulates the capacity of p^{115} RhoGEF to facilitate the dissociation of GDP from Rho and thus increase Rho-dependent signaling.

G-protein α-subunits also possess specific residues that can be covalently modified by bacterial toxins. Cholera toxin catalyzes the transfer of ADP-ribose moiety of NAD to a specific arginine residue in certain α-subunits, whereas pertussis toxin ADP-ribosylates those α-subunits that contain a specific cysteine residue near the carboxy-terminus. Modification of the α-subunit by cholera toxin persistently activates these protein by inhibiting their GTPase activity, whereas pertussis toxin inactives Giα protein and thereby results in the uncoupling of receptor from the effector. G-protein α-subunits are regulated by covalent modifications by fatty acids myristate and palmate. These lipid modifications serve to anchor the subunits to the membrane and increase the interaction with other protein and also increase the affinity of the α-subunit for βγ. In this regard, the myristoylation of Giα is required for adenylyl cyclase inhibition in cell-free assay (Taussig et al. 1993).

G-proteins and Membrane Signaling in Cardiovascular Disease

A number of cardiovascular disease states that eventually result in chronic congestive heart failure are associated with alterations in cardiac performance. Several hormonal factors such as angiotensin II, endothelin, and alterations in signal transduction mechanisms including adenylyl cyclase and phospholipase C (PLC) have been reported to play an important role in the alterations of cardiac performance.

G-proteins and Membrane Signaling in Cardiac Hypertrophy

Altered expression of G-proteins has been reported in several pathophysiological conditions such as hypertension, hypertrophy, heart failure, atherosclerosis, and diabetes (Spiegel et al. 1992; Fleming et al. 1992; Eschenhagen et al. 1993). Cardiac hypertrophy is often associated with augmented intracardiac sympathetic nerve activity and enhanced levels of plasma catecholamine (Siri 1988). Exposure of cardiomyocytes to catecholamines has been reported to alter their functions including heart rate, contractile activity and result in the induction of hypertrophic responses (Laks et al. 1973; Simpson 1983; Zierhut and Zimmer 1989; Thorburn 1994; Bogoyevitch et al. 1996; Yamazaki et al. 1997). The implication of both the β- and α-adrenergic receptor (AR) in the induction of cardiomyocyte hypertrophy has been shown (Laks et al. 1973; Simpson 1983; Zierhut and Zimmer 1989; Thorburn 1994; Bogoyevitch et al. 1996; Yamazaki et al. 1997). Phenylephrine (PHE), an α-AR agonist, induces hypertrophic responses in cardiomyocytes of neonatal rats (Thorburn 1994) whereas the expression of constitutively active α-AR has been reported to induce cardiac hypertrophy in adult mice (Milano et al. 1994). AR agonists signal through different second messengers, stimulation of α_1-AR activates Gqα-mediated PLC signaling pathway and forms inositol triphosphate and diacylglycerol (DAG). DAG activates protein kinase C (PKC) and results in the activation of Raf-1 kinase/extracellular signal regulated protein kinase (ERK) cascade (Karliner et al. 1990; Terzic et al. 1993). PHE has been reported to induce cardiomyocyte hypertrophy through a Ras-Erk-dependent pathway (Thorburn 1994). On the other hand, β-AR signals through the adenylyl cyclase/cAMP/protein kinase A pathway (Morgan and Baker 1991). cAMP/PKA pathway has been reported to inhibit growth factor-stimulated ERK activation in various cell types including Rac-1 cells, adipocytes, smooth muscle cells, COS-7 cells, and Chinese hamster ovary cells (Cook and McCormick 1993; Graves et al. 1993; Burgering et al. 1993; Hordijk et al. 1994; Crespo et al. 1995), whereas cAMP has also been reported to activate ERK and potentiate the effect of growth factors on differentiation and cell proliferation in PC12, Swiss-3T3, and mouse lymphoma cells (Heidemann et al. 1985; Frodin et al. 1994; Faure et al. 1994; Yao et al. 1995). However, in human endothelial cells, Gsα-mediated cAMP elevation and resultant PKA activation has been shown to activate ERKs (Sexl et al. 1997). In addition, β-AR agonists including isoproterenol has been reported to activate ERK and protein synthesis through cAMP/PKA in cardiac myocytes (Bogoyevitch et al. 1996; Yamazaki et al. 1997; Zou et al. 1999). The implication of Gsα as well as Giα protein in isoproterenol-induced cardiomyocyte hypertrophy has also been reported (Zou et al. 1999). However, a decreased expression of Gsα in heart and a decreased β-AR density and reduced sensitivity to the chronotropic effects of isoproterenol have been shown in pigs with volume-overload cardiac hypertrophy (Hammond et al. 1988), whereas the levels of pertussis toxin substrates were unchanged (Hammond et al. 1992). DiFusco and Anand-Srivastava (2000) have demonstrated a reduction in Gsα protein levels and decreased responsiveness of adenylyl cyclase to isoproterenol stimulation in hearts from volume-overload hypertrophic rats. In addition, dogs with pressure-overload cardiac hypertrophy also demonstrated decreased levels of Gsα (Longabaugh et al. 1988). On the other hand, in rats with pressure-overload hypertrophy [left ventricular hypertrophy (LVH)], basal adenylyl cyclase activity and its responsiveness to GTPγS, forskolin, and manganese chloride stimulation were attenuated in hypertrophied left ventricles (Holmer et al. 1996). The functional activity of Gs was reduced in LVH, whereas protein and mRNA expression of Gsα and Giα-2 were not altered (Moalic et al. 1995; Mondry et al. 1995; Bohm et al. 1995). In addition, mRNA levels of type V enzyme were unaltered (Holmer et al. 1996); however, the mRNA and protein levels of β-AR (Moalic et al. 1993; Bohm et al. 1995) and M2-muscarinic receptor were significantly decreased in LVH (Moalic et al. 1993), suggesting that the relative levels of β1-AR mRNA may be correlated negatively with the degree of LVH. However, in hypertensive patients with cardiac hypertrophy, the levels of myocardial Giα-2 were significantly elevated (Bohm et al. 1995). In a minipig model of pressure-overload hypertrophy, a decrease in β-AR density and an increase in antagonist affinity were shown in left ventricular membranes of hypertrophied animals as compared to control. Gsα, as measured by CT-catalyzed ADP-ribosylation, was increased in early hypertrophy and thereafter decreased in late hypertrophy. A similar pattern with Goα was observed by PT-catalyzed ADP-ribosylation, whereas Giα was unaltered. Modifications of Gs functional activity in later hypertrophic stages may be important in the pathogenesis of decompensation from compensated hypertrophy to cardiac failure (Nieto et al. 1993).

Decreased functions of Gsα in femoral arteries (Asano et al. 1988) and decreased levels of Gsα protein and mRNA in failing hearts and in hearts with compensated LVH have been reported (Chen et al. 1991). In addition, adenylyl cyclase activity, both basal and stimulated, was also decreased in these hearts. These decreases in Gsα protein and adenylyl cyclase paralleled the development of LVH. However, Gao et al. (1999) have demonstrated that directed expression of cardiac adenylyl cyclase ACVl results in structurally normal hearts with normal basal heart rate and function and increased responsiveness to catecholamine stimulation. The increased cardiac adenylyl cyclase ACVl expression was associated with increased survival after myocardial infarction (Takahashi et al. 2006), suggesting that cardiac ACVl increases function of the failing heart. Increased levels of Giα without any change in Gsα in failing human hearts due to idiopathic dilated cardiomyopathy have also been reported (Feldman et al. 1988). Similarly, Böhm et al. (1990) have reported increased levels of Giα proteins in human hearts with dilated cardiomyopathy but not with ischemic cardiomyopathy. The expression level of Gsα mRNA in ventricles from cardiomyopathic hamsters has also been shown to be lower than that of the control hamster strain and to decrease as the stage of cardiomyopathy progressed (Katoh et al. 1992). In addition, the increase in Giα was shown to accompany progression of postinfarction remodeling in hypertensive cardiomyopathy (Kouchi et al. 2000). However, Cai et al. (1993) did not observe any changes in Gsα levels in hearts from hypertrophic cardiomyopathic Syrian hamsters at 30 days and 6 months of age as compared to control hamsters, whereas PKC activities were higher in both cardiac membranes and cytosol in hypertrophic cardiomyopathic hamsters at 6 months of age as compared to control hamsters.

The heterotrimeric Gqα protein and associated signaling pathway has also been shown to be implicated in the development and progression of cardiomyocyte hypertrophy and heart failure (Dorn and Force 2005). The hormones such as angiotensin II, endothelin, and phenylephrine that activate Gqα-mediated signaling have also been reported to induce cardiac hypertrophy (Milano et al. 1994; Bogoyevitch and Sugden 1996; Akhtar et al. 1997; Hein et al. 1997). The stimulation of the Gq pathway by Gqα protein-coupled receptor (GPCR) agonists activates a common G-protein (Gqα/G11) which in turn initiates a divergent signaling cascade involving PLCβ activation resulting in the hydrolysis of phosphatidylinositol and producing inositol 1,4,5-triphosphate (IP3) and DAG. DAG activates several PKC isoenzymes which regulate a variety of cellular functions including cell growth and differentiation. In recent years, several groups have examined the role of individual components of Gq-coupled receptor signaling cascade using a variety of experimental approaches. The role of PKC in the development and progression of cardiac hypertrophy has been reported (Takeishi et al. 1998, 2000; Bowling et al. 1999). Recent studies showing that the cardiac overexpression of DAG kinase that terminates DAG signaling and thereby activates PKC prevented Gqα-coupled agonist-induced cardiac hypertrophy in transgenic mice (Arimoto et al. 2006) further supports the implication of PKC in cardiac hypertrophy.

Cardiac overexpression of Gqα in transgenic mice (D'Angelo et al. 1997) has been reported to result in hypertrophy, decreased ventricular function, loss of β-AR inotropic responsiveness without loss of β-AR expression, and induction of classic hypertrophy gene expression profile. Expression of the transgene to a greater extent or by surgical transverse aortic constriction or pregnancy has been shown to result in cardiac failure and death (Adams et al. 1998). The transgenic overexpression of a Gαq dominant negative minigene that has been shown to result in the lack of hypertrophy response to transverse aortic constriction (Akhtar et al. 1998) further supports the implication of Gqα in hypertrophy. The transgenic mice overexpressing Gqα were also shown to exhibit increased expression of Giα and PKCα (Dorn et al. 2000), whereas β-adrenergic receptor kinase (β-ARK) was downregulated and Gsα was unaffected. The increased expression of Giα in these mice was associated with decreased adenylyl cyclase activity stimulated by isoproterenol which may also be due to the uncoupling of β-ARs to the adenylyl cyclase system due to the phosphorylation by increased PKC activity and expression. These results suggest that *in vivo* overexpression of Gqα could also modulate the other signaling pathways involving Giα and Gs which may also contribute to the impaired β-adrenergic function in these mice. In support of this are studies showing that chronic administration of the Gs-coupled receptor agonist isoproterenol induced cardiac hypertrophy associated with downregulation of the signaling pathway, β-AR density and decrease in adenylyl cyclase activity (Zou et al. 1999). Thus, it appears that both Gqα and Gs signals have the potential to mediate alterations in cardiac mass. Furthermore, cardiac overexpression of a constitutively activated $\alpha_{1\beta}$-AR also resulted in cardiac hypertrophy (Milano et al. 1994), whereas overexpression of the Gq-coupled Ang II type 1 (AT1) receptor resulted in more severe cardiomyopathy (Hein et al. 1997; Zhai et al. 2005). Recent studies showing that transgenic mice with cardiac-specific overexpression of AT1 receptor second intracellular loop mutant (AT1-12m) which does not couple to Gqα or Giα, exhibited greater cardiac hypertrophy, cardiac dysfunction and bradycardia but less apoptosis and fibrosis than those overexpressing AT1 receptor, suggested that AT1-mediated other downstream signaling pathways may also induce cardiac hypertrophy with a distinct phenotype through Giα- or Gqα-dependent and -independent mechanism (Zhai et al. 2005). Furthermore, the contribution of vascular Gq-coupled signaling in the development of cardiac hypertrophy has also been shown by using transgenic mice with vascular-specific Gq inhibitor (GqI) expression. These mice exhibited attenuated cardiac hypertrophy in response to Ang II, phenylephrine and serotonin, whereas cardiac GqI peptide expression did not attenuate cardiac hypertrophy (Keys et al. 2002).

G-protein and Membrane Signaling in Ischemia

Alterations in G-protein levels and associated functions, i.e., adenylyl cyclase impairment, have also been reported in myocardial ischemia, which may contribute to the altered pathophysiology of the ischemic heart. Tobise et al. (1991) have reported that prolonged periods of global ischemia did not produce any changes in the levels of Giα-2 and Gsα proteins nor in β-ARs in ischemic heart, whereas forskolin-stimulated adenylyl cyclase activity was depressed. Similary, Van den Ende et al. (1994) have also demonstrated no change in the levels of Giα and Gsα in ischemic heart; however, adenylyl cyclase responsiveness was impaired. Basal adenylyl cyclase activity, as well as stimulations exerted by isoproterenol, guanine nucleotides, NaF, forskolin, and Mn^{2+} were diminished in ischemic heart, and the forskolin- and db-cAMP-induced inotropic responses were virtually abolished after ischemia. On the other hand, global myocardial ischemia for 15 and 50 minutes resulted in the upregulation of β-ARs in the cardiac membranes (Strasser et al. 1990). Isoproterenol- and forskolin-stimulated adenylyl cyclase activities were also augmented after 15 minutes of ischemia and then declined after 30 and 50 minutes of ischemia. Furthermore, Susani et al. (1989) have also demonstrated an attenuation of basal, GTP-, isoproterenol-, NaF-, and forskolin-stimulated adenylyl cyclase activities in heart sarcolemma after 1 hour of myocardial ischemia that was associated with the decrease in Gsα levels as determined by cholera-toxin labeling. Decreased levels of Gsα and Giα proteins and of mRNA have also been reported in ischemic myocardium by *in situ* hybridization (Ohyanagi et al. 1995). In addition, Maisel et al. (1990) have reported a decrease in Gsα in ischemic myocardial sarcolemma within 15 minutes of coronary occlusion. On the other hand, Hammond et al. (1993) have shown that regional myocardial ischemia is associated with a reduction in myocardial β-ARs, whereas adenylyl cyclase activity was unaltered. Giα-2 protein levels were decreased in the ischemic bed, and Gsα levels were increased in endocardial sections of the ischemic bed. The decreased levels of Giα which were associated with decreased inhibition of adenylyl cyclase and increased Gsα in the endocardium may play an important role in the preservation of adrenergic activation in the setting of chronic episodic myocardial ischemia. Similarly, Wolff et al. (1994) have shown that ischemic rabbit myocardium obtained after 30 minutes of coronary artery occlusion resulted in the attenuation of adenylyl cyclase activity stimulated by GppNHp, whereas β-adrenergic agonist (isoproterenol)-stimulated cyclase was preserved and forskolin-stimulated enzyme activity was significantly increased. Based on these results the authors suggested that increased catalytic activity may be responsible for the preservation of β-adrenoceptor-mediated adenylyl cyclase in myocardial ischemia. The receptor-G-protein-adenylyl cyclase system has also been shown to be impaired in cerebral ischemia. Decreased levels of PT-catalyzed ADP-ribosylation of Giα and Goα protein have been reported in cerebral cortical membranes obtained after ischemia of rat brain (Takenaka et al. 1991). In addition, transient ischemia resulted in the augmentation of basal, isoproterenol-, and forskolin-stimulated adenylyl cyclase activity in hippocampal slices at 6 and 24 hours after ischemia (Suyama et al. 1995), whereas no change in the levels of Gα protein was observed and Gβ was decreased for 4 days after ischemia, as determined by immunoblotting. However, PT-catalyzed ADP-ribosylation declined progressively, reaching a significant reduction at 6 hours after ischemia (Suyama et al. 1995). On the other hand, hypoxia ischemia decreased the levels of Gqα and G_{11} in the cerebral cortex and cerebellum of rats but not in striatum 3 and 14 days, respectively, after cardiac arrest, suggesting the implication of phosphoinositidase C-linked signal transduction pathways in the expression of motor dysfunction in rats after cardiac arrest (Jaw et al. 1995).

G-proteins and Membrane Signaling in Hypertension

Alterations in G-protein levels and functions such as altered adenylyl cyclase responsiveness to various agonists have also been demonstrated in cardiovascular and noncardiovascular tissues from genetic as well as experimental hypertensive rats (Anand-Srivastava et al. 1991, 1993; Anand-Srivastava 1992, 1993; Thibault and Anand-Srivastava 1992; Bohm et al. 1993; Li et al. 1994).

An overexpression of Giα-2 and Giα-3 proteins as well as their genes was shown in hearts and aorta from spontaneously hypertensive rats (SHRs), deoxycorticosterone acetate (DOCA)-salt hypertensive rats (HR), N^{ω}-nitro-L-arginine methylester (L-NAME) HR, and 1 kidney 1 clip (1K1C) HR (Anand-Srivastava et al. 1991, 1993; Bohm et al. 1992, 1993; Thibault and Anand-Srivastava 1992; Anand-Srivastava 1993; DiFusco and Anand-Srivastava 1997, 2000; Ge et al. 1999, 2006), whereas Gsα protein and its gene were not altered in SHRs, 1K1C and L-NAME HR, and were decreased in DOCA-salt HRs (Anand-Srivastava et al. 1991; Anand-Srivastava 1992, 1993; Thibault and Anand-Srivastava 1992; DiFusco and Anand-Srivastava 1997, 2000; Ge et al. 1999, 2006). In addition, the levels of Goα in heart were not altered (Anand-Srivastava 1992). Alterations in Gi-protein levels have been shown to be reflected in altered responsiveness of adenylyl cyclase to stimulatory and inhibitory hormones in SHRs, and experimental models of hypertensive rats (Anand-Srivastava et al. 1991; Anand-Srivastava 1992). However, a decreased expression of Giα proteins was also shown in different tissues from a different model of HR including Milan hypertensive rats (MHS) (Clark et al. 1993; Li et al. 1994; Michel et al. 1994; Kanagy and Webb 1996). The VSMC from MHS exhibit enhanced basal adenylyl cyclase activity as compared to control normotensive rats (MNS). The number of β-adrenoceptors and the stimulations exerted by isoproterenol and prostaglandin E_1 (PGE_1) were significantly increased in MHS as compared to MNS. On the other hand, platelets from SHRs (Anand-Srivastava 1993) as well as from hypertensive patients (Marcil et al. 1996) exhibited a decreased expression of Giα-2 and Giα-3 protein as compared to WKY and to normotensive control subjects, respectively, whereas the levels of Gsα protein were not altered. The decreased expression of Giα-2 and Giα-3 was correlated with adenylyl cyclase inhibition by inhibitory hormones. The ANP and Ang II-mediated inhibitions were completely attenuated in platelets from SHRs and hypertensive patients, whereas the stimulatory effects of PGE_1, NECA, and forskolin (FSK) were augmented (Anand-Srivastava 1993; Marcil et al. 1996). However, McLellan et al. (1993) were unable to show any changes in the levels of Gsα, Giα-2, and Gβ in platelets from hypertensive patients as compared to normotensive subjects, whereas an enhanced stimulation of adenylyl cyclase by PGE_1 was observed in hypertensive patients as compared to normotensive subjects. On the other hand, lymphocytes from SHRs (Marcil and Anand-Srivastava 2001) and hypertensive patients (Feldman et al. 1995) showed a decreased responsiveness of adenylyl cyclase to stimulatory hormones, which may be attributed to the alterations in Gs and Gi proteins. The potentiation of stimulatory responses of several hormones on adenylyl cyclase has also been demonstrated in platelets and splenocytic membranes from SHRs (Hamet et al. 1980; Zeng et al. 1991). In addition, antihypertensive drug therapy (a combination of β-blockers, Ca^{2+} channel blocker, ACE inhibitor, etc.) partially restored Giα-2 levels toward normotensive subjects by about 60% to 70%. Furthermore, the enhanced stimulation of adenylyl cyclase by GTPγS, NECA, and PGE_1 was partially corrected by about 50% to 80% in the patients under antihypertensive drug therapy (Marcil et al. 1996). These results suggest that the altered responsiveness of platelet adenylyl cyclase to hormones in hypertension and the normalization of the response with antihypertensive drug therapy could partially be due to the ability of the latter to modulate Giα protein expression. These effects on platelet function may underlie the beneficial effects of antihypertensive agents on some of the complications of hypertension.

However, the levels of Gsα, Giα-1, Giα-2, Giα-3, Goα, and Gβ were also shown to be unaltered in myocardium from SHRs, and adenylyl cyclase activity stimulated by PGE_1, glucagon, and isoproterenol was reduced in SHRs, whereas FSK-stimulated enzyme activity was greater in SHRs as compared to WKY (McLellan et al. 1993). On the other hand, a diminished stimulation of adenylyl cyclase by stimulatory hormones, guanine nucleotides, FSK, and NaF in aorta and heart sarcolemma from SHRs (Anand-Srivastava 1992), renal hypertensive rats (Anand-Srivastava 1988) 1K1C HRs (Ge et al. 1999, 2006), and DOCA-salt HRs (Anand-Srivastava et al. 1993) has been demonstrated The reduction in the hormone receptor binding sites may be one of the possible mechanisms responsible for such an impaired response of hormones (Limas and Limas 1978; Woodcock et al.

1979; Bhalla et al. 1980). However, the decreased stimulation of adenylyl cyclase by dopamine D-1 receptors in the kidney tubules from SHRs was shown to be attributed to the defective coupling and not to the changes in the receptor number (Kinoshita et al. 1989).

Furthermore, the increased levels of Giα were shown to be associated with hypertension and not with hypertrophy, due to the fact that heart and aorta from L-NAME-induced HRs, which do not have cardiac hypertrophy exhibited enhanced levels of Giα-2 and Giα-3 proteins as well as mRNA, whereas the levels of Gsα protein were unaltered (DiFusco and Anand-Srivastava 1997, 2000). The increased levels of Giα-2 and Giα-3 proteins and their mRNA in heart and aorta precede the development of blood pressure in SHRs (Marcil et al. 1997), and in DOCA-salt HRs (Marcil et al. 1998), and suggest that the enhanced levels of Giα proteins which result in the decreased levels of cAMP may be one of the contributing factors in the pathogenesis of hypertension. This was further supported by recent studies showing that the inactivation of Giα protein in prehypertensive rats (2-week-old SHR) by single injection of pertussis toxin (PT) (1.5 μg/100 g body weight) prevented the development of high blood pressure (Figure 1.1) which was associated with PT-induced decreased levels of Giα proteins (Figure 1.2). Furthermore, Triggle and Tabrizchi (1993) have shown that treatment of the SHRs (adult) with PT lowered the blood pressure.

The levels of vasoactive peptides such as angiotensin II (Ang II), endothelin (ET-1), and arginine vasopressin (AVP) as well as growth factors that have been reported to be augmented in various models of hypertension (Morishuta et al. 1992; Trinder et al. 1992; Wahlander et al. 1999; Kagiyama et al. 2002, 2003; Iglarz and Schiffrin 2003; Kirchengast et al. 2005; Shermuly et al. 2005; Jesmin et al. 2006) may be responsible for the enhanced expression of Giα proteins in hypertension. In this regard, a role of Ang II in enhanced expression of Giα protein in SHRs and 1K1C HRs has been suggested by studies showing that captopril, an angiotensin-converting enzyme (ACE) inhibitor treatment of the SHRs and 1K1C HRs that decreased the blood pressure, also restored the enhanced levels of Giα protein to control levels (Figure 1.3). Similarly, the increased blood pressure and enhanced expression of Giα proteins in L-NAME hypertensive rats was also shown to be restored to control levels by losartan, an AT1 receptor antagonist (Figure 1.3), implicating Ang II in increased levels of Giα proteins and increased blood pressure in L-NAME-induced hypertension. These treatments were also shown to restore the diminished stimulation of adenylyl cyclase by stimulatory hormones and enhanced inhibition by inhibitory hormones observed in SHRs, 1K1C and L-NAME HRs (Pandey and Anand-Srivastava 1996; Ge et al. 1999; Hashim and Anand-Srivastava 2004). In addition, infusion of Ang II in rats that increased blood pressure has also been reported to enhance the levels of Giα proteins (Sims et al. 1992). Similarly, nitrendipin and fosinopril treatments have also been reported to have similar effects on Gi proteins and functions in hearts from SHRs (Bohm et al. 1995) and further implicate Ang II in enhanced levels of Giα protein in SHR.

The role of MAP kinase and PI3K signaling as well as oxidative stress in Ang II-induced enhanced levels of Giα proteins has also been reported (Ge and

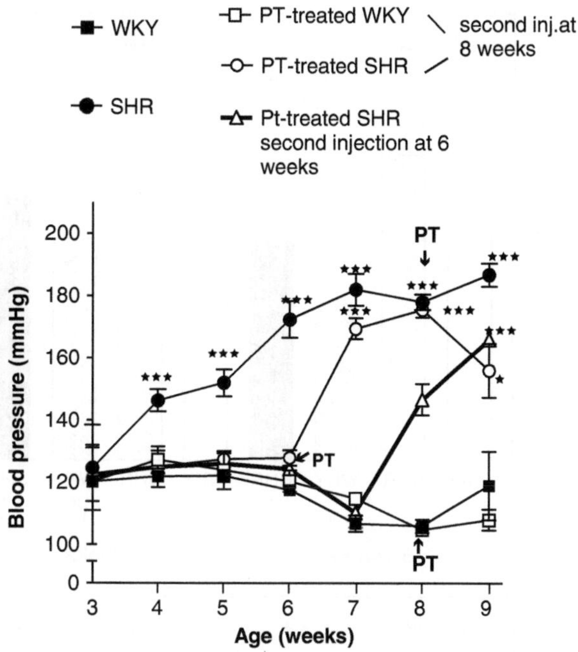

Fig. 1.1 Effect of *in vivo* pertussis toxin (PT) treatment on the development of blood pressure in spontaneously hypertensive rats (SHR) Two-week-old SHR and WKY were injected intraperitoneally with PT (1.5 μg/100 g body weight) in 0.01 M sodium phosphate buffer, pH 7.0, containing 0.05 M NaCl (PT-treated) or vehicle (control WKY and control SHR), as described earlier (Li and Anand-Srivastava 2002) A second injection of PT (1.5 μg/100 g body weight) was given at 6 weeks to one group of PT-treated SHR and at 8 weeks to another group of PT-treated SHR and PT-treated WKY. Blood pressure was monitored weekly as described earlier (Li and Anand-Srivastava 2002). Values are means ± S.E.M. of five or six rats in each group. Reproduced with permission from Li and Anand-Srivastava (2002).

Anand-Srivastava 1998; Li et al. 2007). In addition, studies showing that MEK inhibitor (Figure 1.4) as well as antioxidants such as diphenyleneiodonium (DPI) and *N-acetyl-L-cysteine* (NAC) (Figure 1.5) restored the enhanced levels of Giα proteins in SHR further implicate MAP kinase and oxidative stress in the enhanced expression of Giα protein in SHR. Furthermore, the enhanced phosphorylation of ERK1/2 in SHR was shown to be restored to WKY levels by antioxidants (Figure 1.6) and suggests that enhanced oxidative stress through MAP kinase signaling may contribute to the enhanced expression of Giα protein in SHR (Figure 1.7). The implication of Giα proteins in the regulation of blood pressure was further demonstrated by the studies showing that nitric oxide (NO) donors SNAP and sodium nitroprusside (SNP) that have been reported to decrease blood pressure also attenuated the expression of Giα proteins and associated functions (Bassil and Anand-Srivstava 2006). In addition, cGMP, which is the second messenger of NO action, also decreased the levels of Giα proteins and functions in VSMC (Bassil and Anand-Srivastava 2007). These results indicate that the decreased levels of NO in

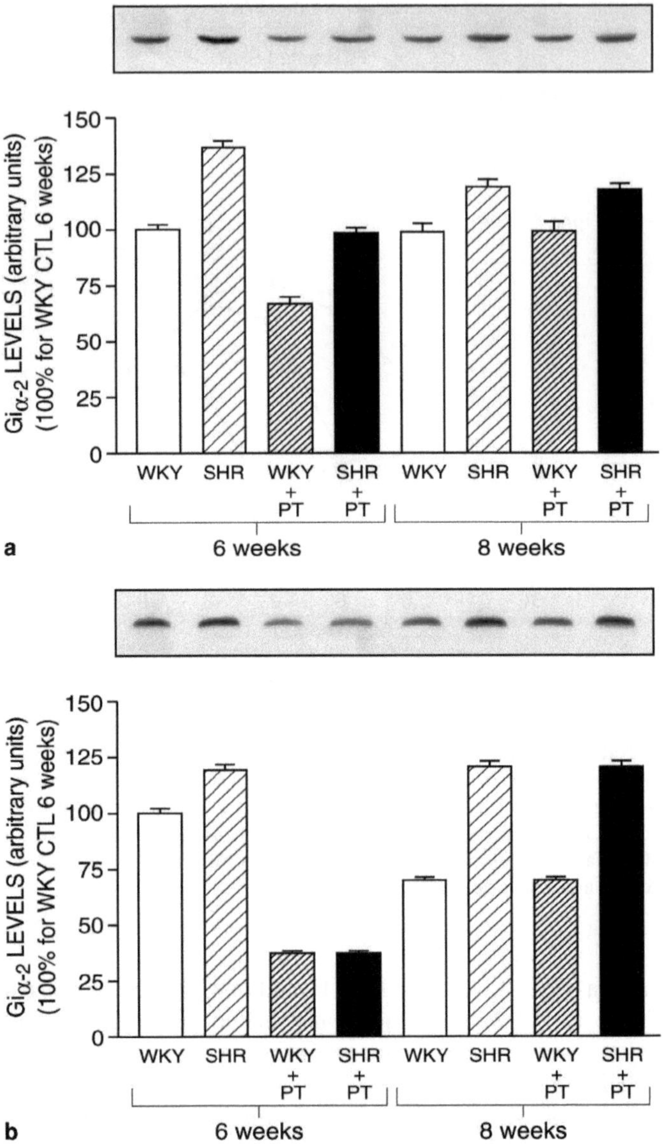

Fig. 1.2 Effect of *in vivo* PT treatment on the levels of Giα-2 and Giα-3 proteins. Upper panels: Heart membrane proteins (20 μg) from 6- and 8-week-old SHR and WKY with or without PT treatment (Li and Anand-Srivastava 2002) were resolved by SDS/PAGE and transferred to nitrocellulose, which was then immunoblotted with antibody AS/7 for Giα-2 (A) or antibody EC/2 for Giα-3 (B). The blots are representative of three or four separate experiments. Lower panels: Quantification of protein bands by densitometric scanning. The results are expressed as percent of WKY control at 6 weeks, which has been taken as 100%. Values are means ± S.E.M. of three or four separate experiments. Reproduced with permission from Li and Anand-Srivastava (2002).

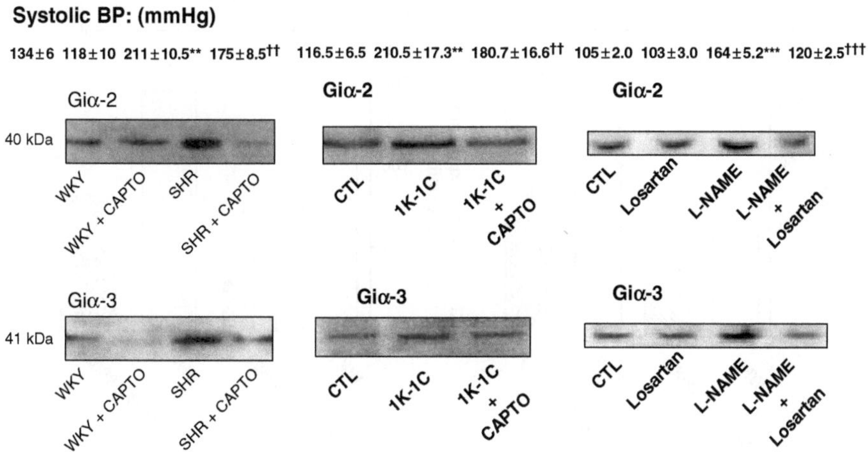

Fig. 1.3 Effect of captopril and losartan treatment on blood pressure and the expression of Giα proteins in different models of hypertensive rats (HR). Twelve-week-old SHR and age-matched WKY rats, 1 kidney 1 clip hypertensive rats (1K-1C HR) were treated with captopril (150 mg/kg body wt/day) as described earlier (Pandey and Anand-Srivastava 1996; Ge et al. 1999), whereas L-NAME-induced hypertensive rats were treated with losartan (10 mg/kg body wt/day) as described earlier (Hashim and Anand-Srivastava 2004). The blood pressure was monitored by the tail cuff method. The expression of Giα-2 and Giα-3 protein in heart from SHR and L-NAME HR and aorta from 1K-1C HR were determined by Western blotting using antidodies AS/7 and EC/1 against Giα-2 and Giα-3 protein, respectively. The blots are representative of three or four separate experiments. $^{**}P < 0.01$, $^{***}P < 0.001$ versus control/WKY, $^{\dagger\dagger}P < 0.01$, $^{\dagger\dagger\dagger}P < 0.001$ versus SHR/1K-1C/L-NAME.

L-NAME hypertensive rats may be responsible for the enhanced expression of Giα proteins and also suggest that NO-induced decreased levels of Giα proteins may represent an additional mechanism through which NO decreases the blood pressure (Figure 1.7).

The levels of Gqα and Gqα mediated signaling molecules have also been shown to be altered in different models of hypertensive rats. The levels of Gqα and G11 mRNA as well as PLC beta were shown to be increased in heart, aorta as well as in kidney from SHR as compared to WKY rats (Chen et al. 2005). The increased expression of Gqα/G11, ERK1/2 as well as PLCβ activity has also been reported in hearts from 1K1C HRs (Bai et al. 2004). The implication of VSMC Gqα signaling in high blood pressure in a renovascular model of hypertension through renal artery stenosis and a genetic model of hypertension using mice with VSM derived hypertension has recently been reported (Harris et al. 2007). These investigators have found that inhibition of Gq signaling by Gq inhibitor GqI peptide as well as by losartan, AT1 receptor antagonist attenuated high blood pressure in both models of hypertension and suggesting that VSM AT1-Gq-coupled receptors may play a critical role in the development of high blood pressure.

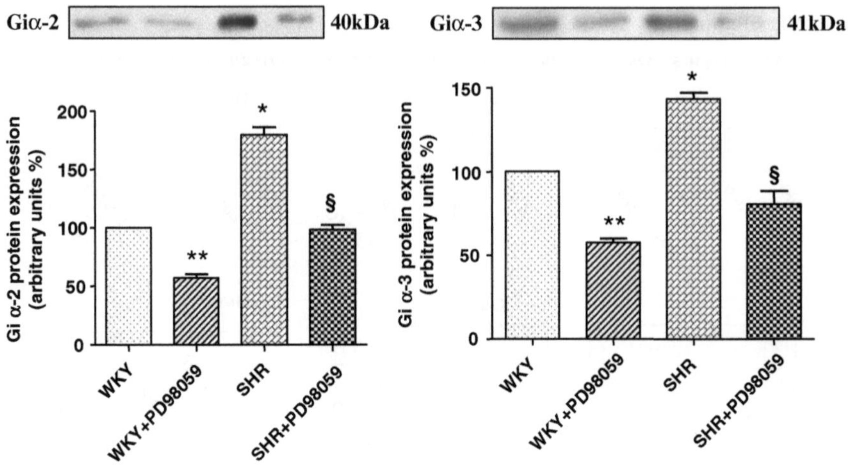

Fig. 1.4 Effect of PD 98059 on Giα-2 and Giα-3 protein expression in VSMC from 12 week-old SHR and age-matched WKY rats. Confluent VSMC from SHR and WKY rats were treated with or without PD 98059 (10 μM) for 24 hours at 37 °C. Membrane proteins (30 μg) were separated and transferred to nitrocellulose, which was then immunoblotted with specific antibodies against Giα-2 (left) and Giα-3 (right) as described earlier (Lappas et al. 2005). The blots are representative of three separate experiments. The graphs in the lower panel show quantification of protein bands by densitometric scanning. The results are expressed as percentage of WKY control which has been taken as 100%. Values are mean ± S.E.M of five separate experiments. $^{*}P < 0.05$, $^{**}P < 0.01$ versus WKY, $^{\S}P < 0.05$ versus SHR. Reproduced with permission from Lappas et al. (2005).

Conclusions

We have discussed the alterations in G-proteins and associated functions in hypertension, hypertrophy, and myocardial ischemia. We have mainly focused on Gi, Gs proteins and Gqα proteins which are implicated in the regulation of adenylyl cyclase/cAMP and PLC/PI turnover signal transduction systems, respectively, that play an important role in the regulation of cardiovascular functions, including vascular tone and reactivity and cell proliferation. The levels of Giα-2 and Giα-3 proteins and mRNA are increased in hearts and aorta from genetic and experimentally induced hypertensive rats, whereas the levels of Gsα are unaltered in genetic and decreased in experimentally-induced hypertensive rats with established hypertrophy. The increased levels of Giα-2 and Giα-3 are associated with increased Gi functions, resulting in greater decreases in cAMP levels, which may partly explain the increased vascular resistance in hypertension. On the other hand, the decreased levels of Gsα and decreased formation of cAMP in hypertension associated with hypertrophy may also contribute to the increased vascular reactivity in hypertension. The increased levels of Giα-2 and Giα-3 may contribute to the pathogenesis of hypertension whereas decreased levels of Gsα may be associated with hypertrophy and not with hypertension. This notion is substantiated by our recent studies, showing that enhanced expression of Giα-2 and Giα-3 proteins and mRNA precede

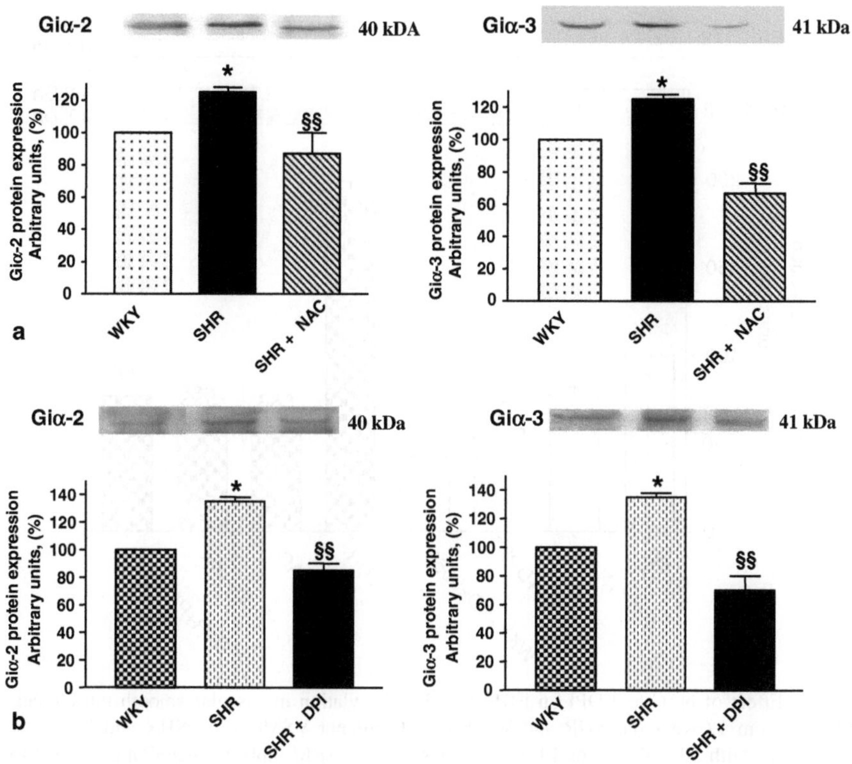

Fig. 1.5 Effect of *N*-acetyl-L-cysteine (NAC) and diphenyleneiodonium (DPI) on Giα-2 and Giα-3 protein expression in vascular smooth muscle cells (VSMC) from 12 week-old SHR and age-matched WKY rats. Confluent VSMC from SHR and WKY rats were treated with 20 mM NAC (A) or 10 μM DPI (B) for 24 hours at 37 °C. Membrane proteins (30 μg) were separated and transferred to nitrocellulose, which was immunoblotted with antibodies AS/7 and EC/1 against Giα-2 (A) and Giα-3 (B), respectively, as described earlier (Lappas et al. 2005). The blots are representative of five separate experiments. The graphs in the lower panel show quantification of protein bands by densitometric scanning. The results are expressed as percentage of WKY control which has been taken as 100%. Values are mean ± S.E.M of five separate experiments. $^{*}P < 0.05$ versus WKY, $^{\S\S}P < 0.01$ versus SHR.

the development of blood pressure. The role of enhanced levels of Giα proteins in the pathogenesis of hypertension was further supported by our studies showing that inactivation of Giα proteins by pertussis toxin treatment in prehypertensive SHR prevented the development of blood pressure. However, the levels of Gsα were decreased only in 15-week-old SHRs with established hypertrophy. Similarly, L-NAME hypertensive rats that do not have cardiac hypertrophy exhibited enhanced expression of Giα-2 and Giα-3 and no changes in Gsα, whereas hypertrophied rats with volume-overload hypertrophy, which do not have hypertension, exhibited decreased levels of Gsα and no augmentation in Giα-2 or Giα-3 proteins. On the other hand, the levels of Gqα/G11 and PLCβ were shown to be augmented in different

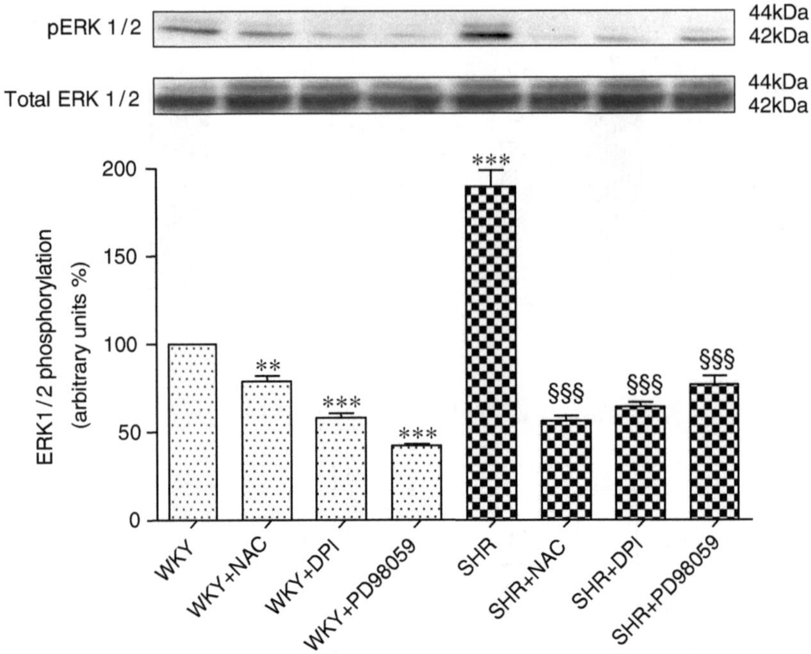

Fig. 1.6 Effect of NAC and DPI on ERK 1/2 phosphorylation in vascular smooth muscle cells (VSMC) from 12 week-old SHR and WKY rats. Confluent VSMC from SHR and WKY rats were treated with 20 mM *N*-acetyl-L-cysteine (NAC) or 10 μM diphenyleneiodonium (DPI) for 24 hours at 37 °C. Cell lysates were immunoblotted by phospho-specif-Tyr^{204}-ERK1/2 antibodies as shown in the top panel. Blots were also analyzed for total ERK1/2 (bottom panel). The blots are representative of three separate experiments. Detection of p-ERK1/2 and total ERK1/2 was performed with chemiluminescence Western blotting detection reagents. WKY levels were taken as 100%. Values are mean ± S.E.M. of three separate experiments. $^{**}P < 0.01$, $^{***}P < 0.001$ versus WKY and $^{\S\S\S}P < 0.001$ versus SHR. Reproduced with permission from Lappas et al. (2005).

models of hypertension. The Gsα levels are also decreased in pressure-overload hypertrophied rat models, whereas the levels of Gqα and PKC are augmented in cardiac hypertrophy. In addition, GPCR activation by Ang II and ET-1 that increase the levels of Gqα also induce cardiac hypertrophy. On the other hand, cardiac ischemia and failure are also associated with decreased levels of Gsα and decreased sensitivity of adenylyl cyclase to stimulatory inputs. The increased levels of Giα are also shown in failing human hearts. Taken together, it can be concluded that decreased formation of cAMP levels, either by increased levels and function of Gi or by decreased levels of Gsα and associated functions, may be responsible for the altered cardiac performance and vascular reactivity in cardiovascular disease, whereas the alterations in Gqα signaling may also play an important role in the induction of cardiac hypertrophy leading to cardiac failure.

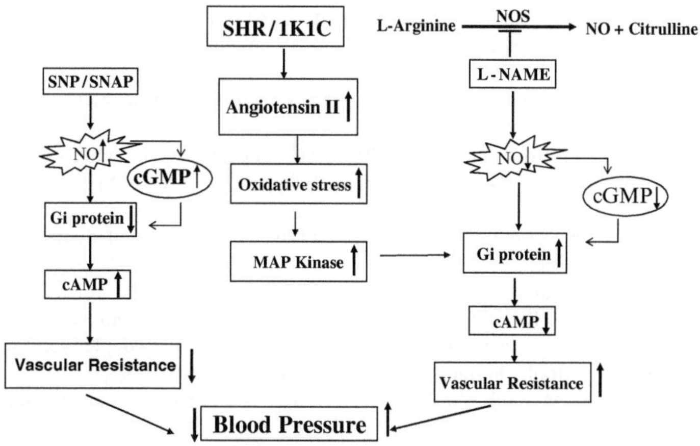

Fig. 1.7 Possible mechanisms involving angiotensin II, oxidative stress and nitric oxide in enhanced Giα protein expression in hypertension. Gi protein expression is enhanced in genetic (SHR) and experimental hypertension including 1 kidney 1 clip (1K1C) and L-NAME-induced hypertension. Inhibition of nitric oxide synthase (NOS) by L-NAME activates renin angiotensin system, and also decreases the level of NO. 1K1C hypertensive rats also exhibit enhanced levels of Ang II. Ang II increases oxidative stress that through increased MAP kinase activity results in enhanced expression of Giα proteins and thereby hypertension. On the other hand, increased levels of NO and cGMP decrease the expression of Giα proteins in VSMC which may be an additional mechanism through which NO decreases blood pressure in L-NAME-induced hypertensive rats.

Acknowledgments The research in the author's laboratory has been supported by the Canadian Institute of Health Research. The author is grateful to Christiane Laurier for her excellent secretarial help.

References

Adams JW, Sakata Y, Davis MG, Sah VP, Wang Y, Liggett SB, Chien KR, Brown JH, Dorn GW 2nd. 1998. Enhanced Gαq signaling: a common pathway mediates cardiac hypertrophy and apoptotic heart failure. Proc Natl Acad Sci USA 95:10140–10145.

Akhtar SA, Milano CA, Shotwell KF, Cho MC, Rockman HA, Lefkowitz RJ, Koch WJ. 1997. Transgenic mice with cardiac overexpression of α1b-adrenergic receptors. J Biol Chem 272:21253–21259.

Akhtar SA, Luttrell RM, Rockman HA, Iaccarino G, Lefkowitz RJ, Joch WJ. 1998. Targeting the receptor-Gq interface to inhibit in vivo pressure overload myocardial hypertrophy. Science 280:574–577.

Anand-Srivsatava MB. 1988. Altered responsiveness of adenylate cyclase to adenosine and other agents in the myocardial sarcolemma and aorta of spontaneously hypertensive rats. Biochem Pharmacol 37:3017–3022.

Anand-Srivstava MB. 1992. Enhanced expression of inhibitory guanine nucleotide regulatory proteins (Giα) in spontaneously hypertensive rats: relationship to adenylate cyclase inhibition. Biochem J 288:79–85.

Anand-Srivastava MB. 1993. Rat platelets from spontaneously hypertensive rats exhibit decreased expression of inhibitory guanine nucleotide regulatory protein: relationship with adenylate cyclase activity. Circ Res 173:1032–1039.

Anand-Srivastava MB. 1997. G-protein and membrane signalling in cardiovascular disease. Heart Failure Rev 2:85–94.

Anand-Srivastava MB, Picard S, Thibault C. 1991. Altered expression of inhibitory guanine nucleotide regulatory proteins (Giα) in spontaneously hypertensive rats. Am J Hypertens 4:840–843.

Anand-Srivastava MB, et al. 1993. DOCA-salt hypertensive rat hearts exhibit altered expression of G-proteins. Am J Hypertens 6:72–75.

Arimoto T, Takeishi Y, Takashi H, Shishido T, Niizeki T, Koyama Y, Shiga R, Nozaki N, Nakajima O, Nishimaru K, Abe J-I, Endoh M, Walsh RA, Goto K, Kubota I. 2006. Cardiac specific overexpression of diacylglycerol kinase prevents Gq protein-coupled receptor agonist-induced cardiac hypertrophy in transgenic mice. Circulation 113:60–66.

Asano M, Masuzawa K, Matsuda T, Asano T. 1988. Reduced function of the stimulatory GTP-binding protein in beta adrenoceptor-adenylate cyclase system of femoral arteries isolated from spontaneously hypertensive rats. J Pharmacol Exp Ther 246:709–718.

Bai H, Wu LP, Xing DG, Liu J, Zhao Ya-Li. 2004. Angiotensin II-induced upregulation of Gα q/11, phospholipase C β3 and extracellular signal regulated kinase 1/2 via aniotensin II type 1 receptor. Chinese Med J 117:88–93.

Bassil M, Anand-Srivstava MB. 2006. Nitric oxide modulates Gi protein expression and adenylyl cyclase signalling in vascular smooth muscle cells. Free Radical Biol Med 41:1162–1173.

Bassil M, Anand-Srivastava MB. 2007. Cyclic GMP modulates the expression of Gi protein and adenylyl cyclase signaling in vascular smooth muscle cells. Cell Biochem Biophys 47:99–108.

Bhalla RC, Sharma RV, Raamanathan S. 1980. Ontogenetic developement of isoproterenol subsensitivity of myocardial adenylate cyclase and beta-adrenergic receptors in spontaneously hypertensive rats. Biochim Biophys Acta 632:497–506.

Bogoyevitch MA, Andersson MB, Gillespie BJ, Clerk A, Glennon PE, Fuller SJ, Sugden PH. 1996. Adrenergic receptor stimulation of the mitogen-activated protein kinase cascade and cardiac hypertrophy. Biochem J 314:115–121.

Bogoyevitch MA, Sugden PH. 1996. The role of protein kinases in adaptational growth of the heart. Int J Biochem Cell Biol 28:1–12.

Bohm M, Gierschik P, Jakos KH, Schnael P, Ungerer M, Erdmann E. 1990. Increase of Gi alpha in human hearts with dilated but not ischemic cardiomyopathy. Circulation 82:1249–1265.

Bohm M, Gierschik P, Knorr A, Larisch K, Weismann K, Erdmann E. 1992. Desensitization of adenylate cyclase and increase of G alpha in cardiac hypertrophy due to acquired hypertension. Hypertension 20:103–112.

Bohm M, Gierschik P, Knorr A, Schmidt U, Weismann K, Erdmann E. 1993. Cardiac adenylyl cyclase, beta-adrenergic receptors, and G proteins in salt-sensitive hypertension. Hypertension 22:715–727.

Bohm M, Kirchmayer R, Erdmann E. 1995. Myocardial Giα alpha-protein levels in patients with hypertensive cardiac hypertrophy, ischemic heart disease and cardiogenic shock. Cardiovasc Res 30:611–618.

Bowling N, Walsh RA, Song G, Estridge T, Sandusky GE, Fouts RL, Mintze K, Pickard T, Roden R, Bristow MR, Sabbah HN, Mizrahi JL, Gromo G, King GL, Vlahos CJ. 1999. Increased protein kinase C activity and expression of Ca^{2+}-sensitive isoforms in the failing human heart. Circulation 99:384–391.

Bray P, Carter A, Simons C, Guo V, Puckett C, Kamholz J, Spiegel A, Nirenberg M. 1986. Human cDNA clones for four species of G alpha's signal transduct protein. Proc Natl Acad Sci USA 83:8893–8897.

Breiwieser GE, Szabo G. 1985. Uncoupling of cardiac muscarinic and beta-adrenergic receptors from ion channels by a guanine nucleotide analogue. Nature 317:538–540.

Brown AM, Birnbauemr L. 1988. Direct G protein gating of ion channels. Am J Physiol 254:H401–H410.

Burgering BM, Pronk GJ,Weeren PC, Chardin P, Bos JL. 1993. cAMP antagonizes p21ras-directed activation of extracellular signal-regulated kinase 2 and phosphorylation of mSos nucleotide exchange factor. EMBO J 12:4211–4220.

Cai JJ, Yu H, Lee HC. 1993. Developmental changes of protein kinase C and Gs alpha in hypertrophic cardiomyopathic hamster hearts. J Lab Clin Med 122:533–541.

Chen LA, Vatner DE, Vatner SF, Hiffinger L, Homcy CJ. 1991. Decreased Gs alpha mRNA levels accompany the fall in Gs and adenylyl cyclase activities in compensated left ventricular hypertrophy. In heart failure, only the impairment in adenylyl cyclase activation progresses. J Clin Invest 87:293–298.

Chen NY, Hu SJ, Dong HT, Li DB, Chen ZK. 2005. Alterations of expression of G protein-inositol phosphates pathway related genes in essential hypertension: experiment with spontaneously hypertensive rats. Zhonghua Yi Za Shi 85:3481–3485.

Clark CJ, Milligan G, Connell JM. 1993.Guanine nucleotide regulatory protein alterations in the Milan hypertensive rat strain. J Hypertens 11:1161–1169.

Cockroft S, Gomperts BD.1985. Role of guanine nucleotide binding protein in the activation of phosphoinositide phosphodiesterase. Nature 315:534–536.

Cook SJ, and McCormick F. 1993. Inhibition by cAMP of Ras-dependent activation of Raf. Science 262:1069–1072.

Crespo P, Cachero TG, Xu N, Gutkind JS. 1995. J Biol Chem 270:25259–25265.

D'Angelo DD, et al. 1997. Transgenic Gαq overexpression induces cardiac contractile failure in mice. Proc Natl Acad Sci USA 94:8121–8126.

DiFusco F, Anand-Srivastava MB. 1997. Nitric oxide synthase inhibition by Nω-nitro-L-arginine methyl ester modulates G-protein expression and adenylyl cyclase activity in rat heart. Am J Hypertens 10:471–475.

DiFusco F, Anand-Srivastava MB. 2000. Enhanced expression of Gi proteins in non-hypertrophic hearts from rats with hypertension induced by L-NAME treatment. J Hypertens 8:1081–1091.

DiFusco F, Hashim S, Anand-Srivastava MB. 2000. Volume overload cardiac hypertrophy exhibits decreased expression of Gsα and not of Giα in heart. Am J Physiol Cell Physiol 279:C990–C998.

Dorn GW II, Force T. 2005. Protein kinase cascades in the regulation of cardiac hypertrophy. J Clin Invest 115:527–537. doi :10.1172/JCI200524178.

Dorn II GW, Tepe NM, Guangyu WU, Atsuko Y, Liggett SB. 2000. Mechanisms of impaired β-adrenergic receptor signaling in Gq-mediated cardiac hypertrophy and ventricular dysfunction. Mol Pharm 57:278–287.

Eschenhagen TG. 1993. Proteins and the heart. Cell Biol Int 17:723–749.

Faure M, Voyno-Yasenetskaya TA, Bourne HR. 1994. cAMP and beta gamma subunits of heterotrimeric G proteins stimulate the mitogen-activated protein kinase pathway in COS-7 cells. J Biol Chem 269:7851–7854.

Feldman AM, Cates AE, Beazey WB, Hershberger RE, Bristow MR, Baughman KL, Baugartner WA, Dop CV. 1988. Increase of the 40,000 mol wt pertussis toxin substrate (G-protein) in the failing human heart. J Clin Invest 82:189–197.

Feldman RD, Jan CM, Chorazyczewski J. 1995. G protein alterations in hypertension and aging. Hypertension 26:725–732.

Fleming JW, Wisler PL, Watanabe AM. 1992. Signal transduction by G proteins in cardiac tissues. Circulation 85:420–433.

Frodin M, Peraldi P, Van Obberghen E. 1994. Cyclic AMP activates the mitogen-activated protein kinase cascade in PC12 cells. J Biol Chem 269:6207–6214.

Gao MH, Lai NC, Roth DM, Zhou J, Zhu J, Anzai T, Dalton N, Hammond HK. 1999. Adenylyl cyclase increases responsiveness to cathecholamine stimulation in transgenomic mice. Circulation 99:1618–1622.

Ge C, Anand-Srivastava MB. 1998. Involvement of phosphatidyl inositol-3-kinase and mitogen-activted protein kinase in AII-mediated enhanced expression of Gi proteins in vascular smooth muscle cells. Biochem Biophys Res Commun 251:570–575.

Ge C, Garcia R, Anand-Srivastava MB. 1999. Altered expression of Gi protein and adenylyl cyclase activity in hearts from one kidney one clip hypertensive rats: effect of captopril. J Hypertens 17:1617–1625.

Ge C, Garcia R, Anand-Srivastava MB. 2006. Enhanced expression of Giα proteins and adenylyl cyclase signalling in aortas from 1 kidney 1 clip hypertensive rats. Can J Physiol Pharmacol 84:739–746.

Gilman A. 1984. G proteins and dual control of adenylyl cyclase. Cell 36:577–579.

Graves LM, Bornfeldt KE, Raines EW, Potts BC, Macdonald SG, Ross R, Krebs EG. 1993. Protein kinase A antagonizes platelet-derived growth factor-induced signaling by mitogen activated protein kinase in human arterial smooth muscle cells. Proc Natl Acad Sci USA 90:10300–10304.

Hamet P, Franks DJ, Adnot S, Coquil JF.1980. Cyclic nucleotides in hypertension. Adv Cyclic Nucleotide Res 12:11–23.

Hammond HK, Ransnas LA, White FC, Bloor CM, Insel PA. 1988. Myocardial β-receptor number, pharmacological responsiveness and Gs in volume overload hypertrophy in PAGE (abstract). Circulation 78(Suppl II):II-163.

Hammond HK, Roth DA, Torsel PA, Ford CE, White FC, Maisel AS, Ziegler MG, Bloor CM, Insel PA. 1992. Myocardial beta-adrenergic receptor expression and signal transduction after chronic volume-overload hypertrophy and circulatory congestion. Circulation 85:269–280.

Hammond HR, Roth DA, McKirnan MD, Ping P. 1993. Regional myocardial down regulation of the inhibitory guanosine triphosphate binding protein (Gi alpha 2) and beta-adrenergic receptos in a porcine model of chronic myocardial ischemia. J Clin Invest 92:2644–2652.

Harris DM, Heather I, Cohn BS, Pesant S, Zhou R-H, Eckhart A. 2007. Vascular smooth muscle Gq signaling is involved in high blood pressure in both induced renal and genetic vascular smooth muscle-derived models of hypertension. Am J Physiol Heart Circ Physiol 1–35. doi:10.1152/ajpheart 00880.2007.

Hart MJ, Jiang X, Kozasa T, Roscoe W, Singer WD, Gilman AG, Sternweis PC, Bollag G. 1998. Direct stimulation of the guanine nucleotide exchange activity of p114 RhoGEF by G α_{13}. Science 280:2112–2114.

Hashim S, Anand-Srivastava MB. 2004. Losartan-induced attenuation of blood pressure in L-NAME hypertensive rats is associated with reversal of the enhanced expression of Giα protein. J Hypertens 22:181–190.

Heidemann SR, Joshi HC, Schechter A, Fletcher JR, Bothwell M. 1985. Tension and compression in the cytoskeleton of PC-12 neurites. II: Quantitative measurements. J Cell Biol 100:915–927.

Hein L, et al. 1997. Overexpression of angiotensin AT1 receptor transgene in the mouse myocardium produces a lethal phenotype associated with myocyte hyperplasia and heart block. Proc Natl Acad Sci USA 94:6391–6396.

Holmer SR, Bruckschlegl G, Schunkert H, Rataj DB, Krower EP, Riegger GA. 1996. Functional activity and expression of the myocardial post-receptor adenylyl cyclase system in pressure-overload hypertrophy in rat. Cardiovasc Res 31:719–728.

Hordijk PL, Verlaan I, Jalink K, van Corven EJ, Moolenaar WH.1994. cAMP abrogates the p21ras-mitogen-activated protein kinase pathway in fibroblasts. J Biol Chem 269:35334–35338.

Iglarz M, Schiffrin EL. 2003. Role of endothelin-1 in hypertension. Curr Hypertens Rep 5:144–148.

Itoh H, Kozasa T, Nagata S, Nakamura S, Katada T, Ui M, Iwai S, Ohtsuka E, Kawasaki H, Susuki K. 1986. Molecular cloning and sequence determination of cDNAs for alpha of the guanine nucleotide-binding proteins Gs, Gi and Go from rat brain. Proc Natl Acad Sci USA 83:3776–3786.

Itoh H, Toyama RI, Kozasa TI, Tsukamoto TI, Matsuoka M, Kaziro Y. 1988. Presence of three distinct molecular species of G protein: a subunit structure of rat cDNA and human genomic DNAs. J Biol Chem 263:6656–6664.

Jaw SP, Su DD, Matsumoto RR, Truong DD.1995. Alteration of brain levels of phosphoinositidase C-linked Gq alpha/G11 alpha proteins and motor function in rats after cardiac arrest. Stroke 26:1067–1070.

Jesnum S, Zaedi S, Maeda S, Togashi H, Yamaguchi I, Goto K, Niyauchi T. 2006. Endothelin antagonism suppresses plasma and cardiac endothelin-1 levels in SHR SPs at the typical hypertensive stage. Exp Biol Med 231:919–924.

Jones DT, Reed RR. 1987. Molecualr cloning of five GTP-binding protein cDNA species from rat olfactory neuroepithelium. J Biol Chem 262:14241–14249.

Kagiyama S, Eguchi S, Frank GD, Inagani T, Zhang YC, Philipps MI. 2002. Angiotensin II-induced cardiac hypertrophy and hypertension are attenuated by epidermal growth factor receptor antisense. Circulation 106:909–912.

Kagiyama S, Qian K, Kagiyama T, Philipps MI. 2003. Antisense to epidermal growth factor receptors prevents the development of left ventricular hypertrophy. Hypertension 41:824–829.

Kanagy NL, Webb RC. 1996. Increased responsiveness and decreased expression of G-proteins in deoxycorticosterone-salt hypertension. Hypertension 27:740–745.

Karliner JS, Kagiya T, Simpson PC. 1990. Effects of pertussis toxin on alpha 1-agonist-mediated phosphatidylinositide turnover and myocardial cell hypertrophy I neonatal rat ventricular myocytes. Experientia (Basel) 46:81–84.

Katoh Y, Komuro I, Yamaguchi H, Yazaki Y. 1992. Molecular mechanism of hypertrophied failing heart abnormalities of the diastolic properties and contractility. Jpn Circ J 56:694–700.

Keys JR, Greene EA, Koch WJ, Eckhart AD. 2002. Gq-coupled receptor agonists mediate cardiac hypertrophy via the vasculature. Hypertension 40:660–666.

Kinoshita S, Ghidhu A, Felder RA. 1989. Defective dopamine-1 receptor adenylate cyclase coupling in the proximal convoluted tubule from the spontaneously hypertensive rat. J Clin Invest 84:1849–1856.

Kirchengast M, Witte K, Stolpe K, Schilling L, Nedvetsky PI, Schmidt HH, Lemmer B. 2005. Effects of chronic endothelin ET(A) receptor blockade on blood pressure and vascular formation of cyclic guanosine 3′5′ monophosphate in spontaneously hypertensive rats. Arzneimittelforschung 55:498–504.

Kouchi I, Zolk O, Jockenhovel F, Itter G, Linz W, Cremers B, Bohm M. 2000. Increase in Giα protein accompanies progression of post-infarction remodelling in hypertensive cardiomyopathy. Hypertension 36:42–47.

Laks MM, Morady F, Swan HJC.1973. Myocardial hypertrophy produced by chronic infusion of subhypertensive doses of norepinephrine in the dog. Chest 64:75–78.

Li P, Zou AP, Al-kayed NJ, Rusch NJ, Harder DR. 1994. Guanine nucleotide-binding proteins in aortic smooth mucle from hypertensive rats. Hypertension 23:914–918.

Li Y, Anand-Srivastava MB. 2002. Inactivation of Gi proteins by pertussis toxin attenuated the development of hypertension in spontaneously hypertensive rats. Circ Res 91:247–254.

Li Y, Lappas G, Anand-Srivastava MB. 2007. Role of oxidative stress in angiotensin II-induced enhanced expression of Gi(alpha) proteins and adenylyl cyclase signaling in A10 vascular smooth muscle cells. Am J Physiol Heart Circ Physiol 292:H1922–H1930.

Limas CJ, Limas C.1978. Reduced number of β-adrenergic receptors in the myocardium of spontaneously hypertensive rats. Biochem Biophys Res Commun 83:710–714.

Litosch I, Wallis C, Fain JN. 1985. 5-Hydroxytryptamine stimulates inositol phosphate production in a cell-free system from blowfly salivary glands. Evidence for a role of GTP in coupling receptor activation to phosphoinositide breakdown. J Biol Chem 260:5464–5471.

Longabaugh JP, Vatner DE, Vatner SF, Homcy CJ. 1988. Decreased stimulatory guanosine triphosphate binding protein in dogs with pressure-overload left ventricular failure. J Clin Invest 81:420–424.

Maisel AS, Ransnas LA, Insel PA. 1990. Beta-adrenergic receptors and the Gs protein in myocardial ischemia and injury. Basic Res Cardiol 85:47–56.

Marcil J, Anand-Srivastava MB. 2001. G protein alterations in lymphocytes from spontaneously hypertensive rats: correlation with adenylyl cyclase. Cardiovasc Res 49:234–243.

Marcil J, Schiffrin EL, Anand-Srivastava MB. 1996. Aberrant adenylyl cyclase/cAMP signal transduction and G protein levels in platelet from hypertensive patients improve with antihypertensive drug therapy. Hypertension 28:83–90.

Marcil J, Thibault C, Anand-Srivastava MB. 1997. Enhanced expression of Gi protein precedes the development of blood pressure in spontaneously hypertensive rats. J Mol Cell Cardiol 29:1008–1022.

Marcil J, de Champlain J, Anand-Srivastava MB. 1998. Overexpression precedes the development of DOCA-salt-induced hypertension: relationship with adenylyl cyclase. Cardiovasc Rev 39:492–505.

McLellan AR, Milligan G, Houslay MD, Connell JM. 1993. G-proteins in essential hypertension: a study of human platelet plasma membranes. J Hypertens 11:543–549.

Michel MC, Farke W, Erdlbrugger W, Philipps T, Broddle OE. 1994. Ontogenesis of sympathetic responsiveness in spontaneously hypertensive rats. II. Renal G proteins in male and female rats. Hypertension 23:653–658.

Milano CA, Dolber PC, Rockman HA, Bond RA, Venable MED, Allen LF, Lefkowitz RJ. 1994. Proc Natl Acad Sci USA 91:10109–10113.

Moalic JM, Bourgeois F, Mainser P, Machida CA, Carre F, Chevalier B, Pilarque P, Swynghedauw B. 1993. Beta-1 adrenergic receptor and G alpha s mRNA in rat heart as a function of mechanical load and thyrosine intoxication. Cardiovasc Res 27:231–237.

Mondry A, Bourgeois F, Carre F, Swynghedauw B, Moalic JM. 1995. Decrease in beta-1-adrenergic and M2 muscarinic receptor mRNA levels and unchanged accumulation of mRNAs coding for G-alpha i-2 and G alpha s proteins in rat cardiac hypertrophy. J Mol Cell Cardiol 27:2287–2294.

Morgan HE, Baker KM. 1991. Cardiac hypertrophy. Mechanical, neural and endocrine dependence. Circulation 83:13–25.

Morishiuta KH, Migaki J, Miyazaki, Ogiharan T. 1992. Possible role of vascular rennin angiotensin system in hypertension and vascular hypertrophy. Hypertension 19:62–67.

Murakami T, Yasuda H. 1988. Rat heart cell membranes contain three substrates for cholera toxin-catalyzed ADP-ribosylation and a single substrate for pertussis toxin-catalyzed ADP-ribosylation. Biochem Biophys Res Commun 138:1355–1361.

Nieto JL, Diaz-Laviade I, Gnillen A, Garcia-Barreno P, Haro A. 1993. Cardiac beta-adrenoceptors, G-proteins and adenylate cyclase regulation during myocardial hypertrophy. Cell Signal 5:169–179.

Ohyanagi M, Yamamoto J, Nakamura K, Shibuyen J, Morita M, Masutani M, Arii T, Iwasaki T. 1995. Messenger RNA for the guanine nucleotide-binding regulatory protein (G-protein) is reduced in the acute ischemic myocardium. J Mol Cell Cardiol 27:1131–1139.

Pandey S, Anand-Srivastava MB. 1996. Modulation of G-protein expression by the angiotensin converting enzyme inhibitor captopril in hearts from spontaneously hypertensive rats: relationship with adenylyl cyclase. Am J Hypertens 9:833–837.

Pfaffinger PJ, Martin JM, Hunter DD, Nathanson NM, Hille B. 1985. GTP-binding proteins couple cardiac muscarinic receptors to a K channel. Nature 317:536–538.

Robishaw JD, Smigel MD, Gilman AG. 1986. Molecular basis for two forms of the G protein that stimulates adenylate cyclase. J Biol Chem 261:9587–9590.

Rodbell M, Krans HM, Pohl SL, Birnbaumer L. 1971. The glucagon-sensitive adenylyl cyclase system in plasma membranes of rat liver: effects of guanylyl nucleotides on binding of ^{125}I-glucagon. J Biol Chem 246:1872–1876.

Sexl V, Mancusi G, Holler C, Gloria-Maercker E, Schutz W, Freissmuth M. 1997. Stimulation of the mitogen-activated protein kinase via the A2A-adenosine receptor in primary human endothelial cells. J Biol Chem 272:5792–5799.

Shermuly RT, Dony E, Ghofrani HA, Pullamcetti S, Savai R, Roth M, Sydykov A, Lai YJ, Weissmann N, Seeger W, Grimminger F. 2005. Reversal of experimental pulmonary hypertension by PDGF inhibition. J Clin Invest 115:2811–2821.

Shi CS, Sinnarajah S, Cho H, Kozasa T, Kehri JH. 2000. $G_{11}\,\alpha$-mediated PYK2 activation. PYK2 is a mediator of $G_{11}\,\alpha$-induced serum response element-dependent transcription. J Biol Chem 275:24470–24476.

Simon MI, Srathmann MP, Gantam N. 1991. Diversity of G proteins in signal transduction. Science 252:802–808.

Simpson PJ. 1983. Norepinephrine stimulated hypertrophy of cultured rat myocardial cells is an alpha-1 adrenergic response. Clin Invest 72:732–738.

Sims C, Ashby K, Douglas JG. 1992. Angiotensin II-induced changes in guanine nucleotide binding and regulatory proteins. Hypertension 19:146–152.

Siri FM. 1988. Sympathetic change during development of cardiac hypertrophy in aortic constricted rats. Am J Physiol 255:H452–H457.

Spiegel AM. 1987. Signal transduction by guanine nucleotide binding proteins. Mol Cell Endocrinol 49:1–16.

Spiegel AM, Shenker A, Weinstein LS. 1992. Receptor-effector coupling by G-proteins: implications for normal and abnormal signal transduction. Endocrinol Rev 13:536–565.

Strasser RH, Krimmer J, Braun-Dullaeus R, Marguentant R, Kubler W. 1990. Dual sensitization of the adrenergic system in early myocardial ischemia: independent regulation of the beta-adrenergic receptors and the adenylyl cyclase. J Mol Cell Cardiol 22:1405–1423.

Strathmann MP, Simon MI. 1991. G alpha 12 and alpha 13 subunits define a fourth class of G protein alpha subunits. Proc Natl Acad Sci USA 88:5582–5586.

Stryer L, Bourne HR. 1986. G-proteins: a family of signal transducers. Annu Rev Cell Biol 2:391–419.

Susani EE, Manders WT, Knight DR, Vatner DE, Vatner SF, Homcy CJ. 1989. One hour of myocardial ischemia decreases the activity of the stimulatory guanine-nucleotide regulatory protein Gs. Circ Res 65:1145–1150.

Suyama K, Saito K, Chen G, Pan PS, Manji HK, Potter WZ. 1995. Alterations in cyclic AMP generation and G-protein subunits following transient ischemia in gerbil hippocampus. J Cereb Blood Flow Metab 15:877–885.

Takahashi, T, Tang T, Lai NC, Roth DV, Rebolledo B, Saito M, Lew WYW, Clopton P, Hammond HK. 2006. Increased cardiac adenylylcyclase expression is associated with increased survival after myocardial infarction. Circulation 114:388–396.

Takeishi Y, Chu G, Kirkpatrick DL, Wakasaki H, Li Z, Kranias EG, King GL, Walsh RA. 1998. *In vivo* phosphorylation of cardiac troponin I by PKCβ2 decreases cardiomyocyte calcium responsiveness and contractility in transgenic mouse heart. J Clin Invest 102:72–78.

Takeishi Y, Ping P, Bolli R, Kirkpatrick DL, Hoit BD, Walsh RA. 2000. Transgenic overexpression of constitutely active protein kinase C-E causes concentric cardiac hypertrophy. Circ Res 86:1218–1223.

Takenaka K, Kanaho Y, Nagata K, Sakai N, Yamada H, Nozawa Y. 1991. Ischemia of rat brain decreases pertussis toxin-catalyzed [^{32}P]ADP-ribosylation of GTP-binding proteins (Gil and Go) in membranes. J Cereb Blood Flow Metab II:155–160.

Tang WJ, Gilman G. 1991. Type-specific regulation of adenylyl cyclase by G protein beta gamma submits. Science 254:1500–1503.

Taussig R, Iniguez-Lluhi JA, Gilman AG. 1993. Inhibition of adenylyl cyclase by Gi alpha. Science 261:218–221.

Terzic A, Puceart M, Vassort G, Vogel SM. 1993. Cardiac alpha 1-adrenoceptors: an overview. Pharmacol Rev 45:147–175.

Thibault C, Anand-Srivastava MB. 1992. Altered expression of G-protein mRNA in spontaneously hypertensive rat hearts exhibit altered expression of G-proteins. FEBS Lett 313:160–164.

Thorburn A.1994. Ras activity is required for phenylephrine-induced activation of mitogen-activated protein kinase in cardiac muscle cells. Biochem Biophys Res Commun 205:1417–1422.

Tobise K, Abe M, Onodera S. 1991. Prolonged period of global ischemia causes no change in the GTP-binding proteins in the isolated perfused heart. Eur J Pharmacol 208:183–187.

Triggle R, Tabrizchi R. 1993. Changes in vascular smooth muscle function in hypertension. Clin Med J 106:250–257.

Trimder D, Philipps RA, Risvani J, Stephenson JM Johnston CI. 1992. Regulation of vasopressin receptor in deoxycorticosterone acetate salt hypertension. Hypertension 20:569–574.

Van den Ende R, Balink HD, ichel MC, VanZwieten PA. 1994. Influence of ischemia and reperfusion on cardiac signal transduction G protein content, adenylyl cyclase activity, cyclic AMP content and forskolin and dibutyryl cyclic AMP-induced inotropy in the rat Langendorff heart. Fundam Clin Pharmacol 8:408–416.

Wahlander H, Wickman A, Isgaard J, Friberg P. 1999. Interaction between renin-angiotensin system and insulin-like growth factor-1 in aorta-caval fistula. Acta Physiol Scand 165:143–154.

Wankerl M, Schwartz K. 1005. Calcium transport proteins in the nonfailing and failing heart: gene expression and function. J Mol Med 73:487–493.

Wedegaertner PB, Wilson PT, Bourne HR. 1995. Lipid modifications of trimeric G proteins. J Biol Chem 270:503–506.

Wickman KD, Iniguez Lluhl JA, Davenport PA, Taussig R, Krapivinsky GB, Linder ME, Gilman AG, Clapham DE. 1994. Recombinant G-protein beta gamma-subunits activate the muscarinic-gated atrial potassium channel. Nature 368:255–257.

Woodcock EA, Farder JW, Johnston CI. 1979. Decreased cardiac beta-adrenergic receptors in deoxycorticosterone-salt and renal hypertensive rats. Circ Res 45:560–565.

Wolff AA, Hines DK, Karliner JS. 1994. Preserved beta-adrenoceptor-mediated adenylyl cyclase activity despite receptor and post-receptor dysfunction in acute myocardial ischemia. Am Heart J 128:542–550.

Wong YH, Conklin BB, Bourne HR. 1992. Gz-mediated hormonal inhibition of cyclic AMP accumulation. Science 255:339–342.

Yamazaki T, Komuro I, Zou Y, Kudoh S, Shiojima I, Hiroi Y, Mizuno T, Aikawa R, Takano H, Yasaki Y. 1997. Norepinephrine induces the raf-1 kinase/mitogen-activated protein kinase cascade through both alpha 1- and beta-adrenoceptors. Circulation 95:1260–1268.

Yao H, Labudda K, Rim C, Capodieci P, Loda M, Stork PJS. 1995. Cyclic adenosine monophosphate can convert epidermal growth factor into a differentiating factor in neuronal cells. J Biol Chem 270:20748–20753.

Yatani A, Brown AM. 1989. Rapid beta-adrenergic modulation of cardiac calcium channel currents by a fast G protein pathway. Science 245: 71–74.

Zeng YY, Benishin CG, Pank PKT. 1991. Increased responsiveness of the adenylate cyclase (AC) signalling system in splenocyte membranes from spontaneously hypertensive rats (SHR) (abstract). FASEB J 5:A1597.

Zhai P, Yamamoto M, Galeotti J, Liu J, Masurekar M, Thaisz J, Irie K, Holle E, Yu, X, Kupershmidt S, Roden DM, Wagner T, Yatani A, Vatner DE, Vatner SF, Sadoshima J. 2005. Cardiac-specific overexpression of AT1 receptor mutant lacking $G\alpha q/G\alpha$ I coupling causes hypertrophy and bradycardia in transgenic mice. J Clin Invest 115:3045–3056.

Zierhut W, Zimmer HG. 1989. Significance of myocardial α- and β-adrenoceptors in catecholamine-induced cardiac hypertrophy. Circ Res 65:1417–1425.

Zou Y, Komuro I, Yamazaki T, Kudoh S, Uozumi H, Kadowaki T, Yazaki Y. 1999. Both Gs and Gi proteins are critically involved in isoproterenol-induced cardiomyocyte hypertrophy. J Biol Chem 274:9760–9770.

Chapter 2
β-Adrenoceptor-Linked Signal Transduction Mechanisms in Congestive Heart Failure

Melissa R. Dent, Tushi Singal, Paramjit S. Tappia, Rajat Sethi, and Naranjan S. Dhalla

Abstract The cardiac β-adrenoceptor (β-AR)-mediated signal transduction system is composed of β_1- and β_2-ARs, stimulatory (G_s) and inhibitory (G_i) guanine nucleotide binding proteins, adenylyl cyclase (AC), and cAMP-dependent protein kinase (PKA). The activation of β_1- and β_2-ARs is known to increase heart function by promoting Ca^{2+}-movements in cardiomyocytes through the stimulation of G_s-proteins, activation of AC and PKA enzymes, and phosphorylation of the target sites. The activation of PKA increases the phosphorylation of some myofibrillar proteins resulting in cardiac relaxation, whereas PKA-mediated phosphorylation of some nuclear proteins results in cardiac hypertrophy. The activation of β_2-AR has also been shown to affect G_i-proteins, stimulate mitogen-activated protein kinase, and increase protein synthesis by enhancing gene expression. β_1- and β_2-ARs as well as AC are thought to be regulated by PKA- and protein kinase C (PKC)-mediated phosphorylations directly; both PKA and PKC also regulate β-AR indirectly through the involvement of β-AR kinase (βARK), β-arrestins, and $G_{\beta\gamma}$-protein subunits. Differences in the extent of defects in the β-AR signaling system have been identified in different types of heart failure to explain the attenuated response of the failing heart to sympathetic stimulation or catecholamine infusion. A decrease in β_1-AR density, an increase in the level of G_i-proteins, and overexpression of βARK are usually associated with heart failure; however, these changes have been shown to be dependent on the type and stage of heart failure as well as region of the heart. Both local and circulating renin–angiotensin systems, sympathetic nervous system, and endothelial cell function appears to regulate the status of β-AR signal transduction pathway in the failing heart. In this article, we highlight alterations in different components and regulators of the β-AR signal transduction pathway and review the biological basis for altered β-AR-mediated signal transduction in heart failure due to different etiologies as well as discuss the pharmacologic blockade of the β-adrenergic system as an approach for the treatment of congestive heart failure.

A. K. Srivastava, M. B. Anand-Srivastava (eds.) *Advances in Biochemistry in Health and Disease.* Signal Transduction in the Cardiovascular System in Health and Disease,

Introduction

Activation of the sympathetic nervous system (SNS) (adrenergic system) in response to a variety of stimuli is essential to maintain homeostasis in a constantly changing environment, and in fact is known to regulate myocardial function on a beat–beat or short-term basis (Lamba and Abraham 2000). The physiological and metabolic responses to sympathetic activation are mediated through the action of endogenous catecholamines, norepinephrine (NE) and epinephrine, on adrenoceptors (ARs) (Stiles et al. 1984; Clark and Cleland 2000; Lamba and Abraham 2000). Based on the pharmacological and molecular structure, ARs are divided into two broad classes, α-ARs and β-ARs; however, this review will focus on the β-adrenergic system and its role in the development of congestive heart failure (CHF). Although α-ARs are also altered in CHF, no effort will be made to deal with this issue at this time. It should be mentioned that β-ARs are of three types—β_1-ARs, β_2-ARs, and β_3-ARs—and these differ significantly with respect to the types of cellular responses they mediate (Dhalla et al. 1977; Stiles et al. 1984; Brodde 1991; Lamba and Abraham 2000). Furthermore, it is repertoire and quantity of different β-ARs that determine the overall response of an organ to the circulating catecholamines. Acute changes in cardiac function are controlled predominantly by β-AR intracellular signaling pathways. The signal transduction pathways triggered by agonist occupancy of β-ARs are key regulators of the heart rate, systolic and diastolic function, as well as myocardial metabolism (Lamba and Abraham 2000). However, biology of the β-AR signaling pathway is altered dramatically in CHF (Clark and Cleland 2000) and in fact, adrenergic over-activity is one of the hallmarks of CHF which is associated with a poor prognosis (Clark and Cleland 2000). Initially, increased β-AR signaling allows the heart to adapt quickly to work loads that may vary, allowing the heart to increase its output within a matter of seconds by increasing the pacemaker frequency and myocardial contractility (Lamba and Abraham 2000). Although the adrenergic drive functions as a control mechanism that maintains cardiac performance at an acceptable level, prolonged activation of the SNS exerts a direct adverse action on the heart and produces deleterious peripheral effects (Clark and Cleland 2000). Apart from this classic role in acute regulation of mechanical and electrical functions of the heart, β-ARs may also be involved in long-term control of myocytes including cell survival and apoptosis under various conditions (Communal and Colucci 2005; Weil and Schunkert 2006). With respect to the development of CHF, what amounts to an initially appropriate compensatory adrenergic response to diminished myocardial performance, eventually results in an inappropriate or maladaptive response. Thus, the β-AR signaling mechanisms associated with CHF of different etiologies will be discussed in this article.

β-AR Pharmacology

It is now well known that the positive inotropic action of catecholamines is primarily mediated by their interaction with β-ARs on the cardiac cell surface (Dhalla et al. 1977; Stiles et al. 1984; Brodde 1991). The availability of selective agonists and antagonists as well as radioligand binding techniques have confirmed the classification of β-AR into β_1-, β_2-, and β_3-ARs (Buxton et al. 1987; Brodde et al. 1989). These have been cloned and the probes derived from these genes have been used to examine the regulation of these receptor proteins (Collins et al. 1981; Tate et al. 1991; Saffitz and Liggett 1992). Such molecular studies have also indicated the existence of a fourth subtype, namely, β_4-ARs. The sequences of β_1- and β_2-AR have a 71% and 54% amino acid identity in the transmembrane domains and in overall sequence, respectively (Searles et al. 1995). The ratio of β_1- to β_2-ARs in the myocardium is about 4:1; this ratio seems to depend on the chamber of the heart as well as species employed for investigation and the type of heart disease. The mammalian heart expresses primarily β_1-ARs (75–85%) and a substantial number of β_2-ARs are also detected in cardiac tissue (Collins et al. 1981; Tate et al. 1991; Saffitz and Liggett 1992). However, the β_2-ARs are mainly expressed in cells such as endothelial cells, fibroblasts, and vascular smooth muscle cells, which are present in the heart. Although the physiological relevance of cardiac β_3-ARs is not well understood, recent evidence suggests that β_3-ARs promote a negative inotropic effect (Tavernier et al. 2003). Since β_3-ARs can modulate relaxation of smooth muscle, the extent to which the β_3-AR-associated negative inotropic effect is direct or secondary to peripheral vasodilation is unknown (Dessy et al. 2004). In contrast, the putative β_4-ARs appear to be akin to β_1- and β_2-ARs in promoting a positive inotropic effect, but their biochemical and pharmacological characteristics are poorly defined (Kohout et al. 2001).

It has been demonstrated that excessive amounts of circulating catecholamines trigger changes in the β-AR system, leading to deterioration of ventricular function (Lamba and Abraham 2000). This is thought to be an adaptive mechanism of the heart to protect compromised myocardium from catecholamine overstimulation. In CHF, due to either idiopathic dilated cardiomyopathy or ischemic heart disease, β_1-ARs selectively undergo downregulation due to uncoupling of the receptors from their respective signaling pathways (Wallukat 2002); these desensitization changes lead to a marked attenuation of the myocardial response to catecholamines. This process of desensitization is initiated by a family of Ser/Thr kinases known as G-protein-coupled receptor kinases (GRKs) that phosphorylate the agonist-coupled G-protein-coupled receptors (GPCRs) (Lamba and Abraham 2000). GPCR desensitization requires not only the kinase activity of GRKs, but also the action of a second protein family, β-arrestins that bind to phosphorylated receptors. β-Arrestin binding subsequently directs the internalization of desensitized GPCRs that can lead to receptor downregulation, or receptor recycling back to the sarcolemmal membrane and stimulation of intracellular signaling pathways (Lamba and Abraham 2000).

Adrenoceptor Signaling in the Heart

The ARs belong to the superfamily of GPCRs, which contain a conserved structure of seven transmembrane α-helices linked by three alternating intracellular and extracellular loops. According to the classic paradigm of GPCR signaling, binding of the ligand to the receptor induces a sequence of conformational changes that result in its coupling to a heterotrimeric G-protein. Activated G-proteins then dissociate into G_{α} and $G_{\beta\gamma}$ subunits, each capable of modulating the activity of a variety of intracellular effector molecules. Thus, receptors that couple to G stimulatory (G_s) or G inhibitory (G_i) proteins modulate the activity of adenylyl cyclase (AC) to generate the second messenger cAMP and subsequently activate cAMP-dependent protein kinase A (PKA). The α subunit (45 to 52 kDa) of G_s-protein was found to be different from that (39 to 41 kDa) of G_i-protein. Furthermore, ADP ribosylation of the α subunit in G_s-protein is catalyzed by cholera toxin whereas that in G_i-protein is catalyzed by pertussis toxin. Studies at the molecular level for both G_s- and G_i-proteins have revealed that genes for these proteins are not co-regulated (Itoh et al. 1988; Kozasa et al. 1988). On the other hand, G_q-coupled receptors including α-ARs stimulate phospholipase C that in turn generates diacylglycerol and inositol 1,4,5-trisphosphate and activates protein kinase C. The nature of the intracellular response to catecholamine stimulation therefore depends not only on the type of activated receptors and their expression levels, but also on the type of G-proteins they couple to and intracellular pathways activated by the various second messengers. Different isoforms of GRKs (GRK2, GRK3, and GRK5) are known to regulate β-ARs in the heart (Brodde 1991). In fact, GRK2 has been shown to play a critical role in the transition of cardiac hypertrophy to heart failure in transgenic mice overexpressing α_{1B}-ARs (Iaccarino et al. 2001).

β-ARs are most commonly associated with regulation of metabolic pathways and have been described to both inhibit and stimulate AC activity. These GPCRs initiate the production of cAMP with subsequent activation of PKA and thus regulate diverse metabolic and functional events (Homcy et al. 1991). PKA activation is a critical step in mediation of the positive inotropic effect of catecholamines through phosphorylation of L-type Ca^{2+} channels in the sarcolemmal membrane and phospholamban in the sarcoplasmic reticulum to regulate Ca^{2+} movements in the cardiomyocytes. Although the exact mode of coupling G-proteins with AC is not clear, both genetic and biochemical evidence indicate that there are multiple forms of AC with a molecular mass in the range of 120 to 150 kDa (Manolopoulos et al. 1995). Of the nine isoforms of AC, the presence of types II to VII has been detected in cardiac tissue but types V and VI are abundant in the mammalian heart (Yu et al. 1995; Sunahara et al. 1996). The catalytic subunit of AC, which is involved in the formation of cAMP from ATP, is activated by cations such as Mn^{2+} as well as by forskolin whereas other agents such as NaF, Gpp(NH)p, cholera toxin, and pertussis toxin are considered to stimulate the enzyme activity through their interaction with G-proteins. Since different hormones including angiotensin II (Ang II), endothelin I, and NE bind to receptors which are coupled to G_q-proteins, molecular targeting of $G_{q\alpha}$-protein in transgenic mouse models has also shown its involvement in both adaptive and maladaptive responses of the heart to stress (Adams et al. 1998; Dorn and Brown 1999; Sabri et al. 2002). Various studies with transgenic mouse models have revealed that specific overexpression of β_1-ARs, β_2-ARs, G_s-proteins, and AC results in an enhanced cardiac function (Milano et al. 1994; Bond et al. 1995; Xiao et al. 1999).

Recent evidence from cell culture and transgenic mouse models suggest distinct differences between β_1- and β_2-ARs in their ability to modulate the process of apoptosis (Milano et al. 1994; Saito et al. 2000; Morisco et al. 2001; Shizukuda and Buttrick 2002). Expression of these receptors in a double knockout mouse model has revealed that stimulation of β_1-ARs is involved in apoptosis whereas that of β_2-ARs elicits cell survival (Zhu et al. 2001). It appears that overexpression of the human β_1-ARs increases expression of proapoptotic proteins like bax (Bisognano et al. 2000). Also, the proapoptotic influence of the β_1-AR pathway has been attributed to the activation of calcineurin via increased intracellular Ca^{2+} through L-type channels; the activation of calcineurin dephosphorylates Bad, permitting it to heterodimerize with the antiapoptotic proteins Bcl-2 and Bcl-xl (Saito et al. 2000). The differential effect of the β-ARs on the induction of apoptosis may be due to the fact that the β_2-AR coupling to $G_{\alpha i}$ is cardioprotective (Iwai-Kanai and Hasegawa 2004). Interestingly, Zhu et al. (2001) have uncovered a connection between the antiapoptotic effects of β_2-ARs and stimulation of a pertussis toxin-sensitive, phophatidylinositol 3-kinase (PI3K) and Akt-PKB pathway, which may be one of the several antiapoptotic pathways. Overexpression of $G_{s\alpha}$-proteins has been reported to increase heart function and produce apoptosis whereas that for $G_{i\alpha}$-proteins has been shown to attenuate β-AR stimulation (Geng et al. 1999; Janssen et al. 2002). In spite of the complexities of transgenic models and some conflicting results, these experiments have provided evidence that the β_1-AR signal transduction is required for maintaining heart function, but overexpression of any of the components of this system can cause cardiac hypertrophy, apoptosis, and heart dysfunction. In addition to the promotion of apoptosis, production of cytotoxicity via Ca^{2+} overload, and increased free radical generation, adrenergic activation is the major stimulus to pathologic hypertrophy (Wallukat 2002). Thus, in view of the critical role of β-AR signaling pathway in influencing cardiac contractility, any change in the components of this system under pathological conditions can be seen to impair signal transduction mechanisms in the myocardium. However, abnormalities in β-ARs, G-proteins, and AC in failing human hearts appear to depend on the etiology of CHF (Bristow et al. 1991; Bohm et al. 1992; Bristow and Feldman 1992; Steinfath et al. 1992).

Altered β-AR Signaling During Various Cardiac Pathologies

Various pathologic factors such as pressure overload (PO) or volume overload (VO), ischemic reperfusion injury, myocardial infarction, and different types of cardiomyopathies are associated with an excessive stimulation of SNS. This sustained adrenergic drive is not only considered to result in the downregulation of β-AR but is also believed to produce a depression in myocardial reserve, Ca^{2+}-cycling, myocardial energetics, in addition to inducing fetal gene program. All of these abnormalities are considered to contribute toward cardiac dysfunction in CHF. A schematic representation of these events is shown in Figure 2.1. However, this article will be focused

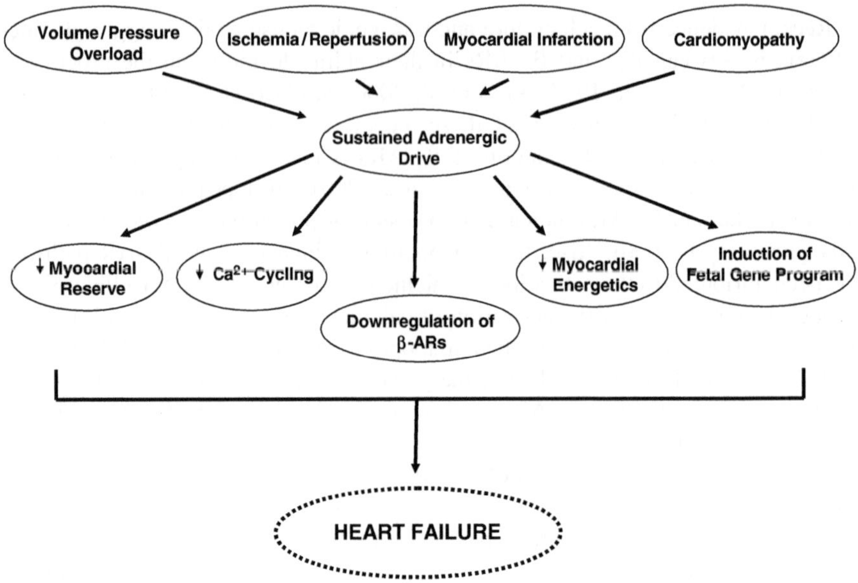

Fig. 2.1 Schematic representation of the mechanisms involved in the development of heart failure due to different etiologies.

on discussion of four components of the β-AR mechanism—β-AR, G-proteins, AC, and PKA—which are known to regulate the major routes of Ca^{2+} entry in the sarcolemmal membrane as well as Ca^{2+}-release from the sarcoplasmic reticular stores in addition to modifying the sensitivity of myofibrils to Ca^{2+} (Figure 2.2). Since changes in β-AR mechanism seem to depend on the type of CHF, it is planned to describe alterations in different components of β-AR systems in different types of heart diseases.

Volume and Pressure Overload Hypertrophy-Induced Changes

Alterations in β_1-AR signaling account for the occurrence of pathologic hypertrophy; this has been shown by reverse remodeling studies using the β-AR blocking agents in the failing human hearts (Frigerio and Roubina 2005). In addition, reversal of fetal gene induction, the molecular hallmark of pathologic hypertrophy, is accompanied by reverse cardiac remodeling in response to β-AR blockers (Lowes et al. 1999). It should be pointed out that cardiac hypertrophy is generally categorized into two broad types: PO-induced hypertrophy and VO-induced hypertrophy. PO occurs in many clinical settings that include hypertension, mitral valve stenosis, and aortic valve stenosis resulting in concentric cardiac remodeling. Thus, an increase in pressure is offset by an increase in the ventricular wall thickness (Carabello 2002). The other type of cardiac hypertrophy due to VO occurs in anemia, heart block, regurgitant mitral or aortic valves, atrial or ventricular septal defects, or other congenital

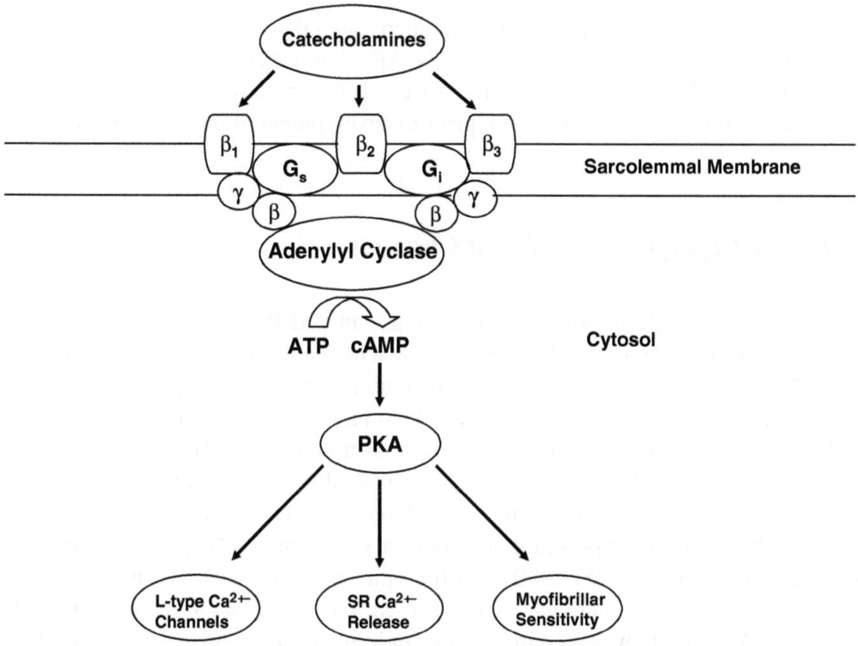

Fig. 2.2 Components of the cardiomyocyte β-adrenergic signaling system and their physiological effects.

diseases, resulting in eccentric cardiac hypertrophy. Dilatation of the left ventricle (LV) chamber occurs via elongation of the surrounding myocytes—the result of sarcomeric replication in series (Carabello 1996). Alterations in β-AR density, AC activity, and G-protein have been identified in cardiac hypertrophy, which normally precedes heart failure due to PO (Galinier et al. 1994; Iaccarino et al. 1999). Further, modification of cardiac AC activities by changes in the G-protein function has been observed in hypertension (Bohm et al. 1993). Changes in the β-AR signaling system have also been shown to be involved in the development of CHF due to both PO and VO. In fact, PO-induced CHF in guinea pigs was associated with an increase in β-AR density without any changes in their affinity (Karliner et al. 1980). A wide variety of changes in the β-AR-linked signal transduction mechanism have also been reported in heart failure induced by rapid pacing and VO (Di Fusco et al. 2000). On the other hand, cardiac hypertrophy and heart failure due to VO induced by aortocaval shunt in rats were associated with hypersensitivity of the myocardium to β-AR stimulation (Wang et al. 2003). We have also shown that the upregulation of the β-AR system as well as changes in the subcellular distribution of regulatory proteins, GRK isoforms, and β-arrestins in the failing hearts due to VO, were partially prevented by treatment of these animals with angiotensin-converting enzyme (ACE) inhibitors and Ang II type 1 receptor (AT_1R) antagonists (Wang et al. 2005). However, such alterations in cardiac hypertrophy and late stages of CHF in this experimental model remain to be examined. Other animal studies of CHF have also shown myocardial β_1-AR expression and increased GRK activity (Anderson et al. 1999). Importantly, increasing levels of β_1-AR often precede the development of overt clinical CHF and may represent a novel early marker for cardiac dysfunction and a potential target for intervention prior to development of end-stage CHF.

Ischemia-Reperfusion-Induced Changes

An increase in β-AR density and an increase in cAMP formation due to catecholamines have been reported in myocardial ischemia due to coronary occlusion in dogs (Mukherjee et al. 1982). Although increased density of β-ARs was also seen in CHF in dogs, this change was associated with a loss of high affinity for these receptors as well as uncoupling of β-ARs from G-proteins (Maisel et al. 1985, 1987; Freissmuth et al. 1987). Other investigators have also observed an increase in the β-AR density in the ischemic myocardium from dogs and calves and the activities of AC in the absence or presence of different stimulants of G_s-proteins were depressed (Vatner et al. 1988, 1990). On the other hand, no changes in the density of β-ARs and basal AC activity were observed, but a depression in isoproterenol-stimulated AC activity was seen in ischemic or hypoxic dog hearts (Freissmuth et al. 1987; Karliner et al. 1989). Ischemic guinea pig hearts showed an increase and a decrease in the β-AR densities in cell surface and cytoplasmic membranes (Maisel et al. 1985, 1987), respectively, whereas opposite results were obtained upon exposing the neonatal rat cardiomyocytes to hypoxia (Rocha-Singh et al. 1991). The changes in ARs and post-receptor mechanisms including changes in mRNA levels due to ischemia/hypoxia seem to depend on the experimental model employed and the degree as well as duration of the reperfusion injury (Will-Shahab et al. 1991; Bernstein et al. 1992; van den Ende et al. 1994; Ohyanagi et al. 1995). Nonetheless, studies from our laboratory have indicated that the ischemia-reperfusion-induced changes in β-AR signal transduction mechanism in the myocardium are mediated through the generation of oxidative stress (Persad et al. 1997, 1998).

Cardiomyopathy-Induced Changes

The AC activities due to the stimulation of β-receptors and G_s-proteins were increased in adriamycin-induced cardiomyopathy in rabbits (Calderone et al. 1991). On the other hand, no alterations in β-AR density, G-proteins, or AC activities were seen in adriamycin-induced cardiomyopathy in rats (Fu et al. 1991). Depressions in β-ARs and AC activities in the absence or presence of various stimulants were noted in catecholamine-induced cardiomyopathy in rats in addition to an increase and loss of G_i- and G_s-proteins, respectively (Meszaros and Levai 1992; Muller et al. 1993; Zhou et al. 1995). Rats with monocrotaline-induced right heart cardiomyopathy showed depressions in β_1-AR density and AC activities in the presence of isoproterenol and Gpp(NH)p without any changes in the absence or presence of NaF and forskolin as well as in the β_2-receptor density; these alterations were chamber-specific (Pela et al. 1990; Yoshie et al. 1994). High level of overexpression of β_2-ARs was also found to cause heart failure in a mouse model of cardiomyopathy, which was prevented by expression of GRK2 (Freeman et al. 2001). Overexpression of GRK2 inhibitor of gene-targeted mice was observed to prevent heart failure and improve cardiac function (Rockman et al. 1998). Conflicting results showing either an increase (Ikegaya et al. 1992) or no change (Kessler et al. 1989; Horackova et al. 1991) in β-AR receptor density have also been reported in hamster cardiomyopathy. However, AC activities in the presence of different stimulants as well as the levels of G_s-protein were found to be depressed in the cardiomyopathic hamster hearts (Panagia et al. 1984; Kessler et al. 1989; Ikegaya et al. 1992; Urasawa et al. 1992), but the basal enzyme activity was normal (Panagia et al. 1984; Kessler et al. 1989) and the level of G_i-protein was increased (Urasawa et al. 1992). No alterations in the levels of mRNA encoding G_s-proteins in cardiomyopathic hamsters were detected (Kessler et al. 1989), while information concerning changes in mRNA specific for AC in failing hearts is still lacking. Increased level of G_i-proteins as well as uncoupling of β_1-AR from AC has been suggested to explain the attenuated responses of cardiomyopathic hamster hearts to catecholamines (Witte et al. 1993; Kawamoto et al. 1994). The work carried out in our laboratory has revealed that changes in the β-ARs, AC, and G-proteins are dependent on the stage of CHF in cardiomyopathic hamsters (Sethi et al. 1994).

Myocardial Infarction-Induced Changes

Several investigators have reported a wide variety of alterations in different components of the β-ARs, G-proteins, and AC system in heart dysfunction in failing human heart as well as in various experimental animal models of heart failure (Dhalla et al. 1997; Wang and Dhalla 2000). Some efforts have been made to understand the mechanisms of attenuated responses of failing hearts to catecholamines; these changes are invariably seen in all types of CHF. Both β-AR density and responsiveness to inotropic stimulation are significantly reduced in failing human hearts (Lamba and Abraham 2000). The loss of cardiac β_1-ARs is critical, since this translates to a larger overall percentage of β_2-ARs and emphasizes their distinct signaling properties. A decrease in the density of β-AR was observed in CHF in dogs with pulmonary artery constriction as well as in rats with myocardial infarction (Dhalla et al. 1992). There was no evidence of any change in β-AR density or isoproterenol-stimulated AC activity in heart failure due to myocardial infarction in rats (Hammond et al. 1993) and dogs (Strasser et al. 1990); however, experiments with rats at two stages of myocardial infarction revealed defects in both β-ARs and postreceptor sites associated with attenuated responses to catecholamines (Sethi and Dhalla 1995). Heart dysfunction in rats with nonocclusive coronary artery constriction without any myocardial infarction was associated with depressions in β-AR density, G_s-protein activity, and isoproterenol-stimulated AC activity (Meggs et al. 1991). CHF due to rapid pacing in dogs was found to decrease β-receptor density, G_s-protein, and AC activity (Marzo et al. 1991; Juneau et al. 1992). Some work employing the molecular biology techniques has shown a decrease in the levels of mRNA specific for β_1-receptors (Bristow and Feldman 1992) and an increase in the levels of mRNA for G_i-protein without any changes in mRNA for G_s-protein in failing human hearts (Eschenhagen et al. 1992).

In spite of the extensive research for identifying defects in the β-AR-mediated signal transduction in failing hearts from patients and experimental animals, several issues remain unresolved. Although marked changes in the pattern of plasma hormones including renin–angiotensin, catecholamines, atrial natriuretic peptide, endothelium-derived relaxing factor as well as endothelin in CHF have been identified (Hodsman et al. 1988; Basu et al. 1996), the exact role of these changes in the genesis of signal transduction abnormalities is far from clear. Some investigators have recently emphasized the importance of local mechanisms such as cardiac renin–angiotensin system (RAS) and NE transport in sympathetic nerve endings in the myocardium, rather than changes in circulating renin–angiotensin and catecholamines, in the development of cardiac dysfunction in heart failure (Bohm et al. 1992, 1995; Yoshikawa et al. 1994; Wollert et al. 1994; Basu et al. 1996; Ganguly et al. 1997). Such local alterations can be seen to explain the differential behavior of the left and right ventricle with respect to changes in adrenergic mechanisms during the development of CHF due to myocardial infarction (Sethi et al. 1997, 1998). Although Yoshida et al. (2001a) failed to observe differences in the isoproterenol-induced response of single cardiomyocytes from left and right ventricles of the 8-week infarcted rats, the concentration (10 nM) of isoproterenol used in this study was too low to elicit any response. Furthermore, the biased selection of single cardiomyocyte employed may have also been another factor for their failure to observe changes. Although Ca^{2+}-handling abnormalities have been reported to occur in hearts failing due to myocardial infarction (Dixon et al. 1990; Afzal and Dhalla 1992; Sethi et al. 2006), the role of these changes in causing an impairment of the signal transduction mechanisms has not been established. In this regard, it should be pointed out that augmented Ca^{2+}-fluxes in the myocardium have been shown to exert a negative regulation of the adrenergic stimulation and the AC activation (Frace et al. 1993; Wang et al. 2002). Furthermore, diltiazem and β-AR antagonists, which are known to regulate Ca^{2+} movements in the myocardium, have been shown to exert beneficial effects with respect to defects in transmembrane signaling due to different pharmacological and pathophysiological interventions (Brodde 1991; Chapados et al. 1992). Likewise, blockade of RAS by ACE inhibitors and AT_1R antagonists was found to prevent changes in β-AR mechanisms in heart failure (Forster et al. 1994; Bohm et al. 1998; Yoshida et al. 2001b; Makino et al. 2003).

Some studies have indicated that alterations in β-AR signal transduction mechanisms were attenuated by different agents, which are known to block β_1-ARs in the heart (Asai et al. 1999; Asano et al. 2001; Liu et al. 2002). In this regard, it should be noted that both SNS and RAS are known to be activated in different types of heart failure. In fact, activation of both SNS (Communal et al. 1998; Iaccarino et al. 1998;

Leineweber et al. 2002) and RAS (Bohm et al. 1998; Bohlender et al. 2001) has been shown to be associated with desensitization of the β-AR mechanisms. Treatment of hypertensive rats with ACE inhibitors and AT_1R antagonists has been reported to normalize the augmented sympathetic activity (K.-Laflamme et al. 1997). Since the activation of RAS was seen before any change in the heart or plasma upon inducing PO (Akers et al. 2000), it appears that the activation of RAS plays a dominant role in changing the sensitivity of failing hearts to β-adrenergic stimulation. Also, we have demonstrated that downregulation of the β-AR G-protein AC system in the LV from the failing heart due to myocardial infarction is attenuated by blockade of the RAS (Sethi et al. 2003). In addition, we have shown that treatment of the infarcted animals with propionyl L-carnitine, a metabolic therapy, not only improved cardiac function but also attenuated defects in the β-adrenergic mechanisms (Sethi et al. 2004). However, whether the desensitization of β-adrenergic mechanisms is due to changes in cardiac gene expression for each of these components, or alterations in the regulatory factors for the β-ARs in the LV of the infarcted animals, remains to be determined. Likewise, nothing is known regarding the molecular mechanisms of hypersensitivity of the right ventricle to catecholamines in the failing hearts due to myocardial infarction.

β-Adrenergic System Blockade

In view of abnormalities in β-AR signal transduction mechanisms during the development of CHF, it is important to discuss the pharmacologic blockade for the β-AR system as an important approach in the treatment of CHF. Over the past 25 years, β-adrenergic blockade has been one of the most successful therapies aimed at attenuating neurohormonal overactivation used in the treatment of patients with CHF (Frigerio and Roubina 2005). β-AR antagonists lead to a decreased risk of death and hospital admission and tend to reverse the adverse effects of prolonged adrenergic stimulation in patients with CHF (Frigerio and Roubina 2005; Chizzola et al. 2006). β-AR blockers for the management of cardiovascular disease are well-established and these agents are widely recommended as important parts of antihypertensive regimens as well as preferred therapies for patients at high risk for coronary heart disease, including those with heart failure. Among three of the most common treatments for heart failure (ACE inhibitors, AT_1 receptor blockers, and β-AR blockers), β-AR blockers have shown a more permanent benefit than that obtained with ACE inhibitors alone. Furthermore, β-AR blockers most likely lower blood pressure and provide target organ protection by mechanisms such as inhibition of RAS, central inhibition of SNS outflow, and slowing of heart rate with a decrease in cardiac output. Efficacy of β-AR blockers appears to be superior to that of ACE inhibitors and AT_1 receptor blockers; however, almost all patients with heart failure enrolled in β-AR blocker clinical trials were already taking ACE inhibitors. Therefore, it is difficult to indicate if β-AR blockade would produce the same results in the absence of RAS antagonism. Recently, the CARMEN (Carvedilol and ACE Inhibitor

Remodeling Mild Heart Failure Evaluation Trial) clinical trial examined the effects of enalapril and carvedilol in patients with mild heart failure (Remme et al. 2004). They found that enalapril alone or carvedilol alone did not cause a significant reduction in LVESV whereas a combined treatment with these agents resulted in a significant reduction in LVESV. In contrast, a previous study comparing carvedilol and captopril showed a significant decrease in LVESV and increased ejection fraction in patients treated with carvedilol only (Khattar et al. 2001).

The three main β-AR blockers that have proven to be of clinical benefit in heart failure are metoprolol, carvedilol, and bisoprolol. These β-AR blockers produce a significant and sustained improvement in ejection fraction and reverse remodeling in addition to reductions in LV sphericity and mitral regurgitation in patients with CHF (Lowes et al. 1999). The newer β-AR blocker used in the treatment of CHF, carvedilol, is a racemic mixture of $R(+)$- and $S(-)$-enantiomers mainly metabolized by cytochrome P (CYP2D6) as well as partially metabolized by CYP1A2 and CYP2C9. Carvedilol is a nonselective β-AR antagonist that leads to systemic arterial vasodilation without reflex tachycardia due to concomitant antagonism of vascular β_1-AR and myocardial β-ARs (DasGupta et al. 1991). Carvedilol is a blocker for β_1-, β_2-, and α_1-ARs raising controversy on whether this type of blocker is superior to selective β_1-AR blockers (Bristow et al. 2003). Therapy of heart failure patients with carvedilol produced a safe and tolerable reduction in heart rate and improved LV function (improved LV systolic shortening fraction and LVEF) within 2 months of treatment. Recent studies have shown that a higher dosage than previously used (75 mg versus 42 mg/day) results in greater benefit in the treatment of heart failure (Chizzola et al. 2006). Carvedilol has also shown to result in improved neuronal adrenergic capture in the myocardial effector cell. Thus, a possible mechanism of action for β-AR antagonists is through improved neuronal reuptake of NE. It is pointed out that AT_1 receptor blockers were found useful in combination with carvedilol in CHF patients (Iwata et al. 2006). Carvedilol is often administered in combination with amiodarone for the treatment of arrhythmias and heart failure (Fukumoto et al. 2005). Many studies have shown a causal relationship between β-AR blockade and reverse remodeling. A small study investigating the effects of metoprolol found that LV function deteriorated after withdrawal of metoprolol in the treatment of heart failure patients and improved LV function after readministration (Waagstein et al. 1989). Furthermore, dose of β-AR blockers has been correlated with the degree of effect on LV volume and ejection fraction (Bristow et al. 1996). Specifically, carvedilol dose was inversely related to mortality (Bristow et al. 1996) and in fact carvedilol therapy and dose were found to be predictors of cardiac size normalization and improved cardiac function. However, the reverse remodeling of the cardiac sympathetic neurons was not associated with an alteration in plasma NE levels. Treatment of heart failure with medication that antagonizes β-ARs tends to reverse the adverse effects of prolonged adrenergic stimuli.

The efficacy of β-AR blockade appears to be superior in achieving reverse remodeling and this is more directly dose-related than that of ACE inhibitors (Frigerio and Roubina 2005). Recently, in a clinical trial examining the effects of metoprolol and atenolol, it was found that patients with heart failure treated with metoprolol experienced an 88% survival rate; survival rate for patients treated with atenolol was 78% as compared to 48% in control groups. This clearly indicates that metoprolol and atenolol have a favorable effect on the survival rate of patients with CHF. Specifically, metoprolol is considerably more effective than atenolol (Celic et al. 2005). In controlled clinical trials, bisoprolol, carvedilol, and metoprolol exerted favorable effects on survival (CIBIS-II 1999; MERIT-HF Study Group 1999; Patrianakos et al. 2005). Bucindolol had little effect on mortality rates and xamoterol was associated with an increased risk of death (The Xamoterol in Severe Heart Failure Study Group 1990; Domanski et al. 2003).

Nebivolol, a third-generation β_1-AR selective blocker, increases endothelial NO release causing peripheral vasodilation. Nebivolol shows both a high degree of selectivity for β_1-ARs and an ability to stimulate endothelial NO production (Patrianakos et al. 2005). In this study Patrianakos et al. (2005) compared the effects of nebivolol and carvedilol on LV function and exercise capacity in patients with mild to moderate heart failure. It was observed that nebivolol is a safe choice with many beneficial effects on systolic and diastolic LV function and exercise capacity after 1 year of treatment. However, nebivolol produced an initial deterioration in exercise capacity, which was not observed with carvedilol treatment (Patrianakos et al. 2005). Thus, it was concluded that carvedilol produces more favorable effects as compared to nebivolol. The treatment of patients with carvedilol produces a faster effect than nebivolol in terms of improved diastolic dysfunction because of the added antagonism of β_2- and α_1-ARs. The effects of carvedilol may be due to restored Ca^{2+}-homeostasis because chronic adrenergic stimulation has detrimental cardiotoxic effects and causes abnormal Ca^{2+}-handling. This may partially explain the observed diastolic restoration and changes in LV filling pattern (Patrianakos et al. 2005). It should be noted that CHF patients exhibit high body mass index (Horwich et al. 2001; Davos et al. 2003) and treatment with β-AR blockers can further increase total body fat mass and total body fat content (Lainscak et al. 2006).

β-AR blockers differ with respect to many pharmacologic properties in β_1/β_2-AR selectivity, intrinsic sympathomimetic activity, and vasodilatory capabilities (Weber 2005). The β-AR blocker class is divided into three generations of antagonists. The first-generation β-AR blockers, such as propanolol, exert blockade on both β_1- and β_2-receptors equally and are therefore termed nonselective β-AR blockers (Weber 2005). The second-generation β-AR blockers (metoprolol, bisoprolol, and atenolol) have a higher affinity binding to β_1-receptors and are referred to as selective β-AR blockers. At higher doses, these selective β-AR blockers may exert some inhibition of β_2-AR as well (Weber 2005). The third-generation β-AR blockers differ from the previous two in their vasodilatory activity. In this class, there are β-AR blockers such as labetalol, which is nonselective with a higher affinity for the β_1-receptor than for the β_2-receptors, whereas carvedilol is a β_1 selective blocker but becomes less selective at higher doses (Weber 2005). Bucindolol, also a third-generation β-AR blocker, is completely nonselective for all ARs. These three third-generation β-AR blockers provide some vasodilatory action through blockade of the α_1-AR thereby regulating endothelial function and vasoconstriction in peripheral blood vessels. The newest β-AR blocker, nebivolol, is a third-generation blocker that has higher β_1-AR selectivity compared with other β-AR blockers in addition to its vasodilatory effects (Weber 2005). Some β-AR blockers (acebutol, penbutolol, pindolol) are capable of both stimulating β-ARs as well as opposing the transmission of SNS signaling. This combination has been shown to attenuate the decreases in heart rate and cardiac output and increases in peripheral vascular resistance associated with β-AR blockade (Weber 2005).

Conclusions

A detailed analysis of the existing literature reveals that changes in β-ARs, G-proteins, and AC depend on the type and stage of heart disease as well as area of the heart and the type of membrane preparations from failing hearts employed for investigations (Pela et al. 1990; Persad et al. 1997; Rockman et al. 1998). The results on β-AR mechanisms in different experimental models as well as in patients with heart failure support the view that alterations in this system depend on the underlying type of the disease (Matsuda et al. 2000; Sayar et al. 2000; Wallukat 2002; Leineweber et al. 2003). The ARs, G-proteins, and AC systems are either unchanged, upregulated, or downregulated in failing myocardium. Furthermore, alterations may occur in one component of the system without changes in the others; such a discrepancy in results seems to depend on the type and stage of the heart disease. Although most of the work in this field has been carried out on myocardial tissues from patients with heart disease (Schotten et al. 2000), it should be recognized

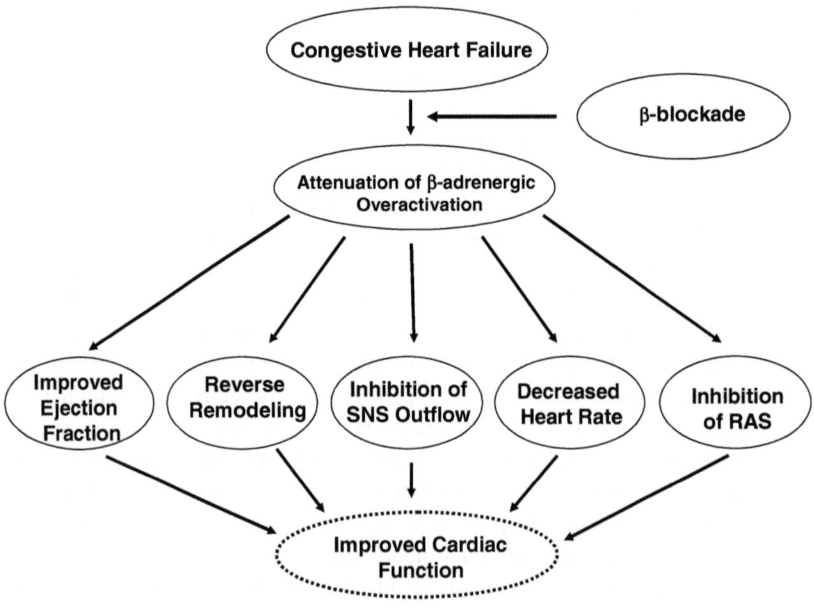

Fig. 2.3 The beneficial mechanisms of action due to β-blockade in congestive heart failure.

that all of these patients were on different cardiac drugs, and thus the results are difficult to interpret in terms of pathophysiological changes in CHF. Nonetheless, it is evident from the foregoing discussion that cardiac dysfunction in patients with CHF is improved upon the blockade of β-AR. This beneficial effect is primarily due to the reduction of the β-adrenergic overdrive which results in reversal of cardiac remodeling, depression in heart rate, and decrease in the overactivity of RAS (Figure 2.3).

Acknowledgments The work reported in this article was supported by a grant from the Heart and Stroke Foundation of Manitoba. M.R.D. is a predoctoral fellow of the Heart and Stroke Foundation of Canada. T.S. is a predoctoral fellow of the Manitoba Health Research Council.

References

Adams, J.W., Sakata, Y., Davis, M.G., Sah, V.P., Wang, Y., Liggett, S.B., Chien, K.R., Brown, J.H., and Dorn, G.W., 2nd. 1998. Enhanced $G_{\alpha q}$ signaling: a common pathway mediates cardiac hypertrophy and apoptotic heart failure. Proc. Natl. Acad. Sci. USA 95:10140–10145.

Afzal, N., and Dhalla, N.S. 1992. Differential changes in left and right ventricular SR calcium transport in congestive heart failure. Am. J. Physiol. Heart Circ. Physiol. 262:H868–H874.

Akers, W.S., Cross, A., Speth, R., Dwoskin, L.P., and Cassis, L.A. 2000. Renin-angiotensin system and sympathetic nervous system in cardiac pressure-overload hypertrophy. Am. J. Physiol. Heart Circ. Physiol. 279:H2797–H2806.

Anand-Srivastava, M.B., de Champlain, J., and Thibault, C. 1993. DOCA-salt hypertensive rat hearts exhibit altered expression of G-proteins. Am. J. Hypertens. 6:72–75.

Anderson, K.M., Eckhart, A.D., Willette, R.N., and Koch, W.J. 1999. The myocardial β-adrenergic system in spontaneously hypertensive heart failure (SHHF) rats. Hypertension 33:402–407.

Asai, K., Yang, G.P., Geng, Y.J., Takagi, G., Bishop, S., Ishikawa, Y., Shannon, R.P., Wagner, T.E., Vatner, D.E., Homcy, C.J., and Vatner, S.F. 1999. β-adrenergic receptor blockade arrests myocyte damage and preserves cardiac function in the transgenic $G_{s\alpha}$ mouse. J. Clin. Invest. 104:551–558.

Asano, K., Zisman, L.S., Yoshikawa, T., Headley, V., Bristow, M.R., and Port, J.D. 2001. Bucindolol, a nonselective β_1- and β_2-adrenergic receptor antagonist, decreases β-adrenergic receptor density in cultured embryonic chick cardiac myocyte membranes. J. Cardiovasc. Pharmacol. 37:678–691.

Basu, S., Sinha, S.K., Shao, Q., Ganguly, P.K., and Dhalla, N.S. 1996. Neuropeptide Y modulation of sympathetic activity in myocardial infarction. J. Am. Coll. Cardiol. 27:1796–1803.

Bernstein, D., Doshi, R., Huang, S., Strandness, E., and Jasper, J.R. 1992. Transcriptional regulation of left ventricular β-adrenergic receptors during chronic hypoxia. Circ. Res. 71:1465–1471.

Bisognano, J.D., Weinberger, H.D., Bohlmeyer, T.J., Pende, A., Raynolds, M.V., Sastravaha, A., Roden, R., Asano, K., Blaxall, B.C., Wu, S.C., Communal, C., Singh, K., Colucci, W., Bristow, M.R., and Port, D.J. 2000. Myocardial-directed overexpression of the human β_1-adrenergic receptor in transgenic mice. J. Mol. Cell. Cardiol. 32:817–830.

Bohlender, J., Hildenbrand, U., Wagner, K.D., Gunther, J., Hempel, P., Schlegel, W.P., Luft, F.C., Krause, E.G., and Bartel, S. 2001. Myocardial adrenergic dysfunction in rats with transgenic, human renin-dependent hypertension. J. Hypertens. 19:1453–1463.

Bohm, M., Gierschik, P., and Erdmann, E. 1992. Quantification of $G_{i\alpha}$-proteins in the failing and nonfailing human myocardium. Basic Res. Cardiol. 87:37–50.

Bohm, M., Gierschik, P., Knorr, A., Schmidt, U., Weismann, K., and Erdmann, E. 1993. Cardiac adenylyl cyclase, β-adrenergic receptors, and G proteins in salt-sensitive hypertension. Hypertension 22:715–727.

Bohm M., La Rosee, K., Schwinger, R.H., and Erdmann, E. 1995. Evidence for reduction of norepinephrine uptake sites in the failing human heart. J. Am. Coll. Cardiol. 25:146–153.

Bohm, M., Zolk, O., Flesch, M., Schiffer, F., Schnabel, P., Stasch, J.P., and Knorr A. 1998a. Effects of angiotensin II type 1 receptor blockade and angiotensin-converting enzyme inhibition on cardiac β-adrenergic signal transduction. Hypertension 31:747–754.

Bohm, M., Ettelbruck, S., Flesch, M., van Gilst, W.H., Knorr, A., Maack, C., Pinto, Y.M., Paul, M., Teisman, A.C., and Zolk, O. 1998b. β-adrenergic signal transduction following carvedilol treatment in hypertensive cardiac hypertrophy. Cardiovasc. Res. 40:146–155.

Bond, R.A., Leff, P., Johnson, T.D., Milano, C.A., Rockman, H.A., McMinn, T.R., Apparsundaram, S., Hyek, M.F., Kenakin, T.P., and Allen, L.F. 1995. Physiological effects of inverse agonists in transgenic mice with myocardial overexpression of the β_2-adrenoceptor. Nature 374:272–276.

Bristow, M.R., and Feldman, A.M. 1992. Changes in the receptor-G protein-adenylyl cyclase system in heart failure from various types of heart muscle disease. Basic Res. Cardiol. 87:15–35.

Bristow, M.R., Anderson, F.L., Port, J.D., Skerl, L., Hershberger, R.E., Larrabee, P., O'Connell, J.B., Renlund, D.G., Volkman, K., Murray, J., et al. 1991. Differences in β-adrenergic neuroeffector mechanisms in ischemic versus idiopathic dilated cardiomyopathy. Circulation 84:1024–1039.

Bristow, M.R., Gilbert, E.M., Abraham, W.T., Adams, K.F., Fowler, M.B., Hershberger, R.E., Kubo, S.H., Narahara, K.A., Ingersoll, H., Krueger, S., Young, S., and Shusterman, N. 1996. Carvedilol produces dose-related improvements in left ventricular function and survival in subjects with chronic heart failure. MOCHA Investigators. Circulation 94:2807–2816.

Bristow, M.R., Feldman, A.M., Adams, K.F., Jr., and Goldstein, S. 2003. Selective versus nonselective β-blockade for heart failure therapy: are there lessons to be learned from the COMET trial? J. Card. Fail. 9:444–453.

Brodde, O.E. 1991. Pathophysiology of the β-adrenoceptor system in chronic heart failure: consequences for treatment with agonists, partial agonists or antagonists? Eur. Heart J. 12 Suppl. F:54–62.

Brodde, O.E., Zerkowski, H.R., Doetsch, N., Motomura, S., Khamssi, M., and Michel, M.C. 1989. Myocardial β-adrenoceptor changes in heart failure: concomitant reduction in β_1- and β_2-adrenoceptor function related to the degree of heart failure in patients with mitral valve disease. J. Am. Coll. Cardiol. 14:323–331.

Buxton, B.F., Jones, C.R., Molenaar, P., and Summers, R.J. 1987. Characterization and autoradiographic localization of β-adrenoceptor subtypes in human cardiac tissues. Br. J. Pharmacol. 92:299–310.

Calderone, A., de Champlain, J., and Rouleau, J.L. 1991. Adriamycin-induced changes to the myocardial β-adrenergic system in the rabbit. J. Mol. Cell. Cardiol. 23:333–342.

Carabello, B.A. 1996. Models of volume overload hypertrophy. J. Card. Fail. 2:55–64.

Carabello, B.A. 2002. Concentric versus eccentric remodeling. J. Card. Fail. 8:S258–S263.

Celic, V., Pencic, B., Dekleva, M., Dimkovic, S., and Kocijancic, M. 2005. [Metoprolol and atenolol in mild-to-moderate chronic heart failure: comparative study]. Srp. Arh. Celok. Lek. 133:242–247.

Chapados, R.A., Gruver, E.J., Ingwall, J.S., Marsh, J.D., and Gwathmey, J.K. 1992. Chronic administration of cardiovascular drugs: altered energetics and transmembrane signaling. Am. J. Physiol. Heart Circ. Physiol. 263:H1576–H1586.

Chizzola, P.R., de Freitas, H.F.G., Marinho, N.V.S., Mansur, J.A., Meneghetti, J.C., and Bocchi, E.A. 2006. The effect of β-adrenergic receptor antagonism in cardiac sympathetic neuronal remodeling in patients with heart failure. Int. J. Cardiol. 106:29–34.

CIBIS-II Investigators and Committees. 1999. The Cardiac Insufficiency Bisoprolol Study II (CIBIS-II): a randomised trial. Lancet 353:9–13.

Clark, A.L., and Cleland, J.G.F. 2000. The control of adrenergic function in heart failure: therapeutic intervention. Heart Fail. Revi. 5:101–114.

Collins, S., Caron, M.G., and Lefkowitz, R.J. 1991. Regulation of adrenergic receptor responsiveness through modulation of receptor gene expression. Annu. Rev. Physiol. 53:497–508.

Communal, C., and Colucci, W.S. 2005. The control of cardiomyocyte apoptosis via the β-adrenergic signaling pathways. Arch. Mal. Coeur Vaisse 98:236–241.

Communal, C., Singh, K., Pimental, D.R., Colucci, W.S. 1998. Norepinephrine stimulates apoptosis in adult rat ventricular myocytes by activation of the beta-adrenergic pathway. Circulation 98: 1329–1334.

DasGupta, P., Broadhurst, P., and Lahiri, A. 1991. The effects of intravenous carvedilol, a new multiple action vasodilatory β-blocker, in congestive heart failure. J. Cardiovasc. Pharmacol. 18:S12–S16.

Davos, C.H., Doehner, W., Rauchhaus, M., Cicoira, M., Francis, D.P., Coats, A.J., Clark, A.L., and Anker, S.D. 2003. Body mass and survival in patients with chronic heart failure without cachexia: the importance of obesity. J. Card. Fail. 9:29–35.

Dessy, C., Moniotte, S., Ghisdal, P., Havaux, X., Noirhomme, P., and Balligand, J.L. 2004. Endothelial β_3-adrenoceptors mediate vasorelaxation of human coronary microarteries through nitric oxide and endothelium-dependent hyperpolarization. Circulation 110:948–954.

Dhalla, N.S., Ziegelhoffer, A., and Harrow, J.A. 1977. Regulatory role of membrane systems in heart function. Can. J. Physiol. Pharmacol. 55:1211–1234.

Dhalla, N.S., Dixon, I.M., Suzuki, S., Kaneko, M., Kobayashi, A., and Beamish, R.E. 1992. Changes in adrenergic receptors during the development of heart failure. Mol. Cell. Biochem. 114:91–95.

Dhalla, N.S., Wang, X., Sethi, R., Das, P.K., and Beamish, R.E. 1997. β-adrenergic linked signal transduction mechanisms in failing hearts. Heart Fail. Rev. 2:55–65.

Di Fusco, F., Hashim, S., and Anand-Srivastava, M.B. 2000. Volume overload cardiac hypertrophy exhibits decreased expression of $G_{s\alpha}$ and not of $G_{i\alpha}$ in heart. Am. J. Physiol. Cell. Physiol. 279:C990–C998.

Dixon, I.M., Lee, S.L., and Dhalla, N.S. 1990. Nitrendipine binding in congestive heart failure due to myocardial infarction. Circ. Res. 66:782–788.

Dixon, I.M., Hata, T., and Dhalla, N.S. 1992. Sarcolemmal calcium transport in congestive heart failure due to myocardial infarction in rats. Am. J. Physiol. Heart Circ. Physiol. 262:H1387–H1394.

Domanski, M.J., Krause-Steinrauf, H., Massie, B.M., Deedwania, P., Follmann, D., Kovar, D., Murray, D., Oren, R., Rosenberg, Y., Young, J., Zile, M., and Eichhorn, E.; BEST Investigators. 2003. A comparative analysis of the results from 4 trials of β-blocker therapy for heart failure: BEST, CIBIS-II, MERIT-HF, and COPERNICUS. J. Card. Fail. 9: 354–363.

Dorn, G.W., 2nd, and Brown, J.H. 1999. G_q signaling in cardiac adaptation and maladaptation. Trends Cardiovasc. Med. 9:26–34.

Eschenhagen, T., Mende, U., Nose, M., Schmitz, W., Scholz, H., Haverich, A., Hirt, S., Doring, V., Kalmar, P., Hoppner, W., et al. 1992. Increased messenger RNA level of the inhibitory G protein α subunit $G_{i\alpha}$-2 in human end-stage heart failure. Circ. Res. 70:688–696.

Forster, C., Naik, G.O., and Larosa, G. 1994. Myocardial β-adrenoceptors in pacing-induced heart failure: regulation by enalapril? Can. J. Physiol. Pharmacol. 72:667–672.

Frace, A.M., Mery, P.F., Fischmeister, R., and Hartzell, H.C. 1993. Rate-limiting steps in the β-adrenergic stimulation of cardiac calcium current. J. Gen. Physiol. 101:337–353.

Freeman, K., Lerman, I., Kranias, E.G., Bohlmeyer, T., Bristow, M.R., Lefkowitz, R.J., Iaccarino, G., Koch, W.J., and Leinwand, L.A. 2001. Alterations in cardiac adrenergic signaling and calcium cycling differentially affect the progression of cardiomyopathy. J. Clin. Invest. 107:967–974.

Freissmuth, M., Schutz, W., Weindlmayer-Gottel, M., Zimpfer, M., and Spiss, C.K. 1987. Effects of ischemia on the canine myocardial β-adrenoceptor-linked adenylate cyclase system. J. Cardiovasc. Pharmacol. 10:568–574.

Frigerio, M., and Roubina, E. 2005. Drugs for left ventricular remodeling in heart failure. Am. J. Cardiol. 96:10L–18L.

Fu, L.X., Bergh, C.H., Hoebeke, J., Liang, Q.M., Sjogren, K.G., Waagstein, F., and Hjalmarson, A. 1991. Effect of metoprolol on activity of β-adrenoceptor coupled to guanine nucleotide binding regulatory proteins in adriamycin-induced cardiotoxicity. Basic Res. Cardiol. 86:117–126.

Fukumoto, K., Kobayashi, T., Komamura, K., Kamakura, S., Kitakaze, M., and Ueno, K. 2005. Stereoselective effect of amiodarone on the pharmacokinetics of racemic carvedilol. Drug Metab. Pharmacokinet. 20:423–427.

Galinier, M., Senard, J.M., Valet, P., Arias, A., Daviaud, D., Glock, Y., Bounhoure, J.P., and Montastruc, J.L. 1994. Cardiac β-adrenoceptors and adenylyl cyclase activity in human left ventricular hypertrophy due to pressure overload. Fundam. Clin. Pharmacol. 8:90–99.

Ganguly, P.K., Dhalla, K.S., Shao, Q., Beamish, R.E., and Dhalla, N.S. 1997. Differential changes in sympathetic activity in left and right ventricles in congestive heart failure after myocardial infarction. Am. Heart J. 133:340–345.

Geng, Y.J., Ishikawa, Y., Vatner, D.E., Wagner, T.E., Bishop, S.P., Vatner, S.F., and Homcy, C.J. 1999. Apoptosis of cardiac myocytes in $G_{s\alpha}$ transgenic mice. Circ. Res. 84:34–42.

Hammond, H.K., Roth, D.A., McKirnan, M.D., Ping, P. 1993. Regional myocardial downregulation of the inhibitory guanosine triphosphate-binding protein (Gi alpha 2) and beta-adrenergic receptors in a poreine model of chronic episodic myocardial ischemia. J. Clin. Invest. 92: 2644–2652.

Hodsman, G.P., Kohzuki, M., Howes, L.G., Sumithran, E., Tsunoda, K., Johnston, C.I. 1998. Neurohormonal responses to chronic myocardial infarction in rats. Circulation 78: 376–381.

Homcy, C.J., Vatner, S.F., and Vatner, D.E. 1991. β-adrenergic receptor regulation in the heart in pathophysiologic states: abnormal adrenergic responsiveness in cardiac disease. Annu. Rev. Physiol. 53:137–159.

Horackova, M., Beresewicz, A., Rowden, G., and Wilkinson, M. 1991. Neurohumoral regulation of excitation-contraction coupling in ventricular myocytes from cardiomyopathic hamsters. Cardiovasc. Res. 25:1023–1034.

Horwich, T.B., Fonarow, G.C., Hamilton, M.A., MacLellan, W.R., Woo, M.A., and Tillisch, J.H. 2001. The relationship between obesity and mortality in patients with heart failure. J. Am. Coll. Cardiol. 38:789–795.

Iaccarino, G., Tomhave, E.D., Lefkowitz, R.J., and Koch, W.J. 1998. Reciprocal *in vivo* regulation of myocardial G protein-coupled receptor kinase expression by β-adrenergic receptor stimulation and blockade. Circulation 98:1783–1789.

Iaccarino, G., Dolber, P.C., Lefkowitz, R.J., and Koch, W.J. 1999. β-adrenergic receptor kinase-1 levels in catecholamine-induced myocardial hypertrophy: regulation by β- but not α_1-adrenergic stimulation. Hypertension 33:396–401.

Iaccarino, G., Keys, J.R., Rapacciuolo, A., Shotwell, K.F., Lefkowitz, R.J., Rockman, H.A., and Koch, W.J. 2001. Regulation of myocardial βARK1 expression in catecholamine-induced cardiac hypertrophy in transgenic mice overexpressing α_{1B}-adrenergic receptors. J. Am. Coll. Cardiol. 38:534–540.

Ikegaya, T., Kobayashi, A., Hong, R.B., Masuda, H., Kaneko, M., and Noboru, Y. 1992. Stimulatory guanine nucleotide-binding protein and adenylate cyclase activities in Bio 14.6 cardiomyopathic hamsters at the hypertrophic stage. Mol. Cell. Biochem. 110:83–90.

Itoh, H., Toyama, R., Kozasa, T., Tsukamoto, T., Matsuoka, M., and Kaziro, Y. 1988. Presence of three distinct molecular species of G_i protein α subunit. Structure of rat cDNAs and human genomic DNAs. J. Biol. Chem. 263:6656–6664.

Iwai-Kanai, E., and Hasegawa, K. 2004. Intracellular signaling pathways for norepinephrine- and endothelin-1-mediated regulation of myocardial cell apoptosis. Mol. Cell. Biochem. 259:163–168.

Iwata, A., Miura, S., Nishikawa, H., Kawamura, A., Matsuo, Y., Sako, H., Kumagai, K., Matsuo, K., and Saku, K. 2006. Significance of combined angiotensin II receptor blocker and carvedilol therapy in patients with congestive heart failure and arginine variant. J. Cardiol. 47:1–7.

Janssen, P.M., Schillinger, W., Donahue, J.K., Zeitz, O., Emami, S., Lehnart, S.E., Weil, J., Eschenhagen, T., Hasenfuss, G., and Prestle, J. 2002. Intracellular β-blockade: overexpression of $G_{\alpha i2}$ depresses the β-adrenergic response in intact myocardium. Cardiovasc. Res. 55:300–308.

Juneau, C., Calderone, A., and Rouleau, J.L. 1992. Myocardial β-adrenergic and mechanical properties in pacing-induced heart failure in dogs. Am. J. Physiol. 262:H1458–H1467.

Karliner, J.S., Barnes, P., Brown, M., and Dollery, C. 1980. Chronic heart failure in the guinea pig increases cardiac α_1- and β-adrenoceptors. Eur. J. Pharmacol. 67:115–118.

Karliner, J.S., Stevens, M.B., Honbo, N., and Hoffman, J.I. 1989. Effects of acute ischemia in the dog on myocardial blood flow, β receptors, and adenylate cyclase activity with and without chronic β blockade. J. Clin. Invest. 83:474–481.

Kawamoto, H., Ohyanagi, M., Nakamura, K., Yamamoto, J., and Iwasaki, T. 1994. Increased levels of inhibitory G protein in myocardium with heart failure. Jpn. Circ. J. 58:913–924.

Kessler, P.D., Cates, A.E., Van Dop, C., and Feldman, A.M. 1989. Decreased bioactivity of the guanine nucleotide-binding protein that stimulates adenylate cyclase in hearts from cardiomyopathic Syrian hamsters. J. Clin. Invest. 84:244–252.

Khattar, R.S., Senior, R., Soman, P., van der Does, R., and Lahiri, A. 2001. Regression of left ventricular remodeling in chronic heart failure: Comparative and combined effects of captopril and carvedilol. Am. Heart J. 142:704–713.

K.-Laflamme, A., Oster, L., Cardinal, R., and de Champlain, J. 1997. Effects of renin-angiotensin blockade on sympathetic reactivity and β-adrenergic pathway in the spontaneously hypertensive rat. Hypertension 30:278–287.

Kohout, T.A., Takaoka, H., McDonald, P.H., Perry, S.J., Mao, L., Lefkowitz, R.J., and Rockman, H.A. 2001. Augmentation of cardiac contractility mediated by the human β_3-adrenergic receptor overexpressed in the hearts of transgenic mice. Circulation 104:2485–2491.

Kozasa, T., Itoh, H., Tsukamoto, T., and Kaziro, Y. 1988. Isolation and characterization of the human G_s α gene. Proc. Natl. Acad. Sci. USA 85:2081–2085.

Lainscak, M., Keber, I., and Anker, S.D. 2006. Body composition changes in patients with systolic heart failure treated with β blockers: a pilot study. Int. J. Cardiol. 106:319–322.

Lamba, S., and Abraham, W.T. 2000. Alterations in adrenergic receptor signaling in heart failure. Heart Fail Rev 5:7–16.

Leineweber, K., Brandt, K., Wludyka, B., Beilfuss, A., Ponicke, K., Heinroth-Hoffmann, I., and Brodde, O.E. 2002. Ventricular hypertrophy plus neurohumoral activation is necessary to alter the cardiac β-adrenoceptor system in experimental heart failure. Circ. Res. 91:1056–1062.

Leineweber, K., Seyfarth, T., Abraham, G., Gerbershagen, H.P., Heinroth-Hoffmann, I., Ponicke, K., and Brodde, O.E. 2003. Cardiac β-adrenoceptor changes in monocrotaline-treated rats: differences between membrane preparations from whole ventricles and isolated ventricular cardiomyocytes. J. Cardiovasc. Pharmacol. 41:333–342.

Lemoine, H., Schonell, H., and Kaumann, A.J. 1988. Contribution of β_1- and β_2-adrenoceptors of human atrium and ventricle to the effects of noradrenaline and adrenaline as assessed with (-)-atenolol. Br. J. Pharmacol. 95:55–66.

Liu, X., Callaerts-Vegh, Z., Evans, K.L., and Bond, R.A. 2002. Chronic infusion of β-adrenoceptor antagonist and inverse agonists decreases elevated protein kinase A activity in transgenic mice with cardiac-specific overexpression of human β_2-adrenoceptor. J. Cardiovasc. Pharmacol. 40:448–455.

Lowes, B.D., Gill, E.A., Abraham, W.T., Larrain, J.R., Robertson, A.D., Bristow, M.R., and Gilbert, E.M. 1999. Effects of carvedilol on left ventricular mass, chamber geometry, and mitral regurgitation in chronic heart failure. Am. J. Cardiol. 83:1201–1205.

Maisel, A.S., Motulsky, H.J., and Insel, P.A. 1985. Externalization of β-adrenergic receptors promoted by myocardial ischemia. Science 230:183–186.

Maisel, A.S., Motulsky, H.J., Ziegler, M.G., and Insel, P.A. 1987. Ischemia- and agonist-induced changes in α- and β-adrenergic receptor traffic in guinea pig hearts. Am. J. Physiol. Heart Circ. Physiol. 253:H1159–H1166.

Makino, T., Hattori, Y., Matsuda, N., Onozuka, H., Sakuma, I., and Kitabatake, A. 2003. Effects of angiotensin-converting enzyme inhibition and angiotensin II type 1 receptor blockade on β-adrenoceptor signaling in heart failure produced by myocardial Infarction in rabbits: reversal of altered expression of β-adrenoceptor kinase and $G_{i\alpha}$. J. Pharmacol. Exp. Ther. 304:370–379.

Manolopoulos, V.G., Liu, J., Unsworth, B.R., and Lelkes, P.I. 1995. Adenylyl cyclase isoforms are differentially expressed in primary cultures of endothelial cells and whole tissue homogenates from various rat tissues. Biochem. Biophys. Res. Commun. 208:323–331.

Marzo, K.P., Frey, M.J., Wilson, J.R., Liang, B.T., Manning, D.R., Lanoce, V., and Molinoff, P.B. 1991. β-adrenergic receptor-G protein-adenylate cyclase complex in experimental canine congestive heart failure produced by rapid ventricular pacing. Circ. Res. 69:1546–1556.

Matsuda, N., Hattori, Y., Akaishi, Y., Suzuki, Y., Kemmotsu, O., and Gando, S. 2000. Impairment of cardiac β-adrenoceptor cellular signaling by decreased expression of $G_{s\alpha}$ in septic rabbits. Anesthesiology 93:1465–1473.

Meggs, L.G., Huang, H., Li, P., Capasso, J.M., and Anversa, P. 1991. Chronic nonocclusive coronary artery constriction in rats. β-adrenoceptor signal transduction and ventricular failure. J. Clin. Invest. 88:1940–1946.

MERIT-HF Study Group. 1999. Effect of metoprolol CR/XL in chronic heart failure: Metoprolol CR/XL Randomised Intervention Trial in Congestive Heart Failure (MERIT-HF). Lancet 353:2001–2007.

Meszaros, J., and Levai, G. 1992. Catecholamine-induced cardiac hypertrophy uncouples β-adrenoceptors from slow calcium channels. Eur. J. Pharmacol. 210:333–338.

Milano, C.A., Allen, L.F., Rockman, H.A., Dolber, P.C., McMinn, T.R., Chien, K.R., Johnson, T.D., Bond, R.A., and Lefkowitz, R.J. 1994. Enhanced myocardial function in transgenic mice overexpressing the β_2-adrenergic receptor. Science 264:582–586.

Morisco, C., Zebrowski, D.C., Vatner, D.E., Vatner, S.F., and Sadoshima, J. 2001. β-adrenergic cardiac hypertrophy is mediated primarily by the β_1-subtype in the rat heart. J. Mol. Cell. Cardiol. 33:561–573.

Mukherjee, A., Bush, L.R., McCoy, K.E., Duke, R.J., Hagler, H., Buja, L.M., and Willerson, J.T. 1982. Relationship between β-adrenergic receptor numbers and physiological responses during experimental canine myocardial ischemia. Circ. Res. 50:735–741.

Muller, F.U., Boheler, K.R., Eschenhagen, T., Schmitz, W., and Scholz, H. 1993. Isoprenaline stimulates gene transcription of the inhibitory G protein alpha-subunit $G_{i\alpha}$-2 in rat heart. Circ. Res. 72:696–700.

Ohyanagi, M., Yamamoto, J., Nakamura, K., Shibuya, J., Morita, M., Masutani, M., Arii, T., and Iwasaki, T. 1995. Messenger RNA for the guanine nucleotide-binding regulatory protein (G protein) is reduced in the acute ischemic myocardium. J. Mol. Cell. Cardiol. 27:1131–1139.

Panagia, V., Singh, J.N., Anand-Srivastava, M.B., Pierce, G.N., Jasmin, G., and Dhalla, N.S. 1984. Sarcolemmal alterations during the development of genetically determined cardiomyopathy. Cardiovasc. Res. 18:567–572.

Patrianakos, A.P., Parthenakis, F.I., Mavrakis, H.E., Diakakis, G.F., Chlouverakis, G.I., and Vardas, P.E. 2005. Comparative efficacy of nebivolol versus carvedilol on left ventricular function and exercise capacity in patients with nonischemic dilated cardiomyopathy. A 12-month study. Am. Heart J. 150:985.

Pela, G., Missale, C., Raddino, R., Condorelli, E., Spano, P.F., and Visioli, O. 1990. β_1- and β_2-receptors are differentially desensitized in an experimental model of heart failure. J. Cardiovasc. Pharmacol. 16:839–846.

Persad, S., Takeda, S., Panagia, V., and Dhalla, N.S. 1997. β-adrenoceptor-linked signal transduction in ischemic-reperfused heart and scavenging of oxyradicals. J. Mol. Cell. Cardiol. 29:545–558.

Persad, S., Panagia, V., and Dhalla, N.S. 1998. Role of H_2O_2 in changing β-adrenoceptor and adenylyl cyclase in ischemia-reperfused hearts. Mol. Cell. Biochem. 186:99–106.

Remme, W., Boccanelli, A., Cline, C., Cohen-Solal, A., Dietz, R., Hobbs, R., Keukelaar, K., Sendon, J.L., Macarie, C., McMurray, J., Rauch, B., Ruzyllo, W., and Zannad, F; SHAPE Study. 2004. Increasing awareness and perception of heart failure in Europe and improving care—rationale and design of the SHAPE Study. Cardiovasc. Drug Ther. 18:153–159.

Rocha-Singh, K.J., Honbo, N.Y., and Karliner, J.S. 1991. Hypoxia and glucose independently regulate the β-adrenergic receptor-adenylate cyclase system in cardiac myocytes. J. Clin. Invest. 88:204–213.

Rockman, H.A., Chien, K.R., Choi, D.J., Iaccarino, G., Hunter, J.J., Ross, J., Jr., Lefkowitz, R.J., and Koch, W.J. 1998. Expression of a β-adrenergic receptor kinase 1 inhibitor prevents the development of myocardial failure in gene-targeted mice. Proc. Natl. Acad. Sci. USA 95:7000–7005.

Sabri, A., Wilson, B.A., and Steinberg, S.F. 2002. Dual actions of the $G_{\alpha q}$ agonist Pasteurella multocida toxin to promote cardiomyocyte hypertrophy and enhance apoptosis susceptibility. Circ. Res. 90:850–857.

Saffitz, J.E., and Liggett, S.B. 1992. Subcellular distribution of β_2-adrenergic receptors delineated with quantitative ultrastructural autoradiography of radioligand binding sites. Circ. Res. 70:1320–1325.

Saito, S., Hiroi, Y., Zou, Y., Aikawa, R., Toko, H., Shibasaki, F., Yazaki, Y., Nagai, R., and Komuro, I. 2000. β-adrenergic pathway induces apoptosis through calcineurin activation in cardiac myocytes. J. Biol. Chem. 275:34528–34533.

Sayar, K., Ugur, M., Gurdal, H., Onaran, O., Hotomaroglu, O., and Turan, B. 2000. Dietary selenium and vitamin E intakes alter β-adrenergic response of L-type Ca-current and β-adrenoceptor-adenylate cyclase coupling in rat heart. J. Nutr. 130:733–740.

Schotten, U., Filzmaier, K., Borghardt, B., Kulka, S., Schoendube, F., Schumacher, C., and Hanrath, P. 2000. Changes of β-adrenergic signaling in compensated human cardiac hypertrophy depend on the underlying disease. Am. J. Physiol. Heart Circ. Physiol. 278:H2076–H2083.

Searles, R.P., Midson, C.N., Nipper, V.J., and Machida, C.A. 1995. Transcription of the rat β_1-adrenergic receptor gene. Characterization of the transcript and identification of important sequences. J. Biol. Chem. 270:157–162.

Sethi, R., and Dhalla, N.S. 1995. Inotropic responses to isoproterenol in congestive heart failure subsequent to myocardial infarction in rats. J. Card. Fail. 1:391–399.

Sethi, R., Bector, N., Takeda, N., Nagano, M., Jasmin, G., and Dhalla, N.S. 1994. Alterations in G-proteins in congestive heart failure in cardiomyopathic (UM-X7.1) hamsters. Mol. Cell. Biochem. 140:163–170.

Sethi, R., Dhalla, K.S., Beamish, R.E., and Dhalla, N.S. 1997. Differential changes in left and right ventricular adenylyl cyclase activities in congestive heart failure. Am. J. Physiol. Heart Circ. Physiol. 272:H884–H893.

Sethi, R., Elimban, V., Chapman, D., Dixon, I.M., and Dhalla, N.S. 1998. Differential alterations in left and right ventricular G-proteins in congestive heart failure due to myocardial infarction. J. Mol. Cell. Cardiol. 30:2153–2163.

Sethi, R., Shao, Q., Takeda, N., and Dhalla, N.S. 2003. Attenuation of changes in G_i-proteins and adenylyl cyclase in heart failure by an ACE inhibitor, imidapril. J. Cell. Mol. Med. 7:277–286.

Sethi, R., Wang, X., Ferrari, R., and Dhalla, N.S. 2004. Improvement of cardiac function and β-adrenergic signal transduction by propionyl L-carnitine in congestive heart failure due to myocardial infarction. Coron. Artery Dis. 15:65–71.

Sethi, R., Saini, H.K., Wang, X., Elimban, V., Babick, A., and Dhalla, N.S. 2006. Differential changes in β-adrenoceptor signal transduction in left and right ventricles of infarcted rats. Can. J. Physiol. Pharmacol. 84:747–754.

Shite, J., Qin, F., Mao, W., Kawai, H., Stevens, S.Y., and Liang, C. 2001. Antioxidant vitamins attenuate oxidative stress and cardiac dysfunction in tachycardia-induced cardiomyopathy. J. Am. Coll. Cardiol. 38:1734–1740.

Shizukuda, Y., and Buttrick, P.M. 2002. Subtype specific roles of β-adrenergic receptors in apoptosis of adult rat ventricular myocytes. J. Mol. Cell. Cardiol. 34:823–831.

Steinfath, M., Danielsen, W., von der Leyen, H., Mende, U., Meyer, W., Neumann, J., Nose, M., Reich, T., Schmitz, W., Scholz, H., et al. 1992. Reduced β_1- and β_2-adrenoceptor-mediated positive inotropic effects in human end-stage heart failure. Br. J. Pharmacol. 105:463–469.

Stiles, G.L., Caron, M.G., and Lefkowitz, R.J. 1984. β-adrenergic receptors: biochemical mechanisms of physiological regulation. Physiol. Rev. 64:661–743.

Strasser, R.H., Krimmer, J., Braun-Dullaeus, R., Marquetant, R., and Kubler, W. 1990. Dual sensitization of the adrenergic system in early myocardial ischemia: independent regulation of the β-adrenergic receptors and the adenylyl cyclase. J. Mol. Cell. Cardiol. 22:1405–1423.

Sunahara, R.K., Dessauer, C.W., and Gilman, A.G. 1996. Complexity and diversity of mammalian adenylyl cyclases. Annu. Rev. Pharmacol. Toxicol. 36:461–480.

Tate, K.M., Briend-Sutren, M.M., Emorine, L.J., Delavier-Klutchko, C., Marullo, S., and Strosberg, A.D. 1991. Expression of three human β-adrenergic-receptor subtypes in transfected Chinese hamster ovary cells. Eur. J. Biochem. 196:357–361.

Tavernier, G., Toumaniantz, G., Erfanian, M., Heymann, M.F., Laurent, K., Langin, D., and Gauthier, C. 2003. β_3-adrenergic stimulation produces a decrease of cardiac contractility *ex vivo* in mice overexpressing the human β_3-adrenergic receptor. Cardiovasc. Res. 59:288–296.

The Xamoterol in Severe Heart Failure Study Group. 1990. Xamoterol in severe heart failure. Lancet. 336:1–6.

Urasawa, K., Sato, K., Igarashi, Y., Kawaguchi, H., Yasuda, H. 1992. A mechanism of catecholamine tolerance in congestive heart failure-alterations in the hormone sensitive adenylyl cyclase system of the heart. Jpn. Circ, J. 56: 456–461.

van den Ende, R., Batink, H.D., Michel, M.C., and van Zwieten, P.A. 1994. Influence of ischaemia and reperfusion on cardiac signal transduction. G protein content, adenylyl cyclase activity, cyclic AMP content, and forskolin and dibutyryl cyclic AMP-induced inotropy in the rat Langendorff heart. Fundam. Clin. Pharmacol. 8:408–416.

Vatner, D.E., Knight, D.R., Shen, Y.T., Thomas, J.X., Jr., Homcy, C.J., and Vatner, S.F. 1988. One hour of myocardial ischemia in conscious dogs increases β-adrenergic receptors, but decreases adenylate cyclase activity. J. Mol. Cell. Cardiol. 20:75–82.

Vatner, D.E., Young, M.A., Knight, D.R., and Vatner, S.F. 1990. β-receptors and adenylate cyclase: comparison of nonischemic, ischemic, and postmortem tissue. Am. J. Physiol. Heart Circ. Physiol. 258:H140–H144.

Waagstein, F., Caidahl, K., Wallentin, I., Bergh, C.H., and Hjalmarson, A. 1989. Long-term β-blockade in dilated cardiomyopathy. Effects of short- and long-term metoprolol treatment followed by withdrawal and readministration of metoprolol. Circulation 80:551–563.

Wallukat, G. 2002. The β-adrenergic receptors. Herz 27:683–690.

Wang, X., and Dhalla, N.S. 2000. Modification of β-adrenoceptor signal transduction pathway by genetic manipulation and heart failure. Mol. Cell. Biochem. 214:131–155.

Wang, X., Wang, J., Takeda, S., Elimban, V., and Dhalla, N.S. 2002. Alterations of cardiac β-adrenoceptor mechanisms due to calcium depletion and repletion. Mol. Cell. Biochem. 232:63–73.

Wang, X., Ren, B., Liu, S., Sentex, E., Tappia, P.S., and Dhalla, N.S. 2003. Characterization of cardiac hypertrophy and heart failure due to volume overload in the rat. J. Appl. Physiol. 94:752–763.

Wang, X., Sentex, E., Saini, H.K., Chapman, D., and Dhalla, N.S. 2005. Upregulation of β-adrenergic receptors in heart failure due to volume overload. Am. J. Physiol. Heart Circ. Physiol. 289:H151–H159.

Weber, M.A. 2005. The role of the new β-blockers in treating cardiovascular disease. Am. J. Hypertens. 18:169S–176S.

Weil, J., and Schunkert, H. 2006. Pathophysiology of chronic heart failure. Clin. Res. Cardiol. 95:1–17.

Will-Shahab, L., Rosenthal, W., Schulze, W., and Kuttner, I. 1991. G protein function in the ischaemic myocardium. Eur. Heart J. 12:135–138.

Witte, K., Olbrich, H.G., Langer, L., and Lemmer, B. 1993. Uncoupling of β_1-adrenoceptors from cardiac adenylyl cyclase in cardiomyopathic and control hamsters. Eur. J. Pharmacol. 247:215–218.

Wollert, K.C., Studer, R., von Bulow, B., and Drexler, H. 1994. Survival after myocardial infarction in the rat. Role of tissue angiotensin-converting enzyme inhibition. Circulation 90:2457–2467.

Xiao, R.P., Avdonin, P., Zhou, Y.Y., Cheng, H., Akhter, S.A., Eschenhagen, T., Lefkowitz, R.J., Koch, W.J., and Lakatta, E.G. 1999. Coupling of β_2-adrenoceptor to G_i proteins and its physiological relevance in murine cardiac myocytes. Circ. Res. 84:43–52.

Yoshida, H., Tanonaka, K., Miyamoto, Y., Abe, T., Takahashi, M., Anand-Srivastava, M.B., and Takeo, S. 2001a. Characterization of cardiac myocyte and tissue β-adrenergic signal transduction in rats with heart failure. Cardiovasc. Res. 50:34–45.

Yoshida, H., Takahashi, M., Tanonaka, K., Maki, T., Nasa, Y., and Takeo, S. 2001b. Effects of ACE inhibition and angiotensin II type 1 receptor blockade on cardiac function and G proteins in rats with chronic heart failure. Br. J. Pharmacol. 134:150–160.

Yoshie, H., Tobise, K., and Onodera, S. 1994. Intraventricular changes in the β-adrenoceptor-adenylate cyclase system of the rat heart with the progress of monocrotaline-induced right ventricular hypertrophy. Jpn. Circ. J. 58:855–865.

Yoshikawa, T., Handa, S., Suzuki, M., and Nagami, K. 1994. Abnormalities in sympathoneuronal regulation are localized to failing myocardium in rabbit heart. J. Am. Coll. Cardiol. 24:210–215.

Yu, H.J., Unnerstall, J.R., and Green, R.D. 1995. Determination and cellular localization of adenylyl cyclase isozymes expressed in embryonic chick heart. FEBS Lett. 374:89–94.

Zhou, Y., Friedman, E., Roberts, J., and Johnson, M.D. 1995. Modulation of aortic and cardiac G protein alpha subunits and their mRNAs during norepinephrine infusion in rats. J. Vasc. Res. 32:16–23.

Zhu, W.Z., Zheng, M., Koch, W.J., Lefkowitz, R.J., Kobilka, B.K., and Xiao, R.P. 2001. Dual modulation of cell survival and cell death by β_2-adrenergic signaling in adult mouse cardiac myocytes. Proc. Natl. Acad. Sci. USA 98:1607–1612.

Chapter 3
Cardiac Gαs and Gαi Modulate Sympathetic Versus Parasympathetic Mechanisms in Hyperhomocysteinemia

Thomas P. Vacek, Karni S. Moshal, Naira Metreveli, Neetu Tyagi, Utpal Sen, Dorothea Rosenberger, and Suresh C. Tyagi

Abstract The increase in sympathetic activity is a compensatory mechanism that provides inotropic support to the heart and peripheral vasoconstriction. Increased sympathetic nerve activity is associated with chronic heart failure (CHF). Increased sympathetic nervous activity activates NMDA receptors and GPCRs, which results in the physiological effect of cardiac arrhythmias. It is clear that GPCR activation feeds into the NMDA receptor signal transduction axis, and heightens the physiological effect of cardiac arrhythmias. Effectors that modulate GPCR activation also physically modify the NMDA receptor via intracellular tyrosine kinases. However, it is unknown whether homocysteine-induced NMDAR activation can backtalk to GPCRs and modulate G-proteins: Gαs, Gαi. This bidirectional crosstalk to GPCRs may be mediated by a number of possible events; perhaps calpain or members of the MAPK cascade that includes ERK 1/2 and MMP-9, or even reactive oxygen species may feed back to GPCRs and potentiate G-proteins. Or, perhaps the influx of calcium from NMDAR activation may somehow activate the GPCR signal transduction axis, including G-proteins. Our specific interest begins in knowing exactly whether homocysteine-induced activation of NMDAR can backtalk to GPCRs and modulate G-proteins—Gαs, Gαi—before beginning to elucidate the players that modulate this kind of backtalk. This review addresses the following: pathophysiology of CHF and sudden cardiac death (SCD), causes of cardiac arrhythmias (electrical and structural remodeling), the relationship between homocysteine and CHF, homocysteine and NMDAR activation, the NMDAR signal cascade, GPCR signal cascade, NMDA-to-GPCR crosstalk, and proposal for homocysteine-induced NMDAR and GPCR bidirectional crosstalk that modulates G-proteins.

Introduction: Autonomic Pathophysiology of CHF and SCD

Heart failure (HF) is a condition in which the ventricles fail to eject adequate quantities of blood to meet the needs of the peripheral organs; common causes of HF

A. K. Srivastava, M. B. Anand-Srivastava (eds.) *Advances in Biochemistry in Health and Disease.* Signal Transduction in the Cardiovascular System in Health and Disease,

are ischemic heart disease, hypertension, and idiopathic dilated cardiomyopathy (Watson et al. 2006). As a result, the body tries to compensate through various neurohumoral systems (Emoto et al. 2001; Watson et al. 2006). For example, the renin–angiotensin system is activated to maintain systemic blood pressure. Hormones can have direct vascular and renal actions as well as have effects on sympathetic nerve activity (Giltay et al. 1998); in addition to angiotensin II (AII) having vasoconstrictive effects, it is also sympathoexcitatory (Watson et al. 2006).

It is well established that increased sympathetic nerve activity is associated with chronic heart failure (CHF) (Porter et al. 1990; Singh 2000; Olshansky 2005; Brodde et al. 2006; Watson et al. 2006). The increase in sympathetic activity is a compensatory mechanism that provides inotropic support to the heart and peripheral vasoconstriction. However, it promotes disease progression and worsens prognosis (Watson et al. 2006). The autonomic nervous system (ANS) is a very complex, balanced system that influences the initiation, termination, and perpetuation of atrial fibrillation (AF), and the AF affects the ANS (Olshansky, 2005). At rest, sympathetic and parasympathetic outflows are related reciprocally: heart failure patients had high sympathetic and low parasympathetic outflows, and healthy subjects had low sympathetic and high parasympathetic outflows (Porter et al. 1990).

One such example of how sympathetic innervation can influence CHF is seen with the paraventricular nucleus (PVN) of the hypothalamus; the PVN is an important site that integrates sympathetic nerve activity. Studies have shown that glutamate invokes excitatory effects on neurons in the PVN through the NMDA receptor (Kurtis and Patel 2003). Microinjection of NMDA into the PVN of rats produced dose-dependent increases in renal sympathetic nerve discharge (RSND), arterial blood pressure (BP), and heart rate (HR) (Kurtis and Patel 2003). Conversely, microinjection of the NMDA receptor antagonist into the PVN caused significant decreases in RSND, BP, and HR only in rats with HF, but very slight changes in sham rats (Kurtis and Patel 2003). Moreover, although glutamate levels in the PVN were not increased significantly in HF compared to sham rats, the expression of NMDA NR subunit was significantly increased (Kurtis and Patel 2003).

Another known mechanism by which the sympathetic system can influence CHF is through agonist binding to α- and β-adrenergic receptors (AR) that are coupled to G-proteins (Singh et al. 2000). β-AR can stimulate apoptosis and the development of dilated cardiomyopathy (Singh et al. 2000). Eight adrenoceptor (AR) subtypes have been identified: α1A, α1B, α1D, α2A, α2B, β1, β2, and β3 AR. Of the nine identified adrenoreceptors, β1 and β2 are the most powerful physiologic mechanism to acutely increase cardiac performance (Singh et al. 2000). The signaling cascade involves variable G-protein expression that could result in cardiac arrhythmia. Both β-AR subtypes couple to the Gs-protein, thereby elevating the intracellular level of cyclic AMP; this causes a positive inotropic and chronotropic effect (Brodde et al. 2006). The beneficial effect of β-AR antagonists on survival in patients with HF suggests that cardiac sympathetic activation is harmful (Watson et al. 2006).

Causes of Cardiac Arrhythmia (Electrical Remodeling, Structural Remodeling)

Remodeling qualifies changes that result in the rearrangement of normally existing structures (Swyngheadauw 1999). Although remodeling does not necessarily define a pathological condition, myocardial remodeling is usually restricted to diseased conditions (Swyngheadauw 1999). Expression of several different proteins, including ion channels, is the likely mechanism that promotes the susceptibility of the human atrium to arrhythmia (Swyngheadauw 1999). Electrical remodeling comes from a change in the electrical properties of the atrium, notably a shortening of the AERP and a loss of rate adaptation (Brundel et al. 2002). AF is also associated with elaborate adaptive and maladaptive changes in tissue and cellular architecture; this type of change is dubbed structural remodeling (Brundel et al. 2002). It is important to note that electrical and structural remodeling are not mutually exclusive since one can evoke the other. For instance, an increase in calcium will certainly have an effect with respect to electrical remodeling; however, it will also trigger calpain and MAPK cascade that leads to MMP-9 activation, a metalloproteinase known to induce structural remodeling.

Calcium and Electrical Remodeling

Calcium plays a crucial role in contraction and pacing within the heart such that an increase or decrease of calcium can result in pathology. It is known that a reduction of extracellular Ca^{2+} concentration causes a reduction in both systolic and diastolic [Ca^{2+}] and the spontaneous firing rate also gradually declines in pacemaker cells (Ju and Allen 1998). In the presence of ryanodine, which interferes with Ca^{2+} release from the sarcoplasmic reticulum, the systolic and diastolic [Ca^{2+}] influx both declined and the firing rate decreased until the cells stopped firing (Ju and Allen 1998). When stimulation rate was increased, there were increases in diastolic Ca^{2+} influx, systolic Ca^{2+} influx, and the Ca^{2+} content of the SR (Frampton et al. 1991). The SR inhibitor ryanodine decreased the size of the Ca^{2+} influx transient, and abolished the increase of Ca^{2+} influx produced by caffeine. In the presence of ryanodine, increasing stimulation rate increased diastolic Ca^{2+} but not systolic Ca^{2+} (Frampton et al. 1991). An important molecular feature of AF is the reduction in L-type Ca^{2+} channel function and protein expression (Brundel et al. 2002). Electrophysiological changes in AF are primarily caused by this reduction in L-type Ca^{2+} channel (Brundel et al. 2002). Secondary to this process, the reduced expression of K^{+} channels may serve to adapt the myocardial cell to the high rate and counteract the shortening of AERP (Brundel et al. 2002). This reduction may serve to protect the cell against a potentially lethal Ca^{2+} overload resulting from the increased activation rate in AF (Hardingham et al. 2001). Protection against cellular stress is achieved at the cost of electrophysiological changes in cardiac function (Brundel et al. 2002).

Gap Junctions and Electrical Remodeling

Electrophysiological remodeling involving gap junctions has been demonstrated in failing hearts and may contribute to intercellular uncoupling, delayed conduction, enhanced arrhythmias, and vulnerability to SCD (Betsuyaku et al. 2006). Gap junctions are clusters of connexin channels, which span the closely opposed plasma membranes forming cell-to-cell pathways (Brundel et al. 2002). The thickness of the endothelium (E)–myocyte (M) cell–cell connection becomes important in the efficiency of transport of endothelial-derived cardioactive agents to the cardiac muscle (Rosenberger et al. 2006). Connexins are permeable to ions and small molecules up to 1 kDa in molecular mass, including second messengers such as inositol triphosphate, cyclic AMP, and calcium (Brundel et al. 2002). Failing human hearts exhibit marked increases in connexin 45 (C × 45) expression in addition to decreases in connexin 43 (C × 43) expression; these changes result in reduced gap junction coupling. Gap junctional connexin proteins (connexin 40 [C × 40], C × 43 are a determinant of myocardial conduction and are implicated in the development of AF (Kanagartnam et al. 2004). The pattern of atrial activation is related to immunoconfocal connexin signal only in the fully remodeled atria of chronic AF. This suggests that intercellular coupling and pattern of atrial activation are interrelated, but only in conjunction with the remodeling of atrial electrophysiology that occurs in chronic AF (Kanagartnam et al. 2004). All patients with AF exhibited a concomitant lateralization of gap junctional proteins C × 43, C × 40, and N-cadherin (the major mechanical junction protein), instead of being confined to the intercalated disKs, as observed in SR (Kanagartnam et al. 2004). The structural correlate of AF comprises extensive concomitant remodeling of mechanical and electrical junctions, reduction of C × 43, and heterogeneous distribution of C × 40. The increase in distance from E to M impairs endothelial-derived NO diffusion mechanism to the cardiac muscle (Rosenberger et al. 2006). These changes, along with fibrosis, may underlie conduction abnormalities (Kostin et al. 2002).

MMP-9 Activation and Structural Remodeling

MMPs are collagenases as well as elastases, but there is a differential turnover rate of collagen and elastin when activated. Activation of certain MMPs will, therefore, alter collagen/elastin ratio, disrupt C × 43, and lead to accumulation of interstitial collagen (fibrosis) between the endothelium and myocyte (Rosenberger et al. 2006). There is an inverse relationship between the increase in heart rate and cardiac elastin/collagen ratio in mice, rats, rabbits, and humans (Rosenberger et al. 2006).Hence, extracellular matrix (ECM) turnover plays an important role in cardiac remodeling. Previous studies from our lab demonstrated an increase in gelatinolytic MMP-2 and -9 activities in endocardial tissue from ischemic cardiomyopathic (ICM) and idiopathic dilated cardiomyopathic (DCM) hearts (Moshal et al. 2005). In chronic volume overload model of CHF in mice, impaired cardiac function is associated with endothelial and myocyte apoptosis (Ovechkin and Tyagi 2005). The mechanism of the phenomena involves activation of MMP-9, leading to cardiac remodeling and E-M uncoupling. MMP-9 has a major role in ventricular remodeling associated with endothelial and myocyte apoptosis (Ovechkin and Tyagi 2005).

Influence of Homocysteine on Cardiac Arrhythmias

Homocysteine Metabolomics and CHF

Homocysteine (Hcy) is a homologue of the naturally occurring amino acid cysteine; it differs in its side chain that contains an additional methylene group before the thiol group (Moshal et al. 2005). There are at least five ways by which Hcy is accumulated in the plasma and tissues as shown in Figure 3.1: (1) a methionine-rich protein diet; (2) methionine demethylation by methyl transferase (MT); (3) Hcy remethylation by methyl synthase (MS) and a vitamin b_{12}/folate/methyl tetrahydrofolate reductase (MTHFR); (4) a heterozygous/homozygous trait for CBS activity and B_6 deficiency; and (5) renovascular stenosis and volume retention. Although Hcy plays a constitutive role in DNA/RNA gene methylation (Tyagi 1999; Yi et al. 2000), hyperhomocysteinemia leads to endothelial damage (McCully 1996; Mujumdar et al. 2001), especially since mammalian endothelial cells lack the CBS enzyme (Finkelstein 1990, 1998). Every 3 mol/liter increase in Hcy level contributes to a 10% increased risk of coronary heart diseases and a 20% increased risk of stroke (Homocysteine Studies Collaboration 2002). A common genetic polymorphism, MTHFR C677T,

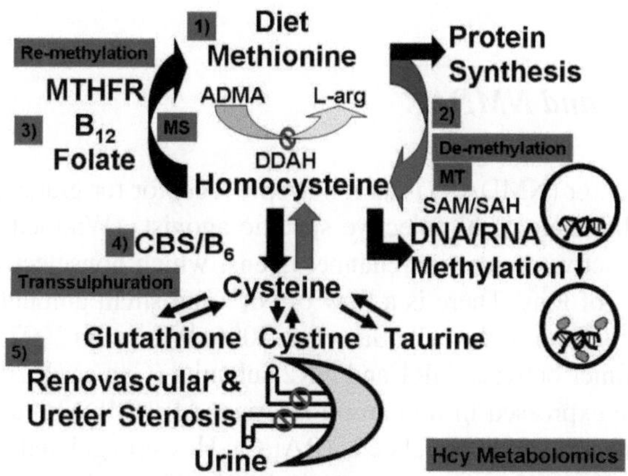

Fig. 3.1 Metabolomics of Hcy. During gene and protein methylation, S-adenosyl homocysteine (SAH) is generated by methyl transferase and methionine. SAH hydrolase (SAHH) generates Hcy. DZA blocks SAHH, otherwise Hcy induces NOS, NADH oxidase and decreases thioredoxin in mitochondria. Hcy inhibits DDAH and increases ADMA causing decrease in NO. Deficiency in MTHFR and CBS increases Hcy. The decrease in kidney ability to filter increases Hcy.

which determines Hcy levels, also has similar effects on heart disease and stroke (Shcherbak et al. 1999; Klerk et al. 2002; Wald et al. 2002; Kaye et al. 2002). The association between this polymorphism and heart disease is unlikely to be confounded by other factors, such as smoking or blood pressure, but influences Hcy levels, suggesting a causal association between Hcy and heart disease or stroke (Davey and Ebrahim 2003). A secondary prevention trial (Clarke and Collins 1998) of folic acid supplementation demonstrated unequivocally that folate and other B-complex vitamins protect against heart disease. Another study demonstrated a beneficial effect on the rate of revascularization (Schnyder et al. 2002). A trial with stroke patients did not demonstrate a robust difference in recurrent stroke associated with a reduction of Hcy levels by 2 mol/liter (Toole et al. 2004). Although the overall risk of heart disease with Hcy may be small, there is evidence of synergism between Hcy and other risk factors such as smoking (Graham et al. 1997; Fallon et al. 2003), hypertension (Fallon et al. 2001), diabetes (Audelin and Genest 2001), and insulin resistance (Fonseca et al. 2003). Plasma Hcy has been suggested as a newly recognized risk factor (Herrmann et al. 2005). It was found that there is a consistent association of plasma Hcy with clinical and echocardiographic measures of CHF; this indicates that there is a relationship between Hcy and the severity of CHF (Herrmann et al. 2005). In fact, hyperhomocysteinemia is common in CHF, and is found to be related to disease severity that creates a metabolic imbalance (Naruszewicz et al. 2006). The metabolic imbalance is evidenced by hyperuricemia, which predicts poor long-term prognosis (Naruszewicz et al. 2006). The elastin/collagen ratio was decreased in Hcy-treated animals. The chronic administration of Hcy increases the heart rate (HR). The withdrawal of Hcy reduces the Hcy-mediated increase in HR. This suggests that Hcy decreases elastin/collagen and increases HR. However, the observations that Hcy reduces the elastin/collagen ratio and increases HR may be associative and not causative (Rosenberger et al. 2006).

Homocysteine and NMDAR

The NMDA receptor (NMDAR) is an ionotropic receptor for glutamate (*N*-methyl-D-aspartate is the name of its selective specific agonist) (Wikipedia 2007). When the NMDAR is activated, the ion channel opens, which nonselectively allows the influx and efflux of ions. There is a flow of Na^+ and small amounts of Ca^{2+} ions into the cell and K^+ out of the cell (Gao et al. 2006; Wikipedia 2007). The NMDAR forms a heterodimer between NR1 and NR2 subunits (Gao et al. 2006). Both NR1 and NR2A were expressed in rat carotid arteries. All NMDAR subunits were expressed in the rat aorta endothelial cells (RAEC); Hcy upregulated NR1 expression and increased cell proliferation (Chen et al. 2005). Hcy-induced NMDAR activation is especially important in calcium flux since an increase in calcium has already been mentioned to induce electrical cardiac remodeling; calcium also plays an important role in signal transduction that results in structural remodeling and interruption of connexins that result in fibrosis. The antagonist to the NMDAR protects against HCy-mediated oxidative toxic effects in neurons, and protects against increase in HR by NMDA analogue, suggesting that Hcy is an agonist to the NMDAR (Rosenberger et al. 2006). Hcy could potentially be a ligand to the NMDAR.

Homocysteine and Activation of NMDAR Signaling Cascade

Calpains are a family of cytosolic cysteine proteinases whose enzymatic activities depend on Ca^{2+} (Khorchid and Ikura 2002). Hcy can activate calpains via activation of the NMDAR, which causes a calcium influx, leading to calcium binding by calpains and calpain activation (Khorchid and Ikura 2002). Members of the calpain family have been shown to function in processes including integrin-mediated cell migration, cytoskeletal remodeling, cell differentiation, and apoptosis (Khorchid and Ikura 2002; Razeghi et al. 2007). Calpain activity is induced during AF and correlates with parameters of ion-channel protein, structural and electrical remodeling. This suggests that calpain activation represents an important mechanism linking calcium overload to cellular adaptation mechanisms in human AF (Brundel et al. 2002). With an increase of calcium from NMDAR activation, there comes an activation of calpain. In primary hippocampal and cerebellar granule (CG) neurons, calcium influx activates calpain and ERK1,2 and increases neurofilament phosphorylation on carboxy-terminal polypeptide sites known to be modulated by ERK1,2 and to be altered in AD. Calpeptin, a cell-permeable calpain inhibitor, blocked ERK1,2. It is concluded that calpains are upstream activators of ERK1,2 signaling (Veeranna et al. 2004). To recapitulate, Hcy induces translocation of active calpain from cytosol to mitochondria, leading to MMP-9 activation, in part, by causing intramitochondrial oxidative burst; studies with pharmacological inhibitors of calpain, ERK and the mitochondrial uncoupler FCCP, suggested that calpain and ERK1/2 are the major events within the Hcy/MMP-9 signal axis. It was concluded that intramitochondrial oxidative stress regulates MMP-9 via ERK1/2 signal cascade (Moshal et al. 2006a, 2006b; Gao et al. 2006).

Influence of GPCRs on Cardiac Arrhythmias

The family of G-protein-coupled receptors (GPCRs) contains a conserved structure of seven transmembrane α-helices. The adrenergic and muscarinic cholinergic receptors are important for the heart because they function in the homeostatic regulation of the cardiovascular system system (Rockman et al. 2002). There are at least eight subtypes of adrenergic receptors that have been cloned; the β1-AR is the most predominant subtype, comprising 75–80% of total β-ARs (Rockman et al. 2002). Interaction of the receptor with an agonist promotes the dissociation of G-proteins into Gαs (stimulatory), Gαi (inhibitory), and G$\beta\gamma$ (inhibitory) subunits. The G-protein subunits then amplify and propagate signals inside the cell by modulating the activity of one or more effector molecules, including adenylyl cyclases, phospholipases, and ion channels. The second messengers cAMP and DAG activate, respectively, protein kinase A (PKA) and protein kinase C (PKC) (Rockman et al. 2002). Acetylcholine, released on parasympathetic stimulation, slows HR through activation of muscarinic receptors on the sinus nodal cells and subsequent opening of atrial muscarinic potassium channel (K ACh) (Gehrmann et al. 2002). Reduction of the amount of functional Gβγ protein caused a pronounced blunting in carbachol-induced bradycardia as well as the increases in time- and frequency-domain indexes of HR variability and baroreflex sensitivity that were observed in wild types (Gehrmann et al. 2002). Data show that the inhibitory G-protein subunit, Gβγ, plays a crucial role for parasympathetic HR control, sinus node automaticity, and atrial arrhythmia vulnerability (Gehrmann et al. 2002). The AT1 receptor mediates downstream signaling mechanisms through Gαq/Gαi-dependent and independent mechanisms, which induce hypertrophy, along with bradycardia (Zhai et al. 2005). Consistent with this result, there is an accelerated cardiomyopathy in mice with overexpression of cardiac Gαsand a missense mutation in the α-myosin heavy chain (Hardt et al. 2002). Moreover, conditional expression of a fabricated Gi-coupled receptor decreased HR by up to 80% (Redfern et al. 1999). Short-term Gi signaling events in the heart include inhibition of adenylyl cyclase and activation of a membrane potassium channel, resulting in a decreased HR (bradycardia) (Redfern et al. 1999, 2000). In a similar design, conditional expression of a Gαi-coupled receptor causes ventricular conduction delay and lethal cardiomyopathy. Likewise, overexpression of cardiac Gαs (stimulatory) exhibited enhanced HR and contractility in rabbits; in this model, however, the transgene rabbit did not develop cardiomyopathy—possibly due to a compensatory increase in Gαi (Nishizawa et al. 2006).

Crosstalk Mechanism of NMDAR and GPCRs in Hcy-Induced Cardiac Arrhythmias

GPCRs have been shown to activate a variety of other signaling complexes, including receptor tyrosine kinases (RTK), cytoskeletal complexes, as well as signaling scaffolds (Lee 2003). Stimulation of the GPCR leads to the activation of the intrinsic tyrosine kinase activity of the RTK, with subsequent binding of adaptor proteins, such as Shc, that regulate downstream signaling pathways (Lee 2003). It is clear, though, that there are multiple pathways capable of transactivating RTKs; GPCRs can also transactivate RTKs via intracellular effectors, such as Gβγ, Src, and Pyk2 (Lee 2003).

Already, we know the physiological response to chronic GPCR activation is cardiac arrhythmia: bradycardia or tachycardia depending on whether the G-protein is stimulatory (Gαs) or inhibitory (Gαi, Gβγ) (Redfern et al. 1999; Gehrmann et al. 2002; Hardt et al. 2002; Zhai et al. 2005; Nishizawa et al. 2006). A stimulatory G-protein results in an increase of intracellular calcium, while an inhibitory

G-protein results in a decrease of intracellular calcium. Receptors that are able to activate PLC (phospholipase C) enzymes cause release of Ca^{2+} from intracellular stores and influence Ca^{2+} entry across the plasma membrane (Werry et al. 2003).

Cardiac arrhythmia also results from NMDA activation that triggers calcium influx, and a cascade of intracellular events (Veeranna et al. 2004; Chen et al. 2005; Moshal et al. 2006a,b). Intuitively, GPCRs utilize the same signal transduction pathway as the NMDAR since upstream activation begins with an influx of calcium. This same influx of calcium can result in activation of calpains, MAPKs (ERK1/2), MMP-9 activation, and cardiac structural remodeling—mirroring the NMDA signal transduction axis (Shah and Catt 2004).

There are other intracellular effectors that are known to be activated by G-proteins that further interact with the NMDAR. Several GPCRs are known to direct tyrosine phosphorylation (Felsch and Cachero 1998). Stimulation of the m1 muscarinic receptor leads to the tyrosine phosphorylation of Pyk2. This phosphorylation of Pyk2 induces two cytosolic proteins (c-src, Grb-2) to bind to Pyk2. By activating an src tyrosine kinase, this begets activation of the MAPK pathway, which further enhances the effect of NMDAR activation (Bi et al. 2000). The enhanced effect of NMDAR activation occurs because the NMDAR signal transduction axis is utilized. Estrogen stimulates tyrosine phosphorylation of NMDAR via an src tyrosine kinase/MAPK pathway (Bi et al. 2000). Activation of tyrosine kinase src potentiates NMDAR currents; this is thought to be necessary for induction of hippocampal long-term potentiation (Zheng et al. 1998). The carboxy terminal domain of the NR2A subunit contains potential tyrosine phosphorylation sites (Zheng et al. 1998). In an unrelated signaling pathway that utilizes neuregulin1 (NRG1), there is a modulation of NMDAR phosphorylation via intracellular tyrosine kinases, Fyn and Pyk2. These kinases stimulate phosphorylation on the NR2B subunit of the NMDAR(Bjarnadottir et al. 2007).

Hypothesis: Backtalk via Hcy-Induced NMDAR Activation Modulates G-Proteins

It is already clear that GPCR activation feeds into the NMDAR signal transduction axis, and heightens the physiological effect of cardiac arrhythmias (Zheng et al. 1998; Felsch and Cachero 1998; Bi et al. 2000; Lee 2003; Shah and Catt 2004). Effectors that modulate GPCR activation also physically modify the NMDAR itself via intracellular tyrosine kinases (Bjarnadottir et al. 2007). This one-way crosstalk from GPCRs to NMDA has already been elucidated. However, it is unknown whether Hcy-induced NMDAR activation can crosstalk to GPCRs and modulate G-proteins: Gαs, Gαi. This feedback to GPCRs may be mediated by a number of possible events; perhaps calpain or members of the MAPK cascade that includes ERK1/2 and MMP-9, or even reactive oxygen species may feed back to GPCRs and potentiate G-proteins. Or, perhaps the influx of calcium from NMDAR activation may somehow activate the GPCR signal transduction axis, including G-proteins. Our specific interest begins in knowing exactly whether Hcy-induced activation of

NMDAR can feed back to GPCRs and modulate G-proteins—Gαs, Gαi—before beginning to elucidate the players that modulate this kind of backtalk. Cardiac tissue will be acquired from wild-type mice and CBS mice that were administered Hcy until cardiomyopathy develops, at which time the mice are sacrificed. Western blot will test for the expression of Gαs and Gαi relative to control mice. We assume that backtalk can only result in amplification of the original pathology since pathology does develop from NMDAR activation (Salter 2003). This pathology results from an increase in calcium; an increase in calcium via the GPCR signal axis results from an increase in expression of Gαs (stimulatory) protein. In a related design using cardiomyocytes, G-protein expression will be determined in wild-type samples, Hcy samples, and Hcy with NMDA blocker samples. This design will allow us to determine whether the Hcy-induced cardiac arrhythmia and potentiation via G-proteins is mediated by NMDAR activation. In addition to acting as a ligand to NMDAR, Hcy also directly activates PAR-4, a GPCR that utilizes G-proteins (Tyagi et al. 2005). The variable expression of G-proteins, then, is not only due to activation of NMDAR by Hcy, but is also due to activation of PAR-4, a GPCR. This, however, will not interfere with testing for variable expression of G-proteins when knocking out the NMDAR in cardiomyocytes to determine if backtalk occurs from NMDAR to GPCR G-proteins. Figure 3.5 illustrates this proposed mechanism along with what is already known about the signal transduction axis.

Constitutive role of Hcy: Hcy plays a significant role in modulating cardiovascular, renal, and brain microvascular function. In the presence of CBS and CGL, Hcy generates H_2S (a gas that relaxes the vessels wall) (Figure 3.2). Therefore, Hcy plays a constitutive role in modulating the cardiovascular function. The rate-limiting steps are the enzymes CBS and CGL in generating H_2S; therefore, the deficiencies in one of these enzymes may lead to cardiovascular dysfunction or in brain to vascular leakage and dementias and strokes.

Load creates hyperhomocysteinemia (HHcy): The chronic pressure or volume overload by hypertension, aortic stenosis, and aortic vena cava fistula increases Hcy by volume retention, in addition, by decreasing the activities of MTHFR, MS, CBS, and CGL enzymes (Figure 3.3). During load Hcy chelates the Cu^{2+} in CytC, Cox,

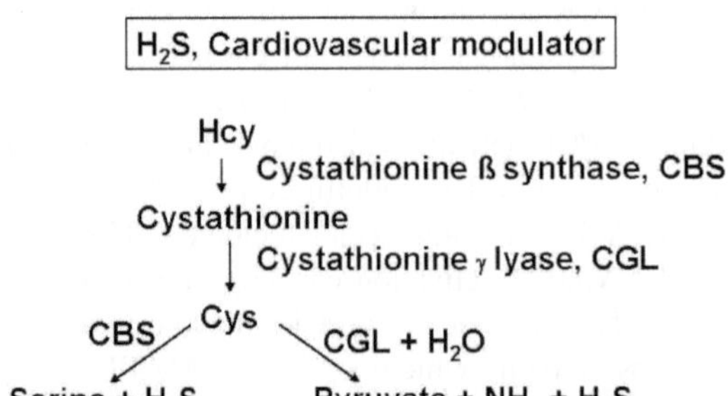

Fig. 3.2 Hcy generates H_2S that relaxes the vascular endothelium. The rate-limiting steps are the CBS and CGL in generation of H_2S. The deficiency in CBS or CGL causes increase in Hcy and leads to vascular dysfunction.

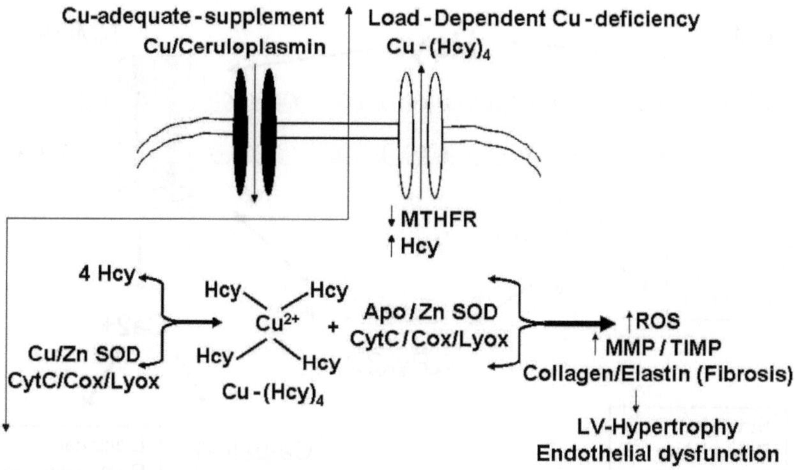

Fig. 3.3 During load the Hcy metabolic enzymes are decreased. This causes increase in Hcy that chelates the metal ions, creating apoenzymes. Metal-Hcy is excreted from the cell. The treatment with copper-dependent ceruloplasmin mitigates the Hcy-dependent decrease in intracellular metalloenzymes.

and lysyl oxidase, therefore deactivates these enzymes and copper is transported out of the cell. The copper supplement by ceruloplasmin ameliorates the load-induced HHcy and copper deficiency.

Latent MMP are activated by Hcy: In the heart most of the resident MMP are in latent form by forming ternary complex between MMP/NO/TIMP. However, during HHcy and oxidative stress by generation of Hcy-peroxynitrite intermediate the TIMP is oxidized and MMP is activated in the milieu of ECM and basement membrane of the endothelium (Figure 3.4).

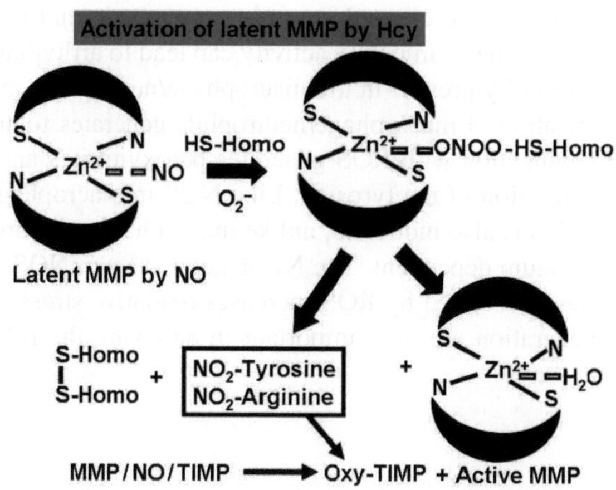

Fig. 3.4 Oxidative stress and increase in reactive oxygen species (ROS) and reactive thiol species (RTS) decreases constitutive NO in MMP/TIMP/NO ternary complex and generates reactive nitrogen species (RNS) and nitrotyrosine. This process oxidizes the TIMP and liberates active MMP.

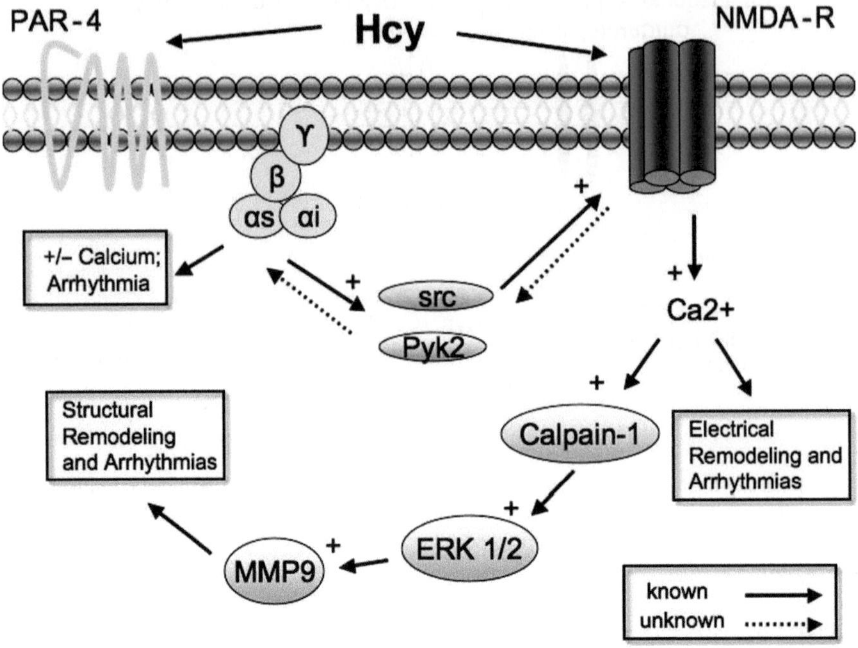

Fig. 3.5 Bidirectional crosstalk. The solid arrows indicate the known activation of molecular species while the dotted arrows indicate one possible model for bidirectional crosstalk. Current research only establishes a one-way crosstalk from GPCRs to NMDA receptor.

Spatial role of NO and mitochondrial NO: There are three nitric oxide synthases (NOS). The endothelial NOS (a calcium-dependent enzyme) is adjacent to the muscle and controls the muscle tone by NO generation; however, decrease in this NOS activity causes muscle dysfunction. The neuronal NOS (a calcium-dependent enzyme) is also close to the muscle and controls sympathetic and parasympathetic tones of the muscle. Imbalance in nNOS activity can lead to arrhythmogenesis. The inducible NOS is primarily present in the macrophage/neutrophils and calcium independent. The activation of macrophage/neutrophils generates robust amounts of NO. This NO in conjunction with ROS generates peroxynitrites and leads to protein oxidation and nitration of the tyrosine. Like NOS in macrophage/neutrophils, the mitochondrial NOS is also inducible; unlike macrophage/neutrophils, the mitochondrial NOS is calcium dependent. The NO generated by mtNOS in conjunction with NADPH oxidase generated by ROS increases oxidative stress. Therefore, the locations of NO generation are very important in annexing the role of NO as to friend or foe.

Summary

Homocysteine binds to GPCRs and NMDAR, which activates separate, though not mutually exclusive, signal transduction pathways; this results in the physiological effect of cardiac arrhythmias. Cardiac arrhythmias result from structural and electrical remodeling that further cause CHF and SCD. G-proteins modulate this effect of cardiac arrhythmias, in part, by increasing or decreasing intracellular calcium. The GPCR signal transduction pathway feeds into the NMDAR signal transduction axis. However, it is currently unknown whether the NMDAR signal transduction pathway can backtalk to GPCRs to modulate G-proteins. Knowing how the pathways feed into each other can help in designing more robust gene therapy drugs to treat cardiac arrhythmias (Praveen et al. 2006).

Acknowledgement This work was supported in part by NIH grants HL-71010, HL-74185, HL-88012, and NS-51568.

References

Audelin MC, Genest J Jr. 2001. Homocysteine and cardiovascular disease in diabetes mellitus. Atherosclerosis 159(2):497–511.

Betsuyaku T, Nnebe N, Sundset R, et al. 2006. Overexpression of cardiac connexin 45 increases susceptibility to ventricular tachyarrhythmias *in vivo*. Am J Physiol Heart Circ Physiol 290:H163–H171.

Bi R, Broutman G, et al. 2000. The tyrosine kinase and mitogen-activated protein kinase pathways mediate multiple effects of estrogen in hippocampus. Proc Natl Acad Sci USA 97:3602–3607.

Bjarnadottir M, Misner D, et al. 2007. Neuregulin1 (NRG1) signaling through fyn modulates NMDA receptor phosphorylation: differential synpatric function in NRG+/− knock-outs compared with wild-type mice. J Neurosci 27:4519–4529.

Brodde O-E, Bruck H, et al. 2006. Cardiac adrenoceptors: physiological and pathophysiological relevance. J Pharmacol Sci 100:323–337.

Brundel B, Ausma J, Gelder I, et al. 2002a. Activation of proteolysis by calpains and structural changes in human paroxysmal and persistent atrial fibrillation. Cardiovasc Res 54:380–389.

Brundel B, Henning R, et al. 2002b. Molecular mechanisms of remodeling in human atrial fibrillation. Cardiovasc Res 54:315–324.

Chen H, Fitzgerald R, Brown A T, et al. 2005. Identification of a homocysteine receptor in the peripheral endothelium and its role in proliferation. J Vasc Surg 41:853–860.

Clarke R, Collins R. 1998. Can dietary supplements with folic acid or vitamin B6 reduce cardiovascular risk? Design of clinical trials to test the homocysteine hypothesis of vascular disease. J Cardiovasc Risk 5:249–55.

Davey SG, Ebrahim S. 2003. Mendelian randomization: can genetic epidemiology contribute to understanding environmental determinants of disease? Int J Epidemiol 32:1–22.

Emoto M, Kanda H, Shoji T, et al. 2001. Impact of insulin resistance and nephropathy on homocysteine in type 2 diabetes. Diabetes Care 24:533–538.

Fallon UB, Elwood P, Ben Shiomo Y, Ubbink JB, Greenwood R, Smith GD. 2001. Homocysteine and ischaemic stroke in men: the Caerphilly study. J Epidemiol Community Health 55:91–96.

Fallon UB, Virtamo J, Young I, et al. 2003. Homocysteine and cerebral infarction in Finnish male smokers. Stroke 34:1359–1363.

Felsch J, Cachero T. 1998. Activation of protein tyrosine kinase Pyk2 by the m1 muscarinic acetylcholine receptor. Proc Natl Acad Sci USA 95:5051–5056.

Finkelstein JD. 1990. Methionine metabolism in mammals. J Nutr Biochem 1:228–237.

Finkelstein JD. 1998. The metabolism of Hcy: pathways and regulation. Eur J Pediatr 157(S-2):S40–S44.
Fonseca VA, Fink UM, Kern PA. 2003. Insulin sensitivity and plasma homocysteine concentrations in non diabetic obese and normal weight subjects. Atherosclerosis 167:105–109.
Frampton JE, Orchard CH, et al. 1991. Diastolic, systolic and sarcoplasmic reticulum [Ca^{2+}] during inotropic interventions in isolated rat myocytes. J Physiol (London) 437:351–375.
Gao X, Xu X, Pang J, et al. 2006. NMDA receptor activation induces mitochondrial dysfunction, oxidative stress and apoptosis in cultured neonatal rat cardiomyocytes. Physiol Res 22:1–25.
Gehrmann J, Meister M, et al. 2002. Impaired parasympathetic heart rate control in mice with a reduction of functional G protein beta-gamma-subunits. Am J Physiol Heart Circ Physiol 282:H445–H456.
Giltay EJ, Hoogeveen EK, Elbers JM, Gooren U, Asscheman H, Stehouwer CD. 1998. Insulin resistance is associated with elevated plasma total homocysteine levels in healthy, non obese subjects. Atherosclerosis 139:197–198.
Graham IM, Daly LE, Refsum HM, et al. 1997. Plasma homocysteine as a risk factor for vascular disease. The European Concerted Action Project [see comments]. JAMA 277:1775–1781.
Hardingham G, Arnold F, et al. 2001. A calcium microdomain near NMDA receptors: on switch for ERK-dependent synapse-to-nucleus communication. Nat Neurosci 4:565–566.
Hardt S, Geng, Y-J, et al. 2002. Accelerated cardiomyopathy in mice with overexpression of cardiac G-s-alpha and a missense mutation in the alpha-myosin heavy chain. Circulation 105:614–620.
Herrmann M, Kindermann I, Muller S, et al. 2005. Relationship of plasma homocysteine with the severity of chronic heart failure. Clin Chem doi: 10.1373/clinchem.2005.049841
Homocysteine. 15 June 2007. <http://en.wikipedia.org/wiki/Homocysteine>
Homocysteine Studies Collaboration. 2002. Homocysteine and risk of ischemic heart disease and stroke: a meta analysis. JAMA 288:2015–2022.
Ju Y-K, Allen D. 1998. Intracellular calcium and Na + -Ca2+ exchange current in isolated toad pacemaker cells. J Physiol (London) 508:153–166.
Kanagartnam P, Cherian A, et al. 2004. Relationship between connexins and atrial activation during human atrial fibrillation. J Cardiovascr Electrophysiol 15:206–213.
Kaye JM, Stanton KG, McCann VJ, Vasikaran VB, Taylors RR, van Bockxmeer FM. 2002. Homocysteine, folate, methylene tetrahydrofolate reductase genotype and vascular morbidity in diabetic subjects. Clin Sci 102:631–637.
Khorchid A, Ikura M. 2002. How calpain is activated by calcium. Nat Struct Biol 9:239–241.
Klerk M, Verhoef P, Clarke R, Blom HJ, Kok FJ, Schouten EG. 2002. MTHFR 677C > T polymorphism and risk of coronary heart disease: A meta analysis. JAMA 288:2023–2031.
Kostin S, Klein G, et al. 2002. Structural correlate of atrial fibrillation in human patients. Cardiovasc Res 54:361–371.
Kurtis Y-F L, Patel K. 2003. Alteration of NMDA NR1 receptors within the paraventricular nucleus of hypothalamus in rats with heart failure. Circ Res 93:990–997.
Lee F S. 2003. Novel crosstalk between G-protein-coupled receptors and NMDA receptors. Exp Neurol 183:269–272.
McCully KS. 1996. Homocysteine and vascular disease. Nat Med 2:386–389.
Moshal K, Tyagi N, et al. 2005. Early induction of matrix metalloproteinase-9 transduces signaling in human heart end stage failure. J Cell Mol Med 9:704–713.
Moshal K, Sen U, Tyagi N, et al. 2006a. Regulation of homocysteine-induced MMP-9 by ERK 1/2 pathway. Am J Physiol Cell Physiol 290:C883–C891.
Moshal K, Singh M, Sen U, et al. 2006b. Homocysteine-mediated activation and mitochondrial translocation of calpain regulates MMP-9 in MVEC. Am J Physiol Heart Circ Physiol 291:H2825–H2835.
Mujumdar VS, Aru GM, Tyagi SC. 2001. Induction of oxidative stress by homocysteine impairs endothelial function. J Cell Biochem 82:491–500.

Naruszewicz M, et al. 2006. Hyperhomocysteinemia in patients with symptomatic chronic heart failure: Prevalence and prognostic importance – pilot study. Atherosclerosis doi: 10.1016/j.atherosclerosis.2006.08.014.

Nishizawa T, Vatner S, et al. 2006. Overexpressed cardiac G-s-alpha in rabbits. J Mol Cell Cardiol 41:44–50.

NMDA Receptor. 15 June 2007. <http://en.wikipedia.org/wiki/NMDA_receptor>

Olshansky B. 2005. Interrelationships between the autonomic nervous system and atrial fibrillation. Prog Cardiovasc Dis 48:57–78.

Ovechkin A, Tyagi N. 2005. Role of matrix metalloproteinase-9 in endothelial apoptosis in chronic heart failure in mice. J Appl Physiol 99:2398–2405.

Porter T, Eckberg D, et al. 1990. Autonomic pathophysiology in heart failure patients. J Clin Invest 85:1362–1371.

Praveen SV, Francis J, et al. 2006. Gene therapy in cardiac arrhythmias. Indian Pacing Electrophysiol J 6:111–118.

Razeghi P, Volpini K, et al. 2007. Mechanical unloading of the heart activates the calpain system. J Mol Cell Cardiol 42:449–452.

Redfern C, Coward P, Degtyarev M, et al. 1999. Conditional expression and signaling of specifically designed Gi-coupled receptor in transgenic mice. Nat Biotechnol 17:165–169.

Redfern C, Degtyarev M, et al. 2000. Conditional expression of a Gi-coupled receptor causes ventricular conduction delay and a lethal cardiomyopathy. Proc Natl Acad Sci USA 97:4826–4831.

Rockman H, Koch W, et al. 2002. Seven-transmembrane-spanning receptors and heart function. Nature 415:206–211.

Rosenberger D, Moshal K, Kartha G, et al. 2006. Arrhythmia and neuronal/endothelial myocyte uncoupling in hyperhomocysteinemia. Arch Physiol Biochem 112:219–227.

Salter M W. 2003. D1 and NMDA receptors hook up: expanding on an emerging theme. Trends Neurosci 26:235–237.

Schnyder G, Roffi M, Flammer Y, Pin R, Hess OM. 2002. Effect of homocysteine lowering therapy with folic acid, vitamin B(12), and vitamin B(6) on clinical outcome after percutaneous coronary intervention: the Swiss Heart Study: a randomized controlled trial. JAMA 288:973–979.

Shah B, Catt K. 2004. GPCR-mediated transactivation of RTKs in the CNS: mechanisms and consequences. Trends Neurosci 27:48–53.

Shcherbak NS, Shutskaya ZV, Sheidina AM, Larionova VI, Schwartz El. 1999. Methylenetetrahydrofolate reductase gene polymorphism as a risk factor for diabetic nephropathy in IDDM patients. Mol Genet Metab 68:375–378.

Singh K, Communal C, et al. 2000. Adrenergic regulation of myocardial apoptosis. Cardiovasc Res 45:713–719.

Swyngheadauw B. 1999. Molecular mechanisms of myocardial remodeling. Physiol Rev 79:215–262.

Toole JF, Malinow MR, Chambless LE, et al. 2004. Lowering homocysteine in patients with ischemic stroke to prevent recurrent stroke, myocardial infarction, and death: the Vitamin Intervention for Stroke Prevention (VISP) randomized controlled trial. JAMA 291:565–575.

Tyagi SC. 1999. Homocysteine and heart disease: Pathophysiology of extracellular matrix. Clin Exp Hypertens 21:181–198.

Tyagi N, Sedoris K, Steed M, et al. 2005. Mechanisms of homocysteine-induced oxidative stress. Am J Physiol Heart Circ Physiol 289:H2649–H2656.

Veeranna, Kaji T, et al. 2004. Calpain mediates calcium-induced activation of the ERK1,2 MAPK pathway and cytoskeletal phosphorylation in neurons. Am J Pathol 265:795–805.

Wald DS, Law M, Morris JK. 2002. Homocysteine and cardiovascular disease: evidence on causality from a meta analysis. BMJ 325:1202.

Watson AMD, Hood SG, May CN. 2006. Mechanisms of sympathetic activation in heart failure. Clin Exp Pharmacol Physiol 33:1269–1274.

Werry T, Wilkinson G, et al. 2003. Mechanisms of cross-talk between G-protein-coupled receptors resulting in enhanced release of intracellular Ca^{2+}. Biochemical Society 374:281–296.

Yi P, Melnyk S, Pogribna M, Pogribny IP, Hine RJ, James SJ. 2000. Increase in plasma homocysteine associated with parallel increase in plasma S-adenosylhomocysteine and lymphocyte DNA hypomethylation. J Biol Chem 275:29318–29323.
Zhai P, Yamamoto M, et al. 2005. Cardiac-specific overexpression of AT1 receptor mutant lacking Galphaq/Galphai coupling causes hypertrophy and bradycardia in transgenic mice. J Clin Invest 115:30453049.
Zheng F, Gingrich MB, et al. 1998. Tyrosine kinase potentiates NMDA receptor currents by reducing tonic zinc inhibition. Nat Neurosci 1:185–191.

Chapter 4
The Role of the Beta-Adrenergic Signal Transduction Pathway in Myocardial Protection

Johan Moolman, Erna Marais, Sonia Genade, Syanda Makaula, and Amanda Lochner

Abstract Beta-adrenergic activation is a major factor causing myocardial damage in the context of ischemia. Activation of the beta-adrenergic signal transduction pathway can, however, also elicit protective responses in the myocardium. Activation of the beta-adrenergic signal transduction pathway participates in the protective effect of ischemic preconditioning, and administration of catecholaminergic agents such as isoproterenol and noradrenaline can elicit pharmacological preconditioning. Protection induced by beta-adrenergic activation (beta-adrenergic preconditioning) elicits both classic (early) and delayed (late) preconditioning and utilizes adenosine to mediate acute protection, while utilizing NO in pharmacological late beta-preconditioning. It is unclear which beta-adrenergic receptor subtype is involved in mediating protection, with evidence for a role of both the "harmful" receptor, $beta_1$, and the $beta_2$ adrenergic receptor. Our own findings suggest cAMP and PKA as the second messengers involved in this ability of beta-adrenergic activation to activate a protective response during the triggering phase of protection, whereas attenuation of activation of p38 MAPK during ischemia is involved in protection against ischemia-mediated necrosis and apoptosis.

Introduction

The deleterious consequence of activation of the beta-adrenergic signal transduction pathway in the context of myocardial ischemia is widely appreciated. However, there is also evidence that beta-adrenergic activation can result in myocardial protection against ischemia. The reason for the differences in beta-adrenergic effects probably relates to the existence of subtypes of beta-adrenergic receptors which are coupled to different signal transduction cascades which can either elicit damaging or protective intracellular responses. $Beta_1$-adrenergic activation results in intracellular calcium overload, oxygen wastage, and radical oxygen species generation, contributing to myocardial necrosis as well as apoptosis. This is the basis of the rationale

A. K. Srivastava, M. B. Anand-Srivastava (eds.) *Advances in Biochemistry in Health and Disease.* Signal Transduction in the Cardiovascular System in Health and Disease,

for beta-adrenergic blockade therapy of patients with ischemic heart disease, both in the setting of acute myocardial infarction and for the management of chronic ischemic heart disease. The protective effects of beta-adrenergic activation have been less well studied, but activation of the $beta_2$-adrenergic signal transduction pathway seems to be involved with protective and antiapoptotic effects.

Ischemic preconditioning refers to the phenomenon whereby a short, transient episode of ischemia elicits endogenous protection against a subsequent sustained episode of ischemia (Murry et al. 1986). The protective effect of preconditioning can be immediate, which is referred to as "acute" or "classical preconditioning" (Yellon and Downey 2003). The protective effect of acute preconditioning is transient and is lost within a few hours but becomes manifest again after 24 hours, called "late preconditioning" or "second window of protection" (Yellon and Downey 2003). Numerous agonists, receptors, and signal transduction pathways have been implicated in the mechanism of ischemic preconditioning (Gross and Gross 2006). We have studied the role of the beta-adrenergic signal transduction pathway in both classical and late preconditioning, and our findings showed it to be an important contributor to the mechanism of ischemic preconditioning against both myocardial infarction (Lochner et al. 1999) and apoptosis (Moolman et al. 2006), and as a pharmacological means of eliciting both classical (Lochner et al. 1999) and late preconditioning (unpublished observations). Furthermore, our investigations suggest adenosine as a mediator of beta-adrenergic preconditioning and implicate attenuation of the p38 MAPK signal transduction pathway in the mechanism of its protection against apoptosis and necrosis.

Agonist/Receptor Interactions and Signal Transduction Pathways in Ischemic Preconditioning

Following the description of the phenomenon of ischemic preconditioning by Murry and coworkers (1986) and the recognition that it was an extremely potent form of protection, experimental investigations strived to elucidate the mechanism. Although metabolic changes such as ATP sparing and glycogen depletion received attention initially, it soon became apparent that the phenomenon was due to an interaction between the agonist adenosine and its receptor (Liu et al. 1991). Subsequently, other agonists and receptor interactions were recognized as either being involved in the mechanism of ischemic preconditioning, or able to mimic it pharmacologically after being administered, as for bradykinin, opioids, adrenergics, and muscarinics (Gross and Gross 2006).

In addition to agonist–receptor interactions the signal transduction pathways involved in the protective effect have been elucidated. In ischemic preconditioning adenosine acts through the A_1, A_{2A}, and A_3 receptor subtypes (the exact importance of each subtype is still being investigated) and activation of PKC, and specifically the $-\varepsilon$ isotype is important in the downstream signal transduction events leading to protection (Ytrehus et al. 1994). The actions of bradykinin are most likely mediated by ligand binding to the B_2 receptor, but the role of the B_1 receptor is still receiving attention. Opioids exert their protective effects through activation of δ and κ opioid receptor subtypes. The receptor subtypes involved by muscarinics are not currently known (Gross and Gross 2006). Investigations into the mechanism of preconditioning have shown that ischemic preconditioning and pharmacological preconditioning with agonists mentioned above converge on common signal transduction pathways or share common mediators of protection. There is good evidence that the salvage kinases, specifically PI3K, ERK, and PKB (Hausenloy et al. 2005), play a pivotal role as final mediators of preconditioning for most ligands involved in preconditioning. These pathways have been suggested to cause inhibition of GSK3β, which inhibits the mitochondrial permeability transition pore (MPTP) (Juhaszova et al. 2004).

Beta-Adrenergic Involvement in Ischemic Preconditioning: Catecholamine Release and Beta-Adrenergic Receptor Activation

Ischemia-mediated release of catecholamines and concomitant increase in tissue cAMP were well described by the time preconditioning was reported (Schömig et al. 1984). The possibility that adrenergic receptors could be involved in preconditioning was therefore raised early on. In initial studies it was reported that alpha-adrenergic receptors were involved in preconditioning (Banerjee et al. 1993). In our own studies, we could not demonstrate any role for alpha-adrenergic activation in ischemic preconditioning, as prazosin, an alpha-1 adrenergic receptor blocker, could not prevent ischemic preconditioning, and no protection was elicited with the alpha-1 adrenergic agonist phenylephrine (Moolman et al. 1996a). The involvement of the beta-adrenergic signal transduction system was suggested by Asimakis et al. (1994) who demonstrated protection against global ischemia by pharmacological preconditioning with noradrenaline and isoproterenol which could be blocked by propranolol. In these studies, no role for the alpha-adrenergic receptor activation could be demonstrated either.

The above observations and findings in our laboratory that cAMP increased in a cyclic fashion at the end of each episode of a three times 5 min (3×5 min) ischemic preconditioning protocol (see Figure 4.1) strongly suggested a role for the beta-adrenergic signal transduction system in ischemic preconditioning (Lochner et al. 1998, 1999), and we undertook a systematic investigation into the changes in this system at different levels during preconditioning as well as during sustained ischemia. The first question we investigated was whether the increase in cAMP seen during the triggering phase was due to catecholamine release. This matter was not clear at that stage, as studies on the release of catecholamines had been done only after periods of 10 min ischemia or longer (Schömig et al. 1984). The work of Strasser et al. (1990) clearly showed that ischemia caused an increase in cAMP. This increase in cAMP was due to increased beta-adrenergic receptor density

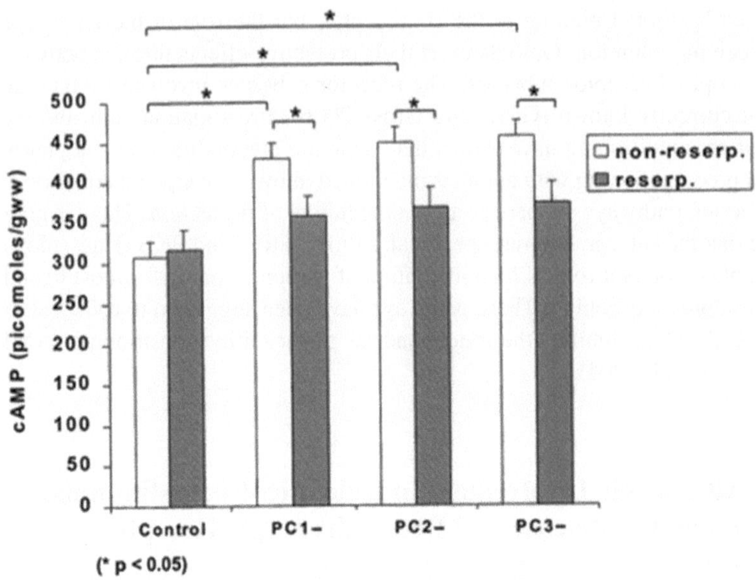

Fig. 4.1 Tissue cAMP levels during a multicycle preconditioning (3×5 min) protocol. PC1-, PC2-, PC3-: samples taken at the end of each 5-min period of the ischemic preconditioning protocol. Reserp.: hearts from animals reserpinized (7 mg/kg) 24 hours before experimentation.

and PKC-mediated sensitization of adenylyl cyclase, and not to catecholamine release as alprenolol could not prevent or decrease the increase in cAMP in response to ischemia, negating a role for catecholamine-mediated activation of the beta-adrenergic receptor (Strasser et al. 1990). However, cAMP changes were only studied at 15 min ischemia or longer, which is significantly longer than the duration of ischemia used in preconditioning. Comparing the ability of myocardium to generate cAMP during a preconditioning protocol of 3 × 5 min ischemia in control and reserpinized rats, we found the increase in cAMP significantly attenuated in the latter group (Figure 4.1) (Lochner et al. 1999). These results suggested catecholamine release and beta-adrenergic activation during the short periods of ischemia as used in preconditioning, to cause the increase in cAMP seen during the preconditioning protocol. These observations suggested the possibility of catecholamine-mediated beta-adrenergic activation as part of the mechanism of preconditioning.

Beta-Adrenergic Activation as Trigger in Ischemic and Pharmacologic Preconditioning

The question arose whether these observed changes in cAMP, and by implication, beta-adrenergic activation, played any role in the cardioprotection of ischemic preconditioning. Using the beta-1 adrenergic blocker alprenolol 7.5×10^{-5} M during the triggering phase, we showed the protective effect of a 3 × 5 min ischaemia preconditioning protocol to be significantly attenuated but not abolished in an isolated perfused working heart model (Lochner et al. 1999). This indicated that although the beta-adrenergic system was involved in ischemic preconditioning, subsequent cardioprotection was not solely dependent on it. In a rabbit model of infarct size, tyramine-induced release of endogenous catecholamines elicited protection comparable to ischemic preconditioning that was alpha-1 adrenergic dependent,(blocked by BE 2254, an alpha-1 adrenergic blocker), but was not affected by beta-adrenergic blockade with propranolol (Thornton et al. 1993). These authors also found that BE 2254 had no effect on ischemic preconditioning, but did not study the effect of beta-adrenergic blockade. They concluded that catecholamines could mimic ischemic preconditioning, but was not involved in the protective effect of ischemic preconditioning, a conclusion that was not fully substantiated, due to the lack of generating data with a beta-adrenergic blocker. Recent findings by Mallet et al. (2006) that preconditioning with intermittent hypoxia in dogs was blunted but not abolished (infarct size decreasing from $38 \pm 6\%$ to $1.1 \pm 0.3\%$ in preconditioned dogs, increased to $27 \pm 3\%$ with the beta-1 blocker metoprolol), are in agreement with ours.

If beta-adrenergic activation played a role in ischemic preconditioning, it was hypothesized that pharmacological activation of the beta-adrenergic receptor should be able to elicit protection against ischemia. Indeed, as alluded to above, Asimakis and coworkers (1994) had demonstrated the ability of isoproterenol to elicit protection against global ischemia using functional recovery as endpoint. We found this protective ability of isoproterenol to be protocol dependent—protection could only be elicited by a single administration of concentrations varying between 10^{-6} M and 10^{-8} M, whereas multiple cycles resulted in severe mechanical failure during reperfusion after exposure to a long period of ischemia (Lochner et al. 1999) (Figure 4.2). Pharmacological preconditioning with beta-adrenergic activation in rat hearts was also demonstrated by Nasa et al. (1997) using norepinephrine (0.25 μM), phenylephrine (10 μM), and isoproterenol (0.25 μM) administered for 2 min, followed by 10 min drug-free perfusion before exposure to ischemia-reperfusion. In this work, left ventricular developed pressure was significantly increased in hearts preconditioned with noradrenaline and isoproterenol but not phenylephrine, and the beta-1 adrenergic blocker timolol abolished the protective effect of norepinephrine, whereas bunazosin, an alpha-1 adrenergic receptor blocker, had no effect. The synthetic beta agonist dobutamine has also been shown to elicit pharmacological preconditioning (Asimakis and Conti 1995).

Pharmacological preconditioning with anaesthetic agents, using similar intracellular mechanisms as ischemic preconditioning, has been described. Recently, Lange et al. (2006) demonstrated involvement of the beta-1 adrenergic pathway in anaesthetic preconditioning elicited with desflurane and sevoflurane as well as ischemic preconditioning in rabbits. Desflurane and sevoflurane reduced infarct size to $34 \pm 2\%$ and $36 \pm 5\%$, respectively, compared to $61 \pm 4\%$ in controls. Anaesthetic preconditioning was abolished by esmolol, a beta-1 adrenergic blocker (infarct size $57 \pm 2\%$ and $52 \pm 4\%$ for desflurane and sevoflurane, respectively), while ischemic preconditioning was attenuated to $40 \pm 4\%$. These findings are important as they

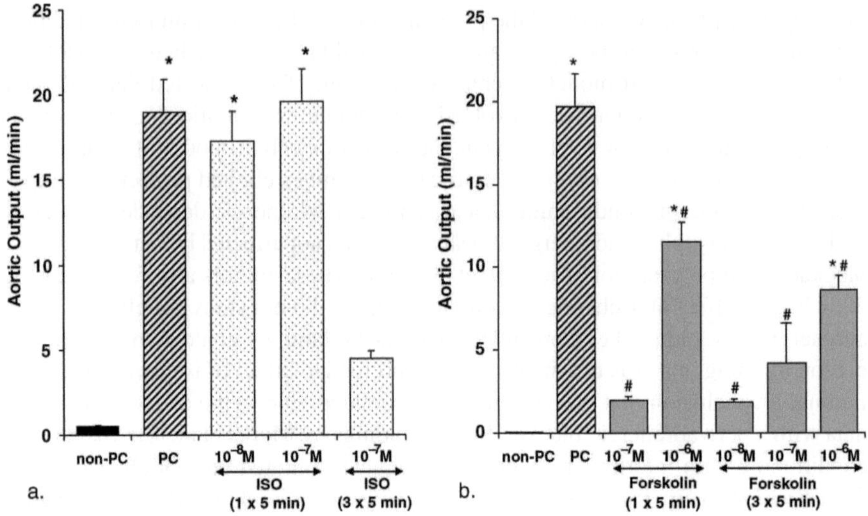

Fig. 4.2 (a) Aortic output during reperfusion after 25 min global ischemia of hearts preconditioned with different concentrations of isoproterenol. (b) Aortic output during reperfusion after 25 min global ischemia of hearts preconditioned with different concentrations of forskolin (10^{-6}–10^{-8} M).

suggest that both ischemic and anaesthetic preconditioning share a mechanism, namely, beta-1 adrenergic signal transduction activation.

PKA and Preconditioning

As stated above, cyclic elevations in tissue cAMP levels during a preconditioning protocol, induced by ischemia or beta-adrenergic stimulation, appear to be pivotal in eliciting cardioprotection. However, the significance of downstream events such as PKA and p38 MAPK activation has not yet been established without a doubt. Using isolated rat hearts, we observed 55% and 87% increases in PKA activation after ischemic (1 × 10 min) or beta-adrenergic preconditioning (forskolin, 0.3×10^{-6} M; 1 × 5 min), respectively (Makaula et al. 2005). Cyclic elevations in PKA activation during a preconditioning protocol were also observed by others (Lan et al. 2005; Inserte et al. 2004).

As in the case of cAMP, activation of PKA during ischemia is potentially harmful, for example, causing phosphorylation and activation of the L-type Ca^{2+} channels, thereby promoting the harmful effects of Ca^{2+} influx into the cardiomyocyte (Tsien 1983; Bunemann et al. 1999) and hyperphosphorylation of the ryanodine receptor to liberate excess amounts of Ca^{2+}, as may occur in heart failure (Marks 2003). The harmful effects of cAMP accumulation during ischemia may be due, at least in part, to activation of PKA. However, activation of PKA has also been linked to cardioprotection. Brief exposure to beta-agonists (Sanada et al. 2004; Lochner et al. 1999), or an adenylyl cyclase activator (Lochner et al. 1999; Makaula et al. 2005) or phosphodiesterase type III inhibitors (Sanada et al. 2001; Nomura et al. 2003), all of which cause rapid activation of PKA, protect the heart against subsequent ischemia, independently of PKC.

The necessity of PKA activation for successful preconditioning is demonstrated by the following: (i) the beta-adrenergic receptor blocker alprenolol partially abolished ischemic preconditioning (Lochner et al. 1999); (ii) landiolol, a short-acting beta-blocker, blunted the infarct size limitation induced by ischemic preconditioning (Sanada et al. 2004); (iii) PKA inhibitors such as H89 (Sanada et al. 2004; Inserte et al. 2004) and Rp-cAMPs (Sanada et al. 2004) blunted ischemic and dibutyryl-cAMP-induced preconditioning.

The putative role of PKA activation after beta-1 adrenergic receptor stimulation with xamoterol was further confirmed by the finding that both atenolol and H89 completely abolished protection (Robinet et al. 2005). These workers also showed that, besides PKA, transduction mechanisms following beta-1-adrenergic receptor stimulation, also involved PI-3K and PKC, with PKA activation occurring prior to PKC.

Possible mechanisms of action of PKA include (i) calpain inhibition, resulting in reduced hydrolysis of structural proteins, reduced sarcolemmal fragility and less cell death (Inserte et al. 2004); (ii) activation of p38 MAPK and its downstream effector HSP27, which, in turn, increase the resistance of the cytoskeleton to conformational changes and fragmentation (Martin et al. 1997); and (iii) phosphorylation of phospholamban, leading to increased Ca^{2+} uptake into the SR, thereby reducing cytosolic Ca^{2+} levels (Sichelschmidt et al. 2003). How these events occurring during a preconditioning protocol translate into cardioprotection during sustained ischemia, remains to be elucidated.

In contrast to the above studies, we recently made the interesting observation that PKA inhibition by H89 when given 5 min before preconditioning with 10 min transient ischemia or forskolin (0.3×10^{-6} M for 5 min) augmented postischemic functional recovery and reduced infarct size (see Figures 4.3 and 4.4). For example, H89 administered for 5 min before 10 min transient ischemia reduced the infarct size from 13.4±1.0% (preconditioning alone) to 7.0±0.93 ($p < 0.01$). In view of its putative harmful effects it appears that PKA activation may offset some of the benefits of incomplete preconditioning (such as 1 × 10 min ischemia). It should also be kept in mind that PKA-independent cAMP pathways may be activated by the preconditioning trigger to promote contractile recovery and to decrease infarct size following ischemia/reperfusion. Many of the cAMP functions previously attributed to PKA, may be dependent on a novel cAMP receptor protein, the guanine nucleotide exchange factor Epac (Kawasaki et al. 1998; Mei et al. 2002). In HEK cells activation of Epac leads to a prosurvival response via phosphatidylinositol-3-kinase-dependent protein kinase B(Akt) activation, while stimulation of PKA inhibits Akt (Mei et al. 2002).

These results suggest that PKA activation during a single-cycle ischemic or pharmacological preconditioning protocol is actually harmful for the preconditioning

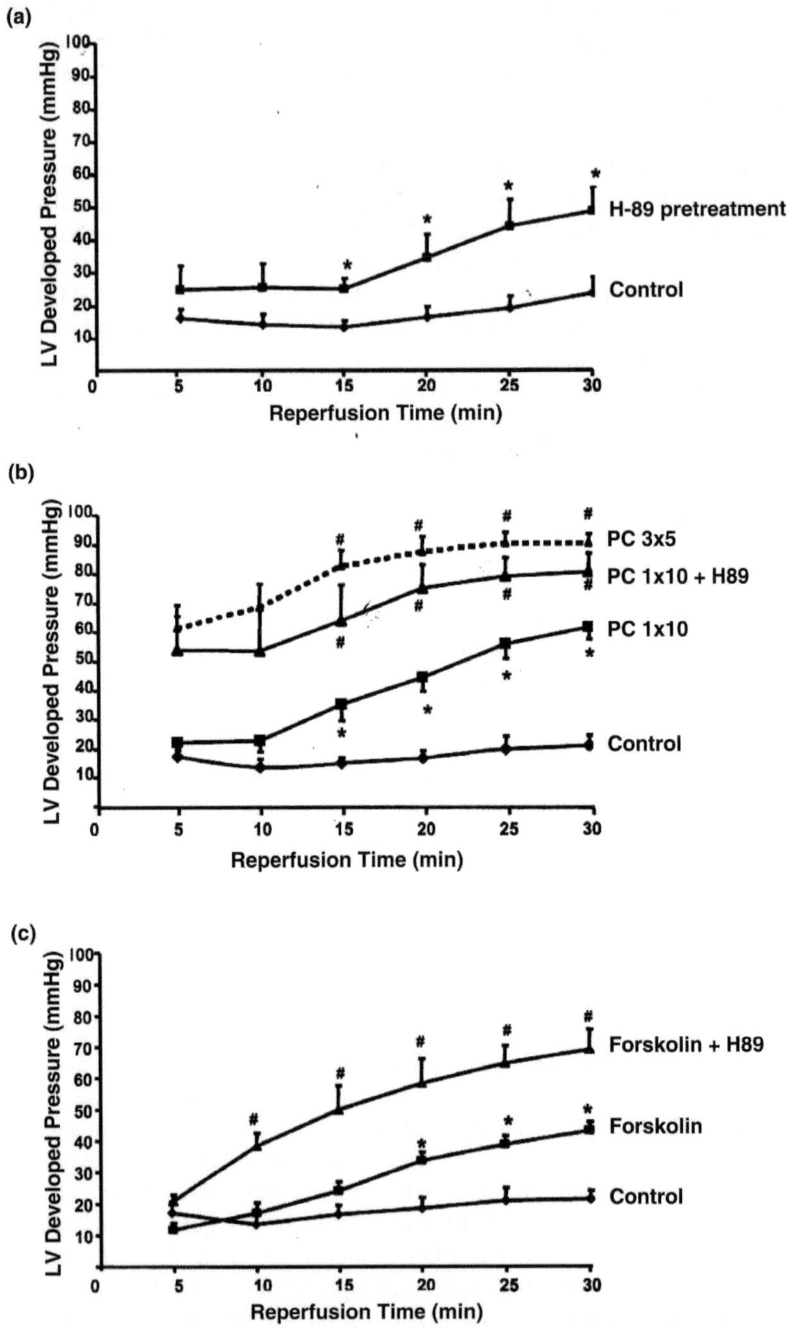

Fig. 4.3 Mechanical performance of retrogradely perfused hearts during reperfusion after 30 min global ischemia. (A) Hearts were pretreated with H89 (2×10^{-6} M) for 10 min before onset of ischemia. (B, C) Hearts were preconditioned with 1×10 or 3×5 min ischemia or forskolin (1×5 min; 0.3×10^{-6} M). H89 (2×10^{-6} M) was administered for 5 min before and during the preconditioning protocols. $^{*}p < 0.05$ versus controls; $\#p < 0.05$ versus PC 1×10 min or forskolin.

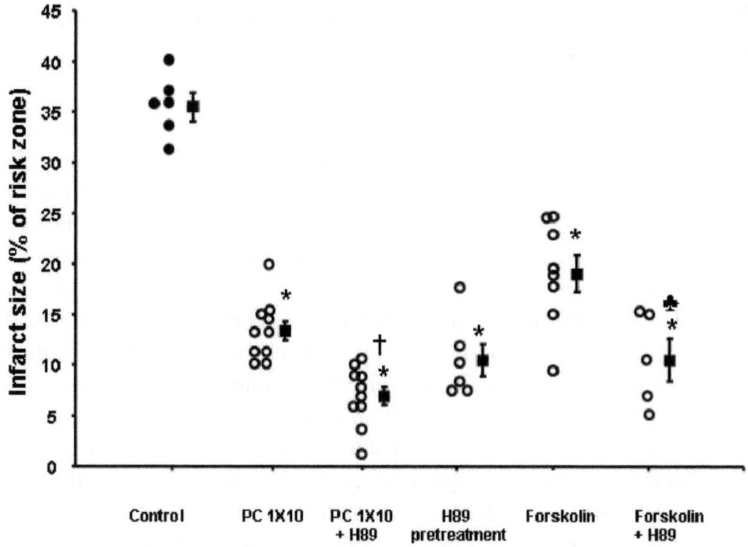

Fig. 4.4 Infarct size as expressed as a percentage of area at risk. Hearts were treated as described in Figure 4.3.

process and that H89, by inhibiting PKA, enhances the effects of suboptimal preconditioning with ischemia or beta-adrenergic stimulation. These observations contradict results obtained by others, as mentioned above, where H89 was observed to abolish or attenuate cardioprotection. Differences in experimental protocol and model (*in vivo* versus *in vitro*) and species may account for the differences observed. For example, Sanada et al. (2004) used anaesthetized, open-chest dogs, while Inserte et al. (2004) used a two-cycle preconditioning protocol, in combination with a very high concentration of H89 (1×10^{-5} M). At this high concentration H89 also significantly affects other kinases (Davis et al. 2000). Robinet et al. (2005), while using the same H89 concentration (1 μM) as ourselves, used a lower concentration of beta-stimulation and a longer period of global ischemia (40 min). It would seem that the experimental protocol plays an important role in our study: we found the beneficial effect of H89 on contractile recovery and infarct size only when it preceded suboptimal single-cycle preconditioning or forskolin administration, thereby allowing scope for the improvement with PKA inhibition and this could be the basis for the differing results obtained.

Finally, the finding that PKA activation during a preconditioning protocol is actually harmful, contradicts the generally accepted assumption that it contributes to triggering the cardioprotective process. Since cAMP is a prerequisite in this process, signaling must occur via an alternative pathway, presumably via the Epac protein. Recent studies from our laboratory showed that Epac is rapidly activated by ischemia as well as by beta-adrenergic preconditioning and is enhanced by simultaneous PKA inhibition. The latter finding confirms the evidence of an alternative beta-adrenergic signaling pathway in the myocardium (unpublished observations).

Possible Mechanisms of Beta-Adrenergic Preconditioning: A Decrease in Tissue AMP During Sustained Ischemia

The role of cAMP accumulation during a long period of ischemia in mediating necrosis and arrhythmias is well recognised (Marks 2003). We asked the question whether the perturbations of the beta-adrenergic system had an effect on cAMP accumulation during the sustained period of ischemia. Indeed, accumulation of intracellular cAMP levels during sustained ischemia was less in preconditioned hearts than in controls, both in rats (Moolman et al. 1996b) (Figure 4.5) and in rabbits (Sandhu et al. 1996, 1997), and interventions such as beta-adrenergic blockade and depletion of endogenous catecholamines by prior reserpine treatment mimicked the effects of preconditioning, causing less cAMP accumulation during ischemia and resulting in functional protection (Moolman et al. 1996b). These observations suggested a decrease in beta-adrenergic signal transduction during sustained ischemia.

This observation raised the question as to the mechanism of the decrease in cAMP, and whether the decrease in cAMP during sustained ischemia was responsible for the protective effect of preconditioning. Our approach was to

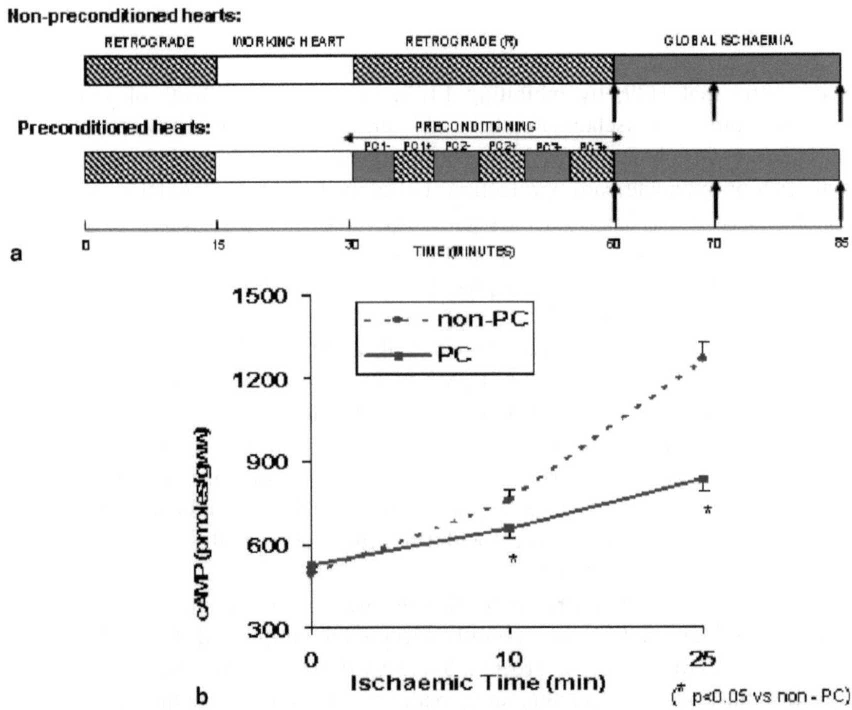

Fig. 4.5 (A) Protocols for non-preconditioned and preconditioned hearts. Arrows indicate times of freeze-clamping. (B) cAMP content of hearts during sustained ischemia. $^{*}p < 0.05$ versus non-PC.

study the state of the beta-adrenergic signal transduction system in terms of beta-adrenergic receptor density and affinity, forskolin-stimulated adenylyl cyclase activity, and PKA activity of the myocardium immediately following the preconditioning protocol, i.e., immediately prior to the sustained ischemia (Lochner et al. 1999). We found an increase in density and affinity of the beta-receptor following a preconditioning protocol of 3 × 5 min ischemia and 5 min reperfusion, B_{max} being increased by 39% and K_d decreased by 35%. Adenylyl cyclase and PKA activities increased significantly with each cycle of preconditioning, but at the end of three episodes of 5 min ischemia and 5 min reperfusion both were reduced. The effect of these changes was assessed by investigating cAMP generation in response to isoproterenol of hearts preconditioned with a 3 × 5 min cycle of ischemia and 5 min reperfusion. cAMP increased significantly in non-preconditioned hearts, but remained unchanged in preconditioned hearts (Figure 4.6), indicating desensitization. cAMP generation in response to forskolin, however, did not differ between preconditioned and non-preconditioned hearts, indicating that the reduced response was at beta-adrenergic receptor level. Our findings are supported by Simonis et al. (2003) who found that beta-adrenergic density increased with repeated cycles, but sensitization of adenylyl cyclase was lost after more than one cycle of 5 min ischemia and reperfusion. These findings suggest reduced responsiveness of the beta-adrenergic signal transduction pathway in preconditioned hearts as the mechanism of reduced cAMP accumulation during the sustained ischemia. However, this seems to differ from rabbits, as Sandhu et al. (1996) found no evidence for reduced responsiveness to isoproterenol in preconditioned rabbit hearts, and based on results obtained with propranolol concluded that reduced cAMP accumulation in preconditioned rabbit hearts was mediated by an attenuated norepinephrine release.

The question remained how reduced cAMP accumulation during sustained ischemia related to the mechanism of ischemic preconditioning. Administration of forskolin to preconditioned hearts resulted in an increase in cAMP during sustained ischemia, but did not abolish protection in our isolated working rat heart model (Moolman et al. 1996b). Sandhu et al. (1996) used NKH477 to activate adenylyl cyclase and increase cAMP in preconditioned hearts during ischemia, and likewise found no loss of protection as measured with infarct size in rabbits. These data strongly suggest that the preconditioning-induced reduction in cAMP accumulation seen during sustained ischemia is a reflection of protection, rather than a causal factor.

p38 MAPK and HSP27

Previous studies from our laboratory suggested a dual role for p38 MAPK in both ischemic and beta-adrenergic preconditioning (Marais et al. 2005): activation of the kinase during the preconditioning protocol had a triggering action, while attenuation of its phosphorylation during sustained ischemia may act as mediator of protection. Significant phosphorylation of cytosolic and myofibrillar HSP27 also occurred

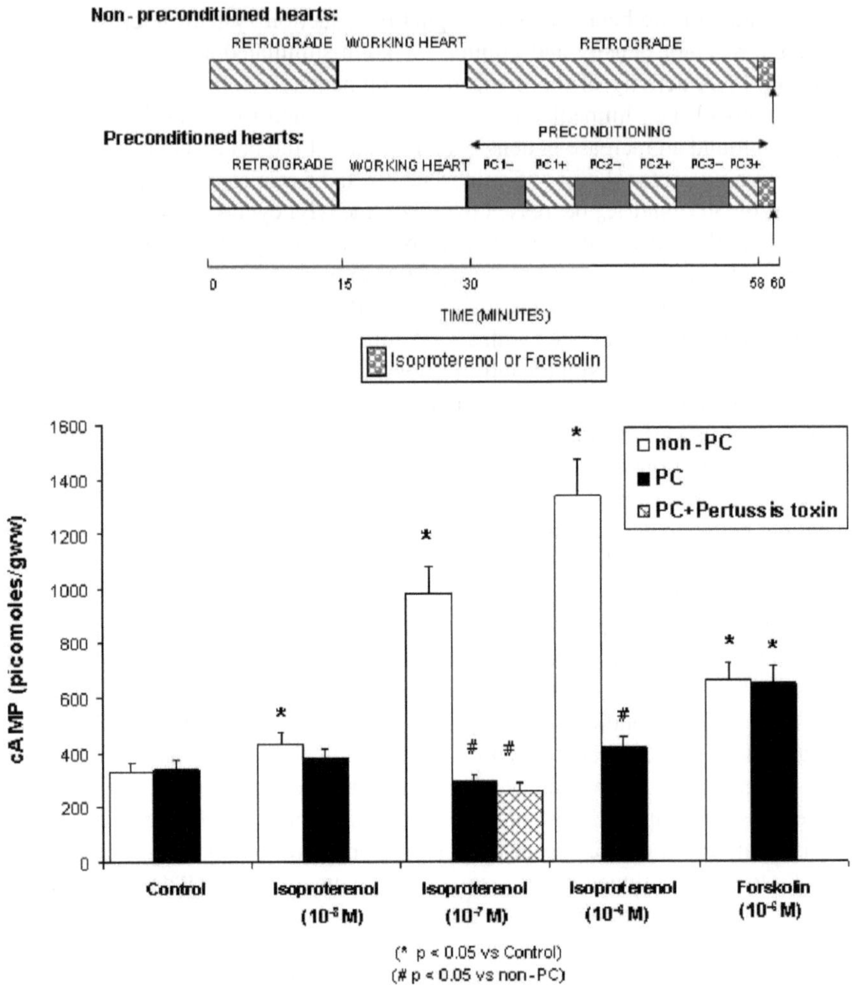

Fig. 4.6 cAMP generation in non-PC and PC hearts in response to isoproterenol and forskolin. Arrows indicate time of administration of the drugs. One series of hearts was reserpinized 24 hours before experimentation. $^{*}p < 0.05$.

during both protocols; this was maintained throughout the sustained ischemic period. According to our results attenuation of p38 MAPK activation and elevation of HSP27 phosphorylation during sustained ischemia are prerequisites for cardioprotection. However, contradictory results regarding the exact role of p38 MAPK have been published (Mocanu et al. 2000; Steenbergen 2002) and the matter warrants further investigation.

The cardioprotective actions of the small heat shock proteins are by now well established (Chi and Karliner 2004). However, how they confer protection is still unclear. Among others, they may act as chaperones (Georgopolous and Welch 1993), stabilize the cytoskeleton (Larsen et al. 1997), or inhibit apoptosis (Rane et al. 2003).

The Role of Adenosine in Mechanism of Beta-Adrenergic Protection

The mechanism of beta-adrenergic preconditioning has not been fully elucidated, but knowledge about it is evolving. The first agonist identified to cause receptor/ligand interaction and to elicit the protective effect of ischemic preconditioning was adenosine (Liu et al. 1991), an effect that was dependent on the activation of protein kinase C (Ytrehus et al. 1994). It is conceivable that beta-adrenergic activation causes demand ischemia, resulting in adenosine production and the downstream activation of its effectors. We have investigated this possibility by perfusing isolated rat hearts with buffer containing adenosine deaminase (ADA), which rapidly breaks down adenosine. Administration of ADA during the triggering period of beta-adrenergic preconditioning (i.e., bracketing the 5-min cycle of isoproterenol administration) abolished the protective effect, suggesting that beta-adrenergic activation functions through adenosine (Figure 4.7) (unpublished observations). Indeed, Thornton et al. (1993) also showed that induction of preconditioning by tyramine-mediated release of endogenous catecholamines in rabbits was blocked by the nonselective adenosine blocker PD115,199, thus proving a role for adenosine in adrenergic-mediated preconditioning. Yabe et al. (1998) subsequently showed that beta-adrenergic preconditioning elicited by isoproterenol was abolished by the

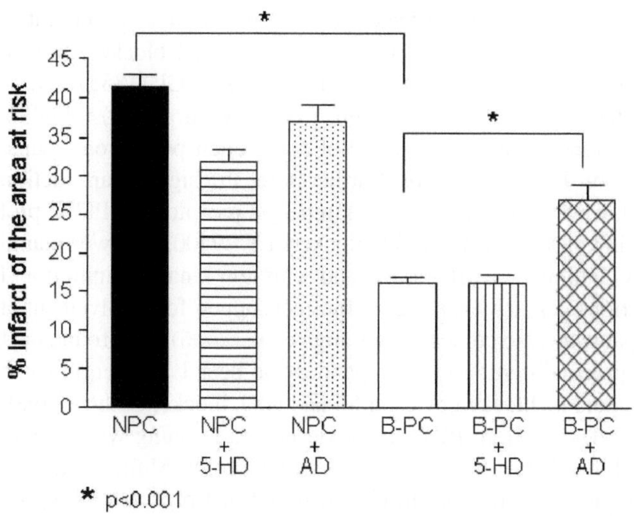

Fig. 4.7 Infarct sizes of beta-adrenergic preconditioned hearts: effects of 5-hydroxy-decanoate (1×10^{-4} M) or adenosine deaminase (0.3 U/ml) pretreatment. The drugs were given for 5 min before and during the administration of isoproterenol (1×10^{-7} M).

PKC inhibitor polymyxin B. If this was true, one would expect beta-adrenergic preconditioning to share another characteristic of adenosine-mediated preconditioning, such as independence from activation of the mitochondrial K_{ATP} channel. Indeed, in our own laboratory, administration of 5-hydroxydecanoate (5-HD) did not have any effect on beta-adrenergic protection, supporting this notion (Figure 4.7) (unpublished observations).

Role of the Beta-Receptor Subtype

In view of the fact that activation of beta-1 and beta-2 adrenergic receptors clearly has a different effect on the response of the myocardium to ischemia (Cross et al. 1999), which even differs between different sexes (Cross et al. 2002), one question which is currently being addressed is which beta-adrenergic subtype is involved in the protective effect of beta-adrenergic preconditioning. Due to the mediating action of adenosine in beta-adrenergic preconditioning, beta-1 adrenergic activation would seem more logical, the rationale being that isoproterenol causes demand ischemia. In view of the known role of the beta-1 adrenergic receptor in adrenergic-mediated myocardial necrosis and apoptosis, this is counterintuitive (Zhu et al. 2001). In view of the opposing effect of beta-2 adrenergic receptor activation on necrosis and apoptosis (Patterson et al. 2004), and the fact that most cell membrane receptors coupled to the G_i protein are able to elicit cardiac protection, the beta-2 adrenergic receptor may be a strong candidate for eliciting beta-adrenergic preconditioning.

Evidence points, however, to a role for the beta-1 adrenergic receptor activation as trigger. Frances et al. (2003) have shown beta-adrenergic preconditioning to be dependent on beta-1 adrenergic receptor activation, as propranolol (nonselective beta-1 blocker) and atenolol (more selective beta-1 blocker) abolished protection, whereas the selective beta-2 adrenergic blocker ICI 118551 had no effect. In a follow-up paper from the same group, Robinet et al. (2005) found that the specific beta-1 adrenergic agonist xamoterol could elicit protection against ischemia, which was abolished by atenolol. Furthermore, the signal transduction pathways implicated downstream of the beta-1 adrenergic receptor are PI3K, protein kinase C, and protein kinase A, as the PI3K blockers LY 294002 and wortmannin, the protein kinase C inhibitor GF 109203X, and a protein kinase A inhibitor (H89) abolished protection elicited by xamoterol. Indirect support for involvement of the beta-1 adrenergic pathway also comes from Lange et al. (2006), who found desflurane and sevoflurane preconditioning to be dependent on beta-1 adrenergic receptor activation, which could be blocked by esmolol and H89, blocker of the second messenger target of beta-1 activation, PKA. Hypoxic preconditioning was likewise found to be attenuated by a beta-1 selective blocker, metoprolol (Mallet et al. 2006). In contrast to these findings, Tong et al. (2005) found that preconditioning could not be elicited with isoproterenol in transgenic beta-2 adrenergic receptor knock-out mice. The exact roles of the beta-1 and beta-2 adrenergic receptors thus remain to be elucidated.

Beta-Adrenergic Preconditioning and Protection Against Apoptosis

The spectrum of protection against ischemia elicited by ischemic preconditioning was initially studied in the context of necrosis and dysrhythmias, and later found to include protection against apoptosis (Piot et al. 1999). The mechanism of the anti-apoptotic effect of ischemic preconditioning involves a host of factors, such as the generation of reactive oxygen species, an altered Bcl-2/Bax ratio and concomitant reduction in cytochrome c release from mitochondria, reduced activation of caspase activity and reduced ceramide production during ischemia (Zhao and Vinten-Johansen 2002). The involvement of altered signal transduction signaling has also been investigated, and p38 mitogen-activated protein kinase (p38 MAPK), known to mediate apoptosis in other cell types, and known to be activated by ischemia, received considerable attention (Ma et al. 1999; Moolman et al. 2006). Since the temporal pattern of JNK and p38 MAPK activation coincided with the appearance of markers of apoptosis during reperfusion (Ma et al. 1999), these kinases appeared likely to play a key role in the signal transduction events causing apoptosis under these conditions. We have described that while p38 MAPK activation during a preconditioning protocol acts as trigger, cardioprotection, whether elicited by ischemic or beta-adrenergic preconditioning, is characterized by attenuation of p38 MAPK

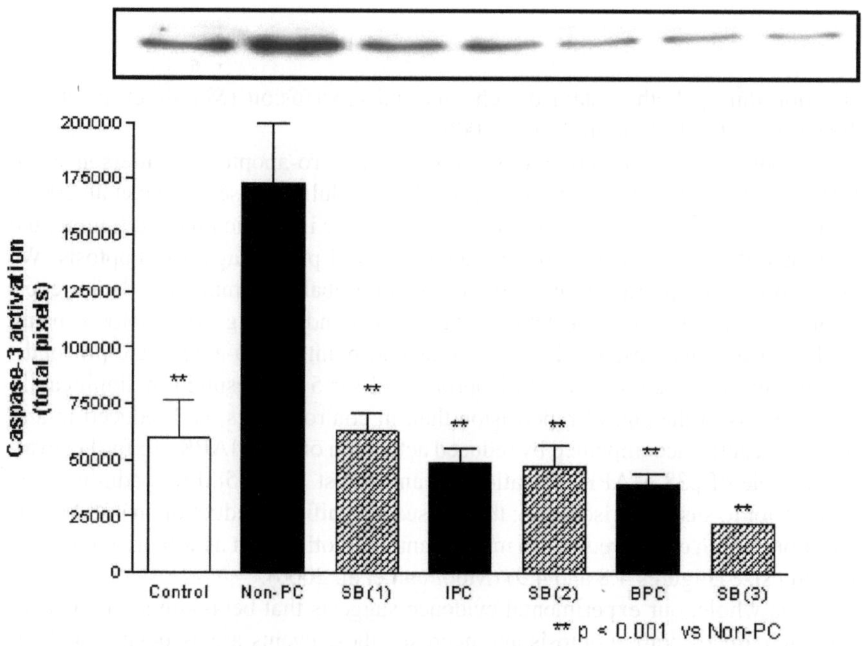

Fig. 4.8 Caspase-3 activity of hearts preconditioned with ischemia (1 × 5 min) or isoproterenol (1 × 5 min; 1 × 10^{-7} M). Samples were taken at end of 30 min reperfusion.

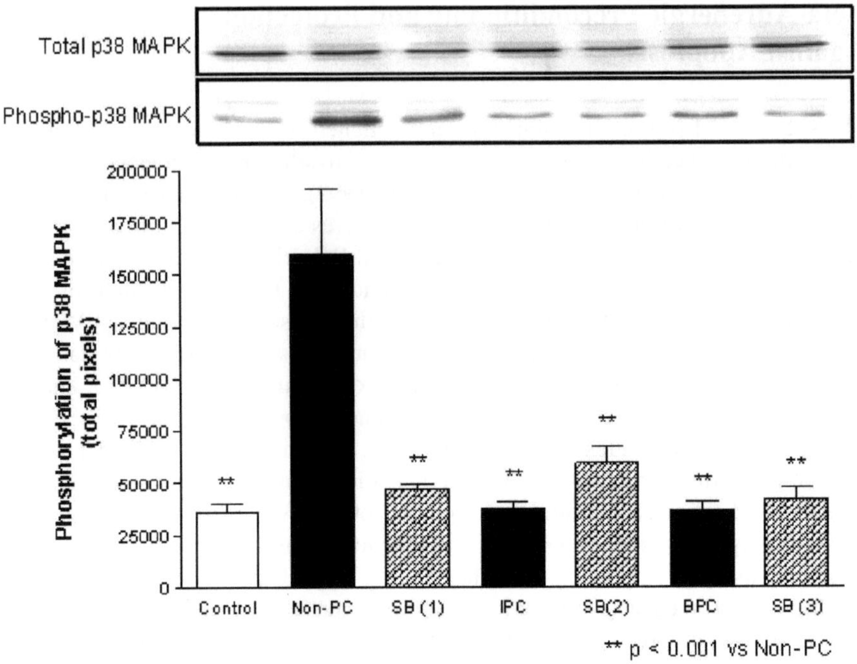

Fig. 4.9 Activation of p38 MAPK in hearts preconditioned with ischemia (1×5 min) or isoproterenol (1×5 min; 1×10^{-7} M). Samples were taken at end of reperfusion.

activation during both sustained ischemia and reperfusion (Marais et al. 2001a, Marais et al. 2001b, Moolman et al. 2006).

Beta adrenergic stimulation was known to be pro-apoptotic (Patterson et al. 2004), an effect attributed to its activation of calmodulin kinase II (Zhu et al. 2001). In view of our finding that isoproterenol could mimic ischemic preconditioning, the question was whether a proapoptotic agonist could protect against apoptosis. We subjected isolated perfused rat hearts to 25 min global ischemia and 30 min reperfusion to study the effect of beta-adrenergic preconditioning on apoptosis, using PARP cleavage and caspase-3 activation as end-points. Beta-adrenergic preconditioning with one cycle of 10^{-7} M isoproterenol for 5 min resulted in significantly less apoptosis at the end of reperfusion than in control hearts, and reduced infarct size significantly, accompanied by reduced activation of p38 MAPK. To further evaluate the role of p38 MAPK activation, its antagonist SB203580 was administered 10 min prior to sustained ischemia: this caused a significant reduction in p38 MAPK activation, which concurred with a marked antiapoptotic effect as well as a reduction in infarct size (Figures 4.8 and 4.9) (Moolman et al. 2006).

On the whole, our experimental evidence suggests that beta-adrenergic preconditioning reduces both apoptosis and necrosis: these events are associated with attenuated activation of p38 MAPK during ischemia and reperfusion. Whether this is the cause or the result of the cardioprotection remains to be established.

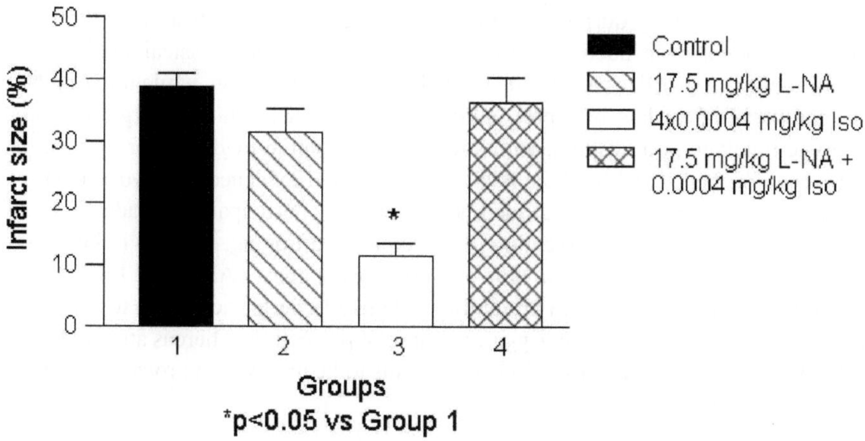

Fig. 4.10 Effect of late beta-adrenergic preconditioning and L-NA pretreatment on infarct size. L-NA: N^{ω}-nitro-L-arginine.

Late Preconditioning with Pharmacological Beta-Adrenergic Preconditioning

Whereas the mechanism of classic preconditioning involved rapid kinase activation, without the production of new peptides, it became clear that late preconditioning had a different mechanism. Bolli (2000) elucidated the important role of NO as final common pathway in late preconditioning. We asked the question whether beta-adrenergic preconditioning could also elicit late preconditioning, and set out to develop a model of beta-adrenergic late preconditioning. Again we found that very small doses of isoproterenol (4×0.0004 mg/kg) administered at 4 hourly intervals could elicit late preconditioning after 24 hours (Figure 4.10). Co-administration of the NOS inhibitor L-NA completely abolished the effects of beta-adrenergic late preconditioning, suggesting a role for NO production in this scenario, as is the case for late preconditioning elicited by ischemia (unpublished observations).

Summary and Conclusions

The role of beta-adrenergic activation as mediator of ischemic damage is undisputed. It is, however, becoming clear that activation of the beta-adrenergic signal transduction pathway can elicit protective responses in the myocardium. Activation of the beta-adrenergic signal transduction pathway occurs, and participates in the protective effect of ischemic preconditioning, although ischemic preconditioning is not dependent on this particular signal transduction pathway. Attenuation of this pathway during sustained ischemia is associated with cardioprotection, as reflected by a reduction in infarct size, apoptosis, and an improvement in functional recovery during reperfusion. Pharmacological activation of the beta-adrenergic signal transduction pathway per se can elicit both classical and late preconditioning, and utilizes adenosine to mediate acute protection, while utilizing NO in pharmacological late beta-preconditioning. The exact beta-adrenergic receptor subtype involved is in dispute, with strong circumstantial evidence for the "harmful" receptor, beta-1, while there is also experimental evidence for involvement of the beta-2 adrenergic receptor. Our findings support participation of adenosine in beta-adrenergic-mediated protection of acute preconditioning, and NO in late beta-adrenergic-mediated protection. Our own findings suggest cAMP and PKA as the second messengers involved in this ability of beta-adrenergic activation to activate a protective response during the triggering phase of protection, whereas attenuation of activation of p38 MAPK during ischemia seems to be involved in protection against ichemia-mediated necrosis and apoptosis.

Acknowledgments The work described was supported by the Medical Research Council of South Africa, the Harry Crossley Fund, Stellenbosch University, and the Cape Heart Institute.

References

Asimakis, G.K., Inners-McBride, K., Conti, V.R., and Yang, C.J. 1994. Transient beta adrenergic stimulation can precondition the rat heart against post-ischaemic contractile dysfunction. Cardiovasc. Res. 28:1726–1734.

Asimakis, G.K., and Conti, V.R. 1995. Preconditioning with dobutamine in the isolated rat heart. Life Sci. 57:177–187.

Banerjee, A., Locke-Winter, C., Rogers, K., Mitchell, M.B., Brew, E.C., Cairns, C., Bensard, D., and Harken, A.H. 1993. Preconditioning against myocardial dysfunction after ischemia and reperfusion by an alpha-adrenergic mechanism. Circ. Res. 73:656–670.

Bolli, R. 2000. The late phase of preconditioning. Circ. Res. 87:972–983.

Bunemann, M., Gerhardstein, B.L., Gao, T., and Hosey, M.M. 1999. Functional regulation of L-type calcium channels via protein kinase A-mediated phosphorylation of the beta (2) subunit. J. Biol. Chem. 274:33851–33854.

Chi, N.C., and Karliner, J.S. 2004. Molecular determinants of responses to myocardial ischaemia-reperfusion injury: focus on hypoxia-inducible and heat shock factors. Cardiovasc. Res. 61:437–447.

Cross, H.R., Steenbergen, C., Lefkowitz, R.J., Koch, W.J., and Murphy, E. 1999. Overexpression of the cardiac beta(2)-adrenergic receptor and expression of a beta-adrenergic receptor kinase-1 (betaARK1) inhibitor both increase myocardial contractility but have differential effects on susceptibility to ischemic injury. Circ. Res. 85:1077–1084.

Cross, H.R., Murphy, E., Koch, W.J., and Steenbergen, C. 2002. Male and female mice overexpressing the beta(2)-adrenergic receptor exhibit differences in ischemia/reperfusion injury: role of nitric oxide. Cardiovasc. Res. 53:662–671.

Davis, S., Reddy, H., Caivano, M., and Cohen, P. 2000. Specificity and mechanism of some commonly used protein kinase inhibitors. Biochem. J. 351:95–105.

Frances, C., Nazeyrollas, P., Prevost, A., Moreau, F., Pisani, J., Davani, S., Kantelip, J.P., and Millart, H. 2003. Role of beta 1- and beta 2-adrenoceptor subtypes in preconditioning against myocardial dysfunction after ischemia and reperfusion. J. Cardiovasc. Pharmacol. 41:396–405.

Georgopolous, C., and Welch, W. 1993. Role of major heat shock proteins as molecular chaperones. Annu. Rev. Cell Biol. 9:601–635.

Gross, E.R., and Gross, G.J. 2006. Ligand triggers of classical preconditioning and postconditioning. Cardiovasc. Res. 70:212–221.

Hausenloy, D.J., Tsang, A., Mocanu, M.M., and Yellon, D.M. 2005. Ischemic preconditioning protects by activating prosurvival kinases at reperfusion. Am. J. Physiol. Heart Circ. Physiol. 288:H971–976.

Inserte, J., Garcia-Dorado, D., Ruiz-Meana, M., Agullo, L., Pina, P., and Soler-Soler, J. 2004. Ischemic preconditioning attenuates calpain-mediated degradation of structural proteins through a protein kinase A-dependent mechanism. Cardiovasc. Res. 64:105–114.

Juhaszova, M., Zorov, D.B., Kim, S.H., Pepe, S., Fu, Q., Fishbein, K.W., Ziman, B.D., Wang, S., Ytrehus, K., Antos, C.L., Olson, E.N., and Sollott, S.J. 2004. Glycogen synthase kinase-3beta mediates convergence of protection signaling to inhibit the mitochondrial permeability transition pore. J Clin. Invest. 113:1535–1549.

Kawasaki, H., Springett, G.M., Mochizuki, N., Toki, S., Nakaya, M., Matsuda, M., Housman, D.E., and Graybiel, A.M. 1998. A family of cAMP-binding proteins that directly activate Rap 1. Science 282:2275–2279.

Lan, X., Wang, J., and Zhang, Y. 2005. Role of beta-adrenergic signal transduction pathway on myocardial ischemic preconditioning of rats. J. Huazhong Univ. Sci. Technol. Med. Sci. 25:709–711.

Lange, M., Smul, T.M., Blomeyer, C.A., Redel, A., Klotz, K.N., Roewer, N., and Kehl, F. 2006. Role of the beta1-adrenergic pathway in anesthetic and ischemic preconditioning against myocardial infarction in the rabbit heart *in vivo*. Anesthesiology 105:503–510.

Larsen, J.K., Yamboliev, I. A., Weber, L.A., and Gerthoffer, W.T. 1997. Phosphorylation of the 27-kDa heat shock protein via p38 MAP kinase and MAPKAP kinase in smooth muscle. Am. J. Physiol. 273:L930–L940.

Liu, G.S., Thornton, J., Van Winkle, D.M., Stanley, A.W.H., Olsson, R.A., and Downey, J.M. 1991. Protection against infarction afforded by preconditioning is mediated by A1 adenosine receptors in rabbit heart. Circulation 84:350–356.

Lochner, A., Genade, S., Tromp, E., Opie, L., Moolman, J., Thomas, S., and Podzuweit, T. 1998. Role of cyclic nucleotide phosphodiesterases in ischemic preconditioning. Mol. Cell. Biochem. 186:169–175.

Lochner, A., Genade, S., Tromp, E., Podzuweit, T., and Moolman, J.A. 1999. Ischemic preconditioning and the beta-adrenergic signal transduction pathway. Circulation. 100:958–966.

Ma, X.L., Kumar, S., Gao, F., Louden, C.S., Lopez, B.L., Christopher, T.A., Wang, C., Lee, J.C., Feuerstein, G.Z., and Yue, T.L. 1999. Inhibition of p38 mitogen activated protein kinase decreases cardiomyocyte apoptosis and improves cardiac function after myocardial ischaemia and reperfusion. Circulation 99:1685–1691.

Makaula, S., Lochner, A., Genade, S., Sack, M.N., Awan, M.M., and Opie, L.H. 2005. H-89, a non-specific inhibitor of protein kinase A, promotes post-ischaemic cardiac contractile recovery and reduces infarct size. J. Cardiovasc. Pharmacol. 45:341–347.

Mallet, R.T., Ryou, M.G., Williams, A.G., Jr., Howard, L., and Downey, H.F. 2006. beta(1)-Adrenergic receptor antagonism abrogates cardioprotective effects of intermittent hypoxia. Basic Res. Cardiol. 101:436–446.

Marais, E., Genade, S., Strijdom, H., Moolman, J.A., and Lochner, A. 2001a. p38 MAPK activation triggers pharmacologically-induced beta-adrenergic preconditioning, but not ischaemic preconditioning. J. Mol. Cell. Cardiol. 33:2157–2177.

Marais, E., Genade, S., Huisamen, B., Strijdom, J.G., Moolman, J.A., and Lochner, A. 2001b. Activation of p38 MAPK induced by a multi-cycle ischaemic preconditioning protocol is associated with attenuated p38 MAPK activity during sustained ischaemia and reperfusion. J. Mol. Cell. Cardiol. 33:769–778.

Marais, E., Genade, S., Salie, R., Huisamen, B., Maritz, S., Moolman, J.A., and Lochner, A. 2005. The temporal relationship between p38 MAPK and HSP27 activation in ischaemic and pharmacological preconditioning. Basic Res. Cardiol. 100: 35–47.

Marks, A.R. 2003. Calcium and the heart, a question of life and death. J. Clin. Invest.111:597–600.

Martin, J.L., Mestril, R., Hilal-Dandan, R., Brunton, L.L., and Dillmann, W.H. 1997. Small heat shock proteins and protection against ischemic injury in cardiac myocytes. Circulation 96:4343–4348.

Mei, F.C., Qiao, J., Tsygankova, O.M., Meinkoth, J.L., Quilliam, L.A., and Cheng, X. 2002. Differential signaling of cyclic AMP: opposing effects of exchange protein directly activated by cyclic AMP and cAMP-dependent protein kinase on protein kinase B activation J. Biol. Chem. 277:11497–11504.

Mocanu, M., Baxter, G.F., Yue, Y., Critz, S.D., and Yellon, D.M. 2000. The p38 MAPK inhibitor, SB 293580, abrogates ischaemic preconditioning in rat heart but timing of administration is critical. Basic Res. Cardiol. 95:472–478.

Moolman, J.A., Genade, S., Tromp, E., and Lochner, A. 1996a. No evidence for mediation of ischemic preconditioning by alpha 1-adrenergic signal transduction pathway or protein kinase C in the isolated rat heart. Cardiovasc. Drugs Ther. 10:125–136.

Moolman, J.A., Genade, S., Tromp, E., and Lochner, A. 1996b. A comparison between ischemic preconditioning and anti-adrenergic interventions: cAMP, energy metabolism and functional recovery. Basic Res. Cardiol. 91:219–233.

Moolman, J.A., Hartley, S., Van Wyk, J., Marais, E., and Lochner, A. 2006. Inhibition of myocardial apoptosis by ischaemic and beta-adrenergic preconditioning is dependent on p38 MAPK. Cardiovasc. Drugs Ther. 20:13–25.

Murry, C.E., Jennings, R.B., and Reimer, K.A. 1986. Preconditioning with ischemia: a delay of lethal cell injury in ischemic myocardium. Circulation 74:1124–1136.

Nasa, Y., Yabe, K., and Takeo, S. 1997. Beta-adrenoceptor stimulation-mediated preconditioning-like cardioprotection in perfused rat hearts. J. Cardiovasc. Pharmacol. 29:436–443.

Nomura, Y., Horimoto, H., Mieno, S., Nakahara, K., Okawa, H., Yoshida, M., and Shinjiro, S. 2003. Repetitive preischemic infusion of phosphodiesterase III inhibitor olprinone elicits cardioprotective effects in the failing heart after myocardial infarction. Mol. Cell. Biochem. 248:179–184.

Patterson, A.J., Zhu, W., Chow, A., Agrawal, R., Kosek, J., Xiao, R.P., and Kobilka, B. 2004. Protecting the myocardium: a role for the beta2 adrenergic receptor in the heart. Crit. Care Med. 32:1041–1048.

Piot, C.A., Martini, J.F., Bui, S.K., and Wolfe, C.L. 1999. Ischemic preconditioning attenuates ischemia/reperfusion-induced activation of caspases and subsequent cleavage of poly(ADP-ribose) polymerase in rat hearts *in vivo*. Cardiovasc. Res. 44:536–542.

Rane, M.J., Pan, Y., Singh, S., Powel, D.W., Wu, R., Cummins, T., Chen, Q., McLeish, K.R., and Klein, J.B. 2003. Heat shock protein 27 controls apoptosis by regulating Akt activation. J. Biol. Chem. 278:27828–27835.

Robinet, A., Hoizey, G., and Millart, H. 2005. PI 3-kinase, protein kinase C, and protein kinase A are involved in the trigger phase of beta1-adrenergic preconditioning. Cardiovasc. Res. 66:530–542.

Sanada, S., Kitakaze, M., Papst, P.J. Asanuma, H., Node, K., Takashima, S., Asakura, M., Ogita, H., Liao, Y., Sakata, Y., Ogai, A., Fukushima, T., Yamada, J., Shinozaki, Y., Kuzuya, T., Mori, H., Terada, N., and Hori, M. 2001. Cardioprotective effect afforded by transient exposure to phosphodiesterase III inhibitors. The role of protein kinase A and p35 mitogen-activated protein kinase. Circulation 104:705–710.

Sanada, S., Asanuma, H., Tsukamoto, O., Minamino, T., Node, K., Takashima, S., Fukushima, T., Ogai, A., Shinozaki, Y., Fujita, M., Hirata, A., Okuda, H., Shimokawa, H., Tomoike, H., Hori, M., and Kitakaze, M. 2004. Protein kinase A as another mediator of ischemic preconditioning independent of protein kinase C. Circulation 110:51–57.

Sandhu, R., Thomas, U., Diaz, R.J., and Wilson, G.J. 1996. Effect of ischemic preconditioning of the myocardium on cAMP. Circ. Res. 78:137–147.

Sandhu, R., Diaz, R.J., Mao, G.D., and Wilson, G.J. 1997. Ischemic preconditioning: differences in protection and susceptibility to blockade with single-cycle versus multicycle transient ischemia. Circulation 96:984–995.

Schömig, A., Dart, A.M., Dietz, R., Mayer, E., and Kubler, W. 1984. Release of endogenous catecholamines in the ischemic myocardium of the rat. Part A: Locally mediated release. Circ. Res. 55:689–701.

Sichelschmidt, O.J., Hahnefeld, C., Hohlfeld, T., Herberg, F.W., and Schrör, K. 2003. Trapidil protects ischemic hearts from reperfusion injury by stimulating PKA II activity. Cardiovasc. Res. 58:602–610.

Simonis, G., Weinbrenner, C., and Strasser, R.H. 2003. Ischemic preconditioning promotes a transient, but not sustained translocation of protein kinase C and sensitization of adenylyl cyclase. Basic Res. Cardiol. 98:104–113.

Steenbergen, C. 2002. The role of p38 mitogen-activated protein kinase in myocardial ischemia/reperfusion injury; relationship to ischemic preconditioning. Basic Res. Cardiol. 97:276–285.

Strasser, R.H., Marquetant, R., and Kubler, W. 1990. Adrenergic receptors and sensitization of adenylyl cyclase in acute myocardial ischemia. Circulation 82(3 Suppl.):II23–29.

Thornton, J.D., Daly, J.F., Cohen, M.V., Yang, X.M., and Downey, J.M. 1993. Catecholamines can induce adenosine receptor-mediated protection of the myocardium but do not participate in ischemic preconditioning in the rabbit. Circ. Res. 73:649–655.

Tong, H., Bernstein, D., Murphy, E., and Steenbergen, C. 2005. The role of beta-adrenergic receptor signaling in cardioprotection. FASEB J. 19:983–985.

Tsien, R.W. 1983. Calcium channels in excitable membranes. Annu. Rev. Physiol. 45:341–358.

Yabe, K., Ishishita, H., Tanonaka, K., and Takeo, S. 1998. Pharmacologic preconditioning induced by beta-adrenergic stimulation is mediated by activation of protein kinase C. J. Cardiovasc. Pharmacol. 32:962–968.

Yellon, D.M., and Downey, J.M. 2003. Preconditioning the myocardium: from cellular physiology to clinical cardiology. Physiol. Rev. 83:1113–1151.

Ytrehus, K., Liu, Y., and Downey, J.M. 1994. Preconditioning protects ischemic rabbit heart by protein kinase C activation. Am. J. Physiol. 266:H1145–H1152.

Zhao, Z.Q., and Vinten-Johansen, J. 2002. Myocardial apoptosis and ischemic preconditioning. Cardiovasc. Res. 55:438–455.

Zhu, W.Z., Zheng, M., Koch, W.J., Lefkowitz, R.J., Kobilka, B.K., and Xiao, R.P. 2001. Dual modulation of cell survival and cell death by beta(2)-adrenergic signaling in adult mouse cardiac myocytes. Proc. Natl. Acad. Sci. USA 98:1607–1612.

Chapter 5
Angiotensin II Signaling in Vascular Physiology and Pathophysiology

Niels Engberding and Kathy K. Griendling

Abstract Initially recognized as a physiologic regulator of blood pressure and body fluid homeostasis, angiotensin (Ang) II has now been shown in innumerable experiments and clinical studies to contribute to the development and maintenance of cardiovascular disease. Dissection of its signaling mechanisms over the past decades has led to the discovery of several novel concepts, such as tissue-specific metabolism of Ang peptides. Identification and cloning of the various receptors through which Ang II acts on almost all tissues has led to the development of specific pharmacologic inhibitors with proven clinical benefit in patients with cardiovascular disorders. Work on the G-protein-coupled Ang II Type 1 receptor has demonstrated that different receptors interact through oligomerization, compartmentalization, and transactivation, and may explain how Ang II can activate G-protein-independent pathways. Unraveling the downstream effects of Ang II in specific cell types corroborates the importance of the cellular redox state on certain signaling pathways. Finally, the effects of Ang II on cell function and phenotype, such as the expression of inflammatory cytokines and receptors promoting the recruitment of inflammatory cells into vascular tissues, have indicated its role in local inflammation as a general pathogenetic basis of cardiovascular disease. The recognition of Ang II as a contributor to such fundamental pathophysiologic mechanisms, which are believed to be a common pathway for diverse cardiovascular risk factors like hypertension and diabetes, has greatly advanced our knowledge of pathologic signaling in vascular tissues and may help to eventually define novel targets for pharmacologic interventions.

The Renin–Angiotensin System (RAS)

Classic RAS

The canonical renin–angiotensin system (RAS) is a circulating hormonal system that controls systemic Ang II production. The octapeptide Ang II was first

A. K. Srivastava, M. B. Anand-Srivastava (eds.) *Advances in Biochemistry in Health and Disease.* Signal Transduction in the Cardiovascular System in Health and Disease,

discovered in the 1940s as the actual pressor substance in animal models of renal hypertension, and is considered the primary effector molecule of the RAS, regulating blood pressure by vasoconstriction and fluid homeostasis (Braun-Menendez et al. 1940; Page and Helmer 1940). Generation of Ang II occurs in two enzymatic steps. The aspartyl protease renin, which is produced by specialized juxtaglomerular smooth muscle cells in the kidneys in response to reduced renal perfusion pressure or salt depletion, catalyzes the first and rate-limiting reaction. By splitting a Leu-Val peptide bond in humans or a Leu-Leu bond in other species, it releases the final ten N-terminal amino acids from its only known substrate, angiotensinogen (AGT). AGT is a liver-derived α_2-globulin that belongs to the family of serine-protease inhibitors (serpines), and its concentration is the major determinant of systemic renin activity. The peptide fragment resulting from cleavage of AGT, Ang I, is the inactive precursor of Ang II. Ang II is generated by cleavage of two C-terminal amino acids (His-Leu) from Ang I by the dipeptidyl carboxypeptidase angiotensin-converting enzyme (ACE) that is anchored to the luminal surface of the endothelium as a transmembranous ectoenzyme and is particularly abundant in the pulmonary vasculature (Caldwell et al. 1976).

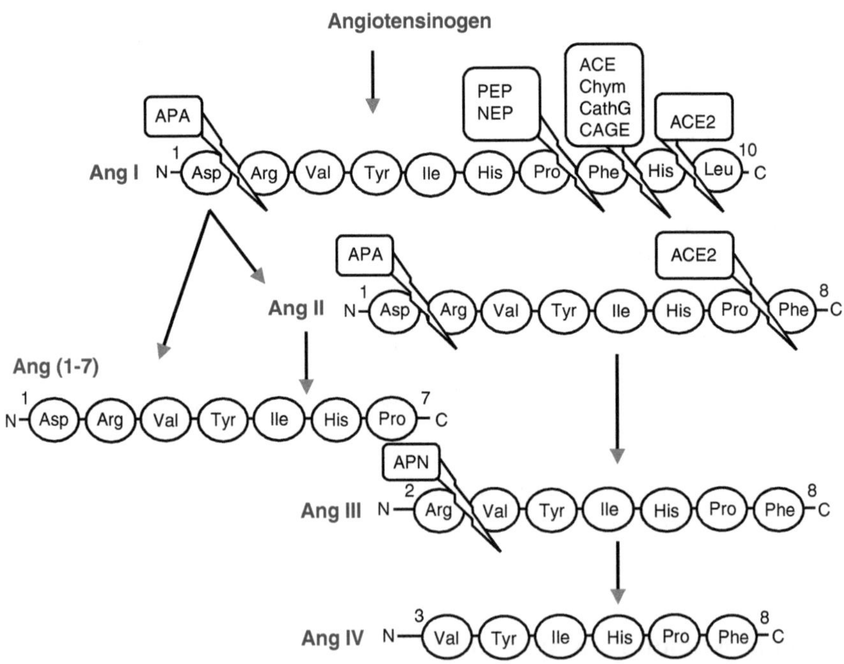

Fig. 5.1 Metabolism of angiotensin peptides. Various peptidases regulate the production and degradation of diverse angiotensin peptides, accounting for tightly regulated agonist and antagonist functions systemically and locally. ACE, angiotensin-converting enzyme; Ang, angiotensin; APA, aminopeptidase A; APN, aminopeptidase N; CAGE, chymostatin-sensitive angiotensin II-generating enzyme; Chym, chymase; CathG, cathepsin G; NEP, neutral endopeptidase; PEP, prolylendopeptidase.

Recent advances in the metabolism of angiotensin peptides have revealed novel functional peptides, which challenge the view of Ang II as the primary effector of the RAS (Figure 5.1). Aminopeptidase A can split off Asp from Ang II, yielding the less potent vasoconstrictor Ang III (Ang 2–8), which can further be transformed to Ang IV (Ang 3–8) by aminopeptidase N. Another angiotensin peptide, Ang 1–7, which is derived either directly from Ang I by tissue-endopeptidases or from Ang II by ACE2, acts as a vasodilator (Ferrario and Iyer 1998). ACE2 is an ACE homologue that acts as a carboxypeptidase by selectively removing the C-terminal Phe from Ang II. Through its contribution to Ang II degradation as well as Ang 1–7 generation, ACE2 has received much attention as an opposing factor of Ang II action. To highlight the heterogeneous functions of the various RAS components, ACE2 has also been recognized as a functional receptor for the SARS coronavirus, mediating its cell intrusion (Kuhn et al. 2004).

Local Tissue RAS

The vascular wall possesses the ability to independently generate Ang II, suggesting local paracrine/autocrine functions of Ang II. Except for renin, all components of the RAS have been found in the vascular wall. The rate-limiting step of local Ang II generation is the conversion of Ang I to Ang II, which can be catalyzed by a number of enzymes in addition to ACE, namely, chymase, cathepsin G, and chymostatin-sensitive Ang II-generating enzyme (CAGE). Several reninlike enzymes that release Ang I from AGT, such as cathepsin D, tonins, tissue plasminogen activator, and other aspartyl proteases, have been proposed as a means to mediate the local RAS independent of kidney-derived renin. While there is no convincing evidence for local synthesis of renin (Hilgers et al. 2001), tissue levels of renin become undetectable within 48 hours after bilateral nephrectomy (Thurston et al. 1979). This suggests accumulation of renin in tissues. Renin uptake into cardiovascular tissues has been reported to result in sustained Ang II effects in local vascular beds (Müller et al. 1998). Furthermore, a functional renin receptor has recently been found on mesangial cells that enhances the catalytic activity of renin upon binding (Nguyen et al. 2002). This receptor specifically binds renin and its precursor prorenin, which may explain why local concentrations of renin (e.g., in interstitial fluid) are oftentimes higher than predicted by simple diffusion from the plasma (Danser et al. 1994).

Angiotensin II Receptors

The cellular effects of Ang II are mediated by distinct high-affinity cell surface receptors. Two subtypes can be discriminated by specific nonpeptide antagonists. Additionally, exposure to the sulfhydryl-reducing dithiothreitol (DTT) inactivates Ang II Type 1 receptors (AT_1Rs), whereas Ang II Type 2 receptors (AT_2Rs) remain intact (Chiu et al. 1989). The AT_1R was initially cloned from rat aortic smooth muscle and bovine adrenal cells (Murphy et al. 1991; Sasaki et al. 1991), while the AT_2R was cloned from a rat pheochromocytoma cell line and from fetal rat tissue (Kambayashi et al. 1993; Mukoyama et al. 1993).

Additional receptor subtypes (e.g., AT_3R and AT_4R) have been described pharmacologically. Based on binding studies in the brain, the AT_4R was originally defined as the binding site for the hexapeptide Ang IV. Later on, the AT_4R was found to be identical to insulin-regulated aminopeptidase (IRAP), a protein first identified as membrane-bound metalloprotease in insulin-sensitive GLUT4 vesicles of fat and muscle cells (Keller et al. 1995). Ang IV binds and inhibits IRAP and enhances cognitive functions in experimental animals (Chai et al. 2004).

The human AT_1R and AT_2R are single-copy genes, but rodents have two AT_1R gene (AGTR1) loci (AGTR1a on chromosome 17; AGTR1b on chromosome 2) and accordingly express two AT_1R isoforms ($AT_{1A}R$ and $AT_{1B}R$) that differ in 18 amino acids, mainly in the carboxy-terminal region. Overall, they are 94% homologous and pharmacologically indistinguishable (Iwai and Inagami 1992). $AT_{1A}Rs$ are expressed predominantly in vascular smooth muscle, endothelial cells, liver, lung, kidney, brain, ovary, and testis, whereas $AT_{1B}Rs$ occur mainly in the adrenal and anterior pituitary gland (Burson et al. 1994). The human AT_1R is $\sim$ 95% homologous to the rodent isoforms (Curnow et al. 1992).

AT_1Rs

Most vascular Ang II effects known to date are mediated by AT_1Rs, the best-characterized Ang II receptor so far. The AT_1R can be blocked by biphenylimidazoles including losartan (DuP 753), valsartan, and candesartan, and is mainly expressed in vascular smooth muscle cells (VSMCs), but also in heart, lung, liver, adrenal cortex, kidney, and brain. Several reports show that Ang II has effects on endothelial cells, but not always do they seem to be mediated by known Ang II receptors (Vaughan et al. 1995). Aside from rodent endothelial cells, which clearly express $AT_{1A}Rs$, AT_1Rs have also been found on bovine and porcine endothelial cells. In primary cultured human arterial umbilical endothelial cells, AT_1Rs have been detected by binding studies; however, the receptors rapidly disappear *in vitro*. Interestingly, a low-affinity binding site persists even in later cell passages, which is not inhibitable by specific receptor blockers (Ko et al. 1997). On the other hand, AT_1Rs have not been found on human endothelial cells using immunohistochemistry (Allen et al. 2000).

Structure and Genomics

The AT_1R is a seven-transmembrane-domain rhodopsinlike peptidergic G-protein-coupled receptor (GPCR) that upon ligand binding activates heterotrimeric

G-proteins to direct subsequent signaling events. The AT_1R has 359 amino acids (MW ~50 kDa). Structurally, it consists of four extracellular (N-terminus and three connecting loops), four intracellular (three connecting loops and C-terminus), and seven α-helical transmembrane domains. Extracellular glycosylation sites are in the N-terminus (Asn^4) and the second extracellular loop (Asp^{176} and Asn^{188}) (Desarnaud et al. 1993). Two disulfide bonds between extracellular loops one and two and between extracellular loop three and the N-terminus stabilize the tertiary structure of the receptor. The N-terminus as well as the first and third extracellular loops contain the epitopes for peptide binding. The binding of nonpeptide antagonists is independent from these epitopes, suggesting distinct modes of interaction for peptide and nonpeptide ligands (Hjorth et al. 1994). Several conserved residues, Asp^{74}, Tyr^{215}, and Tyr^{292}, are important in G-protein binding and activation (Marie et al. 1994). The cytoplasmic tail can be phosphorylated in the basal and Ang II-stimulated state at its numerous Ser and Thr residues, but also at the few Tyr residues at positions 302, 312, 319, and 339 (Kai et al. 1994). Among these the Asp-Pro-Leu-Phe-Tyr ($NPLFY^{302}$) sequence is a variant of the highly conserved tyrosine-containing NPXY motif found in the cytoplasmic tail of many receptor tyrosine kinases (RTKs), in which it is linked to coated pit-mediated receptor internalization. However, in the AT_1R the NPLFY motif is not linked to receptor internalization. Instead, the extra Phe^{301} seems to be important for agonist binding (Hunyady et al. 1995). Tyr^{319} is also part of a functionally important motif: YIPP. This sequence is similar to motifs in the platelet-derived growth factor receptor (PDGF-R) and epidermal growth factor receptor (EGF-R), which, when phosphorylated, are important in SH2-domain coupling to those receptors. In the AT_1R, Tyr^{319} mediates interaction with JAK2 (Ali et al. 1997), a member of the Janus family kinases (JAK), and is also important for EGF-R transactivation.

The human AGTR1 is located on chromosome 3q21–25 and spans about 60 kb, including five exons and four introns. Exon sizes range from 59 to 2014 bp with exon 5 being the largest and the only coding exon, while the first four exons encode the 5′ untranslated region (UTR) (Guo et al. 1994). Several splice variants with different exon composition have been described. The open reading frame of the AGTR1 spans 1080 bp. AGTR1 polymorphisms have been described as a potential link to vascular morbidity. The single nucleotide polymorphism (SNP) A1166C in the 3′ UTR is more common in hypertensive patients than normotensive controls (Bonnardeaux et al. 1994). In addition, patients with coronary artery disease (CAD) who are homozygous for the A1166C mutated allele (CC genotype) have increased AT_1R sensitivity (van Geel et al. 2000). Some other SNPs in the AGTR1 have been implicated in the genetics of hypertension, but much work needs to be done to evaluate the association of other AGTR1 SNPs with cardiovascular disease.

Regulation of Surface Expression

The density of Ang II receptors on VSMCs is a central determinant of the cellular sensitivity to Ang II. Thus, factors that regulate AGTR1 expression have a marked

Table 5.1 Regulation of AT_1R expression in VSMCs

Upregulation	Downregulation
LDL	Angiotensin II
Insulin	Epidermal growth factor
Insulin-like growth factor-1	Fibroblast growth factor
Progesterone	Platelet-derived growth factor
Erythropoietin	α-Thrombin
Interleukin-1α	ATP
Interleukin-6	Interferon-γ
TNF-α	Nitric oxide
C-reactive protein	Reactive oxygen species
Glucocorticoids	HMG-CoA reductase inhibitors
Sodium chloride	PPAR-γ agonists
Hypoxia	Estrogen
Hyperglycemia	Vitamin A
	Thyroid hormone
	All-*trans* retinoic acid
	Forskolin
	Isoproterenol

LDL, low-density lipoprotein; ATP, adenosine triphosphate; TNF, tumor necrosis factor; HMG-CoA, 3-hydroxy-3-methyl-glutaryl-coenzyme A; PPAR, peroxisome proliferator-activated receptor.

effect on cardiovascular function. Among these are cytokines, hormones, growth factors, and vasoactive agents (Table 5.1). In vascular tissues, chronic Ang II exposure itself induces downregulation of the AT_1R in a negative feedback fashion (Nickenig et al. 2000). Both mRNA and protein are significantly downregulated after 2–6 hours of Ang II exposure *in vitro* (Lassegue et al. 1995) and *in vivo* (Gunther et al. 1980). Various pathophysiological conditions affect AT_1R regulation. For example, hypercholesterolemia leads to AT_1R overexpression (Strehlow et al. 2000). This may explain why hyperlipidemia is frequently associated with hypertension. Intriguingly, oxidized low-density lipoprotein (oxLDL), which has been linked to early atherosclerotic events, increases AT_1R expression on endothelial cells (Li et al. 2000), but in contrast to unmodified LDL, it does not affect AT_1R expression on VSMCs (Nickenig et al. 1997). 3-Hydroxy-3-methylglutaryl coenzyme A (HMG-CoA) reductase inhibitors (statins), which are potent cholesterol-lowering drugs, have been shown to reduce AT_1R expression by reducing the half-life of AT_1R mRNA. This mechanism is dependent on their inhibitory effect on geranylgeranylation and may therefore explain lipid-lowering independent (pleiotropic) effects of statins on the vasculature (Wassmann et al. 2001). Proinflammatory cytokines increase AT_1R expression in vascular tissues (Sasamura et al. 1997), whereas nitric oxide (NO) downregulates AGTR1 transcription (Ichiki et al. 1998).

Stability of AT_1R mRNA depends on specific binding motifs for polysomal proteins in the 3′ UTR immediately upstream of the polyadenylation tract (Nickenig et al. 2001). There are two putative polyadenylation sites in the 3′ UTR as well as six AUUUA motifs (Furuta et al. 1992). Mutation of the AU-rich element (3′ UTR 2159–2175) decreases AT_1R mRNA decay, while AU-repeat mRNA-binding proteins promote AT_1R mRNA destabilization. In addition to the primary base composition, the secondary structure of the mRNA, displaying a stem-loop sequence, seems to be necessary for interaction with degrading mRNA-binding proteins (Berger et al. 2005). Moreover, the 5′ UTR sequence has been found to be involved in posttranscriptional regulation. Although in all known splice variants the open reading frame is not affected, variants in the 5′ leader sequence contribute to tissue-specific changes in AGTR1 expression at least in part by altering rates of mRNA translation (Elton and Martin 2003; Zhang et al. 2004).

Initiation of mRNA synthesis is another control point of AGTR1 expression. Aside from binding sites for activator proteins AP-1 and AP-2, and cyclic AMP regulatory element (CRE), there are binding sites for the eukaryotic transcription factor Sp1 in the ATGR1a promoter, including a GC-box-related sequence within the $-58/-34$ region (Takeuchi et al. 1993). Direct binding of the activated peroxisome proliferator-activated receptor (PPAR)-γ to Sp1 causes inhibition of AGTR1a transcription (Sugawara et al. 2001). Another GC-box-related sequence is located at $-98/-79$ of the AGTR1 promoter in rats and humans (Zhao et al. 2000). Both Sp1 binding sites are additively involved in driving basal AGTR1 expression (Kambe et al. 2004). It has also been suggested that increased levels of Sp1 proteins in the hypothalamus contribute to the hypertensive phenotype in spontaneously hypertensive rats (SHRs) via increased AT_1R expression (Kubo et al. 2003). In proximal tubular cells, a *cis*-acting GAGA-box at $-161/-149$ has also been identified as an important regulator for basal and growth factor-induced AGTR1 transcription, suggesting that the AGTR1 may possess alternative initiation sites (Wyse et al. 2000).

Oligomerization

Although a single AT_1R is fully functional, emerging data show that AT_1Rs are capable of forming homo- and heteromers with other GPCRs. Homodimerization of the AT_1R on monocytes correlates with enhanced Ang II signaling and monocyte adhesion to the endothelium (AbdAlla et al. 2004). Coexpression of nonfunctional AT_1R mutants abrogates coupling to G-proteins, whereas recruitment of β-arrestins and subsequent stimulation of the mitogen-activated protein kinase (MAPK) pathway remains intact suggesting, that oligomerization of intact AT_1Rs is particularly pertinent for G-protein signaling (Hansen et al. 2004). Receptors that dimerize with the AT_1R include the bradykinin B_2 receptor, the β_2-adrenergic receptor, the dopamine D_1 receptor, and the AT_2R (AbdAlla et al. 2000, 2001; Barki-Harrington et al. 2003; Zeng et al. 2003). The AT_1R/B_2 heterodimers confer increased Ang II sensitivity, and an increased number of these heterodimers has been clinically correlated with preeclampsia (Quitterer et al. 2004). It has not been conclusively determined if oligomerization is affected by ligand binding, since some data suggest that it may occur even before surface expression (Hansen et al. 2004). Oligomerization may differentially couple the AT_1R to specific downstream signaling proteins, or change receptor ligand binding characteristics or receptor trafficking patterns, all of which may integrate information from multiple receptors and modulate Ang II sensitivity of cells.

Internalization and Trafficking

A characteristic consequence of AT_1R activation is cellular desensitization to Ang II. Within 10 minutes, the receptor is internalized via clathrin-coated pits into endosomes. The intracellular Ang II/AT_1R complex continues to signal until its inactivation, which occurs either by degradation after fusion with lysosomes or by dephosphorylation with subsequent recycling to the cell surface membrane (Hunyady et al. 2002). Approximately 25% of the internalized receptors are recycled back to the cell membrane. Internalization and cellular processing of the Ang II/AT_1R complex itself may have a specific role in signal transduction by engaging the complex with second messenger systems other than those at the cell membrane. The principal mechanism depends on Ser- and Thr-phosphorylation of the activated AT_1R C-terminus mediated by G-protein-related kinases (GRKs). Phosphorylation in the region from Thr^{332} to Ser^{338} facilitates interaction with β-arrestins (Qian et al. 2001), which impair further G-protein activation and subsequently promote receptor endocytosis (Kule et al. 2004). β-Arrestins recruit non-receptor tyrosine kinases (NRTKs) like c-Src and other adapter proteins to the cell membrane to form complexes that orchestrate the internalization process (Fessart et al. 2005). After receptor phosphorylation by GRKs 5/6, β-arrestins can direct signaling events toward the MAPK pathway independent of G-proteins (Kim et al. 2005b). For instance, mutant AT_1Rs incapable of G-protein coupling have been found to activate ERK via β-arrestin-2 (Wei et al. 2003). Of interest, hypertension per se upregulates GRK5 in VSMCs *in vivo*, suggesting that modulation of GRK5 levels may be an adaptive cellular means to autoregulate Ang II sensitivity (Ishizaka et al. 1997). After endocytosis, the AT_1R remains tightly bound to β-arrestin and travels to an early sorting endosome, from which the complex fuses either to a lysosome or to a perinuclear recycling endosome (Gaborik and Hunyady 2004). Sorting of the receptor–ligand complex between intracellular membrane organelles is an important trafficking mechanism. It has been shown that AT_1Rs preferentially traffic to Rab5-positive endosomes (Seachrist et al. 2002). Rab GTPases regulate intracellular vesicle transport and fusion. Furthermore, Rab5 and β-arrestin binding to the AT_1R C-terminus appear to mediate retention of the complex in early endosomes, thereby preventing recycling and degradation (Dale et al. 2004). This could represent a mechanism to prolong Ang II effects intracellularly. Nuclear accumulation of the AT_1R has also been described with effects on cellular proliferation, and it has been speculated that the AT_1R directly participates in transcriptional regulation (Cook et al. 2006).

Interestingly, direct microinjection of Ang II into VSMCs with concomitant extracellular AT_1R blockade has been demonstrated to generate an intracellular calcium increase. However, concomitant microinjection of an AT_1R blocker abolished this Ang II effect (Haller and Luft 1998). This confirms a functional coupling of

Ang II and AT_1R intracellularly as seen in the internalization process described above, but direct actions of free intracellular Ang II, which either has escaped from the internalized AT_1R–Ang II–ligand complex or has been synthesized *de novo*, have also been hypothesized. The functional relevance of this mechanism *in vivo* remains to be determined.

Microdomains and Lipid Rafts

It has become apparent that the lipid bilayer cell membrane has distinct regional characteristics with different functions particularly related to cell signaling. Regions that are predominantly composed of sphingolipids and cholesterol in the outer lipid layer, called lipid rafts, are regions of concentrated signaling proteins such as G-proteins, suggesting that the lipid composition surrounding a GPCR may influence receptor function. Caveolae are special invaginated lipid rafts characterized by the presence of caveolin (Cav)-1, -2, and -3, which are of particular importance for Ang II signaling. Upon stimulation with Ang II, the AT_1R rapidly (<2 min) translocates laterally to Cav-enriched membrane fractions, and directly interacts with Cav-1, promoting the assembly of a Cav scaffolding domain (Ishizaka et al. 1998). This process requires a functional cytoskeleton and cAbl, an actin-binding NRTK, to direct proper translocation of AT_1R into lipid rafts with subsequent Rac1 and NADPH oxidase activation, which eventually mediates VSMC hypertrophy (Zuo et al. 2005). It is possible that the AT_1R complex is also internalized at these noncoated caveolae.

AT_2R

The AT_2R has 363 amino acids (MW ~44 kDa) with only ~30% homology to the AT_1R sequence. The AGTR2 is located on the X chromosome. It is mainly expressed in uterine smooth muscle, brain, ovary, adrenal medulla, heart, and fetal mesenchyme, and is specifically antagonized by tetrahydroimidazopyridines like PD123319 and PD123177 (EXP655) (Wharton et al. 1998). Its wide expression in fetal tissues supports the concept that it has modulatory functions during embryonic development; however, AT_2R surface expression increases under certain pathological conditions. For instance, in heart failure patients the ratio of AT_2R to AT_1R in the heart increases (Tsutsumi et al. 1998). Additionally, it has been reported that AT_2R is reexpressed in vascular inflammation and injury (Akishita et al. 2000). Interestingly, gene targeting of the AT_2R blocks cardiac hypertrophy and fibrosis in mice with Ang II-induced hypertension (Ichihara et al. 2001). The AT_2R is also present on bovine pulmonary endothelial cells where it increases endothelial NO production and counterbalances AT_1R-mediated vasoconstriction in the setting of hypoxia (Olson et al. 2004). In contrast to the AT_1R, the AT_2R does not undergo receptor internalization (Hunyady et al. 2004). The third intracellular loop seems to carry the unique features of the AT_2R in regards to G-protein interaction (Lehtonen et al. 1999). AGTR2 expression is also regulated by multiple extracellular factors. Interestingly, Ang II infusion upregulates vascular AT_2R expression in mesenteric arteries of rats (Bonnet et al. 2001).

Signaling Pathways of the AT_1R

Contraction

Ang II activates a complex series of signaling events that are temporally and spatially tightly controlled. Some second messenger systems are activated within seconds, while others are delayed and persist for more than an hour (Griendling et al. 1986). Upon Ang II binding to VSMCs, the AT_1R activates G-proteins of the pertussis toxin-insensitive subfamilies: $G\alpha_q$ and $G\alpha_{12/13}$. Within 30 sec of Ang II stimulation, a rapid and transient activation of phosphatidylinositol-specific membrane-bound phospholipase C (PLC)-β1 leads to the production of inositol 1,4,5-trisphosphate (IP_3) and diacylglycerol (DAG). $G\alpha_{q/11}$ and $G\alpha_{12}$ proteins as well as their respective Gβγ subunits couple the AT_1R with PLC-β1 (Figure 5.2). Subsequently (≥30 sec), PLC-γ activation accounts for the majority

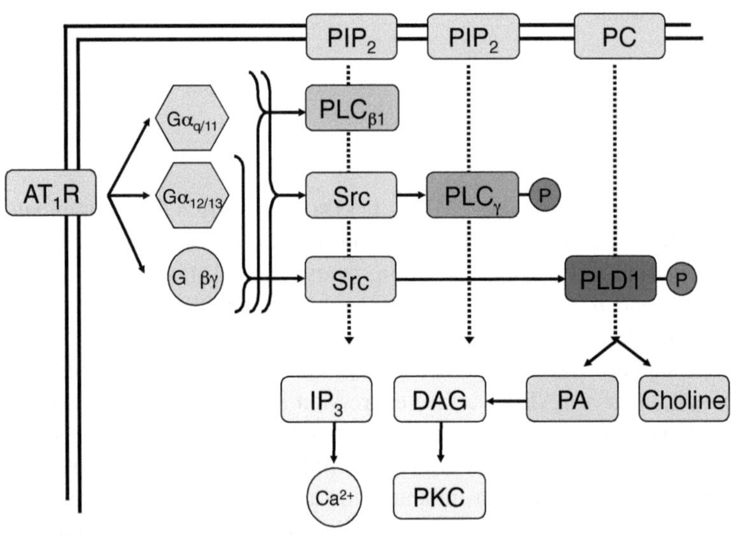

Fig. 5.2 Early G-protein signaling events at the AT_1R. Phospholipases C and D are sequentially activated by heterotrimeric G-protein subunits to produce important second messengers such as IP_3 and DAG. See text for details. AT_1R, angiotensin II type 1 receptor; DAG, diacylglycerol; IP_3, inositol 1,4,5-trisphosphate; PA, phosphatidic acid; PC, phosphatidylcholine; PIP_2, phosphatidylinositol-4,5-bisphosphate; PKC, protein kinase C; PLC, phospholipase C; PLD, phospholipase D.

of IP_3 production. PLC-γ is activated in a tyrosine kinase-dependent manner and returns to baseline activity in 10 min (Ushio-Fukai et al. 1998b). Some evidence exists to suggest the NRTK Src as the responsible tyrosine kinase in this step (Haendeler et al. 2003). IP_3 rapidly diffuses to the endoplasmic reticulum (ER) where it binds to IP_3 receptors to release calcium (Ca^{2+}) into the cytosol. In addition to IP_3-triggered release of Ca^{2+} from intracellular Ca^{2+} stores, sustained Ang II-induced vasoconstriction depends on transmembrane Ca^{2+} influx (Ruan and Arendshorst 1996). The production of the membrane-bound DAG by Ang II is biphasic. The initial peak at 15 sec results from hydrolysis of phosphoinositides by PLC, whereas the second phase, which has its maximum at 5 min, results from phospholipase D (PLD) activation. $G\alpha_{12}$ and $G\beta\gamma$ activate PLD through c-Src and RhoA-dependent mechanisms. PLD remains active for at least 1 hour, splitting phosphatidylcholine into choline and phosphatidic acid (PA) (Ushio-Fukai et al. 1999a), which is subsequently converted to DAG. For this prolonged phase of DAG production, internalization of the ligand-bound AT_1R complex is required, implying that PLD activation occurs spatially separated from the plasma membrane. DAG, in conjunction with phosphatidylserine and Ca^{2+}, subsequently activates protein kinase C (PKC). The early increase in intracellular Ca^{2+} stimulates the Ca^{2+}/calmodulin-dependent myosin light chain kinase to phosphorylate Ser^{19} of the myosin regulatory light chain (MLC) with subsequent VSMC contraction through MLC interaction with actin. This process is importantly modulated by the Ca^{2+} sensitivity of the myofilaments, which is largely determined by Rho-associated kinase (ROCK)-mediated inhibition of MLC phosphatase (Uehata et al. 1997).

In addition, Ang II rapidly phosphorylates and activates PLA_2 to produce arachidonic acid (Rao et al. 1994). Cytochrome P450 metabolites of arachidonic acid, such as 20-hydroxyeicosatetraenoic acids (20-HETEs), have been implicated in vasoconstriction in various vascular beds (Roman 2002).

Controlled relaxation following contraction is vital to the dynamic regulation of blood vessel diameter and hence flow. Termination of Ang II signaling is achieved by regulator of G-protein signaling-2 (RGS2). Ang II upregulates RGS2 mRNA in a PKC-dependent manner thereby providing functional negative feedback (Grant et al. 2000). RGS2-deficient mice have dramatically increased blood pressure due to a prolonged response to Ang II in resistance vessels, as evidenced by the fact that their hypertension is rapidly reversed by Ang II blockade (Heximer et al. 2003).

Cell Growth

Tissue growth is characterized by cellular hypertrophy with increased protein translation, and cellular hyperplasia (proliferation) with increased DNA synthesis, cell cycle progression, and presumably inhibition of cell death pathways. This latter response eventually depends on regulation of gene expression by transcription factors. Although extracellular signals regulating cell growth have classically been

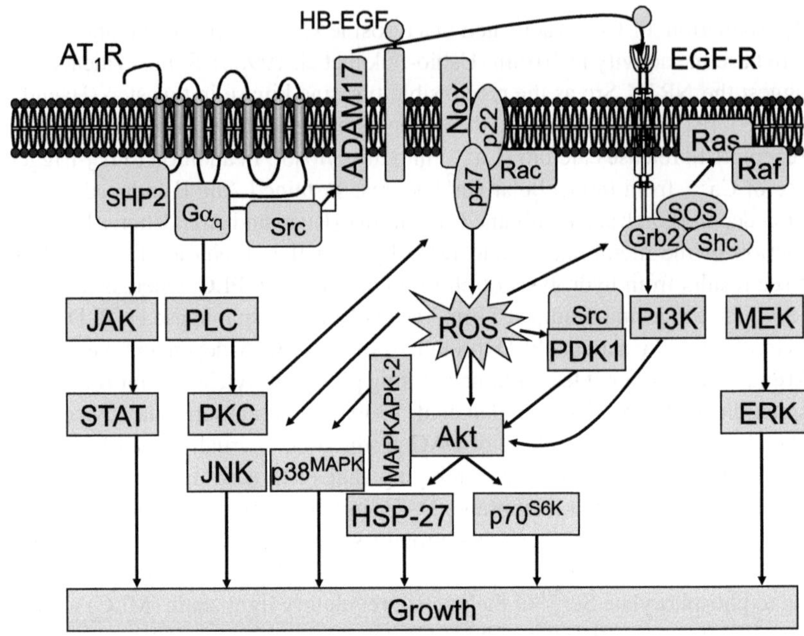

Fig. 5.3 Angiotensin II stimulates vascular smooth muscle cell growth.Growth-promoting effects of Ang II are largely mediated by EGF-R transactivation and stimulation of tyrosine kinase signaling cascades. See text for details. ADAM, a disintegrin and metalloproteinase; AT_1R, angiotensin II type 1 receptor; EGF-R, epidermal growth factor receptor; HB-EGF, heparin-binding epidermal growth factor; HSP, heat shock protein; JAK, Janus kinase; JNK, c-Jun NH_2-terminal kinase; PKC, protein kinase C; PI3K, phosphatidylinositol 3-kinase; PDK1, 3-phosphoinositide-dependent kinase-1; ROS, reactive oxygen species.

considered to be mediated by growth factors and cytokines with activation of RTKs and NRTKs, Ang II has been found to be a potent growth stimulus for VSMCs (Figure 5.3). This is reflected by an increase in cell volume and protein content (Berk et al. 1989; Geisterfer et al. 1988), whereas a proliferative effect of Ang II has been found to be dependent on the cellular milieu (e.g., autocrine production of transforming growth factor-β1 can suppress the hyperplastic response) (Gibbons et al. 1992). Within 1 min of Ang II exposure, numerous cytosolic proteins have been found to be tyrosine phosphorylated (Molloy et al. 1993). Furthermore, the growth-promoting effects of Ang II were found to be abrogated by tyrosine kinase inhibitors (Leduc et al. 1995). It is now well accepted that Ang II rapidly activates the MAPK family, including c-Jun NH_2-terminal kinase (JNK), also termed stress-activated protein kinase, extracellular signal-regulated kinases (ERK, $p42/44^{MAPK}$)1/2, and $p38^{MAPK}$, which are considered upstream mediators of transcription factors involved in VSMC growth.

ERK1/2 activation by Ang II, which can be blocked by PLC inhibition, intracellular Ca^{2+} chelation, and tyrosine kinase inhibition, eventually proceeds through a Ras-dependent pathway (Eguchi et al. 1996). One intermediate tyrosine kinase between AT_1R and Ras is Src (Schieffer et al. 1996). Growth factor receptors, which are typically RTKs such as EGF-R, characteristically propagate their signal to the small GTPase Ras via the GTP-exchange factor son of sevenless (SOS), which forms multimeric signaling complexes with the receptor and adapter proteins such as Grb2 and Shc. Importantly, it has been demonstrated that the atypical PKC-ζ mediates Ang II activation of Ras, which seems to be a unique aspect of the Ang II-induced response as compared to other growth factors (Liao et al. 1997). Ang II induces EGF-R activation by an intracellular Ca^{2+} – and reactive oxygen species (ROS)-dependent mechanism, which has been termed EGF-R transactivation (Eguchi et al. 1998; Ushio-Fukai et al. 2001), and the EGF-R then acts as a scaffold, mediating a number of downstream signaling molecules including phosphatidylinositol 3-kinase (PI3-K) and Akt. Src recruits another NRTK of the focal adhesion kinase (FAK) family, called proline-rich tyrosine kinase (Pyk) 2, and after assembling a signaling complex, activates subtype ADAM17 of the *a d*isintegrin *a*nd *m*etalloprotease (ADAM) family to release cell surface-bound EGF, which subsequently activates its receptor (Eguchi et al. 1999; Ohtsu et al. 2006; Prenzel et al. 1999) and Ras. An important consequence of Ang II-induced ERK1/2 activation in conjunction with PI3-K activity is an increase of protein translation by PHAS-1-eukaryotic initiation factor-4E (Rocic et al. 2003).

Interestingly, EGF-R transactivation is only required for ERK1/2 and $p38^{MAPK}$ activation, but not for JNK activation (Eguchi et al. 2001). A parallel and independent mechanism of ERK1/2 activation by Ang II occurs via β-arrestin-2 as mentioned above (Wei et al. 2003). Apoptosis signaling-related kinase (ASK) 1 is required for JNK and $p38^{MAPK}$ activation (Tobiume et al. 2001), while JNK activation also depends on Rho/ROCK activation and subsequent rac stimulation (Ohtsu et al. 2005). Activation of $p38^{MAPK}$ by Ang II leads to the stimulation of MAPKAPK-2, heat shock protein-27 (HSP-27), and Akt (Taniyama et al. 2004), the latter of which is crucial for VSMC hypertrophy.

JAKs, which are classically activated by cytokine receptors, are upstream controllers of the transcription factors signal transducers and activators of transcription (STATs), which in turn regulate transcription of early growth response genes such as *c-fos, c-myc*, and *c-jun* (Horvath and Darnell 1997). JAK2 binding to Tyr^{319} of the AT_1R is facilitated by SH2 domain-containing tyrosine phosphatases (SHPs) (Marrero et al. 1998).

Several aspects of the growth-promoting pathways have a special characteristic of being ROS-dependent, which means that enzyme activity is susceptible to oxidation. In fact, many signaling proteins, such as transcription factors and protein tyrosine phosphatases (PTPs), rely on reduced Cys residues for activity. Ang II via its AT_1R generates ROS via a membrane-bound multisubunit protein complex, called NADPH oxidase (Nox) (Griendling et al. 1994). Arachidonic acid metabolites play an important role in Nox activation (Zafari et al. 1999). ROS, in turn, modify critical signaling cascades such as $p38^{MAPK}$ (Ushio-Fukai et al. 1998a). Several different isoforms of Nox have now been identified that differ in subunit composition, subcellular localization, and probably type of ROS produced (Bedard and Krause 2007; Hilenski et al. 2004). Different ROS may have distinct vascular effects. For instance, vascular hypertrophy in Ang II-mediated hypertension is mediated by VSMC-derived H_2O_2, while a main effect of superoxide ($O_2^{\cdot -}$) is to inactivate antiatherogenic NO (Rajagopalan et al. 1996; Zhang et al. 2005c).

Tissue growth also occurs through expansion of noncellular components. Chronic Ang II stimulates collagen production (Kato et al. 1991). This observation is frequently referred to as transformation of VSMCs into a synthetic phenotype, which then produce proteoglycans, glycosaminoglycans, collagen type 1, and fibronectin, all of which are constituents of the extracellular matrix. The synthetic phenotype of VSMCs is predominant in hypertensive animals, whereas their normotensive littermates contain more VSMCs of the contractile phenotype (Fukuda et al. 1999). At the same time, VSMCs modulate extracellular matrix degradation by activation of matrix metalloproteinases (MMPs). Extracellular matrix degradation is a prerequisite for cell migration on the one hand, but on the other it has been implicated in plaque instability.

Cell Migration

Ang II-induced VSMC migration is strongly implicated in atherosclerosis and restenosis after vascular injury, because abnormal accumulation of VSMCs in the vessel intima entails movement out of the media (Prescott et al. 1991). It was suggested that VSMCs migrate from the media to the intima as a result of a PDGF gradient (Okuda et al. 1995). *In vitro* studies further confirmed that Ang II is also capable of stimulating VSMCs to migrate, although less potently than PDGF (Bell and Madri 1990), and that NO antagonizes this effect (Dubey et al. 1995). A role for Ang II in the neointima formation of restenosis was suggested based on studies showing that tissue ACE activity is elevated after vessel injury and that ACE inhibition can prevent restenosis (Rakugi et al. 1994), although the relevance of this observation to human disease has been questioned.

Ang II induces VSMC migration by recruiting growth pathways such as the EGF-R–MAPK pathway (Saito et al. 2002; Xi et al. 1999). Additionally, cell migration requires dynamic reorganization of the cytoskeleton and focal adhesion complexes, a process that involves Src activity in VSMCs (Ishida et al. 1999). Ang II stimulates the phosphorylation of FAK, Pyk2, and paxillin (Eguchi et al. 1999; Leduc and Meloche 1995; Okuda et al. 1995), which regulate focal adhesion dynamics. Together with Src, Pyk2 also leads to tyrosine phosphorylation of PDK1, which is critical for focal adhesion assembly (Taniyama et al. 2003). JNK has emerged as key mediator of cell migration through its interaction with these cytoskeleton components and focal adhesion-associated proteins (Huang et al. 2003). The JNK pathway can be activated by the small G-protein Rho and its effector ROCK through PKC-δ and Pyk2 (Ohtsu et al. 2005). Ang II itself is a weak migratory factor, but importantly, it enhances the ability of VSMCs to migrate toward PDGF-BB, possibly by inducing focal adhesions (Dubey et al. 1995).

Apoptosis

A specific suicide capability, apoptosis, allows tissues to regulate their own cell number, which is an essential process in atherosclerosis and restenosis (Isner et al. 1995). Indeed, many growth-promoting factors have been found to also suppress cell death, which may contribute to the deleterious accumulation of VSMCs in the intimal space. Ang II, via its AT_1R, directly inhibits cGMP-induced apoptosis in VSMC (Pollman et al. 1996) by upregulating an enzyme that degrades cGMP, phosphodiesterase 5A (PDE5A) (Kim et al. 2005a). Another mechanism by which Ang II exerts antiapoptotic effects is via stimulation of Noxes to generate H_2O_2, which in turn stimulates antiapoptotic MAPK and Akt activity (Ushio-Fukai et al. 1999b). Conversely, in addition to its antiapoptotic function, Ang II can cause a delayed apoptosis in the epithelioid-shaped subtype of VSMC, which has been suggested to contribute to locally heterogeneous plaque weakening and rupture (Bascands et al. 2001). Apoptosis alone does not cause inflammation in normal arteries, but in atherosclerotic lesions it does, leading to plaque vulnerability (Clarke et al. 2006). Thus, apoptosis is considered both beneficial and detrimental.

Proapoptotic actions of Ang II seem to be mediated substantially by AT_2R activation. It has been demonstrated *in vivo* that AT_2Rs mediate negative vascular remodeling by inducing VSMC apoptosis (Yamada et al. 1998). While ACE inhibitors (ACE-I) block this effect, AT_1R blockers (ARBs) do not (Tea et al. 2000). Interestingly, overexpression of the AT_2R alone stimulates apoptosis in the absence of Ang II (Miura and Karnik 2000). It has been found that PTPs like SHP-1 and MAPK phosphatase (MKP)-1 are linked to the AT_2R via Gi protein coupling to the unique third intracellular loop of the receptor (Lehtonen et al. 1999); however, the signaling pathways linked to the AT_2R have not been well characterized yet.

Inflammation

Numerous studies have shown that VSMCs are capable of producing cytokines in the vessel wall. Cytokines are considered the main modulators of inflammatory events during atherogenesis. Ang II, like inflammatory cytokines, activates proinflammatory transcription factors (Brasier et al. 2000). The resulting inflammatory gene products can be grossly subdivided into adhesion molecules, cytokines, and acute phase reactants. The prototypical Ang II-inducible transcription factors, such as AP-1, STATs, and nuclear factor-κB (NF-κB), are also known cytokine-inducible transcription factors. Ang II dose-dependently stimulates interleukin-6 (IL-6) production in cultured VSMC, which is accompanied by activation of NF-κB. Antioxidants reverse this effect, indicating a role for ROS (Kranzhofer et al. 1999). Ang II increases IL-6 expression as early as 15 min, peaking at 1 h and falling thereafter, whereas NF-κB activation is biphasic with peaks at 15 min and 24 h (Han et al. 1999). Cell cycle regulatory genes, like *AP-1* and *c-myc*, are stimulated via the AP-1 transcription factor (Naftilan et al. 1989; Taubman et al. 1989). Ang II also stimulates transcription of intercellular and vascular adhesion molecules (ICAM,

VCAM) in a NF-κB-dependent manner (Tummala et al. 1999). The link between AT_1R and NF-κB activation involves direct phosphorylation of the p65 subunit of NF-κB by ribosomal S6 kinase (RSK) and by IKK, and both pathways appear to be redox-dependent (Zhang et al. 2005a,b).

In addition to NF-κB, Ang II-induced vascular inflammation is critically regulated by the ETS family of transcription factors such as Ets-1. Systemic administration of Ang II to $Ets^{-/-}$ mice has been associated with a marked reduction in medial hypertrophy and recruitment of inflammatory cells into the vascular wall compared to that in wild-type mice, independent of the blood pressure effect (Zhan et al. 2005).

It is therefore likely that the proinflammatory milieu in atherosclerotic lesions is at least in part a consequence of Ang II-induced inflammatory gene expression. Ang II-stimulated production of VCAM-1 and chemokine monocyte chemotactic protein-1 (MCP-1) enhances leukocyte accumulation in the vessel wall. MCP-1 is thought to function locally by establishing a chemical gradient to attract monocytes and T lymphocytes expressing the CC chemokine receptor-2 (CCR2). It has been found that C-reactive protein (CRP), a known systemic acute phase reactant, is synthesized by VSMCs in response to inflammatory cytokines and Ang II (Peng et al. 2006). In turn, CRP upregulates AT_1R expression, creating a vicious cycle of proinflammatory signals. Furthermore, CRP has been established as useful biomarker for cardiovascular risk stratification.

Physiological Vascular Functions

The main direct physiological function of Ang II in the vasculature is acute regulation of vascular tone, thereby controlling blood flow to various vascular beds. The classic RAS functions as an adaptive mechanism to maintain circulatory homeostasis, regulating blood pressure acutely by vasoconstriction and chronically by releasing aldosterone and expansion of intravascular volume. Effects of Ang II outside of the vasculature include modulation of sympathetic tone and dipsogenic stimulation to maintain vegetative autoregulatory functions of the vasculature. In the adult vessel, VSMCs are present in the quiescent contractile phenotype until an insult triggers a change in phenotype. Ang II seems to mediate this transformation, which may be adaptive and beneficial in the short term, but in the long term sustains unfavorable and unstable vessel architecture prone to increase the likelihood for adverse clinical events.

Pathophysiologic Functions

Hypertension

Patients with essential hypertension can be divided into subgroups with low and high plasma renin activity, with distinct pathophysiological features. Low-renin or

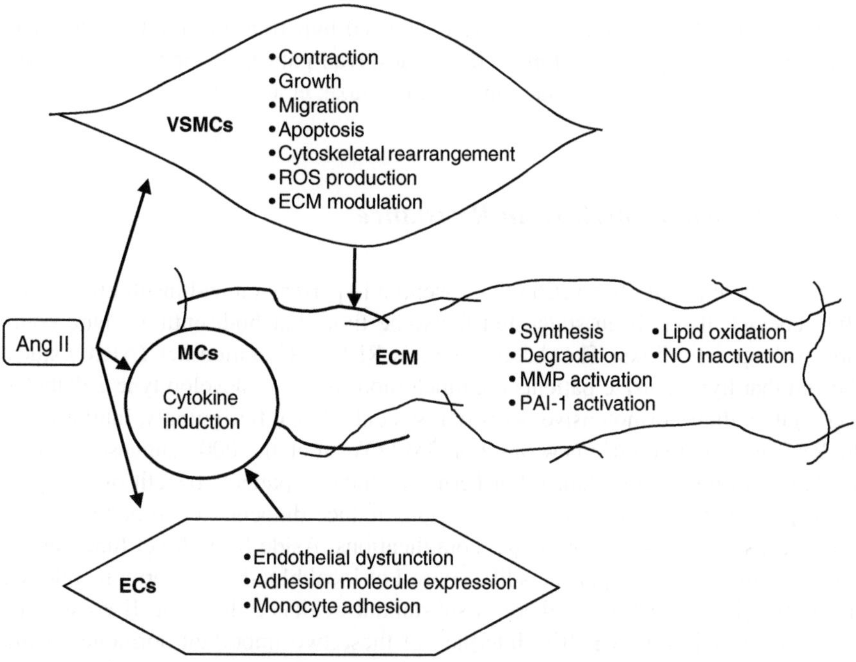

Fig. 5.4 Angiotensin II mediates multiple physiological and pathophysiological responses in the vessel wall.The vessel wall undergoes numerous structural and functional changes mediated by the AT_1R, all of which are considered proatherogenic once physiological thresholds are crossed. See text for details. Ang II, angiotensin II; EC, endothelial cell; MC, monocyte/macrophage; MMP, matrix metalloprotease; NO, nitric oxide; PAI-1, plasminogen activator inhibitor-1. ECM, extracellular matrix; ROS, reactive oxygen species.

renin-independent hypertension is characterized by salt-sensitivity, good response to diuretics, and familial aggregation (Fisher et al. 2002). Blockage of the RAS is most effective in hypertensives with high plasma renin activity, whereas it has only limited effectiveness in low-renin states. Patients with hypertension have a multitude of structural and functional disturbances affecting all compartments and cell types of the vascular wall. In particular, increased media thickness has been consistently found in vessels of hypertensive animals and can be attenuated by Ang II inhibition (Rizzoni et al. 1998). This increased media thickness is due in part to hypertrophy of the VSMCs themselves, but also to an increased production of extracellular matrix (Lee 1987). As described above, Ang II has the potential to increase both of these components by its effects on VSMC growth and synthetic capacity (Figure 5.4). In addition to these trophic effects on the vessel wall implicated in chronic hypertension, prolonged contraction of resistance arteries (as illustrated above for the RGS_2^{-} mouse) and endothelial dysfunction are additional mechanisms of hypertension linked to RAS activity. Ang II-induced $O_2^{\cdot -}$ from VSMC Nox enzymes reacts with endothelial-derived NO leading to impaired endothelial-dependent vasorelaxation (Rajagopalan et al. 1996). Aside from the vessel wall, Ang II also activates Noxes in various other tissues. Recently, it has been shown that activation of circulating T cells is essential for Ang II-induced hypertension, implying that circulating cells may provide a link to this widespread pattern of Nox activation and ROS production in various tissues and organs (Guzik et al. 2006).

Metabolic Syndrome/Insulin Resistance

Based on the clinical observation that essential hypertension and insulin resistance often coexist, it has been suggested for some time that both entities share common pathogenetic links. The Atherosclerosis Risk in Communities (ARIC) study showed that hypertensive patients are much more likely to develop type 2 diabetes (DM2) than their normotensive peers (Gress et al. 2000). Intriguingly, inhibition of the RAS appears to reduce new onset of DM2 (Yusuf et al. 2000), suggesting a role for Ang II in insulin resistance. Furthermore, diabetic patients benefit overproportionally from RAS inhibition as these agents reduce diabetic nephropathy as well as microvascular and macrovascular complications. Aside from direct functions of Ang II on the endocrine pancreas like control of local blood flow, hormone release, and prostaglandin synthesis, there is substantial evidence that Ang II cross-talks to insulin signaling in VSMCs. Interplay of these two important hormones at the level of VSMCs is an attractive hypothesis for accelerated atherosclerosis found in patients with DM2 or insulin resistance. Observations made *in vitro* suggest that Ang II stimulation of VSMCs leads to serine phosphorylation of the insulin receptor substrate-1, which has been implicated in its premature degradation (Taniyama et al. 2005). Of note, Ang II modulates insulin signaling at multiple levels by serine phosphorylating not only insulin receptor substrate-1, but also the insulin receptor and PI3-K (Folli et al. 1997). In addition, Ang II downregulates the vascular expression of PPARγs (Tham et al. 2002), which are nuclear hormonal receptors and transcription factors that promote beneficial effects on lipid metabolism, insulin sensitivity, and atherosclerosis development. Moreover, they exhibit anti-inflammatory properties by negatively modulating inflammatory gene expression. It is noteworthy that certain ARBs have been shown to directly and AT_1R-independently activate PPARγs in adipocytes, which has been suggested as a mechanism for the insulin-sensitizing/antidiabetic effect of certain ARBs (Schupp et al. 2004).

Atherosclerosis

The development of atherosclerotic lesions involves migration and proliferation of VSMCs. Essentially all of the above-mentioned chronic effects of Ang II on the vasculature are therefore considered atherogenic. While the initiating event for atherogenesis remains elusive, it is clear that a subsequent change in VSMC phenotype maintains disease progression. As opposed to the concentric medial thickening observed in hypertension, atherosclerosis results from the buildup of focal or diffuse lipid-laden and fibroproliferative plaques in the vessel intima with subsequent narrowing of the vessel lumen. Prior to formation of a plaque, endothelial dysfunction and upregulation of adhesion molecules occurs in atherosclerosis-prone vascular segments (Nakashima et al. 1998). Ang II induces expression of adhesion molecules promoting monocyte invasion into the vasculature. In addition, Ang II not only induces expression of the oxLDL receptor (LOX-1) on macrophages with subsequent macrophage activation, but it also provides the basis for LDL oxidation by stimulating the production of ROS by VSMCs (Griendling et al. 2000). Not surprisingly, Ang II infusion into $apoE^{-/-}$ mice significantly enhances atherosclerotic lesion progression independent of its effect on blood pressure (Weiss et al. 2001). Conversely, Ang II inhibition is known to reduce atherosclerotic lesion formation (Daugherty et al. 2000).

Furthermore, Ang II has been found to contribute to VSMC senescence, which has been implicated in the pathogenesis of atherosclerosis (Kunieda et al. 2006). Cell senescence may promote plaque instability, since loss of VSMCs leads to transformation into a rupture-prone plaque with a thin fibrous cap over the lipid-rich core (Geng and Libby 2002).

Conclusion

Increased RAS activity has emerged as a key mechanism in the pathogenesis of highly prevalent vascular diseases such as hypertension and atherosclerosis. It is therefore vital to understand mechanisms of local Ang II production, Ang II responsiveness, as well as its signaling pathways in normal and disturbed vascular cell function, particularly in VSMCs, the most abundant cell type in the vessel wall. The cellular events initiated by Ang II binding to its AT_1R are complex, with distinct temporal and spatial characteristics, and are far from completely understood. Moreover, cross-talk between the RAS and other metabolic systems such as lipoproteins and insulin may explain common pathogenetic links for cardiovascular diseases. Thus, advancing our knowledge of Ang II signaling in cardiovascular disease bears enormous potential in identifying novel targets for pharmacologic intervention.

References

AbdAlla, S., Lother, H., and Quitterer, U. 2000. AT1-receptor heterodimers show enhanced G-protein activation and altered receptor sequestration. Nature 407:94–98.

AbdAlla, S., Lother, H., Abdel-tawab, A. M., et al. 2001. The angiotensin II AT2 receptor is an AT1 receptor antagonist. J Biol Chem 276:39721–39726.

AbdAlla, S., Lother, H., Langer, A., et al. 2004. Factor XIIIA transglutaminase crosslinks AT1 receptor dimers of monocytes at the onset of atherosclerosis. Cell 119:343–354.

Akishita, M., Horiuchi, M., Yamada, H., et al. 2000. Inflammation influences vascular remodeling through AT2 receptor expression and signaling. Physiol Genomics 2:13–20.

Ali, M. S., Sayeski, P. P., Dirksen, L. B., et al. 1997. Dependence on the motif YIPP for the physical association of Jak2 kinase with the intracellular carboxyl tail of the angiotensin II AT1 receptor. J Biol Chem 272:23382–23388.

Allen, A. M., Zhuo, J., and Mendelsohn, F. A. 2000. Localization and function of angiotensin AT1 receptors. Am J Hypertens 13:31S–38S.

Barki-Harrington, L., Luttrell, L. M., and Rockman, H. A. 2003. Dual inhibition of beta-adrenergic and angiotensin II receptors by a single antagonist: a functional role for receptor-receptor interaction *in vivo*. Circulation 108:1611–1618.

Bascands, J. L., Girolami, J. P., Troly, M., et al. 2001. Angiotensin II induces phenotype-dependent apoptosis in vascular smooth muscle cells. Hypertension 38:1294–1299.

Bedard, K., and Krause, K. H. 2007. The NOX family of ROS-generating NADPH oxidases: Physiology and pathophysiology. Physiol Rev 87:245–313.

Bell, L., and Madri, J. A. 1990. Influence of the angiotensin system on endothelial and smooth muscle cell migration. Am J Pathol 137:7–12.

Berger, A., Stierkorb, E., and Nickenig, G. 2005. The role of the AUUUUA hexamer for the post-transcriptional regulation of the AT1 receptor mRNA stability. Biochem Biophys Res Commun 330:805–812.

Berk, B. C., Vekshtein, V., Gordon, H. M., et al. 1989. Angiotensin II-stimulated protein synthesis in cultured vascular smooth muscle cells. Hypertension 13:305–314.

Bonnardeaux, A., Davies, E., Jeunemaitre, X., et al. 1994. Angiotensin II type 1 receptor gene polymorphisms in human essential hypertension. Hypertension 24:63–69.

Bonnet, F., Cooper, M. E., Carey, R. M., et al. 2001. Vascular expression of angiotensin type 2 receptor in the adult rat: influence of angiotensin II infusion. J Hypertens 19:1075–1081.

Brasier, A. R., Jamaluddin, M., Han, Y., et al. 2000. Angiotensin II induces gene transcription through cell-type-dependent effects on the nuclear factor-kappaB (NF-kappaB) transcription factor. Mol Cell Biochem 212:155–169.

Braun-Menendez, E., Fasciolo, J. C., Leloir, L. F., et al. 1940. The substance causing renal hypertension. J Physiol (London) 98:283–298.

Burson, J. M., Aguilera, G., Gross, K. W., et al. 1994. Differential expression of angiotensin receptor 1A and 1B in mouse. Am J Physiol 267:E260–E267.

Caldwell, P. R., Seegal, B. C., Hsu, K. C., et al. 1976. Angiotensin-converting enzyme: vascular endothelial localization. Science 191:1050–1051.

Chai, S. Y., Fernando, R., Peck, G., et al. 2004. The angiotensin IV/AT4 receptor. Cell Mol Life Sci 61:2728–2737.

Chiu, A. T., McCall, D. E., Nguyen, T. T., et al. 1989. Discrimination of angiotensin II receptor subtypes by dithiothreitol. Eur J Pharmacol 170:117–118.

Clarke, M. C., Figg, N., Maguire, J. J., et al. 2006. Apoptosis of vascular smooth muscle cells induces features of plaque vulnerability in atherosclerosis. Nat Med 12:1075–1080.

Cook, J. L., Mills, S. J., Naquin, R., et al. 2006. Nuclear accumulation of the AT1 receptor in a rat vascular smooth muscle cell line: effects upon signal transduction and cellular proliferation. J Mol Cell Cardiol 40:696–707.

Curnow, K. M., Pascoe, L., and White, P. C. 1992. Genetic analysis of the human type-1 angiotensin II receptor. Mol Endocrinol 6:1113–1118.

Dale, L. B., Seachrist, J. L., Babwah, A. V., et al. 2004. Regulation of angiotensin II type 1A receptor intracellular retention, degradation, and recycling by Rab5, Rab7, and Rab11 GTPases. J Biol Chem 279:13110–13118.

Danser, A. H., van Kats, J. P., Admiraal, P. J., et al. 1994. Cardiac renin and angiotensins. Uptake from plasma versus in situ synthesis. Hypertension 24:37–48.

Daugherty, A., Manning, M. W., and Cassis, L. A. 2000. Angiotensin II promotes atherosclerotic lesions and aneurysms in apolipoprotein E-deficient mice. J Clin Invest 105:1605–1612.

Desarnaud, F., Marie, J., Lombard, C., et al. 1993. Deglycosylation and fragmentation of purified rat liver angiotensin II receptor: application to the mapping of hormone-binding domains. Biochem J 289(Pt 1):289–297.

Dubey, R. K., Jackson, E. K., and Luscher, T. F. 1995. Nitric oxide inhibits angiotensin II-induced migration of rat aortic smooth muscle cell. Role of cyclic-nucleotides and angiotensin1 receptors. J Clin Invest 96:141–149.

Eguchi, S., Matsumoto, T., Motley, E. D., et al. 1996. Identification of an essential signaling cascade for mitogen-activated protein kinase activation by angiotensin II in cultured rat vascular smooth muscle cells. Possible requirement of Gq-mediated p21ras activation coupled to a Ca2+/calmodulin-sensitive tyrosine kinase. J Biol Chem 271:14169–14175.

Eguchi, S., Numaguchi, K., Iwasaki, H., et al. 1998. Calcium-dependent epidermal growth factor receptor transactivation mediates the angiotensin II-induced mitogen-activated protein kinase activation in vascular smooth muscle cells. J Biol Chem 273:8890–8896.

Eguchi, S., Iwasaki, H., Inagami, T., et al. 1999. Involvement of PYK2 in angiotensin II signaling of vascular smooth muscle cells. Hypertension 33:201–206.

Eguchi, S., Dempsey, P. J., Frank, G. D., et al. 2001. Activation of MAPKs by angiotensin II in vascular smooth muscle cells. Metalloprotease-dependent EGF receptor activation is required for activation of ERK and p38 MAPK but not for JNK. J Biol Chem 276:7957–7962.

Elton, T. S., and Martin, M. M. 2003. Alternative splicing: a novel mechanism to fine-tune the expression and function of the human AT1 receptor. Trends Endocrinol Metab 14:66–71.

Ferrario, C. M., and Iyer, S. N. 1998. Angiotensin-(1–7): a bioactive fragment of the renin-angiotensin system. Regul Pept 78:13–18.

Fessart, D., Simaan, M., and Laporte, S. A. 2005. c-Src regulates clathrin adapter protein 2 interaction with beta-arrestin and the angiotensin II type 1 receptor during clathrin-mediated internalization. Mol Endocrinol 19:491–503.

Fisher, N. D., Hurwitz, S., Jeunemaitre, X., et al. 2002. Familial aggregation of low-renin hypertension. Hypertension 39:914–918.

Folli, F., Kahn, C. R., Hansen, H., et al. 1997. Angiotensin II inhibits insulin signaling in aortic smooth muscle cells at multiple levels. A potential role for serine phosphorylation in insulin/angiotensin II crosstalk. J Clin Invest 100:2158–2169.

Fukuda, N., Hu, W. Y., Satoh, C., et al. 1999. Contribution of synthetic phenotype on the enhanced angiotensin II-generating system in vascular smooth muscle cells from spontaneously hypertensive rats. J Hypertens 17:1099–1107.

Furuta, H., Guo, D. F., and Inagami, T. 1992. Molecular cloning and sequencing of the gene encoding human angiotensin II type 1 receptor. Biochem Biophys Res Commun 183:8–13.

Gaborik, Z., and Hunyady, L. 2004. Intracellular trafficking of hormone receptors. Trends Endocrinol Metab 15:286–293.

Geisterfer, A. A., Peach, M. J., and Owens, G. K. 1988. Angiotensin II induces hypertrophy, not hyperplasia, of cultured rat aortic smooth muscle cells. Circ Res 62:749–756.

Geng, Y. J., and Libby, P. 2002. Progression of atheroma: a struggle between death and procreation. Arterioscler Thromb Vasc Biol 22:1370–1380.

Gibbons, G. H., Pratt, R. E., and Dzau, V. J. 1992. Vascular smooth muscle cell hypertrophy vs. hyperplasia. Autocrine transforming growth factor-beta 1 expression determines growth response to angiotensin II. J Clin Invest 90:456–461.

Grant, S. L., Lassegue, B., Griendling, K. K., et al. 2000. Specific regulation of RGS2 messenger RNA by angiotensin II in cultured vascular smooth muscle cells. Mol Pharmacol 57:460–467.

Gress, T. W., Nieto, F. J., Shahar, E., et al. 2000. Hypertension and antihypertensive therapy as risk factors for type 2 diabetes mellitus. Atherosclerosis Risk in Communities Study. N Engl J Med 342:905–912.

Griendling, K. K., Minieri, C. A., Ollerenshaw, J. D., et al. 1994. Angiotensin II stimulates NADH and NADPH oxidase activity in cultured vascular smooth muscle cells. Circ Res 74:1141–1148.

Griendling, K. K., Rittenhouse, S. E., Brock, T. A., et al. 1986. Sustained diacylglycerol formation from inositol phospholipids in angiotensin II-stimulated vascular smooth muscle cells. J Biol Chem 261:5901–5906.

Griendling, K. K., Sorescu, D., and Ushio-Fukai, M. 2000. NAD(P)H oxidase: role in cardiovascular biology and disease. Circ Res 86:494–501.

Gunther, S., Gimbrone, M. A., Jr., and Alexander, R. W. 1980. Regulation by angiotensin II of its receptors in resistance blood vessels. Nature 287:230–232.

Guo, D. F., Furuta, H., Mizukoshi, M., et al. 1994. The genomic organization of human angiotensin II type 1 receptor. Biochem Biophys Res Commun 200:313–319.

Guzik, T.J., Hoch, N.E., Brown, K.A., et al. 2007. Role of the T cell in the genesis of angiotensin II induced hypertension and vascular dysfunction. J Exp Med 204:2449–2460.

Haendeler, J., Yin, G., Hojo, Y., et al. 2003. GIT1 mediates Src-dependent activation of phospholipase Cgamma by angiotensin II and epidermal growth factor. J Biol Chem 278:49936–49944.

Haller, H., and Luft, F. C. 1998. Angiotensin II acts intracellularly in vascular smooth muscle cells. Basic Res Cardiol 93(Suppl 2):30–36.

Han, Y., Runge, M. S., and Brasier, A. R. 1999. Angiotensin II induces interleukin-6 transcription in vascular smooth muscle cells through pleiotropic activation of nuclear factor-kappa B transcription factors. Circ Res 84:695–703.

Hansen, J. L., Theilade, J., Haunso, S., et al. 2004. Oligomerization of wild type and nonfunctional mutant angiotensin II type I receptors inhibits galphaq protein signaling but not ERK activation. J Biol Chem 279:24108–24115.

Heximer, S. P., Knutsen, R. H., Sun, X., et al. 2003. Hypertension and prolonged vasoconstrictor signaling in RGS2-deficient mice. J Clin Invest 111:1259.

Hilenski, L. L., Clempus, R. E., Quinn, M. T., et al. 2004. Distinct subcellular localizations of Nox1 and Nox4 in vascular smooth muscle cells. Arterioscler Thromb Vasc Biol 24:677–683.

Hilgers, K. F., Veelken, R., Muller, D. N., et al. 2001. Renin uptake by the endothelium mediates vascular angiotensin formation. Hypertension 38:243–248.

Hjorth, S. A., Schambye, H. T., Greenlee, W. J., et al. 1994. Identification of peptide binding residues in the extracellular domains of the AT1 receptor. J Biol Chem 269:30953–30959.

Horvath, C. M., and Darnell, J. E. 1997. The state of the STATs: recent developments in the study of signal transduction to the nucleus. Curr Opin Cell Biol 9:233–239.

Huang, C., Rajfur, Z., Borchers, C., et al. 2003. JNK phosphorylates paxillin and regulates cell migration. Nature 424:219–223.

Hunyady, L., Bor, M., Baukal, A. J., et al. 1995. A conserved NPLFY sequence contributes to agonist binding and signal transduction but is not an internalization signal for the type 1 angiotensin II receptor. J Biol Chem 270:16602–16609.

Hunyady, L., Baukal, A. J., Gaborik, Z., et al. 2002. Differential PI 3-kinase dependence of early and late phases of recycling of the internalized AT1 angiotensin receptor. J Cell Biol 157:1211–1222.

Hunyady, L., Gaborik, Z., Shah, B. H., et al. 2004. Structural determinants of agonist-induced signaling and regulation of the angiotensin AT1 receptor. Mol Cell Endocrinol 217:89–100.

Ichihara, S., Senbonmatsu, T., Price, E., Jr., et al. 2001. Angiotensin II type 2 receptor is essential for left ventricular hypertrophy and cardiac fibrosis in chronic angiotensin II-induced hypertension. Circulation 104:346–351.

Ichiki, T., Usui, M., Kato, M., et al. 1998. Downregulation of angiotensin II type 1 receptor gene transcription by nitric oxide. Hypertension 31:342–348.

Ishida, T., Ishida, M., Suero, J., et al. 1999. Agonist-stimulated cytoskeletal reorganization and signal transduction at focal adhesions in vascular smooth muscle cells require c-Src. J Clin Invest 103:789–797.

Ishizaka, N., Alexander, R. W., Laursen, J. B., et al. 1997. G protein-coupled receptor kinase 5 in cultured vascular smooth muscle cells and rat aorta. Regulation by angiotensin II and hypertension. J Biol Chem 272:32482–32488.

Ishizaka, N., Griendling, K. K., Lassegue, B., et al. 1998. Angiotensin II type 1 receptor: relationship with caveolae and caveolin after initial agonist stimulation. Hypertension 32:459–466.

Isner, J. M., Kearney, M., Bortman, S., et al. 1995. Apoptosis in human atherosclerosis and restenosis. Circulation 91:2703–2711.

Iwai, N., and Inagami, T. 1992. Identification of two subtypes in the rat type I angiotensin II receptor. FEBS Lett 298:257–260.

Kai, H., Griendling, K. K., Lassegue, B., et al. 1994. Agonist-induced phosphorylation of the vascular type 1 angiotensin II receptor. Hypertension 24:523–527.

Kambayashi, Y., Bardhan, S., Takahashi, K., et al. 1993. Molecular cloning of a novel angiotensin II receptor isoform involved in phosphotyrosine phosphatase inhibition. J Biol Chem 268:24543–24546.

Kambe, T., Kinjyo, N., Hiruki, H., et al. 2004. Basal transcriptional regulation of rat AT1 angiotensin II receptor gene expression. Clin Exp Pharmacol Physiol 31:96–100.

Kato, H., Suzuki, H., Tajima, S., et al. 1991. Angiotensin II stimulates collagen synthesis in cultured vascular smooth muscle cells. J Hypertens 9:17–22.

Keller, S. R., Scott, H. M., Mastick, C. C., et al. 1995. Cloning and characterization of a novel insulin-regulated membrane aminopeptidase from Glut4 vesicles. J Biol Chem 270:23612–23618.

Kim, D., Aizawa, T., Wei, H., et al. 2005a. Angiotensin II increases phosphodiesterase 5A expression in vascular smooth muscle cells: a mechanism by which angiotensin II antagonizes cGMP signaling. J Mol Cell Cardiol 38:175–184.

Kim, J., Ahn, S., Ren, X. R., et al. 2005b. Functional antagonism of different G protein-coupled receptor kinases for beta-arrestin-mediated angiotensin II receptor signaling. Proc Natl Acad Sci USA 102:1442–1447.

Ko, Y., Glodny, B., Stier, S., et al. 1997. Angiotensin type-1 (AT1) receptor gene expression in primarily cultured human arterial umbilical endothelial cells. Biochem Pharmacol 53:417–421.

Kranzhofer, R., Schmidt, J., Pfeiffer, C. A., et al. 1999. Angiotensin induces inflammatory activation of human vascular smooth muscle cells. Arterioscler Thromb Vasc Biol 19:1623–1629.

Kubo, T., Kinjyo, N., Ikezawa, A., et al. 2003. Sp1 decoy oligodeoxynucleotide decreases angiotensin receptor expression and blood pressure in spontaneously hypertensive rats. Brain Res 992:1–8.

Kuhn, J. H., Li, W., Choe, H., et al. 2004. Angiotensin-converting enzyme 2: a functional receptor for SARS coronavirus. Cell Mol Life Sci 61:2738–2743.

Kule, C. E., Karoor, V., Day, J. N., et al. 2004. Agonist-dependent internalization of the angiotensin II type one receptor (AT1): role of C-terminus phosphorylation in recruitment of beta-arrestins. Regul Pept 120:141–148.

Kunieda, T., Minamino, T., Nishi, J., et al. 2006. Angiotensin II induces premature senescence of vascular smooth muscle cells and accelerates the development of atherosclerosis via a p21-dependent pathway. Circulation 114:953–960.

Lassegue, B., Alexander, R. W., Nickenig, G., et al. 1995. Angiotensin II down-regulates the vascular smooth muscle AT1 receptor by transcriptional and post-transcriptional mechanisms: evidence for homologous and heterologous regulation. Mol Pharmacol 48:601–609.

Leduc, I., and Meloche, S. 1995. Angiotensin II stimulates tyrosine phosphorylation of the focal adhesion-associated protein paxillin in aortic smooth muscle cells. J Biol Chem 270:4401–4404.

Leduc, I., Haddad, P., Giasson, E., et al. 1995. Involvement of a tyrosine kinase pathway in the growth-promoting effects of angiotensin II on aortic smooth muscle cells. Mol Pharmacol 48:582–592.

Lee, R. M. 1987. Structural alterations of blood vessels in hypertensive rats. Can J Physiol Pharmacol 65:1528–1535.

Lehtonen, J. Y., Daviet, L., Nahmias, C., et al. 1999. Analysis of functional domains of angiotensin II type 2 receptor involved in apoptosis. Mol Endocrinol 13:1051–1060.

Li, D., Saldeen, T., Romeo, F., et al. 2000. Oxidized LDL upregulates angiotensin II type 1 receptor expression in cultured human coronary artery endothelial cells: the potential role of transcription factor NF-kappaB. Circulation 102:1970–1976.

Liao, D. F., Monia, B., Dean, N., et al. 1997. Protein kinase C-zeta mediates angiotensin II activation of ERK1/2 in vascular smooth muscle cells. J Biol Chem 272:6146–6150.

Marie, J., Maigret, B., Joseph, M. P., et al. 1994. Tyr292 in the seventh transmembrane domain of the AT1A angiotensin II receptor is essential for its coupling to phospholipase C. J Biol Chem 269:20815–20818.

Marrero, M. B., Venema, V. J., Ju, H., et al. 1998. Regulation of angiotensin II-induced JAK2 tyrosine phosphorylation: roles of SHP-1 and SHP-2. Am J Physiol 275:C1216–C1223.

Miura, S., and Karnik, S. S. 2000. Ligand-independent signals from angiotensin II type 2 receptor induce apoptosis. EMBO J 19:4026–4035.

Molloy, C. J., Taylor, D. S., and Weber, H. 1993. Angiotensin II stimulation of rapid protein tyrosine phosphorylation and protein kinase activation in rat aortic smooth muscle cells. J Biol Chem 268:7338–7345.

Mukoyama, M., Nakajima, M., Horiuchi, M., et al. 1993. Expression cloning of type 2 angiotensin II receptor reveals a unique class of seven-transmembrane receptors. J Biol Chem 268:24539–24542.

Müller, D. N., Fischli, W., Clozel, J. P., et al. 1998. Local angiotensin II generation in the rat heart: role of renin uptake. Circ Res 82:13–20.

Murphy, T. J., Alexander, R. W., Griendling, K. K., et al. 1991. Isolation of a cDNA encoding the vascular type-1 angiotensin II receptor. Nature 351:233–236.

Naftilan, A. J., Pratt, R. E., and Dzau, V. J. 1989. Induction of platelet-derived growth factor A-chain and c-myc gene expressions by angiotensin II in cultured rat vascular smooth muscle cells. J Clin Invest 83:1419–1424.

Nakashima, Y., Raines, E. W., Plump, A. S., et al. 1998. Upregulation of VCAM-1 and ICAM-1 at atherosclerosis-prone sites on the endothelium in the ApoE-deficient mouse. Arterioscler Thromb Vasc Biol 18:842–851.

Nguyen, G., Delarue, F., Burckle, C., et al. 2002. Pivotal role of the renin/prorenin receptor in angiotensin II production and cellular responses to renin. J Clin Invest 109:1417–1427.

Nickenig, G., Sachinidis, A., Seewald, S., et al. 1997. Influence of oxidized low-density lipoprotein on vascular angiotensin II receptor expression. J Hypertens Suppl 15:S27–S30.

Nickenig, G., Strehlow, K., Baumer, A. T., et al. 2000. Negative feedback regulation of reactive oxygen species on AT1 receptor gene expression. Br J Pharmacol 131:795–803.

Nickenig, G., Michaelsen, F., Muller, C., et al. 2001. Post-transcriptional regulation of the AT1 receptor mRNA. Identification of the mRNA binding motif and functional characterization. Faseb J 15:1490–1492.

Ohtsu, H., Mifune, M., Frank, G. D., et al. 2005. Signal-crosstalk between Rho/ROCK and c-Jun NH2-terminal kinase mediates migration of vascular smooth muscle cells stimulated by angiotensin II. Arterioscler Thromb Vasc Biol 25:1831–1836.

Ohtsu, H., Dempsey, P. J., Frank, G. D., et al. 2006. ADAM17 mediates epidermal growth factor receptor transactivation and vascular smooth muscle cell hypertrophy induced by angiotensin II. Arterioscler Thromb Vasc Biol 26:e133–137.

Okuda, M., Kawahara, Y., Nakayama, I., et al. 1995. Angiotensin II transduces its signal to focal adhesions via angiotensin II type 1 receptors in vascular smooth muscle cells. FEBS Lett 368:343–347.

Olson, S., Oeckler, R., Li, X., et al. 2004. Angiotensin II stimulates nitric oxide production in pulmonary artery endothelium via the type 2 receptor. Am J Physiol 287:L559–L568.

Page, I. H., and Helmer, O. M. 1940. A crystalline pressor substance (angiotonin) resulting from the reaction between renin and renin-activator. J Exp Med 71:29–42.

Peng, N., Liu, J. T., Gao, D. F., et al. 2006. Angiotensin II-induced C-reactive protein generation: Inflammatory role of vascular smooth muscle cells in atherosclerosis. Atherosclerosis.

Pollman, M. J., Yamada, T., Horiuchi, M., et al. 1996. Vasoactive substances regulate vascular smooth muscle cell apoptosis. Countervailing influences of nitric oxide and angiotensin II. Circ Res 79:748–756.

Prenzel, N., Zwick, E., Daub, H., et al. 1999. EGF receptor transactivation by G-protein-coupled receptors requires metalloproteinase cleavage of proHB-EGF. Nature 402:884–888.

Prescott, M. F., Webb, R. L., and Reidy, M. A. 1991. Angiotensin-converting enzyme inhibitor versus angiotensin II, AT1 receptor antagonist. Effects on smooth muscle cell migration and proliferation after balloon catheter injury. Am J Pathol 139:1291–1296.

Qian, H., Pipolo, L., and Thomas, W. G. 2001. Association of beta-arrestin 1 with the type 1A angiotensin II receptor involves phosphorylation of the receptor carboxyl terminus and correlates with receptor internalization. Mol Endocrinol 15:1706–1719.

Quitterer, U., Lother, H., and Abdalla, S. 2004. AT1 receptor heterodimers and angiotensin II responsiveness in preeclampsia. Semin Nephrol 24:115–119.

Rajagopalan, S., Kurz, S., Munzel, T., et al. 1996. Angiotensin II-mediated hypertension in the rat increases vascular superoxide production via membrane NADH/NADPH oxidase activation. Contribution to alterations of vasomotor tone. J Clin Invest 97:1916–1923.

Rakugi, H., Kim, D. K., Krieger, J. E., et al. 1994. Induction of angiotensin converting enzyme in the neointima after vascular injury. Possible role in restenosis. J Clin Invest 93:339–346.

Rao, G. N., Lassegue, B., Alexander, R. W., et al. 1994. Angiotensin II stimulates phosphorylation of high-molecular-mass cytosolic phospholipase A2 in vascular smooth-muscle cells. Biochem J 299(Pt 1):197–201.

Rizzoni, D., Porteri, E., Piccoli, A., et al. 1998. Effects of losartan and enalapril on small artery structure in hypertensive rats. Hypertension 32:305–310.

Rocic, P., Jo, H., and Lucchesi, P. A. 2003. A role for PYK2 in ANG II-dependent regulation of the PHAS-1-eIF4E complex by multiple signaling cascades in vascular smooth muscle. Am J Physiol Cell Physiol 285:C1437–C1444.

Roman, R. J. 2002. P-450 metabolites of arachidonic acid in the control of cardiovascular function. Physiol Rev 82:131–185.

Ruan, X., and Arendshorst, W. J. 1996. Calcium entry and mobilization signaling pathways in ANG II-induced renal vasoconstriction *in vivo*. Am J Physiol 270:F398–F405.

Saito, S., Frank, G. D., Motley, E. D., et al. 2002. Metalloprotease inhibitor blocks angiotensin II-induced migration through inhibition of epidermal growth factor receptor transactivation. Biochem Biophys Res Commun 294:1023–1029.

Sasaki, K., Yamano, Y., Bardhan, S., et al. 1991. Cloning and expression of a complementary DNA encoding a bovine adrenal angiotensin II type-1 receptor. Nature 351:230–233.

Sasamura, H., Nakazato, Y., Hayashida, T., et al. 1997. Regulation of vascular type 1 angiotensin receptors by cytokines. Hypertension 30:35–41.

Schieffer, B., Paxton, W. G., Chai, Q., et al. 1996. Angiotensin II controls p21ras activity via pp60c-src. J Biol Chem 271:10329–10333.

Schupp, M., Janke, J., Clasen, R., et al. 2004. Angiotensin type 1 receptor blockers induce peroxisome proliferator-activated receptor-gamma activity. Circulation 109:2054–2057.

Seachrist, J. L., Laporte, S. A., Dale, L. B., et al. 2002. Rab5 association with the angiotensin II type 1A receptor promotes Rab5 GTP binding and vesicular fusion. J Biol Chem 277:679–685.

Strehlow, K., Wassmann, S., Bohm, M., et al. 2000. Angiotensin AT1 receptor over-expression in hypercholesterolaemia. Ann Med 32:386–389.

Sugawara, A., Takeuchi, K., Uruno, A., et al. 2001. Transcriptional suppression of type 1 angiotensin II receptor gene expression by peroxisome proliferator-activated receptor-gamma in vascular smooth muscle cells. Endocrinology 142:3125–3134.

Takeuchi, K., Alexander, R. W., Nakamura, Y., et al. 1993. Molecular structure and transcriptional function of the rat vascular AT1a angiotensin receptor gene. Circ Res 73:612–621.

Taniyama, Y., Weber, D. S., Rocic, P., et al. 2003. Pyk2- and Src-dependent tyrosine phosphorylation of PDK1 regulates focal adhesions. Mol Cell Biol 23:8019–8029.

Taniyama, Y., Ushio-Fukai, M., Hitomi, H., et al. 2004. Role of p38 MAPK and MAPKAPK-2 in angiotensin II-induced Akt activation in vascular smooth muscle cells. Am J Physiol Cell Physiol 287:C494–C499.

Taniyama, Y., Hitomi, H., Shah, A., et al. 2005. Mechanisms of reactive oxygen species-dependent downregulation of insulin receptor substrate-1 by angiotensin II. Arterioscler Thromb Vasc Biol 25:1142–1147.

Taubman, M. B., Berk, B. C., Izumo, S., et al. 1989. Angiotensin II induces c-fos mRNA in aortic smooth muscle. Role of Ca2+ mobilization and protein kinase C activation. J Biol Chem 264:526–530.

Tea, B. S., Der Sarkissian, S., Touyz, R. M., et al. 2000. Proapoptotic and growth-inhibitory role of angiotensin II type 2 receptor in vascular smooth muscle cells of spontaneously hypertensive rats *in vivo*. Hypertension 35:1069–1073.

Tham, D. M., Martin-McNulty, B., Wang, Y. X., et al. 2002. Angiotensin II is associated with activation of NF-kappaB-mediated genes and downregulation of PPARs. PhysiolGenomics 11:21–30.

Thurston, H., Swales, J. D., Bing, R. F., et al. 1979. Vascular renin-like activity and blood pressure maintenance in the rat. Studies of the effect of changes in sodium balance, hypertension and nephrectomy. Hypertension 1:643–649.

Tobiume, K., Matsuzawa, A., Takahashi, T., et al. 2001. ASK1 is required for sustained activations of JNK/p38 MAP kinases and apoptosis. EMBO Rep 2:222–228.

Tsutsumi, Y., Matsubara, H., Ohkubo, N., et al. 1998. Angiotensin II type 2 receptor is upregulated in human heart with interstitial fibrosis, and cardiac fibroblasts are the major cell type for its expression. Circ Res 83:1035–1046.

Tummala, P. E., Chen, X. L., Sundell, C. L., et al. 1999. Angiotensin II induces vascular cell adhesion molecule-1 expression in rat vasculature: A potential link between the renin-angiotensin system and atherosclerosis. Circulation 100:1223–1229.

Uehata, M., Ishizaki, T., Satoh, H., et al. 1997. Calcium sensitization of smooth muscle mediated by a Rho-associated protein kinase in hypertension. Nature 389:990–994.

Ushio-Fukai, M., Alexander, R. W., Akers, M., et al. 1998a. p38 mitogen-activated protein kinase is a critical component of the redox-sensitive signaling pathways activated by angiotensin II. Role in vascular smooth muscle cell hypertrophy. J Biol Chem 273:15022–15029.

Ushio-Fukai, M., Griendling, K. K., Akers, M., et al. 1998b. Temporal dispersion of activation of phospholipase C-beta1 and -gamma isoforms by angiotensin II in vascular smooth muscle cells. Role of alphaq/11, alpha12, and beta gamma G protein subunits. J Biol Chem 273:19772–19777.

Ushio-Fukai, M., Alexander, R. W., Akers, M., et al. 1999a. Angiotensin II receptor coupling to phospholipase D is mediated by the betagamma subunits of heterotrimeric G proteins in vascular smooth muscle cells. Mol Pharmacol 55:142–149.

Ushio-Fukai, M., Alexander, R. W., Akers, M., et al. 1999b. Reactive oxygen species mediate the activation of Akt/protein kinase B by angiotensin II in vascular smooth muscle cells. J Biol Chem 274:22699–22704.

Ushio-Fukai, M., Griendling, K. K., Becker, P. L., et al. 2001. Epidermal growth factor receptor transactivation by angiotensin II requires reactive oxygen species in vascular smooth muscle cells. Arterioscler Thromb Vasc Biol 21:489–495.

van Geel, P. P., Pinto, Y. M., Voors, A. A., et al. 2000. Angiotensin II type 1 receptor A1166C gene polymorphism is associated with an increased response to angiotensin II in human arteries. Hypertension 35:717–721.

Vaughan, D. E., Lazos, S. A., and Tong, K. 1995. Angiotensin II regulates the expression of plasminogen activator inhibitor-1 in cultured endothelial cells. A potential link between the renin-angiotensin system and thrombosis. J Clin Invest 95:995–1001.

Wassmann, S., Laufs, U., Baumer, A. T., et al. 2001. Inhibition of geranylgeranylation reduces angiotensin II-mediated free radical production in vascular smooth muscle cells: involvement of angiotensin AT1 receptor expression and Rac1 GTPase. Mol Pharmacol 59:646–654.

Wei, H., Ahn, S., Shenoy, S. K., et al. 2003. Independent beta-arrestin 2 and G protein-mediated pathways for angiotensin II activation of extracellular signal-regulated kinases 1 and 2. Proc Natl Acad Sci USA 100:10782–10787.

Weiss, D., Kools, J. J., and Taylor, W. R. 2001. Angiotensin II-induced hypertension accelerates the development of atherosclerosis in apoE-deficient mice. Circulation 103:448–454.

Wharton, J., Morgan, K., Rutherford, R. A., et al. 1998. Differential distribution of angiotensin AT2 receptors in the normal and failing human heart. J Pharmacol Exp Ther 284:323–336.

Wyse, B. D., Linas, S. L., and Thekkumkara, T. J. 2000. Functional role of a novel cis-acting element (GAGA box) in human type-1 angiotensin II receptor gene transcription. J Mol Endocrinol 25:97–108.

Xi, X. P., Graf, K., Goetze, S., et al. 1999. Central role of the MAPK pathway in ang II-mediated DNA synthesis and migration in rat vascular smooth muscle cells. Arterioscler Thromb Vasc Biol 19:73–82.

Yamada, T., Akishita, M., Pollman, M. J., et al. 1998. Angiotensin II type 2 receptor mediates vascular smooth muscle cell apoptosis and antagonizes angiotensin II type 1 receptor action: an *in vitro* gene transfer study. Life Sci 63:PL289–PL295.

Yusuf, S., Sleight, P., Pogue, J., et al. 2000. Effects of an angiotensin-converting-enzyme inhibitor, ramipril, on cardiovascular events in high-risk patients. The Heart Outcomes Prevention Evaluation Study Investigators. N Engl JMed 342:145–153.

Zafari, A. M., Ushio-Fukai, M., Minieri, C. A., et al. 1999. Arachidonic acid metabolites mediate angiotensin II-induced NADH/NADPH oxidase activity and hypertrophy in vascular smooth muscle cells. Antioxidants Redox Signal 1:167–179.

Zeng, C., Luo, Y., Asico, L. D., et al. 2003. Perturbation of D1 dopamine and AT1 receptor interaction in spontaneously hypertensive rats. Hypertension 42:787–792.

Zhan, Y., Brown, C., Maynard, E., et al. 2005. Ets-1 is a critical regulator of Ang II-mediated vascular inflammation and remodeling. J Clin Invest 115:2508–2516.

Zhang, Y., Ji, H., Fabucci, M. E., et al. 2004. Translational control of the rat angiotensin type 1a receptor by alternative splicing. Gene 341:93–100.

Zhang, L., Cheng, J., Ma, Y., et al. 2005a. Dual pathways for nuclear factor kappaB activation by angiotensin II in vascular smooth muscle: phosphorylation of p65 by IkappaB kinase and ribosomal kinase. Circ Res 97:975–982.

Zhang, L., Ma, Y., Zhang, J., et al. 2005b. A new cellular signaling mechanism for angiotensin II activation of NF-kappaB: An IkappaB-independent, RSK-mediated phosphorylation of p65. Arterioscler Thromb Vasc Biol 25:1148–1153.

Zhang, Y., Griendling, K. K., Dikalova, A., et al. 2005c. Vascular hypertrophy in angiotensin II-induced hypertension is mediated by vascular smooth muscle cell-derived H_2O_2. Hypertension 46:732–737.

Zhao, X., Martin, M. M., and Elton, T. S. 2000. Basal level transcriptional regulation of the human angiotensin II type 1 receptor gene. Biochim Biophys Acta 1494:181–184.

Zuo, L., Ushio-Fukai, M., Ikeda, S., et al. 2005. Caveolin-1 is essential for activation of Rac1 and NAD(P)H oxidase after angiotensin II type 1 receptor stimulation in vascular smooth muscle cells: role in redox signaling and vascular hypertrophy. Arterioscler Thromb Vasc Biol 25:1824–1830.

Chapter 6
Angiotensin II Signaling: Cell Growth and Apoptosis

Stephen W. Schaffer and Mahmood S. Mozaffari

Abstract Angiotensin II mediates both growth and death pathways in the young developing heart and the failing heart. Although stimulation of both processes by the same agent might seem counterproductive, the cell is capable of specifically regulating the process that will dominate. The secret behind the specificity of the response lies in the characteristics of the plethora of signaling pathways initiated by angiotensin II. It is generally felt that alterations in these pathways play a pivotal role in both the normal development of the young heart and the onset of pathology in the failing heart. Although the recent introduction of transgenic animals has proven valuable in assessing the contribution of specific pathways in the regulation of growth and apoptosis, more information is required on the interplay between these pathways and mediators of growth and apoptosis.

Introduction

Angiotensin II is an octapeptide, which was initially described as a potent vasoconstrictor agent. However, its functions have since been expanded to include regulation of cell growth, inflammation, electrolyte and water balance, hormone secretion, sympathetic nervous system activity, differentiation, and apoptosis. The discovery that it is produced both systemically and locally was instrumental in establishing a pivotal role for the peptide in several disease states, including hypertension, coronary heart disease, myocarditis, congestive heart failure, atherosclerosis, and nephrosclerosis.

One of the most widely studied and pathologically relevant actions of angiotensin II is the stimulation of cell growth. Because of the complexity of the cellular hypertrophic process, it has been a major challenge to map out the trophic signaling pathways of angiotensin II. Not only is the response to angiotensin II cell specific, but it also depends on the intensity and duration of numerous second messenger signals, the distinct pathways initiated by the two angiotensin II receptors (AT_1 and AT_2), and the size and location of the various angiotensin II pools. Also influencing the

A. K. Srivastava, M. B. Anand-Srivastava (eds.) *Advances in Biochemistry in Health and Disease.* Signal Transduction in the Cardiovascular System in Health and Disease,

trophic signal is cross-talk between the angiotensin II receptor and effectors such as transforming growth factor-β1 (TGFβ1), epidermal growth factor (EGF), and insulinlike growth factor (IGF-1) (Touyz and Schiffrin 2000). *In vivo*, the complexity of the angiotensin II response is compounded by difficulties in separating the growth-promoting actions of angiotensin II from effects caused by elevated afterload pressure. Moreover, AT_{1A} receptor knockout mice develop pressure overload-induced hypertrophy (Hamawaki et al. 1998). Because of these complications, the debate surrounding the importance of angiotensin II's trophic activity *in vivo* has been mired in controversy (Yamazaki and Yazaki 1997; Dostal and Baker 1999; Lijnen and Petrov 1999; Bader 2002). As a result, many of the recent studies have focused on the identity of key angiotensin II signaling pathways that regulate growth, survival, and apoptosis (Touyz and Schiffrin 2000; Touyz 2007). The present review examines several cell-specific signaling pathways initiated by angiotensin II.

AT_1 Receptor Trophic Signaling in Cardiomyocytes and Vascular Smooth Muscle Cells

One of the early events in angiotensin II signaling within the cardiomyocyte is the activation and translocation of protein kinase C-δ and protein kinase C-ε (Schunkert et al. 1995; Ruf et al. 2002), events that arise from G-protein-coupled activation of phospholipase C and D (Sadoshima and Izumo 1993) (Figure 6.1). Downstream from activated protein kinase C is extracellular signal-regulated kinase (ERK), whose inhibition by PD98059 prevents angiotensin II-induced protein synthesis in the cardiomyocyte (Ruf et al. 2002). ERK1/2 is also activated by angiotensin II in aortic smooth muscle cells and neonatal cardiomyocytes, but the period of enhanced activity is more prolonged in the adult ventricular cardiomyocyte (Giasson and Meloche 1995; Sadoshima et al. 1995; Ruf et al. 2002). Nonetheless, exposure of all three cell types to angiotensin II leads to an extended period of p70S6 kinase activation and eventually to cellular hypertrophy. In a related study it was shown that a c-Raf/MEK/ERK/p70S6 kinase pathway is also initiated by the protein kinase C activator, 12-*O*-tetradecanoylphorbol-13-acetate, in the cardiomyocyte (Iijima et al. 2002). These data support the notion that ERK1/2 and p70S6 kinase are coupled in the cardiomyocyte (Sadoshima et al., 1995) (Figure 6.1). However, two lines of evidence indicate that the activation of ERK1/2 and p70S6 kinase involve distinct angiotensin II signaling pathways in vascular smooth muscle cells (Giasson and Meloche 1995; Hafizi et al. 2004). First, the time course for ERK1/2 and p70S6 kinase activation is more consistent with the existence of distinct pathways of activation. Second, sustained ERK1/2 activation is required for cell cycle progression and mitogenicity. Yet ERK1/2 activation in human vascular smooth muscle cells is transient.

Godeny and Sayeski (2006a,b) recently addressed the cause underlying the transient nature of ERK activation in vascular smooth muscle cells. They found that ERK activation was mediated by either an Src-dependent or a PKCζ-dependent signaling pathway. A major difference between the two pathways is that PKCζ signaling initiates ERK1/2 nuclear translocation while the Src-dependent pathway activates ERK1/2 in the cytoplasm (Figure 6.2). The most robust and sustained

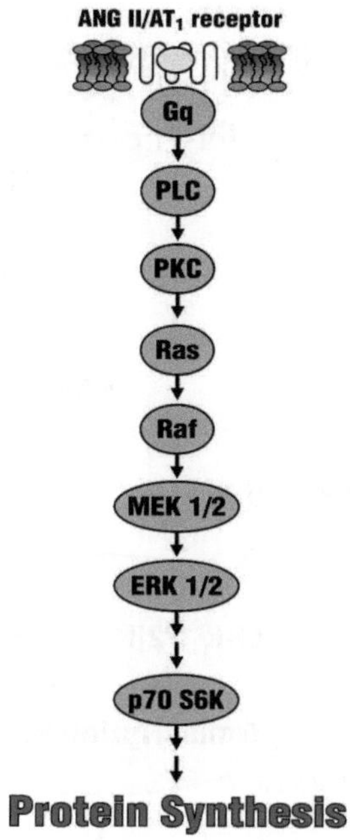

Fig. 6.1 Proposed scheme underlying the stimulation of protein synthesis by the AT_1/ERK/p70S6K pathway. The interaction of angiotensin II (ANG II) with the AT_1 receptor leads to the rapid activation of phospholipase C (PLC), which in turn generates diacylglycerol, an activator of protein kinase C (PKC). Downstream from PKC is Ras, which activates Raf. Upon the phosphorylation of mitogen-activated protein kinase kinase (MEK1/2) the intermediate targets, p42 and p44 kDa MAP kinases (ERK1/2), are phosphorylated and activated. In the cardiomyocyte, the subsequent activation of p70S6 kinase (p70S6K) is required for angiotensin II-mediated stimulation of protein synthesis.

activation of ERK1/2 and the resulting transcription of c-fos was noted when both the PKCζ and Src pathways were simultaneously stimulated.

Two distinct mechanisms of ERK1/2 activation have also been observed in cardiac fibroblasts and COS7 cells exposed to angiotensin II (Seta et al. 2002). Like the observation of Godeny and Sayeski (2006a,b), Seta et al. (2002) found that the G-protein/PKC pathway led to nuclear translocation of ERK1/2 while the G-protein-independent Src/Ras-dependent pathway facilitated the localization of the active form of ERK1/2 to the cytoplasm. Based on immunocytochemical staining and immunoprecipitation experiments, Aplin et al. (2007a,b) proposed that the form of ERK1/2 activated by the Src/Ras pathway is anchored in the cytoplasm by the β-arrestin2 scaffolding complex (Figure 6.3). According to their hypothesis the ERK-β-arrestin2 complex contributes to angiotensin II signaling by activating p90RSK and promoting cell proliferation. However, that interpretation is inconsistent with a study by Zhai et al. (2005), who compared the Gq/PKC-dependent and Src-dependent pathways in normal mice and transgenic mice that either overexpressed the AT_1 wild-type receptor (Tg-WT mice) or contained a mutation in the second intracellular loop of the AT_1 receptor (Tg-i2m mice) to selectively activate the Gq-independent mechanism. As predicted, hearts of Tg-WT mice exhibited

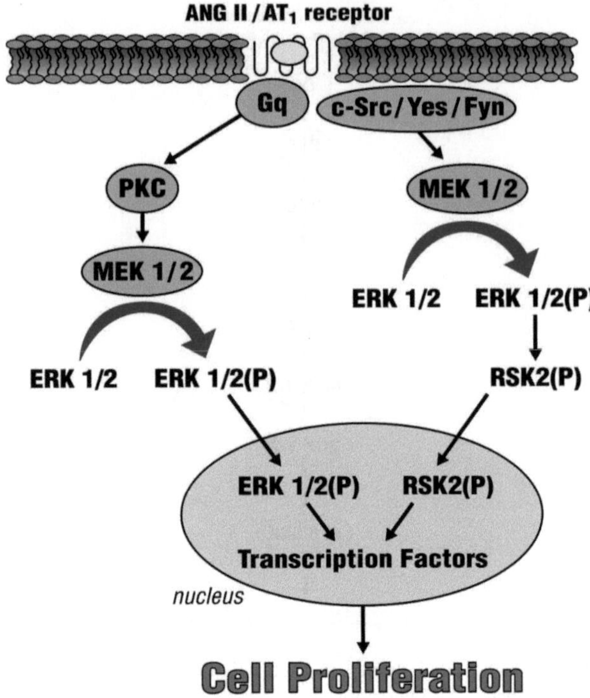

Fig. 6.2 Proposed scheme for stimulation of cell proliferation by the Gq/PKC-dependent and the G-protein-independent pathways of ERK1/2 activation. In one of the pathways, the interaction of angiotensin II with the AT_1 receptor is coupled through a G_q protein to the activation of protein kinase C (PKC), which in turn activates mitogen-activated protein kinase kinase (MEK1/2), the upstream kinase responsible for the phosphorylation of ERK1/2. The PKC-dependent pathway results in the translocation of ERK1/2 to the nucleus, where it upregulates certain transcription factors. A second angiotensin II-mediated pathway is G-protein independent and results in the activation of Src/Yes/Fyn, followed by MEK1/2 and ERK1/2. Since ERK1/2 activated by the G-protein-independent pathway remains sequestered in the cytoplasm, its actions are mediated by RSK2, a serine protein kinase that phosphorylates the S6 ribosomal protein and stimulates protein synthesis.

elevations in nuclear phosphorylated ERK content, enhanced translocation of PKCε, and greater association of Gq in the cytoplasm in comparison to either the normal or Tg-i2m mouse heart. By contrast, the active, phosphorylated form of Src was elevated more in the Tg-i2m mouse heart than in the other two groups. Therefore, the specific development of a hypertrophic cardiomyopathy in the Tg-i2m mouse suggests the involvement of the G-protein-independent/Src-dependent pathway in the development of myocardial hypertrophy, a conclusion inconsistent with the conclusions of Aplin et al. (2007b). Also inconsistent with the work of Aplin et al. (2007b) is the finding that chronic angiotensin II infusion induced greater hypertrophy in the Tg-i2m mouse than in wild-type control mice (Zhai et al. 2005).

Angiotensin II signaling in vascular smooth muscle cells and cardiac fibroblasts is also mediated through EGF receptor transactivation (Figure 6.4). In fact,

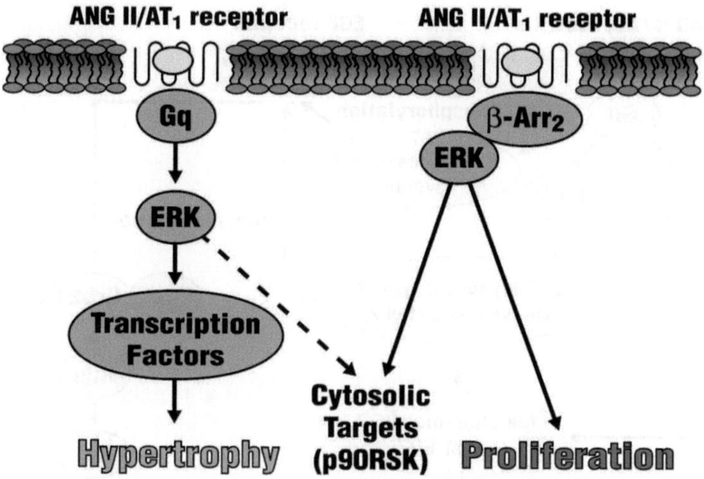

Fig. 6.3 Proposed model for the Gq-dependent and β-arrestin2-dependent pathways of ERK1/2 activation. In the Gq-dependent pathway, ERK1/2 translocates to the nucleus while activation of ERK1/2 by the β-arrestin2 scaffold leads to the sequestration of active ERK1/2 in the cytoplasm, where it activates RSK (see Figure 6.2). The proposed scheme maintains that the Gq-dependent pathway leads to cellular hypertrophy while cytosolic ERK1/2 promotes cell proliferation.

Bokemeyer et al. (2000) have reported that the selective EGF receptor antagonist, AG1478, blocks angiotensin II-induced stimulation of ERK1/2, Akt, and protein synthesis in vascular smooth muscle cells. Moreover, angiotensin II is unable to stimulate ERK1/2 in NIH3T3 cells lacking EGF receptors, but induces significant ERK1/2 activation in NIH3T3 cells containing the EGF receptor (Bokemeyer et al. 2000). A key step in the transactivation cascade is the rapid phosphorylation of the EGF receptor (Eguchi and Inagami 2000). In vascular smooth muscle cells, the phosphorylation event is followed by the formation of the Shc-Grb2-SOS complex, which in turn leads to the activation of Ras and Raf. Farther downstream from Ras and Raf are MEK1/2 and ERK1/2. Touyz et al. (2003) proposed that NADPH oxidase may also lie downstream from the EGF receptor. A recent review by Clempus and Griendling (2006) discusses the importance of NADPH oxidase in the activation of protein kinases that promote cell growth.

The upstream reactions leading to EGF receptor transactivation continue to be actively investigated. Recent studies have focused on four mediators of the transactivation reaction. First, the EGF receptor can be independently activated by a calcium ionophore, leading to the hypothesis that a Ca^{2+}-dependent tyrosine kinase plays a role in the transactivation reaction (Eguchi et al. 1998; Murasawa et al. 1998a; Frank et al. 2001). However, not all studies support a role for Ca^{2+} or Ca^{2+}-dependent tyrosine kinases in the transactivation process. Wang et al. (1999, 2000) found that the nitric oxide donor, *t*-nitroso-*N*-acetylpenicillamine, inhibits angiotensin II-induced elevation in $[Ca^{2+}]_i$ and activation of the Ca^{2+}-sensitive tyrosine kinase, Pyk2, without affecting EGF transactivation. Moreover, *N*-acetylcysteine has been

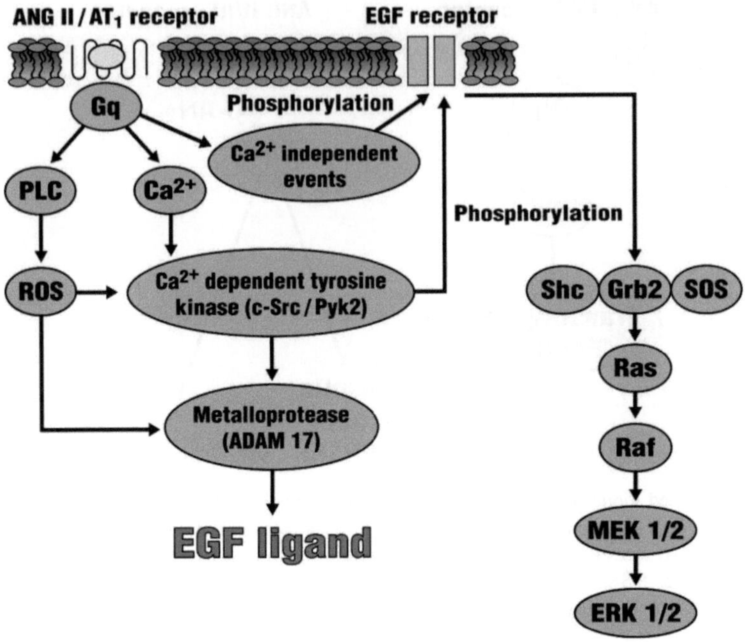

Fig. 6.4 Proposed scheme for AT_1-mediated transactivation of the EGF receptor and the subsequent activation of ERK1/2. The interaction of angiotensin II with the AT_1 receptor stimulates several pathways that result in the transactivation of the EGF receptor. In one pathway, phospholipase C (PLC) is activated, which in turn promotes the production of reactive oxygen species (ROS). AT_1 receptor activation also elevates $[Ca^{2+}]_i$, contributing to the stimulation of a Ca^{2+}-dependent tyrosine kinase, of which Src and proline-rich tyrosine kinase-2 (Pyk2) are the preferred targets. Both ROS and the Ca^{2+}-dependent tyrosine kinase activate a metalloprotease, presumably ADAM17, which proteolytically activates the EGF ligand. The Ca^{2+}-dependent tyrosine kinase, as well as a Ca^{2+}-independent kinase, also phosphorylate the EGF receptor. The activation of the EGF receptor triggers a cascade leading to the activation of ERK1/2. EGF receptor transactivation also proceeds via a G-protein-independent pathway (see text).

found to prevent transactivation of the EGF receptor in the cardiac fibroblast without blocking angiotensin II-mediated elevations in $[Ca^{2+}]_i$. Finally, the Ca^{2+} chelator, BAPTA, has only a marginal influence on angiotensin II-mediated phosphorylation of the EGF receptor (Wang et al. 2000). The debate surrounding the roles of Ca^{2+} and the Ca^{2+}-dependent tyrosine kinases in the transactivation pathway has invariably been linked, with some investigators reporting a requirement of elevated $[Ca^{2+}]_i$ and Ca^{2+}-dependent tyrosine kinase activity for EGF receptor phosphorylation (Ushio-Fukai et al. 2001) while others observe an uncoupling between the two phenomena (Eguchi et al. 2000).

Based on the evidence that several antioxidants inhibit angiotensin II-mediated phosphorylation of the EGF receptor (Ushio-Fukai et al. 2001) while H_2O_2 induces its phosphorylation, it has been postulated that oxidative stress is an early and necessary event in the transactivation of the EGF receptor (Zhang et al. 2005). A likely site of reactive oxygen species (ROS) action is the inhibition of a protein tyrosine phosphatase, leading to the activation of Src and proline-rich tyrosine kinase-2 (Pyk-2). Indeed, transactivation is blocked by the tyrosine kinase inhibitor, genistein, the Src kinase inhibitor, PP1, and by overexpression of kinase-inactive c-Src (Frank et al. 2001; Ushio-Fukai et al. 2001). Moreover, the phosphorylation of Src by angiotensin II is partially inhibited by various antioxidants (Frank et al. 2001). Ushio-Fukai et al. (2001) have confirmed that Src is a key link between ROS and EGF receptor phosphorylation.

Mifune et al. (2005) have recently implicated ROS, Ca^{2+}, and a metalloproteinase mediator in angiotensin II-mediated transactivation of the EGF receptor. They found that angiotensin II-induced EGF receptor transactivation was blocked in cells exposed to the heparin-binding EGF inhibitor, CRM197, or the metalloproteinase inhibitor, 2*R*[(biphenylsulfonyl)amino]-*N*-hydroxy-3-phenylpropionamide. Since the proteolytic activation of heparin-binding EGF was inhibited in angiotensin II-treated COS7 cells by the intracellular Ca^{2+} antagonist, TMB-8, and the antioxidant, *N*-acetylcysteine, they concluded that the elevation in $[Ca^{2+}]_i$ and ROS were required for the formation of the EGF ligand. Because transfection of COS7 cells with dominant negative ADAM 17 (a disintegrin and metalloprotease 17) inhibited angiotensin II-induced EGF receptor transactivation, it seemed likely that a proteolytic step was involved in EGF receptor transactivation. Taken together, the data suggested that angiotensin II-mediated elevations in $[Ca^{2+}]_i$ and ROS are responsible for the activation of a metalloproteinase that increases the levels of a ligand involved in EGF receptor activation. The contribution of ROS and angiotensin II-mediated EGF receptor transactivation is consistent with the pivotal role of ROS generation in hypertrophic signaling (Nakamura et al. 1998; Bendall et al. 2002; Zhang et al. 2005; Clempus and Griendling 2006).

Zhai et al. (2006) have provided convincing evidence for a fourth mechanism of EGF transactivation, one which involves the interaction of the AT_1 receptor with the EGF receptor. A key finding was that transgenic mice (Tg-Y319F), which were incapable of properly forming the AT_1-EGF receptor interaction, did not undergo EGF receptor transactivation and did not significantly develop cardiac hypertrophy *in vivo*. More importantly, the transgenic mice did not develop cardiac hypertrophy following chronic infusion of angiotensin II despite the normal response of G-protein-coupled reactions, such as the translocation of PKC, the phosphorylation of JNK and p38 MAP kinase, and the activation of STAT3. These findings suggest that the G-protein-independent pathway is a major contributor to hypertrophy in the stressed myocardium. Indeed, introduction of dominant negative EGF receptor into the heart of a normal mouse produces a phenotype that resembles the Tg-Y319F mouse (Zhai et al. 2006).

AT_1 Receptor Signaling in Cardiac Fibroblasts

Angiotensin II increases the levels of some of the same hypertrophic effectors in the cardiac fibroblast as in the cardiomyocyte, but the signaling pathways that lead to their elevation are different in the two cell types. According to Zou et al. (1998), angiotensin II-mediated activation of ERK1/2 is pertussis toxin sensitive in the rat cardiac fibroblast, suggesting the involvement of a pertussin toxin-sensitive G_i protein in angiotensin II signaling. Zou et al. (1998) also found that the AT_1 receptor of the cardiac fibroblast is coupled to the Src family of tyrosine kinases, which among other actions phosphorylates Shc to facilitate the association of Shc with Grb2. Since angiotensin II-mediated ERK1/2 activation in fibroblasts is attenuated by the overexpression of dominant negative Ras, the small GTP-binding protein appears to be another key intermediate in the ERK1/2 signal cascade. Based on cytokine signaling, it is not surprising that Raf-1 is also required for the activation of ERK1/2 in cardiac fibroblasts. Thus, angiotensin II signaling in the fibroblast resembles signaling initiated by many cytokines and growth factors, in that SOS and Ras associate with the Shc-Grb2 complex and activate Ras. Once activated, Ras stimulates Raf-1, which in turn activates dual protein kinase MEK, the immediate upstream activator of ERK1/2.

In contrast to the cardiomyocyte, Akt/p70S6 kinase signaling is reportedly absent in the angiotensin II-treated cardiac fibroblast (Olson et al. 2005). Nonetheless, the trophic properties of angiotensin II in both cell types are blocked by the antioxidant, reserveratrol, which appears to interfere with the activation of ERK1/2 signaling (Zou et al. 1998; Olson et al. 2005). Despite evidence that the ERK1/2 pathway plays a central role in rat cardiac fibroblast growth, not all studies consider ERK1/2 an important mediator of angiotensin II-linked growth. Hou et al. (2000) failed to detect an effect of either pertussis toxin or inhibition of ERK1/2 on angiotensin II-induced stimulation of protein synthesis. Another surprising finding in the Hou study was the observation that the Ca^{2+}-sensitive protein kinase C inhibitor, Go6976, prevented angiotensin II-mediated [^{3}H]leucine incorporation; neither Zou et al. (1998) nor Murasawa et al. (1998b) observed an effect of the general protein kinase C antagonist, calphostin C, on angiotensin II-induced activation of ERK1/2. The difference of opinion surrounding the roles of ERK1/2 and protein kinase C in the trophic activity of angiotensin II might be related to species differences, as Hou et al. (2000) used human cardiac fibroblasts rather than the more widely used rat fibroblast. Alternatively, the dose of angiotensin II used might explain the difference in response to angiotensin II. While the concentration of angiotensin II used in the study of Zou et al. (1998) was 1 μM, 10 nM angiotensin II was used in the study by Hou et al. (2000). Particularly noteworthy, in this regard, is the observation that the MEK inhibitor, PD098059, reduced angiotensin II-induced protein synthesis by an insignificant 17%. Had Hou et al. (2000) increased the number of experiments performed to $n > 3$ or elevated the concentration of angiotensin II 100-fold, a significant effect of PD098059 on angiotensin II-induced protein synthesis might have been observed.

Activation of the Jak/STAT Pathway by Angiotensin II

A growth-linked signaling pathway that is stimulated by angiotensin II in neonatal cardiomyocytes, rat aortic smooth muscle cells and cardiac fibroblasts is the

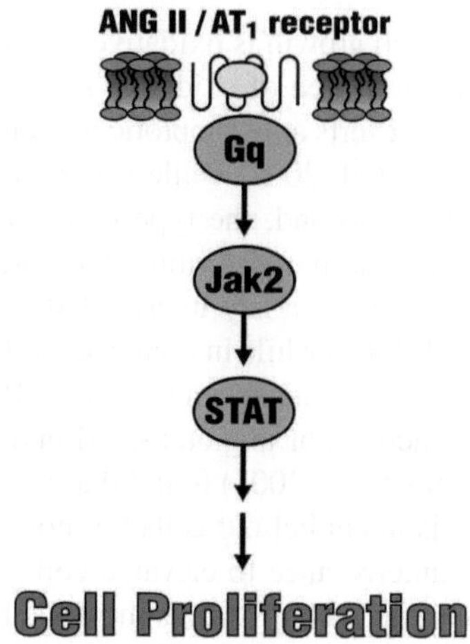

Fig. 6.5 Activation of the Jak/STAT pathway by angiotensin II. Binding of angiotensin II to the AT_1 receptor stimulates, via a G protein, the phosphorylation of janus family kinases (JAK). Jak2 participates in the phosphorylation and activation of signal transducers and activators of transcription (STAT) family members, which translocate to the nucleus and upregulate growth-linked genes.

Jak/STAT pathway (Dostal et al. 1997). This pathway was initially identified as a major cytokine signal transduction pathway, but has subsequently been implicated in cardiac hypertrophy (Dostal et al. 1997). Baker and coworkers (Dostal et al. 1997; Booz et al. 2002) have proposed the existence of multiple pathways for the activation of the Jak/STAT pathway in the cardiac myocyte and fibroblast (Figure 6.5). The major AT_1-linked pathway involves the activation of a tyrosine kinase, followed by the phosphorylation and activation of Jak proteins. Upon phosphorylation by Jak2, members of the STAT (signal transducers and activators of transcription) family translocate to the nucleus, forming a complex that upregulates growth-linked genes, such as the immediate early genes (c-fos, c-jun, junB, Egr-1, and c-myc). While this pathway was once touted as an important mechanism in angiotensin II signaling, enthusiasm for this pathway waned when it was discovered that angiotensin II promotes protein synthesis in the intact adult heart without the induction of c-fos or c-jun (Schunkert et al. 1995). However, more recent data link the activation of the Jak/STAT pathway to the etiology of heart failure (Booz et al. 2002) and hypertension-induced remodeling of the vasculature (Touyz and Schiffrin 2000).

Factors Regulating Angiotensin II-Mediated Apoptosis and Cell Growth

Angiotensin II, like several of the cytokines, is capable of triggering both apoptosis and protein synthesis. Indeed, angiotensin II modulates both cell death and proliferation in the young developing heart (Mel'nikova et al. 2006). It also contributes to the development of heart failure by promoting hypertrophy, fibrosis, and apoptosis (Schroeder et al. 2006). While the initiation of death and growth pathways by the same agent seems counterproductive, several scenarios could account for the dual activity. First, it is likely that a major trigger for angiotensin II-mediated apoptosis and cell growth is oxidative stress. Yet the dependence of apoptosis and cell growth on the degree of oxidative stress likely differs. In endothelial cells, mild oxidative stress exerts anti-apoptotic actions through the upregulation of thioredoxin-1 (Haendeler et al. 2004) while severe oxidative stress triggers apoptosis (Dimmeler et al. 1997). Second, the type of reactive species formed in response to angiotensin II treatment may determine the outcome of the treatment. In the cardiomyocyte, peroxynitrite is a key intermediate in the apoptotic cascade (Lin et al. 1995; Grishko et al. 2003) while in human endothelial cells superoxide appears crucial, as nitric oxide protects against angiotensin II-induced apoptosis (Dimmeler et al. 1997). Third, the actions of angiotensin II and oxidative stress may be phenotype specific. Bascands et al. (2001) found that in aortic smooth muscle, angiotensin II induces apoptosis in epithelioid cells but not in spindle cells. Fourth, apoptosis may represent a counterbalance to elevated cell growth during stress. The interplay between apoptosis and cell growth is most likely temporally related, as evidenced by the increase in smooth muscle cell apoptosis following regression of vascular hypertrophy in spontaneously hypertensive rats treated with antihypertensive therapy (deBlois et al. 1997). Fifth, the competition between growth and apoptosis may be receptor specific (AT_1 versus AT_2). Most of angiotensin II's well-known effects, such as modulation of blood pressure, fluid/electrolyte balance, cellular proliferation, hormone release, and drinking behavior, represent actions of the AT_1 receptor. Much less is known about the physiological function of the AT_2 receptor. However, recent evidence suggests that the AT_2 receptor may play a role in cellular differentiation, antigrowth (although in some cases growth), and apoptosis (Inagami and Senbonmatsu 2001; Gendron et al. 2003; Kaschina and Unger 2003; Joehren et al., 2004). Since the expression of the AT_2 receptor is low in adults but elevated during fetal and early postnatal life, it has been proposed that counteracting effects of the AT_1 and AT_2 receptors may influence development and remodeling (Yamada et al. 1998). However, it is also possible that physiological function in the adult may be influenced by competition between the AT_1 and AT_2 receptors, particularly in cells such as the cardiomyocyte in which the levels of both the AT_1 and AT_2 receptors are normally low. Sixth, the concentration of angiotensin II in the vicinity of the receptor may regulate the balance between proliferation, survival, and apoptosis. In rat neonatal cardiomyocytes, apoptosis is initiated at concentrations of angiotensin II as low as 1 nM (Grishko et al. 2003) while its trophic activity is generally seen at higher concentrations (Sadoshima and Izumo 1995). Seventh, the interaction between angiotensin II and other effectors may determine the response to angiotensin II. According to Pollman et al. (1996) the countervailing balance between nitric oxide and angiotensin II regulates the survivability of the vascular smooth muscle cell. There is also evidence that angiotensin II-induced apoptosis is associated with enhanced generation of TGF-β (Ding et al. 2002; Bhaskaran et al. 2003; Schroeder et al. 2006) and specific growth factors (Eguchi and Inagami 2000). However, the effect of angiotensin II on the activation state of Akt appears to be cell specific. While Li et al. (2006) report that angiotensin II promotes the dephosphorylation of Akt in human vascular muscle cells, in the cardiomyocyte angiotensin II activates the PI 3-kinase/Akt pathway (Wenzel et al. 2006). Therefore, it is not surprising that the levels and activity of several regulatory factors modulate the actions of angiotensin II.

Regulation of Apoptosis by Angiotensin II

Sun and Weber (1998) were the first investigators to propose a role for angiotensin II in cellular apoptosis. Today, it is generally accepted that angiotensin II, acting through either the AT_1 or AT_2 receptor, induces or regulates apoptosis in a variety of cell types, including neonatal and adult cardiomyocytes, human umbilical venous endothelial cells, primary lung alveolar epithelial cells, human coronary artery endothelial cells, rat pheochromocytoma PC12W cell, fibroblast R3T3 cell line, SA and AV nodal cells, pancreatic acinar cells, glomerular epithelial cells, fetal adrenal cells, ovarian granulosa cells, cultured neurons, and vascular smooth muscle cells. However, due to its progrowth properties, angiotensin II may also exhibit antiapoptotic activity. In cardiac fibroblasts, angiotensin II inhibits interleukin-1β and diethylenetriamine NONOate-induced apoptosis, an effect associated with the activation of Akt (Tian et al. 2003). A similar outcome has been reported for serum-deprived vascular endothelial cells exposed to angiotensin II (Ohashi et al. 2004). Although Akt activation may represent the most common cause of angiotensin II-mediated cytoprotection, Grammatopoulos et al. (2004) have implicated AT_2-mediated reductions in caspase 3 activity in the attenuation of azide-induced apoptosis of mouse cortical neurons by angiotensin II.

Mechanisms Underlying AT_2-Mediated Apoptosis

Yamada et al. (1996) provided the initial evidence that the AT_2 receptor is capable of mediating apoptosis. Using cells (PC12W and R3T3) that express abundant AT_2 receptors but few AT_1 receptors, they found that antisense oligonucleotides directed against MAP kinase phosphatase 1 inhibited AT_2-directed ERK1/2 dephosphorylation and apoptosis. In a follow-up study, Horiuchi et al. (1997) found that the antisense oligonucleotide of MAP kinase phosphatase 1 enhanced the phosphorylation of ERK1/2 and Bcl-2 by nerve growth factor, thereby activating ERK1/2 and preventing Bcl-2 degradation (Dimmeler et al. 1999). A logical conclusion was that the AT_2 receptor promoted apoptosis and antagonized the cell survival actions of nerve growth factor by activating MAP kinase phosphatase 1 (Figure 6.6). In a related study, Roessig et al. (2002) reported that the enzyme induced in human endothelial cells by angiotensin II was MAP kinase phosphatase-3 rather than MAP kinase phosphatase 1; the net result of AT_2 stimulation in human endothelial cells was the dephosphorylation of ERK1/2 and the degradation of Bcl-2. In NIE-115 cells, AT_2 receptor-mediated activation of a protein phosphatase 2A also resulted in the dephosphorylation of ERK1/2 (Huang et al. 1996). Similarly, AT_2 receptor

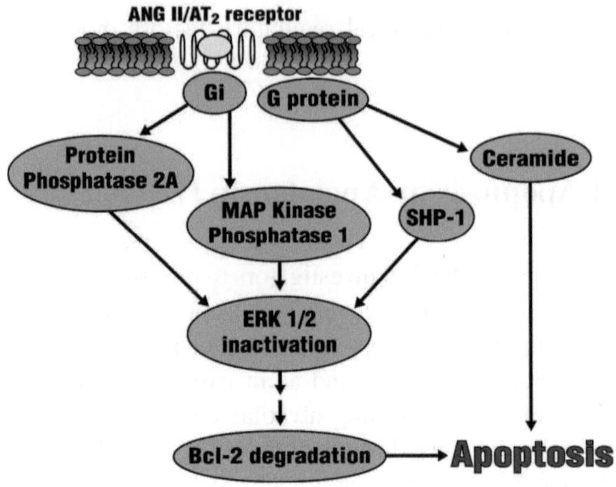

Fig. 6.6 Signaling pathways involved in AT_2-mediated apoptosis. The stimulation of the AT_2 receptor leads to a slow elevation in cellular ceramide, a proapoptotic messenger. The more rapid AT_2 response results in the activation of a protein phosphatase, which has been identified as either a protein phosphatase 2A or MAP kinase phosphatase 1. The activated phosphatase catalyzes the dephosphorylation and inactivation of ERK1/2. Because ERK1/2 maintains the viability of Bcl-2, the inactivation of ERK1/2 leads to the dephosphorylation and degradation of the antiapoptotic factor.

stimulation was found to initiate apoptosis in cultured neurons through the activation of a serine/threonine phosphatase (Shenoy et al. 1999).

A role for the AT_2 receptor in the regulation of Bcl-2 content is also supported by the work of Suzuki et al. (2002), who found that the number of apoptotic vascular smooth muscle cells in injured arteries of AT_2 null mice was reduced despite the elevation in Bcl-2 and Bcl-x_L mRNA content and the decline in Bax mRNA content. While AT_2-dependent changes in Bcl-2 family member activity are generally seen within a few hours of receptor activation, a later event initiated by the AT_2 receptor is the activation of a ceramide pathway, which is followed by caspase 3 activation and the onset of apoptosis (Lehtonen et al. 1999). Since attenuation of nerve growth factor-mediated activation of ERK1/2 activity by the MEK inhibitor, PD098059, has been found to have no influence on ceramide levels, the authors concluded that AT_2 stimulates at least two distinct proapoptotic pathways, the activation of MAP kinase phosphatase 1 and the elevation in ceramide levels (Figure 6.6). Modulation of ERK1/2 activity also underlies the phenotype switch in vascular smooth muscle cells deficient in AT_2 receptor (Akishiti et al. 1999).

Another AT_2 pathway linked to apoptosis leads to a stimulation of a soluble protein tyrosine phosphatase, SHP-1, an enzyme that associates with insulin receptor substrate (IRS)-2 (Cui et al. 2002). Overexpression of a dominant negative form of SHP-1 in PC12W cells has been found to attenuate AT_2 receptor-mediated inhibition of insulin signaling. Since insulin activates Akt, it is interesting that the mechanism of apoptosis in the angiotensin II-treated PC12W cell involves the dephosphorylation and inactivation of Akt. In NIE-115 neuroblastoma cells, angiotensin II-mediated activation of SHP-1 phosphatase is associated with the inactivation of ERK1/2 (Bedecs et al. 1997) (Figure 6.6).

Other possible mechanisms contributing to AT_2-mediated apoptosis include the initiation of a pathway involving both nitric oxide and p53 (Bonnet et al. 2001) and another pathway requiring p38 MAP kinase (Miura and Karnik 2000). There is also evidence that AT_2 activation is coupled to the dephosphorylation of STAT (Steckelings et al. 2005). Recently, Miura and coworkers (Miura and Karnik 2000; Miura et al. 2005) reported that the AT_2 receptor is capable of inducing apoptosis in R3T3 cells without the need of a ligand.

Mechanisms Underlying AT_1-Mediated Apoptosis

Less information is available on the mechanism underlying AT_1-mediated apoptosis. There are at least two apoptotic pathways initiated by the AT_1 receptor. One occurs secondary to the upregulation of $TGF\beta_1$, an event that requires the concurrent and early elevation of transcription factors, AP-1 and GATA. Recently, Schroeder et al. (2006) showed that angiotensin II increased Smad protein binding, an effect blocked by a $TGF\beta_1$-specific antibody. Reductions in Smad binding activity achieved by the introduction of Smad decoy oligonucleotides into cells largely prevented angiotensin II-mediated apoptosis, implicating $TGF\beta_1$-mediated upregulation of Smad in the mechanism of angiotensin II-mediated apoptosis (Figure 6.7).

The other angiotensin II-mediated apoptotic pathway is directly mediated by the AT_1 receptor without the contribution of a secondary factor. Grishko et al. (2003) published a "unifying hypothesis" designed to clarify the divergent views of AT_1 signaling. Since both scavengers of ROS and inhibition of NADPH oxidase prevent angiotensin II-mediated apoptosis, they proposed that the generation of ROS by NADPH oxidase is an early event in angiotensin II signaling (Figure 6.8). Based on the work of Griendling and coworkers (Seshiah et al. 2002; Clempus and Griendling 2006), they concluded that angiotensin II-mediated activation of protein kinase C is responsible for the stimulation of NADPH oxidase activity. This idea has received support from the work of Li et al. (1999) and Papp et al. (2002), who found that inhibition of protein kinase C attenuates angiotensin II-mediated apoptosis.

Another central element of the Grishko hypothesis is the upregulation of inducible nitric oxide synthase (iNOS). Since the induction of iNOS is often associated with extensive nitric oxide formation, it is commonly a toxic event. However, nitric oxide formation is not detrimental in all cell types. Dimmeler et al. (1997) found that nitric oxide protects human endothelial cells against angiotensin II-mediated apoptosis. A key variable appears to be the amount of nitric oxide produced. In the cardiomyocyte, sufficient levels of nitric oxide are generated to cause the formation of toxic levels of peroxynitrite, while in endothelial cells, ROS, rather than reactive nitrogen species, appear to be the toxic mediators of apoptosis.

One of the consequences of either ROS or reactive nitrogen species generation is the oxidation of mitochondrial and nuclear DNA. In the angiotensin II-treated

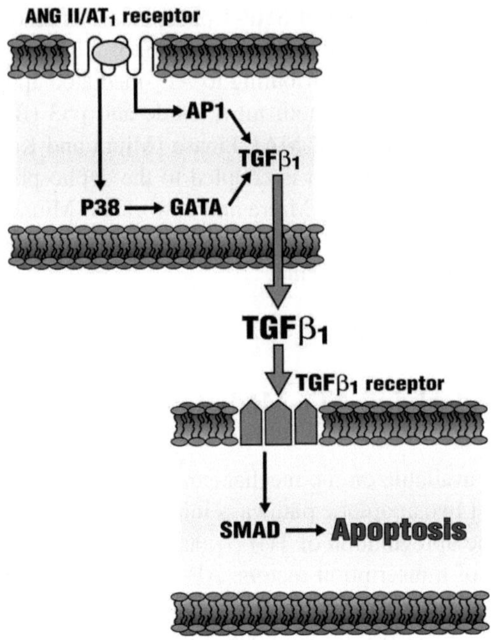

Fig. 6.7 Proposed model for TGFβ1-linked mediation of apoptosis by angiotensin II. In one cardiomyocyte, angiotensin II promotes the upregulation of TGFβ1, which is released into the intracellular space where it interacts with the TGFβ1 receptor of a second cardiomyocyte. Levels of Smad proteins rise in the second cardiomyocyte, causing it to undergo apoptosis.

cardiomyocyte both types of DNA damage may occur (Grishko et al. 2003). In many cell types, nuclear DNA damage has been shown to activate kinases that phosphorylate p53, an oncogene that regulates the levels of Bcl-2 and Bax (Miyashita et al. 1994). In accordance with this sequence of events, Grishko et al. (2003) found that in the neonatal cardiomyocyte, angiotensin II activates p53, causes a transient decrease in Bcl-2 levels, and elevates Bax content. While these results are consistent with the idea that nuclear DNA damage might contribute to angiotensin II-mediated apoptosis, the elevation in the Bax/Bcl-2 ratio in the Grishko study was not large and likely insufficient to cause mitochondrial disruption. Nonetheless, several investigators (Pierzchalski et al. 1997; Diep et al. 2002; Ding et al. 2002) have attributed angiotensin II-mediated apoptosis to the rise in the Bax/Bcl-2 ratio, although the size of the Bax increase in their studies was significantly greater than that reported by Grishko et al. (2003).

Concern over the role of Bax in angiotensin II-mediated apoptosis has led to a modified version of the Grishko hypothesis (Figure 6.8). The modified hypothesis was influenced by the discovery that angiotensin II-induced apoptosis proceeds via the mitochondrial permeability transition (MPT) (Ricci et al. 2005). Although Bcl-2 family members can modulate the activity of the MPT pore (Shimizu et al. 1999; Rostovtseva et al. 2004), regulation of the MPT pore by oxidative stress may be

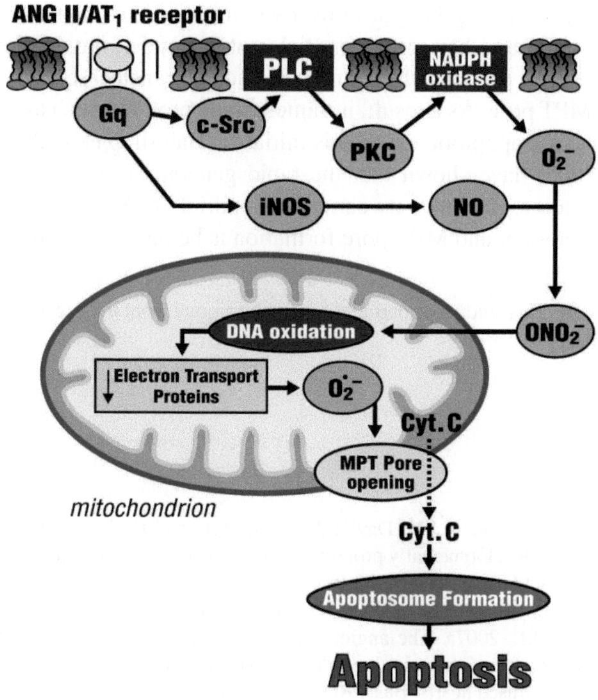

Fig. 6.8 Proposed scheme for the proapoptotic activity of the stimulated AT_1 receptor. Interaction of angiotensin II with the AT_1 receptor triggers a cascade, which leads to the activation of NADPH oxidase and the subsequent generation of superoxide. Angiotensin II also activates inducible nitric oxide synthase, iNOS, a major source of nitric oxide (NO). The two reactants combine to form peroxynitrite (ONO_2^-). After entering the mitochondria, peroxynitrite oxidizes DNA, which interferes with the expression of mitochondrial proteins and the activity of the electron transport. As a result, electron transport slows and an imbalance develops between the generation of reducing equivalents and their utilization by the electron transport chain. ATP production slows, but more importantly, electrons are diverted from the electron transport chain to oxygen forming superoxide. The antioxidant defense system is overwhelmed and mitochondrial proteins undergo oxidation. Upon oxidation of adenine nucleotide translocase, the mitochondrial permeability transition (MPT) pore forms, allowing cytochrome c (cyt c) release. In the cytosol, cytochrome c combines with Apaf-1 and ATP to form the apoptosome, a complex that initiates the caspase cascade.

a more important contributor to angiotensin II-mediated apoptosis. Halestrap et al. (2004) have shown that oxidative modification of adenine nucleotide translocase facilitates MPT pore formation. A major question has been the steps upstream from MPT pore formation. In the modified hypothesis, one of the major steps upstream from MPT pore formation is the slowing of flux through the electron transport chain. According to Ricci et al. (2005), angiotensin II mediates mitochondrial DNA damage, which induces the downregulation of mitochondrially encoded electron transport proteins. It has been assumed that the decreased expression of the electron transport components invariably slows flux of reducing equivalents through the electron transport chain, enhancing the diversion of electrons to oxygen. The resulting formation of superoxide is thought to overwhelm the antioxidative defense system of the mitochondria and key mitochondrial proteins undergo oxidation. Oxidative injury to one of these proteins, the adenine nucleotide translocase, facilitates the formation of the MPT pore. As a result, intramembrane proteins, such as cytochrome c, are released and the apoptotic cascade is initiated (Halestrap et al. 2004). Although Ricci et al. (2005) have shown that the rapid generation of superoxide precedes MPT pore formation, the link between mitochondrial DNA damage, mitochondrial superoxide generation, and MPT pore formation is being actively investigated.

Acknowledgement This study was partially funded by a grant from the American Heart Association (0755288B).

References

Akishita, M., Ito, M., Lehtonen, J.Y., Daviet, L., Dzau, V.J., and Horiuchi, M. 1999. Expression of the AT_2 receptor developmentally programs extracellular signal-regulated kinase activity and influences fetal vascular growth. J. Clin. Invest. 103:63–71.

Aplin, M., Christensen, G.L., Schneider, M., Heydorn, A., Gammeltoft, S., Kjolbye, A.L., Sheikh, S.P., and Hansen, J.L. 2007a. The angiotensin type 1 receptor activates extracellular signal-related kinases 1 and 2 by G protein-dependent and -independent pathways in cardiac myocytes and Langendorff-perfused hearts. Basic Res. Pharmacol. Toxicol. 100:289–295.

Aplin, M., Christensen, G.L., Schneider, M., Heydorn, A., Gammeltoft, S., Kjolbye, A.L., Sheikh, S.P., and Hansen, J.L. 2007b. Differential extracellular signal-regulated kinases 1 and 2 activation by the angiotensin type 1 receptor supports distinct phenotypes of cardiac myocytes. Basic Res. Pharmacol. Toxicol. 100:296–301.

Bader, M. 2002. Role of the local renin-angiotensin system in cardiac damage: a minireview focusing on transgenic animal models. J. Mol. Cell. Cardiol. 34:1455–1462.

Bascands, J.-L., Girolami, J.-P., Troly, M., Escargueil-Blanc, I., Nazzal, D., Salvayre, R., and Blaes, N. 2001. Angiotensin II induces phenotype-dependent apoptosis in vascular smooth muscle cells. Hypertension 38:1294–1299.

Bedecs, K., Elbaz, N., Sutren, M., Masson, M., Susini, C., Strosberg, A.D., and Nahmias, C. 1997. Angiotensin II type 2 receptors mediate inhibition of mitogen-activated protein kinase cascade and functional activation of SHP-1 tyrosine phosphatase. Biochem. J. 325:449–454.

Bendall, J.K., Cave, A.C., Heymes, C., Gall, N., and Shah, A.M. 2002. Pivotal role of a $gp91^{phox}$-containing NADPH oxidase in angiotensin II-induced cardiac hypertrophy in mice. Circulation 105:293–296.

Bhaskaran, M., Reddy, K., Radhakrishnan, N., Franki, N., Ding, G., and Singhal, P.C. 2003. Angiotensin II induces apoptosis in renal proximal tubular cells. Am. J. Physiol. 284:F955–F965.

Bokemeyer, D., Schmi, U., and Kramer, H.J. 2000. Angiotensin II-induced growth of vascular smooth muscle cells requires an Src-dependent activation of the epidermal growth factor receptor. Kidney Int. 58:549–558.

Bonnet, F., Cao, Z., and Cooper, M.E. 2001. Apoptosis and angiotensin II: Yet another renal regulatory system? Exp. Nephrol. 9:295–300.

Booz, G.W., Day, J.N.E., and Baker, K.M. 2002. Interplay between the cardiac renin angiotensin system and JAK-STAT signaling: Role in cardiac hypertrophy, ischemia/reperfusion dysfunction, and heart failure. J. Mol. Cell. Cardiol. 34:1443–1453.

Clempus, R.E., and Griendling, K.K. 2006. Reactive oxygen species signaling in vascular smooth muscle cells. Cardiovasc. Res. 71:216–225.

Cui, T.-X., Nakagami, H., Nahmias, C., Shiuchi, T., Takeda-Matsubara, Y., Li, J.-M., Wu, L., Iwai, M., and Horiuchi, M. 2002. Angiotensin II subtype 2 receptor activation inhibits insulin-induced phosphoinositide 3-kinase and Akt and induces apoptosis in PC12W cells. Mol. Endocrinol. 16:2113–2123.

deBlois, D., Tea, B.-S., Dam, T.-V., Tremblay, J., and Hamet, P. 1997. Smooth muscle apoptosis during vascular regression in spontaneously hypertensive rats. Hypertension 29:340–349.

Diep, Q.N., El Mabrouk, M., Yue, P., and Schiffrin, E.L. 2002. Effect of AT_1 receptor blockade on cardiac apoptosis in angiotensin II-induced hypertension. Am. J. Physiol. 282:H1635–H1641.

Dimmeler, S., Rippmann, V., Weiland, U., Haendeler, J., and Zeiher, A.M. 1997. Angiotensin II induces apoptosis of human endothelial cells: Protective effect of nitric oxide. Circ. Res. 81:970–976.

Dimmeler, S., Breitschopf, K., Haendeler, J., and Zeiher, A.M. 1999. Dephosphorylation targets Bcl-2 for ubiquitinin-dependent degradation: a link between apoptosis and the proteasome pathway. J. Exp. Med. 189:1815–1822.

Ding, G., Reddy, K., Kapasi, A.A., Franki, N., Gibbons, N., Kasinath, B.S., and Singhal, P.C. 2002. Angiotensin II induces apoptosis in rat glomerular epithelial cells. Am. J. Physiol. 283:F173–F180.

Dostal, D.E., and Baker, K.M. 1999. The cardiac renin-angiotensin system: conceptual, or a regulator of cardiac function? Circ. Res. 85:643–650.

Dostal, D.E., Hunt, R.A., Kule, C.E., Bhat, G.J., Karoor, V., McWhinney, C.D., and Baker, K.M. 1997. Molecular mechanisms of angiotensin II in modulating cardiac function: intracardiac effects and signal transduction pathways. J. Mol. Cell. Cardiol. 29:2893–2902.

Eguchi, S., and Inagami, T. 2000. Signal transduction of angiotensin II type 1 receptor through receptor tyrosine kinase. Regul. Peptides 91:13–20.

Eguchi, S., Numaguchi, K., Iwasaki, H., Matsumoto, T., Yamakawa, T., Utsunomiya, H., Motley, E.D., Kawakatsu, H., Owada, K.M., Hirata, Y., Marumo, F., and Inagami, T. 1998. Calcium-dependent epidermal growth factor receptor transactivaiton mediates the angiotensin II-induced mitogen-activated protein kinase activation in vascular smooth muscle cells. J. Biol. Chem. 273:8890–8896.

Frank, G.D., Eguchi, S., Inagami, T., and Motley, E.D. 2001. N-acetylcysteine inhibits angiotensin II-mediated activation of extracellular signal-regulated kinase and epidermal growth factor receptor. Biochem. Biophys. Res. Commun. 280:1116–1119.

Gendron, L., Payet, M.D., and Gallo-Payet, N. 2003. The angiotensin type 2 receptor of angiotensin II and neuronal differentiation: from observations to mechanisms. J. Mol. Endocrinol. 31:359–372.

Giasson, E., and Meloche, S. 1995. Role of p70 S6 protein kinase in angiotensin II-induced protein synthesis in vascular smooth muscle cells. J. Biol. Chem. 270:5225–5231.

Godeny, M.D., and Sayeski, P.P. 2006a. ANG II-induced cell proliferation is dually mediated by c-Src/Yes/Fyn-regulated ERK1/2 activation in the cytoplasm and PKCζ-controlled ERK1/2 activity within the nucleus. Am. J. Physiol. 291:C1297–1307.

Godeny, M.D., and Sayeski, P.P. 2006b. ERK1/2 regulates ANG II-dependent cell proliferation via cytoplasmic activation of RSK2 and nuclear activation of elk1. Am. J. Physiol. 291:C1308–1317.

Grammatopoulos, T.N., Morris, K., Bachar, C., Moore, S., Andres, R., and Weyhenmeyer, J.A. 2004. Angiotensin II attenuates chemical hypoxia-induced caspase-3 activation in primary cortical neuronal cultures. Brain Res. Bull. 62:297–303.

Grishko, V., Pastukh, V., Solodushko, V., Gillespie, M., Azuma, J., and Schaffer, S. 2003. Apoptotic cascade initiated by angiotensin II in neonatal cardiomyocytes: Role of DNA damage. Am. J. Physiol. 285:H2364–2372.

Haendeler, J., Tischler, V., Hoffmann, J., Zeiher, A.M., and Dimmeler, S. 2004. Low doses of reactive oxygen species protect endothelial cells from apoptosis by increasing thioredoxin-1 expression. FEBS Lett. 577:427–433.

Hafizi, S., Wang, X., Chester, A.H., Yacoub, M.H., and Proud, C.G. 2004. ANG II activates effectors of mTOR via PI3-K signaling in human coronary smooth muscle cells. Am. J. Physiol. 287:H1232–H1238.

Halestrap, A.P., Clarke, S.J., and Javadov, S.A. 2004. Mitochondrial permeability transition pore opening during myocardial reperfusion—a target for cardioprotection. Cardiovasc. Res. 61:372–385.

Hamawaki, M., Coffman, T.M., Lashus, A., Koide, M., Zile, M.R., Oliverio, M.I., DeFreyte, G., Cooper, G. 4th, and Carabello, B.A. 1998. Pressure-overload hypertrophy is unabated in mice devoid of AT_{1A} receptors. Am. J. Physiol. 274:H868–H873.

Horiuchi, M., Hayashida, W., Kambe, T., Yamada, T., and Dzau, V.J. 1997. Angiotensin type 2 receptor dephosphorylates Bcl-2 by activating mitogen-activated protein kinase phosphatase-1 and induces apoptosis. J. Biol. Chem. 272:19022–19026.

Hou, M., Pantev, E., Moeller, S., Erlinge, D., and Edvinsson, L. 2000. Angiotensin II type 1 receptors stimulate protein synthesis in human cardiac fibroblasts via a Ca^{2+}-sensitive PKC-dependent tyrosine kinase pathway. Acta Physiol. Scand. 168:301–309.

Huang, X.C., Richards, E.M., and Sumners, C. 1996. Mitogen-activated protein kinases in rat brain neuronal cultures are activated by angiotensin II type 1 receptors and inhibited by angiotensin II type 2 receptors. J. Biol. Chem. 271:15635–15641.

Iijima, Y., Laser, M., Shiraishi, H., Willey, C.D., Sundaravadivel, B., Xu, L., McDermott, P.J., and Kuppuswamy, D. 2002. c-Raf/MEK/ERK pathway controls protein kinase C-mediated p70S6K activation in adult cardiac muscle cells. J. Biol. Chem. 277:23065–23075.

Inagami, T., and Senbonmatsu, T. 2001. Dual effects of angiotensin II type 2 receptor on cardiovascular hypertrophy. Trends Cardiovasc. Med. 11:324–328.

Joehren, O., Dendorfer, A., and Dominiak, P. 2004. Cardiovascular and renal function of angiotensin II type-2 receptors. Cardiovasc. Res. 62:460–467.

Kaschina, E., and Unger, T. 2003. Angiotensin AT1/AT2 receptors: Regulation, signaling and function. Blood Pressure 12:70–88.

Lehtonen, J.Y.A., Horiuchi, M., Daviet, L., Akishita, M., and Dzau, V.J. 1999. Activation of the de novo biosynthesis of sphingolipids mediates angiotensin II type 2 receptor-induced apoptosis. J. Biol. Chem. 274:16901–16906.

Li, D., Yang, B., Philips, M.I., and Mehta, J.L. 1999. Proapoptotic effects of ANG II in human coronary artery endothelial cells: role of AT_1 receptor and PKC activation. Am. J. Physiol. 276:H786–H792.

Li, Y., Song, Y.-H., Mohler, J., and Delafontaine, P. 2006. ANG II induces apoptosis of human vascular smooth muscle via extrinsic pathway involving inhibition of Akt phosphorylation and increased FasL expression. Am. J. Physiol. 290:H2116–H2123.

Lijnen, P., and Petrov, V. 1999. Renin-angiotensin system, hypertrophy and gene expression in cardiac myocytes. J. Mol. Cell. Cardiol. 31:949–970.

Lin, K.T., Xue, J.Y., Nomen, M., Spur, B., and Wong, P.Y. 1995. Peroxynitrite-induced apoptosis in HL-60 cells. J. Biol. Chem. 270:16487–16490.

Mel'nikova, N.P., Timoshin, S.S., Jivotova, E.Y., Pelliniemi, L.J., Jokinen, E., and Abdelwahid, E. 2006. Angiotensin II activates apoptosis, proliferation and protein synthesis in the left heart ventricle of newborn albino rats. Int. J. Cardiol. 112:219–222.

Mifune, M., Ohtsu, H., Suzuki, H., Nakashima, H., Brailoiu, E., Dun, N.J., Frank, G.D., Inagami, T., Higashiyama, S., Thomas, W.G., Eckhart, A.D., Dempsey, P.J., and Eguchi, S. 2005. G protein coupling and second messenger generation are indispensable for metalloprotease-dependent, heparin-binding epidermal growth factor shedding through angiotensin II type-1 receptor. J. Biol. Chem. 280:26592–26599.

Miura, S.-I., and Karnik, S.S. 2000. Ligand-independent signals from angiotensin II type 2 receptor induce apoptosis. EMBO J. 19:4026–4035.

Miura, S.-I., Karnik, S.S., and Saku, K. 2005. Constitutively active homo-oligomeric angiotensin II type 2 receptor induces cell signaling independent of receptor conformation and ligand stimulation. J. Biol. Chem. 280:18237–18244.

Miyashita, T., Krajewski, S., Krajewska, M., Wang, H.G., Lin, H.K., Liebermann, D.A., Hoffmann, B., and Reed, J.C. 1994. Tumor suppressor p53 is a regulator of bcl-2 and bax gene expression *in vitro* and *in vivo*. Oncogene 9:1799–1805.

Murasawa, S., Mori, Y., Nozawa, Y., Gotoh, N., Shibuya, M., Masaki, H., Maruyama, K., Tsutsumi, Y., Moriguchi, Y., Shibasaki, Y., Tanaka, Y., Iwasaka, T., Inada, M., and Matsubara, H. 1998a. Angiotensin II type 1 receptor-induced extracellular signal-regulated protein kinase activation is mediated by Ca^{2+}/calmodulin-dependent transactivation of epidermal growth factor receptor. Circ. Res. 82:1338–1348.

Murasawa, S., Mori, Y., Nozawa, Y., Masaki, H., Maruyama, K., Tsutsumi, Y., Moriguchi, Y., Shibasaki, Y., Tanaka, Y., Iwasaka, T., Inada, M., and Matsubara, H. 1998b. Role of calcium-sensitive tyrosine kinase Pyk2/CAKβ/RAFTK in angiotensin II-induced Ras/ERK signaling. Hypertension 32:668–675.

Nakamura, K., Fushimi, K., Kouchi, H., Mihara, K., Miyazaki, M., Ohe, T., and Namba, M. 1998. Inhibitory effects of antioxidants on neonatal rat cardiac myocyte hypertrophy induced by tumor necrosis factor-α and angiotensin II. Circulation 98:794–799.

Ohashi, H. Takagi, H., Oh, H., Suzuma, K., Suzuma, I., Miyamoto, N., Uemura, A., Watanabe, D., Murakami, T., Sugaya, T., Fukamizu, A., and Honda, Y. 2004. Phosphatidylinositol-3-kinase/Akt regulates angiotensin II-induced inhibition of apoptosis in microvascular endothelial cells by governing surviving expression and suppression of caspase-3 activity. Circ. Res. 94:785–793.

Olson, E.R., Naugle, J.E., Zhang, X., Bomser, J.A., and Meszaros, J.G. 2005. Inhibition of cardiac fibroblast proliferation and myofibroblast differentiation by resveratrol. Am. J. Physiol. 288:H1131–H1138.

Papp, M., Li, X., Zhuang, J., Wang, R., and Uhal, B.D. 2002. Angiotensin receptor subtype AT_1 mediates alveolar epithelial cell apoptosis in response to ANG II. Am. J. Physiol. 282:L713–L718.

Pierzchalski, P., Reiss, K., Cheng, W., Cirielli, C., Kajstura, J., Nitahara, J.A., Rizk, M., Capogrossi, M.C., and Anversa, P. 1997. p53 induces myocyte apoptosis via the activation of the renin-angiotensin system. Exp. Cell Res. 234:57–65.

Pollman, M.J., Yamada, T., Horiuchi, M., and Gibbons, G.H. 1996. Vasoactive substances regulate vascular smooth muscle cell apoptosis. Circ. Res. 79:748–756.

Ricci, C., Pastukh, V., and Schaffer, S. 2005. Involvement of the mitochondrial permeability transition pore in angiotensin II-mediated apoptosis. Exp. Clin. Cardiol. 10:160–164.

Roessig, L., Hermann, C., Haendeler, J., Assmus, B., Zeiher, A.M., and Dimmeler, S. 2002. Angiotensin II-induced upregulation of MAP kinase phosphatase-3 mRNA levels mediates endothelial cell apoptosis. Basic Res. Cardiol. 97:1–8.

Rostovtseva, T.K., Antonsson, B., Suzuki, M., Youle, R.J., Colombini, M., and Bezrukov, S.M. 2004. Bid, but not bax, regulates VDAC channels. J. Biol. Chem. 279:13575–13583.

Ruf, S., Piper, H.M., and Schlueter, K.-D. 2002. Specific role for the extracellular signal-regulated kinase pathway in angiotensin II- but not phenylephrine-induced cardiac hypertrophy *in vitro*. Pfluegers Arch. Eur. J. Physiol. 443:483–490.

Sadoshima, J.-I., and Izumo, S. 1993. Signal transduction pathways of angiotensin II-induced c-fos gene expression in cardiac myocytes *in vitro*: roles of phospholipid-derived second messengers. Circ. Res. 73:424–438.

Sadoshima, J.-I., and Izumo, S. 1995. Rapamycin selectively inhibits angiotensin II-induced increase in protein synthesis in cardiac myocytes *in vitro*: potential role of 70-kD S6 kinase in angiotensin II-induced cardiac hypertrophy. Circ. Res. 77:1040–1052.

Sadoshima, J., Qiu, Z., Morgan, J.P., and Izumo, S. 1995. Angiotensin II and other hypertrophic stimuli mediated by G protein-coupled receptors activate tyrosine kinase, mitogen-activated protein kinase, and 90-kD S6 kinase in cardiac myocytes: The critical role of Ca^{2+}-dependent signaling. Circ. Res. 76:1–15.

Schroeder, D., Heger, J., Piper, H.M., and Euler, G. 2006. Angiotensin II stimulates apoptosis via TGF-β1 signaling in ventricular cardiomyocytes of rat. J. Mol. Med. 84:975–983.

Schunkert, H., Sadoshima, J.-I., Cornelius, T., Kagaya, Y., Weinberg, E.O., Izumo, S., Riegger, G., and Lorell, B.H. 1995. Angiotensin II-induced growth responses in isolated adult rat hearts: evidence for load-independent induction of cardiac protein synthesis by angiotensin II. Circ. Res. 76:489–497.

Seshiah, P.N., Weber, D.S., Rocic, P., Valppu, L., Taniyama, Y., and Griendling, K.K. 2002. Angiotensin II stimulation of NAD(P)H oxidase activity: upstream mediators. Circ. Res. 91:406–413.

Seta, K., Nanamori, M., Modrall, J.G., Neubig, R.R., and Sadoshima, J. 2002. AT1 receptor mutant lacking heterotrimeric G protein coupling activates the Src-Ras-ERK pathway without nuclear translocation of ERKs. J. Biol. Chem. 277:9268–9277.

Shenoy, U.V., Richards, E.M., Huang, X.C., and Sumners, C. 1999. Angiotensin II type 2 receptor-mediated apoptosis of cultured neurons from newborn rat brain. Endocrinology 140:500–509.

Shimizu, S., Narita, M., and Tsujimoto, Y. 1999. Bcl-2 family proteins regulate the release of apoptogenic cytochrome c by the mitochondrial channel VDAC. Nature 399:483–487.

Steckelings, U.M., Kaschina, E., and Unger, T. 2005. The AT2 receptor—A matter of love and hate. Peptides 26:1401–1409.

Sun, Y., and Weber, K.T. 1998. Cardiac remodeling by fibrous tissue: role of local factors and circulating hormones. Ann. Med. 30(Suppl. 1):3–8.

Suzuki, J., Iwai, M., Nakagami, H., Wu, L., Chen, R., Sugaya, T., Hamada, M., Hiwada, K., and Horiuchi, M. 2002. Role of angiotensin II-regulated apoptosis through distinct AT_1 and AT_2 receptors in neointimal formation. Circulation 106:847–853.

Tian, B., Liu, J., Bitterman, P., and Bache, R.J. 2003. Angiotensin II modulates nitric oxide-induced cardiac fibroblast apoptosis by activation of Akt/PKB. Am. J. Physiol. 285:H1105–H1112.

Touyz, R.M., and Schiffrin, E.L. 2000. Signal transduction mechanisms mediating the physiological and pathophysiological actions of angiotensin II in vascular smooth muscle cells. Pharmacol. Rev. 52:639–672.

Touyz, R.M., Cruzado, M., Tabet, F., Yao, G., Salomon, S., and Schiffrin, E.L. 2003. Redox-dependent MAP kinase signaling by Ang II in vascular smooth muscle cells: role of receptor tyrosine kinase transactivation. Can. J. Physiol. Pharmacol. 81:159–167.

Ushio-Fukai, M., Griendling, K.K., Becker, P.L., Hilenski, L., Halleran, S., and Alexander, R.W. 2001. Epidermal growth factor receptor transactivation by angiotensin II requires reactive oxygen species in vascular smooth muscle cells. Arterioscler. Thromb. Vasc. Biol. 21:489–495.

Wang, D., Yu, X., and Brecher, P. 1999. Nitric oxide inhibits angiotensin II-induced activation of the calcium-sensitive tyrosine kinase proline-rich tyrosine kinase 2 without affecting epidermal growth factor receptor transactivation. J. Biol. Chem. 274:24342–24348.

Wang, D., Yu, X., Cohen, R.A., and Brecher, P. 2000. Distinct effects of N-acetylcysteine and nitric oxide on angiotensin II-induced epidermal growth factor receptor phosphorylation and intracellular Ca^{2+} levels. J. Biol. Chem. 275:12223–12230.

Wang, X.P., Zhang, R., Wu, K., Wu, L., and Dong, Y. 2004. Angiotensin II mediates acinar cell apoptosis during the development of rat pancreatic fibrosis by AT1R. Pancreas 29:264–270.

Wenzel, S., Abdallah, Y., Helmig, S., Schaefer, C., Piper, H.M., and Schlueter, K.-D. 2006. Contribution of PI 3-kinase isoforms to angiotensin II- and α-adrenoceptor-mediated signalling pathways in cardiomyocytes. Cardiovasc. Res. 71: 352–362.

Yamada, T., Horiuchi, M., and Dzau, V.J. 1996. Angiotensin II type 2 receptor mediates programmed cell death. Proc. Natl. Acad. Sci. USA 93:156–160.

Yamada, T., Akishita, M., Pollman, M.J., Gibbons, G.H., Dzau, V.J., and Horiuchi, M. 1998. Angiotensin II type 2 receptor mediates vascular smooth muscle cell apoptosis and antagonizes angiotensin II type 1 receptor action: an *in vitro* gene transfer study. Life Sci. 63:PL289–PL295.

Yamazaki, T., and Yazaki, Y. 1997. Is there major involvement of the renin-angiotensin system in cardiac hypertrophy? Circ. Res. 81:639–642.

Zhai, P., Yamamoto, M., Galeotti, J., Liu, J., Masurekar, M., Thaisz, J., Irie, K., Holle, E., Yu, X., Kupershmidt, S., Roden, D.M., Wagner, T., Yatani, A., Vatner, D.E., Vatner, S.F., and Sadoshima, J. 2005. Cardiac-specific overexpression of AT_1 receptor mutant lacking $G_{\alpha q}/G_{\alpha I}$

coupling causes hypertrophy and bradycardia in transgenic mice. J. Clin. Invest. 115:3045–3056.

Zhai, P., Galeotti, J., Liu, J., Holle, E., Yu, X., Wagner, T., and Sadoshima, J. 2006. An angiotensin II type 1 receptor mutant lacking epidermal growth factor receptor transactivation does not induce angiotensin II-mediated cardiac hypertrophy. Circ. Res. 99:528–536.

Zhang, Y., Griendling, K.K., Dikalova, A., Owens, G.K., and Taylor, W.R. 2005. Vascular hypertrophy in angiotensin II-induced hypertension is mediated by vascular smooth muscle cell-derived H_2O_2. Hypertension 46:732–737.

Zou, Y., Komuro, I., Yamazaki, T., Kudoh, S., Aikawa, R., Zhu, W., Shiojima, I., Hiroi, Y., Tobe, K., Kadowaki, T., and Yazaki, Y. 1998. Cell type-specific angiotensin II-evoked signal transduction pathways: critical roles of $G_{\beta\gamma}$ subunit, Src family, and Ras in cardiac fibroblasts. Circ. Res. 82:337–345.

Part II
Redox Signaling

Chapter 7
Thioredoxin Signaling in the Ischemic Heart

Dipak K. Das and Shyamal K. Goswami

Abstract Thioredoxins are a class of small redox-regulating proteins that play a crucial role against various oxidative stress-inducible degenerative diseases. A recent study demonstrated that in the ischemic reperfused myocardium, thioredoxin-1 (Trx-1) level is reduced. Upon adaptation to ischemic stress by preconditioning with repeated cyclic episodes of small duration of ischemia and reperfusion, there was an increased expression of Trx-1. Inhibition of Trx-1 expression resulted in reduced postischemic ventricular recovery and increased myocardial infarct size in the preconditioned heart. Corroborating these findings, transgenic mouse hearts overexpressing Trx-1 were resistant to ischemic reperfusion injury as compared to the hearts from wild-type mice. Thus, it appears that thioredoxin plays a crucial role in cardioprotection induced by preconditioning.

Introduction

Reactive oxygen species (ROS) play a crucial role in the pathophysiology of ischemic heart diseases by causing electrical disturbances, cardiac dysfunction, and cell death (Chen et al. 2007; Kevin et al. 2005; Huang et al. 2003; Turoczi et al. 2003; Yang et al. 2003; Nishida et al. 2000; Nishida et al. 1998). Cardiac arrhythmias and stunning are the hallmarks of ROS-mediated cellular injuries during the reperfusion of the ischemic myocardium (Huang et al. 2003; Nishida et al. 1998). Evidence in favor of ROS being the mediator of ischemia reperfusion injury includes observations that antioxidants and free radical scavengers can ameliorate those effects (Di Lisa and Bernardi 2006; Zorov et al. 2006; Petrosillo et al. 2005; Hangaishi et al. 2000; Tanguy et al. 1996). A number of intracellular sources like mitochondrial respiratory chains and microsomes (Gutierrez et al. 2006; Honda et al. 2005), cyclooxygenase and lipooxygenase pathways (Shinmura et al. 2002; Valen et al. 1993), xanthine oxidase (Kozlovski et al. 2006), circulating elements in the blood and polymorphonuclear leukocytes may generate ROS (Lantos et al. 2001). The ROS can be directly detected by using electron spin resonance (ESR) (Lecour et al.

A. K. Srivastava, M. B. Anand-Srivastava (eds.) *Advances in Biochemistry in Health and Disease.* Signal Transduction in the Cardiovascular System in Health and Disease,

1998; Tortolani et al. 1993) or high-performance liquid chromatography (HPLC) (Cordis et al. 1994, 1998a) and indirectly by measuring the extent of lipid peroxidation (Tosaki et al. 1993), conjugated diene formation (Zhang et al. 2006), or DNA strand breaks (Cordis et al. 1998a). Extensive studies both *in vitro* and *in vivo* have clearly demonstrated that ROS damages virtually all kinds of biomolecules, viz., proteins, lipids, carbohydrates, and nucleotides (Skibska et al. 2006; Babusikova et al. 2004). If ROS are formed in a close vicinity to DNA, it can bring about a change in DNA structure resulting in mutation or cytotoxicity (Kujoth et al. 2005).

While ROS have long been known for their deleterious effects, recent studies indicate that they can also function as the second messengers, inducing events as diverse as cell differentiation, proliferation, tissue repair, wound healing, cardioprotection, etc. (Koneru et al. 2007; Piccoli et al. 2007; auf dem Keller et al. 2006; Chiarugi and Fiaschi 2006; Das et al. 2006; Das and Maulik 2005; Laderoute and Webster 1997). Also, the generation of various reactive oxygen species is closely associated with that of reactive nitrogen species, and they are often collectively considered for their biological effects (Botting and Vane 1989). Following the seminal discovery in the late 1980s that nitric oxide (NO) can act as a signaling molecule (Botting and Vane 1989), a number of other reactive species, especially hydrogen peroxide, have emerged as key signal transducers (Kim et al. 2007). This review will focus on the cell signaling by ROS in the ischemic reperfused myocardium and provide evidence that such signaling activity leads to the repair of the injured cardiomyocytes.

Historical Background

The concept of ROS being a second messenger goes back to the nineteenth century. Charles Darwin's book *The Origin of Species* published in 1859 described that "white blue eyed cats are deaf due to the defect in neuronal development because of the absence of melanin." We now know that melanin is a stable free radical, involved in redox signaling. Oxidation of melanin results in the generation of redox-active tautomer followed by melanosomal redox cycling, leading to the activation of transcription factors and the antiapoptotic phenotype of the melanocytes (Meyskens et al. 2004). In 1979, Proctor et al. suggested that active oxygen metabolites act as specific intermediary transmitter substances for a variety of biological processes including inflammation, fibrosis, and possibly neurotransmission (Proctor et al. 1979). Bochner et al. 1984 probably provided the first direct evidence of redox signaling by showing that AppppA and related adenylated nucleotides are synthesized as a consequence of oxidative stress (Bochner et al. 1984). Direct evidence of ROS functioning as the second messenger(s) in the ischemic reperfused myocardium was first provided by Das et al. (1999) by demonstrating that ROS can make the ischemic heart tolerant to subsequent ischemia and reperfusion (ischemic preconditioning) by upregulating the defense system.

Rapidly accumulating evidence indicates that cellular changes associated with myocardial ischemia and reperfusion are redox-regulated. ROS that are generated in the ischemic reperfused myocardium affect the cellular events in virtually all cellular compartments at the regulatory level. Several redox-sensitive anti- and proapoptotic transcription factors including NFκB and AP-1 progressively and steadily increase in heart during ischemia and reperfusion (Das et al. 1995). The redox signaling also involves multiple kinases including MAP kinases and protein kinase C (Sato et al. 2000; Maulik et al. 1998).

Thioredoxins and Glutaredoxins—The Redox-Sensing Proteins

Several important redox proteins such as thioredoxin (Trx) and glutaredoxin (Grx) play a crucial role in maintaining redox homeostasis in the ischemic myocardium (Turoczi et al. 2003; Isowa et al. 2000). In heart, the intracellular environment is maintained at a reduced state by high levels of glutathione and a number of oxidoreductases, including Trx and Grx, which act as thiol-disulfide redox buffers (Figure 7.1). These thiol proteins are responsible for sensing the cellular redox status and maintain it in a nonequilibrium steady state. During ischemia-reperfusion, due to the generation of ROS, the intracellular environment shifts toward the oxidized state. Thioredoxins are small proteins with redox-active half-cysteine residues (Cys-Gly-Pro-Cys) present in their active center and serving as electron donors for various enzymes (Powis and Monfort 2001; Nakamura et al. 1997; Holmgren and Bjornstedt 1995; Holmgren 1989; Holmgren 1985; Holmgren and Luthman 1978). In mammals, there are two thioredoxin genes, viz., *Trx 1* and *2*, which are ubiquitously expressed but differentially localized. Trx-1 is cytosolic and is involved in

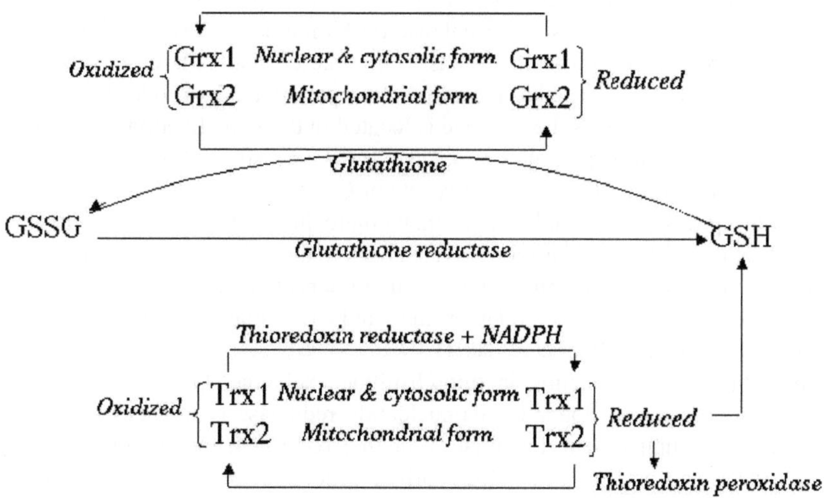

Fig. 7.1 Thioredoxin and glutathione system in mammalian heart.

numerous redox events including the supply of reducing equivalents to thioredoxin peroxidases, ribonucleotide reductase, and the regulation of transcription factor activities (Lillig and Holmgrenm 2007; Chae et al. 1994). Trx-1 also binds to a variety of cellular proteins including Ref-1, ASK 1, PKC, and NFκB and inhibits apoptosis (Zhou et al. 2007; Chen et al. 2006). It possesses free radical scavenging property (Mitsui et al. 1992), acts as a reducing cofactor (Kuster et al. 2006), and can protect cells from H_2O_2-induced oxidative damage (Haendeler et al. 2004). Most of the antioxidant functions of Trx-1 are mediated via thioredoxin peroxidases, which belong to the antioxidant protein family peroxiredoxin. Thioredoxin peroxidases 1 and 2 possess antioxidant property, can scavenge ROS, and reduce apoptosis with a mechanism distinct from that of Bcl-2 (Zhang et al. 1997). Thioredoxin peroxidases are also present in red blood cells and are responsible for protecting them from oxidant injury.

Trx-1 is induced in tissues or cells subjected to a wide variety of stresses including oxidative stress, UV radiation, ionizing irradiation, heavy metals, and inflammation (Tanaka et al. 1997; Ohira et al. 1994). A recent study showed that pretreatment of mice with a sublethal dose of LPS increased Trx concentrations that played an important role in the development of endotoxin tolerance via inhibition of ROS (Sano et al. 2002). Trx is also induced in the tissues subjected to ischemia and reperfusion or ischemic preconditioning (PC; Turoczi et al. 2003), while more recent study has indicated PC-mediated phosphorylation of Akt and nuclear translocation of Trx-1 (Malik et al. 2006). Immunohistochemistry revealed that after translocating into the nucleus, Trx-1 associates with Ref-1, resulting in the increased DNA binding of transcription factor NFκB. PC-mediated phosphorylation of Akt and the activation of NFκB were abolished when the expression of Trx 1 was abrogated by shRNA-Trx-1, suggestng the essentiality of Trx-1 in mediating those effects.

Like thioredoxins, glutaredoxins (Grx) are also conserved in evolution and are involved in the modulation of cellular redox. Glutaredoxins are thioltransferases and mammals have two Grx genes, viz., *Grx 1* and *2*. Although Grx-1 was initially considered to be a cytosolic enzyme, recent results of immunolabeling experiments have shown it to be present in the nucleus as well (Kaarteenaho-Wiik et al. 2004). The second Grx gene, i.e., *Grx 2*, has recently been reported and it encodes two proteins (resulting from alterative splicing); one is located in the mitochondria and the other in the nucleus (Lundberg et al. 2001). Like Trx, Grx also catalyzes thiol/disulfide exchange (Lillig et al. 2005). The active site of Grx [Cys-Pro-Tyr (Phe)-Cys] is oxidized during this process and is regenerated thereafter by the reducing equivalents of GSH. The oxidized GSH is then recycled back to the reduced form by glutathione reductase. Since GSH is the major intracellular nonprotein thiol, it forms the majority of mixed disulfides under oxidative stress and Grx-1 plays a predominant role in reversing this process (Beer et al. 2004).

In addition to its function in thiol/disulfide exchange, Grx-1 also serves as an alternative electron donor to ribonucleotide reductase (Fernandes and Holmgren 2004), participates in deiodination of thyroxine to triiodothyronine (Takagi et al. 1989), and acts as a dehydroascorbate reductase for regenerating ascorbic acid (Washburn and Wells 1989). However, despite the potential role of Grx in antioxidant defense mechanism, studies on its antioxidant function have been exceptionally limited (Pimentel et al. 2006; Fernandes and Holmgren 2004). Recombinant Grx is capable of protecting cultured cerebellar granule neurons from dopamine-induced apoptosis, and the protection is mediated through the activation of the Ras/PI3K/Ref-1/Akt/NFκB pathway (Daily et al. 2001a,b). Thus, it is tempting to speculate that this Ras/PI3K/Ref-1/Akt/NFκB pathway may play an important role in the redox regulation by Trx-1 and/or Grx. Although both Trx-1 and Grx-1 are involved in the regeneration of oxidatively damaged proteins, they have different substrate preferences. While Grx-1 reactivates proteins containing mixed disulfides (Prot-SS and $Prot_1$-SS-$Prot_2$) more efficiently, Trx-1 preferentially regenerates protein sulfenic or sulfinic acid (Yoshitake et al. 1994).

Thioredoxins and Glutaredoxins in Mammalian Hearts

Both the variants of Trx and Grx are present in mammalian hearts and it is believed that they have similar but different roles in modulating cellular redox (Fernandes and Holmgren 2004; Berndt et al. 2006). Recent studies have indicated a correlation between plasma thioredoxin concentrations and severity of heart failure (Jekell et al. 2004). In patients having myocarditis and cardiomyopathies with active necrotic stage, myocardial expression of Trx is elevated in association with DNA damage (Nimata et al. 2003). In another related study, increased Trx expression was found in the skin biopsies (obtained during the cardiac catheterization) of 29 out of 35 patients with congestive heart failure, but none of the 8 control subjects (Miyamoto et al. 2003). Overexpression of Trx-1 in heart attenuates adriamycin-induced cardiotoxicity (Shioji et al. 2002). Also, treatment with recombinant human Trx-1 suppressed cardiomyocyte injury following adrenomycin treatment (Shioji et al. 2003).

Ventricular Remodeling with Thioredoxin

Myocardial infarction is followed by ventricular remodeling that includes both structural and metabolic reorganization and cardiac hypertrophy (Brown et al. 2006; Bujak and Frangogiannis 2007; Frangogiannis 2006). Although at early stages, ventricular remodeling occurs as an adaptive response that is beneficial to the heart, at a later stage it becomes maladaptive, leading to heart failure (Brown et al. 2006; Ho and Seidman 2006). Cardiac hypertrophy is also induced by other pathological conditions, under mechanical stress; and is accompanied by ventricular dysfunction and extensive fibrosis (Ho and Seidman 2006). ROS has been attributed to both early and sustained stages of hypertrophy and cardiac remodeling (Takimoto and Kass 2007; Tsutsui et al. 2006). Antioxidant therapy following experimental myocardial infarction prevents ventricular remodeling and heart failure (Matsushima et al. 2006).

As a general rule, a small amount of ROS elicits adaptive response while massive generation of ROS leads to myocyte loss and heart failure (Tsutsui et al. 2006; Takimoto and Kass 2007). In agreement, recent studies have indicated that suppression of endogenous Trx-1 (by overexpression of dominant negative Trx-1) induces cardiac hypertrophy and exacerbates hypertrophic response induced by thoracic aortic constriction (Yamamoto et al. 2003). Also, in Tg-Trx-1 mice, aortic constriction fails to induce hypertrophy (Yamamoto et al. 2003). The ability of Trx-1 to suppress hypertrophy was due to the inhibition of RAS-MAPK signaling cascade. Another recent study has indicated that Trx-1 attenuates metabolic remodeling by enhancing PPARγ coactivator-1α (PGC-1α), thereby protecting against hypertrophy (Ago et al. 2006).

Trx and Grx being the key redox sensors, their functions are modulated by ROS. While at a lower dose, ROS can induce their expression, thereby eliciting an adaptive response; massive generation of ROS inactivates these proteins by posttranslational modifications (Kondo et al. 2006). In heart failure patients, a significant correlation between the serum concentration of Trx-1 and the severity of the disease has also been documented (Jekell et al. 2004). The increased Trx-1 activity is likely to be due to the adaptive response during the failure to compensate for the increased ROS activities.

Thioredoxins/Glutaredoxins in Ischemia/Reperfusion Injury

Trx-1 is an important component of the defense against cardiac injury. Oxidized thioredoxin is released into plasma of the patients undergoing cardiopulmonary bypass surgery (Nakamura et al. 1998). Nitrative inactivation of thioredoxin-1 has been attributed to postischemic myocardial apoptosis (Tao et al. 2006b). Endurance training by swimming is accompanied by a reduction of ischemia/reperfusion-induced oxidative stress with a concomitant increase in thioredoxin reductase, resulting in protection against myocardial ischemia/reperfusion injury (Kihlstrom 1990). Human thioredoxin attenuates hypoxia-reoxygenation injury of murine endothelial cells in a thiol-free condition, suggesting a novel redox signaling pathway in protecting the myocardium (Isowa et al. 2000). Recombinant human Trx-1 improves lung function and reduces cellular injury of the lung subjected to ischemia and reperfusion (Yokomise et al. 1996). Transgenic mice overexpressing Trx-1 in brain showed smaller infarct after cerebral artery occlusion (Takagi et al. 1999). The brains from the patients with Alzheimer's disease demonstrated reduced amount of Trx-1 with the concomitant increase in thioredoxin reductase (Lovell et al. 2000). Increased Trx-1 expression was reported in the epidermal cells of the skin exposed to sun (Sachi et al. 1995) and for the patients with HIV (Nakamura et al. 1996).

The protective role of Trx-1 in myocardial ischemia/reperfusion injury is further substantiated from the observations that mouse hearts overexpressing Trx-1 were resistant to ischemic injury and ventricular fibrillation (Tao et al. 2006b; Das and Maulik 2003; Aota et al. 1996). Transgenic overexpression of Trx-1 leads to reduced myocardial infarct size and improved left ventricular performance in the postischemic myocardium. Measurements of plasma thioredoxin during open-heart surgery demonstrated elevated levels of Trx in arterial plasma during reperfusion of the postcardioplegic heart of the patients (Nakamura et al. 1998). Trx increases during surgical preparation for cardiopulmonary bypass, but decreases during the bypass due to the release of oxidized Trx into the plasma.

While ischemia/reperfusion reduces the amount of Trx-1, preconditioning induces thioredoxin activity (Malik et al. 2006; Chiueh et al. 2005; Das and Maulik 2003). Increased expression of Trx-1 is due to the adaptive response triggered by PC, as Trx-1 overexpression nicely corroborates with cardioprotective abilities of adaptation. Consistent with these findings, inhibition of Trx-1 abolished cardioprotective effects of PC with a concomitant increase in number of apoptotic myocytes in the myocardium (Malik et al. 2006). These results were further corroborated with the findings that Trx-1 inhibition also abolishes the oxidative stress lowering ability of PC (Das 2004), suggesting a crucial role of Trx in PC-mediated redox signaling. Thioredoxin also induces MnSOD and other antioxidants, which play a crucial role in myocardial defense against ischemic reperfusion (Das and Maulik 2006; Das et al. 1997). Taken together, it is hypothesized that Trx (and Grx) may regulate the "redox switch," which changes ischemia/reperfusion-induced "death signal" into PC-mediated "survival signal (Das and Maulik 2003).

Thioredoxins in Diseased Hearts

Thioredoxin and Hypertension

Impaired expression of Trx-1 in spontaneously hypertensive (SHR) and stroke-prone SHR (SHRSP) rat hearts have been documented (Tanito et al. 2004). Induction of thioredoxin was further reduced upon angiotensin II treatment of peripheral blood mononuclear cells isolated from those animals as compared to control Wistar Kyoto rats. Reduced expression of thioredoxin during hypoxia and reoxygenation of cortical neurons isolated from stroke-prone spontaneously hypertensive rats has also been reported (Yamagata et al. 2000). In another related study, serum Trx concentration in the hypertensive patients were found to be significantly higher than that of normal patients (Miwa et al. 2005).

Thioredoxin and Diabetes

Several reports correlating levels of thioredoxin with diabetes also exist in the literature. According to one study, patients with glucose intolerance have high levels of thioredoxin (Miyamoto et al. 2005). In another study, in patients with type 2 diabetes, serum thioredoxin levels were elevated (Kakisaka et al. 2002). A more recent study demonstrated that Trx-2 was suppressed in the aorta from rats with 2 weeks of diabetes, suggesting a novel role for Trx-2 in protecting cells against high ambient glucose (Liang and Pietrusz 2007). When Trx-2 expression was knocked down by siRNA in human umbilical vein endothelial cells, high ambient glucose elicited substantial cellular damage, increased cytochrome c, lipid peroxidation, and oxidized glutathione; whereas in the presence of Trx-2, such damages were significantly lower. In another related study, after treatment with cisplatin, (a thioredoxin inhibitor), severe hyperglycemic hyperosmolar derangement was observed in a patient without previous history of diabetic mellitus (Arnold et al. 2004). Diabetic rat heart shows upregulation of thioredoxin along with significant reduction in thioredoxin reductase activity (Li et al. 2005).

Thioredoxin and Atherosclerosis

Recent studies have also suggested a correlation between thioredoxin and atherosclerosis. Using DNA microarray analysis, Hagg et al. demonstrated that human macrophages respond to oxidized low-density lipoprotein (oxLDL) by activating thioredoxin systems—thioredoxin, thioredoxin reductase 1, and glutathione reductase—suggesting that these enzymes might play a role in the cellular defense against oxidized LDL and development of atherosclerosis (Hagg et al. 2006). According to another study, lack of antioxidant glutathione peroxidase-1 does not increase atherosclerosis in mice fed a high-fat diet, but reduces the amount of glutaredoxin-1 and increases glutaredoxin-2 (de Haan et al. 2006). Upregulation of thioredoxin reductase in human atherosclerotic plaques has also been reported (Furman et al. 2004).

Summary and Conclusion

Thioredoxin appears to play an important role in the diseased hearts by inducing its own upregulation to compensate for the stress. Trx-1 suppresses cardiac hypertrophy and heart failure and ameliorates myocardial ischemia reperfusion injury. Trx-1 also moderates hypertension and diabetes. A recent study suggests that thioredoxin and its interacting protein TXNIP are key components of biomechanical signal transduction and potentially novel regulators of TNF signaling and inflammation in endothelial cells (Yoshioka et al. 2006). Future studies using either Trx gene therapy or an intervention to increase Trx protein will reveal the usefulness of Trx in the-clinical arena.

Acknowledgement This study was supported by National Institutes of Health grants NIH HL 56803, HL 22559, and HL 33889.

References

Ago, T., Yeh, I., Yamamoto, M., Schinke-Braun, M., Brown, J.A., Tian, B., and Sadoshima, J. 2006. Thioredoxin1 upregulates mitochondrial proteins related to oxidative phosphorylation and TCA cycle in the heart. Antioxid. Redox Signal. 8:1635–1650.

Aota, M., Matsuda, K., Isowa, N., Wada, H., Yodoi, J., and Ban, T. 1996. Protection against reperfusion-induced arrhythmias by human thioredoxin. J. Cardiovasc. Pharmacol. 27:727–732.

Arnold, N.B., Ketterer, K., Kleeff, J,, Friess, H., Buchler, M.W,, and Korc, M. 2004. Thioredoxin is downstream of Smad7 in a pathway that promotes growth and suppresses cisplatin-induced apoptosis in pancreatic cancer. Cancer Res. 64:3599–3606.

auf dem Keller, U., Kumin, A., Braun, S., and Werner, S. 2006. Reactive oxygen species and their detoxification in healing skin wounds. J. Invest. Dermatol. Symp. Proc. 11:106–111.

Babusikova, E., Kaplan, P., Lehotsky, J., Jesenak, M., and Dobrota, D. 2004. Oxidative modification of rat cardiac mitochondrial membranes and myofibrils by hydroxyl radicals. Gen. Physiol. Biophys. 23:327–335.

Beer, S.M., Taylor, E.R., Brown, S.E., Dahm, C.C., Costa, N.J., Runswick, M.J., and Murphy, M.P. 2004. Glutaredoxin 2 catalyzes the reversible oxidation and glutathionylation of mitochondrial membrane thiol proteins: implications for mitochondrial redox regulation and antioxidant defence. J. Biol. Chem. 279:47939–47951.

Berndt, C., Lillig, C.H., and Holmgren, A. 2007. Thiol-based mechanisms of the thioredoxin and glutaredoxin systems: implications for diseases in the cardiovascular system. Am. J. Physiol. Heart Circ. Physiol. 292:H1227–H1236.

Bochner, B.R., Lee, P.C., Wilson, S.W., Cutler, C.W., and Ames, B.N. 1984. AppppA and related adenylylated nucleotides are synthesized as a consequence of oxidative stress. Cell 37:225–232.

Botting, R., and Vane, J.R. 1989. The receipt and dispatch of chemical messengers by endothelial cells. Prog. Clin. Biol. Res. 301:1–11.

Brown, J.H., Del Re, D.P., and Sussman, M.A. 2006. The Rac and Rho hall of fame: a decade of hypertrophic signaling hits. Circ. Res. 98:730–732.

Bujak, M., and Frangogiannis, N.G., 2007. The role of TGF-beta signaling in myocardial infarction and cardiac remodeling. Cardiovasc. Res. 74:184–195.

Chae, H.Z., Chung, S.J., and Rhee, S.G. 1994. Thioredoxin-dependent peroxide reductase from yeast. J. Biol. Chem. 269:27670–27678.

Chen, Q., Camara, A.K., Stowe, D.F., Hoppel, C.L., and Lesnefsky, E.J. 2007. Modulation of electron transport protects cardiac mitochondria and decreases myocardial injury during ischemia and reperfusion. Am. J. Physiol. Cell Physiol. 292:C137–C147.

Chen, Y., Cai, J., and Jones, D.P. 2006. Mitochondrial thioredoxin in regulation of oxidant-induced cell death. FEBS Lett. 580:6596–6602.

Chiarugi, P., and Fiaschi, T. 2006. Redox signalling in anchorage-dependent cell growth. Cell Signal. 19:672–682.

Chiueh, C.C., Andoh, T., and Chock, P.B. 2005. Induction of thioredoxin and mitochondrial survival proteins mediates preconditioning-induced cardioprotection and neuroprotection. Ann. N.Y. Acad. Sci. 1042:403–418.

Cordis, G.A., Bagchi, D., Maulik, N., and Das, D.K. 1994. High-performance liquid chromatographic method for the simultaneous detection of malonaldehyde, acetaldehyde, formaldehyde, acetone and propionaldehyde to monitor the oxidative stress in heart. J. Chromatogr. 661:181–191.

Cordis, G.A., Das, D.K., and Riedel, W. 1998a. High-performance liquid chromato-graphic peak identification of 2,4-dinitrophenylhydrazine derivatives of lipid peroxidation aldehydes by photodiode array detection. J. Chromatogr. 798:117–123.

Cordis, G.A., Maulik, G., Bagchi, D., Riedel, W., and Das, D.K. 1998b. Detection of oxidative DNA damage to ischemic reperfused rat hearts by 8-hydroxydeoxyguanosine formation. J. Mol. Cell. Cardiol. 30:1939–1944.

Daily, D., Vlamis-Gardikas, A., Offen, D., Mittelman, L., Melamed, E., Holmgren, A., and Barzilai, A. 2001a. Glutaredoxin protects cerebellar granule neurons from dopamine-induced apoptosis by dual activation of the ras-phosphoinositide 3-kinase and jun N-terminal kinase pathways. J. Biol. Chem. 276:21618–21626.

Daily, D., Vlamis-Gardikas, A., Offen, D., Mittelman, L., Melamed, E., Holmgren, A., and Barzilai, A. 2001. Glutaredoxin protects cerebellar granule neurons from dopamine-induced apoptosis by activating NF-kappa B via Ref-1. J. Biol. Chem. 276:1335–1344.

Das, D.K. 2004. Thioredoxin regulation of ischemic preconditioning. Antioxid. Redox Signal. 6:405–412.

Das, D.K., and Maulik, N. 2003. Preconditioning potentiates redox signaling and converts death signal into survival signal. Arch. Biochem. Biophys. 420:305–311.

Das, D.K., and Maulik, N. 2005. Mitochondrial function in cardiomyocytes: target for cardioprotection. Curr. Opin. Anaesthesiol. 18:77–82.

Das, D.K., and Maulik, N. 2006. Cardiac genomic response following preconditioning stimulus. Cardiovasc. Res. 70:254–263.

Das, D.K., Maulik, N., and Moraru, I.I. 1995. Gene expression in acute myocardial stress. Induction by hypoxia, ischemia, hyperthermia and oxidative stress. J. Mol. Cell. Cardiol. 27:181–193,

Das, D.K., Maulik, N., Sato, M., and Ray, P.S. 1999. Reactive oxygen species function as second messenger during ischemic preconditioning of heart. Mol. Cell. Biochem. 196:59–67.

Das, K.C., Lewis-Molock, Y., and White, C.W. 1997. Elevation of manganese superoxide dismutase gene expression by thioredoxin. Am. J. Respir. Cell. Mol. Biol. 17:713–726.

Das, S., Otani, H., Maulik, N., and Das, D.K. 2006. Redox regulation of angiotensin II preconditioning of the myocardium requires MAP kinase signaling. J. Mol. Cell. Cardiol. 41:248–255.

de Haan, J.B., Witting, P.K., Stefanovic, N., Pete, J., Daskalakis, M., Kola, I., Stocker, R., and Smolich, J.J. 2006. Lack of the antioxidant glutathione peroxidase-1 does not increase atherosclerosis in C57BL/J6 mice fed a high-fat diet. J. Lipid Res. 47:1157–1167.

Di Lisa, F., and Bernardi, P. 2006. Mitochondria and ischemia-reperfusion injury of the heart: fixing a hole. Cardiovasc. Res. 70:191–199.

Fernandes, A.P., and Holmgren, A. 2004. Glutaredoxins: glutathione-dependent redox enzymes with functions far beyond a simple thioredoxin backup system. Antioxid. Redox Signal. 6:63–74.

Frangogiannis, N.G. 2006. The mechanistic basis of infarct healing. Antioxid, Redox Signal. 8:1907–1939.

Furman, C., Rundlof, A.K., Larigauderie, G., Jaye, M., Bricca, G., Copin, C., Kandoussi, A.M., Fruchart, J.C., Arner, E.S., and Rouis, M. 2004. Thioredoxin reductase 1 is upregulated in atherosclerotic plaques: specific induction of the promoter in human macrophages by oxidized low-density lipoproteins. Free Radical Biol. Med. 37:71–85.

Gutierrez, A.M., Reboredo, G.R., Mosca, S.M., and Catala, A. 2006. A low degree of fatty acid unsaturation leads to high resistance to lipid peroxidation in mitochondria and microsomes of different organs of quail (Coturnix coturnix japonica). Mol. Cell. Biochem. 282:109–115.

Haendeler, J., Tischler, V., Hoffmann, J., Zeiher, A.M., and Dimmeler, S. 2004. Low doses of reactive oxygen species protect endothelial cells from apoptosis by increasing thioredoxin-1 expression. FEBS Lett. 577:427–433.

Hagg, D., Englund, M.C., Jernas, M., Schmidt, C., Wiklund, O., Hulten, L.M., Ohlsson, B.G., Carlsson, L.M., Carlsson, B., and Svensson, P.A. 2006. Oxidized LDL induces a coordinated up-regulation of the glutathione and thioredoxin systems in human macrophages. Atherosclerosis 185:282–289.

Hangaishi, M., Ishizaka, N., Aizawa, T., Kurihara, Y., Taguchi, J., Nagai, R., Kimura, S., and Ohno, M. 2000. Induction of heme oxygenase-1 can act protectively against cardiac ischemia/reperfusion *in vivo*. Biochem. Biophys. Res. Commun. 279:582–588.

Ho, C.Y., and Seidman, C.E. 2006. A contemporary approach to hypertrophic cardio-myopathy. Circulation 113:e858–e862.

Holmgren, A. 1985. Thioredoxin. Annu. Rev. Biochem. 54:237–271.

Holmgren, A. 1989. Thioredoxin and glutaredoxin systems. J. Biol. Chem. 264:13963–13966.

Holmgren, A., and Bjornstedt, M. 1995. Thioredoxin and thioredoxin reductase. Methods Enzymol. 252:199–208.

Holmgren, A., and Luthman, M. 1978. Tissue distribution and subcellular localization of bovine thioredoxin determined by radioimmunoassay. Biochemistry 17:4071–4077.

Honda, H.M., Korge, P., and Weiss, J.N. 2005. Mitochondria and ischemia/reperfusion injury. Ann. N.Y. Acad. Sci. 1047:248–258.

Huang, C.H., Vatner, S.F., Peppas, A.P., Yang, G., and Kudej, R.K. 2003. Cardiac nerves affect myocardial stunning through reactive oxygen and nitric oxide mechanisms. Circ. Res. 93:866–873.

Isowa, N., Yoshimura, T., Kosaka, S., Liu, M., Hitomi, S., Yodoi, J., and Wada, H. 2000. Human thioredoxin attenuates hypoxia-reoxygenation injury of murine endothelial cells in a thiol-free condition. J. Cell. Physiol. 182:33–40.

Jekell, A., Hossain, A., Alehagen, U., Dahlstrom, U., and Rosen, A. 2004. Elevated circulating levels of thioredoxin and stress in chronic heart failure. Eur. J. Heart Fail. 6:883–890.

Kaarteenaho-Wiik, R., Saily, M., Sormunen, R., Paakko, P., Holmgren, A., Soini, Y., and Kinnula, V.L. 2004. Expression of glutaredoxin is highly cell specific in human lung and is decreased by transforming growth factor-beta *in vitro* and in interstitial lung diseases *in vivo*. Hum. Pathol. 35:1000–1007.

Kakisaka, Y., Nakashima, T., Sumida, Y., Yoh, T., Nakamura, H., Yodoi, J., and Senmaru, H. 2002. Elevation of serum thioredoxin levels in patients with type 2 diabetes. Horm. Metab. Res. 34:160–164.

Kevin, L.G., Novalija, E., and Stowe, D.F. 2005. Reactive oxygen species as mediators of cardiac injury and protection: the relevance to anesthesia practice. Anesth. Analgesia 101:1275–1287.

Kihlstrom, M. 1990. Protection effect of endurance training against reoxygenation-induced injuries in rat heart. J. Appl. Physiol. 68:1672–1688.

Kim, Y., Lee, S.H., and Ho, W.K. 2007. Hydrogen peroxide selectively increases TREK-2 currents via myosin light chain kinases. Front. Biosci. 12:1642–1650.

Kondo, N., Nakamura, H., Masutani, H., and Yodoi, J. 2006. Redox regulation of human thioredoxin network. Antioxid. Redox Signal. 8:1881–1890.

Koneru, S., Penumathsa, S., Thirunavukkarasum, M., Samuel, S.M., Zhan, L., Han, Z., Maulik, G., Das, D.K., and Maulik, N. 2007. Redox regulation of ischemic preconditioning is mediated by the differential activation of caveolins and their association with Enos and Glut-4. Am. J. Physiol. Heart Circ. Physiol. 292:H2060–H2072.

Kozlovski, V.I., Olszanecki, R., and Chlopicki, S. 2006. Free radicals generated by xanthine/xanthine oxidase system augment nitric oxide synthase (NOS) and cyclooxygenase (COX)-independent component of bradykinin-induced vasodilatation in the isolated guinea pig heart. Pharmacol. Rep. 58:405–412.

Kujoth, G.C., Hiona, A., Pugh, T.D., Someya, S., Panzer, K., Wohlgemuth, S.E., Hofer, T., Seo, A.Y., Sullivan, R., Jobling, W.A., Morrow, J.D., Van Remmen, H., Sedivy, J.M., Yamasoba, T., Tanokura, M., Weindruch, R., Leeuwenburgh, C., and Prolla, T.A. 2005. Mitochondrial DNA mutations, oxidative stress, and apoptosis in mammalian aging. Science 309:481–484.

Kuster, G.M., Siwik, D.A., Pimentel, D.R., and Colucci, W.S. 2006. Role of reversible, thioredoxin-sensitive oxidative protein modifications in cardiac myocytes. Antioxid. Redox Signal. 8:2153–2159.

Laderoute, K.R., and Webster, K.A. 1997. Hypoxia/reoxygenation stimulates Jun kinase activity through redox signaling in cardiac myocytes. Circ. Res. 80:336–344.

Lantos, J., Temes, G., Gobolos, L., Jaberansari, M.T., and Roth, E. 2001. Is peripheral blood a reliable indicator of acute oxidative stress following heart ischemia and reperfusion? Med. Sci. Monitor. 7:1166–1170.

Lecour, S., Baouali, A.B., Maupoil, V., Chahine, R., Abadie, C., Javouhey-Donzel, A., Rochette, L., and Nadeau, R. 1998. Demonstration of the production of oxygen-centered free radicals during electrolysis using E.S.R. spin-trapping techniques: effects on cardiac function in the isolated rat heart. Free Radical Biol. Med. 24:573–579.

Li, X., Xu, Z., Li, S., and Rozanski, G.J., 2005. Redox regulation of Ito remodeling in diabetic rat heart. Am. J. Physiol. Heart Circ. Physiol. 288:H1417–H1424.

Liang, M., and Pietrusz, J.L., 2007. Thiol-related genes in diabetic complications: a novel protective role for endogenous thioredoxin 2. Arterioscler. Thromb. Vasc. Biol. 27:77–83.

Lillig, C.H., and Holmgren, A. 2007. Thioredoxin and related molecules—from biology to health and disease. Antioxid. Redox Signal. 9:25–47.

Lillig, C.H., Berndt, C., Vergnolle, O., Lonn, M.E., Hudemann, C., Bill, E., and Holmgren, A. 2005. Characterization of human glutaredoxin 2 as iron-sulfur protein: a possible role as redox sensor. Proc. Natl. Acad. Sci. USA 10:8168–8173.

Lovell, M.A., Chengsong, X., Gabbita, S.P., and Markesbery, W.R. 2000. Decreased thioredoxin and increased thioredoxin reductase levels in Alzheimer's disease brain. Free Radical Biol. Med. 28:418–427.

Lundberg, M., Johansson, C., Chandra, J., Enoksson, M., Jacobsson, G., Ljung, J., Johansson, M., and Holmgren, A. 2001. Cloning and expression of a novel human glutaredoxin (Grx2) with mitochondrial and nuclear isoforms. J. Biol. Chem. 276:26269–26275.

Malik, G., Gorbounov, N., Das, S., Gurusamy, N., Otani, H., Maulik, N., Goswami, S., and Das, D.K. 2006. Ischemic preconditioning triggers nuclear translocation of thioredoxin and its interaction with Ref-1 potentiating a survival signal through the PI-3-kinase-Akt pathway. Antioxid. Redox Signal. 8:2101–2109.

Matsushima, S., Ide, T., Yamato, M., Matsusaka, H., Hattori, F., Ikeuchi, M., Kubota, T., Sunagawa, K., Hasegawa, Y., Kurihara, T., Oikawa, S., Kinugawa, S., and Tsutsui, H. 2006. Overexpression of mitochondrial peroxiredoxin-3 prevents left ventricular remodeling and failure after myocardial infarction in mice. Circulation 113:1779–1786.

Maulik, N., Yoshida, T., Zu, Y.L., Sato, M., Banerjee, A., and Das, D.K. 1998. Ischemic preconditioning triggers tyrosine kinase signaling: A potential role for MAPKAP kinase 2. Am. J. Physiol. 275:H1857–H1864.

Meyskens, F.L., Jr., Farmer, P.J., and Anton-Culver, H. 2004. Etiologic pathogenesis of melanoma. A unifying hypothesis for the missing attributable risk. Clin. Cancer Res. 10:2581–2583.

Mitsui, A., Hirakawa, T., and Yodoi, J. 1992. Reactive oxygen-reducing and protein-refolding activities of adult T cell leukomia-derived factor/human thioredoxin. Biochem. Biophys. Res. Commun. 186:1220–1226.

Miwa, K., Kishimoto, C., Nakamura, H., Makita, T., Ishii, K., Okuda, N., Yodoi, J., and Sasayama, S. 2005. Serum thioredoxin and alpha-tocopherol concentrations in patients with major risk factors. Circ. J. 69:291–294.

Miyamoto, M., Kishimoto, C., Shioji, K., Lee, J.D., Shimizu, H., Ueda, T., and Yodoi, J. 2003. Cutaneous arteriolar thioredoxin expression in patients with heart failure. Circ. J. 67:116–118.

Miyamoto, S., Kawano, H., Hokamaki, J., Soejima, H., Kojima, S., Kudoh. T., Nagayoshi, Y., Sugiyama, S., Sakamoto, T., Yoshimura, M., Nakamura, H., Yodoi, J., and Ogawa, H. 2005. Increased plasma levels of thioredoxin in patients with glucose intolerance. Intern. Med. 44:1127–1132.

Nakamura, H., Derosa, S., Roedere, M., Anderson, M.T., and Dubs, J.G. 1996. Elevation of plasma thioredoxin levels in HIV infected individuals. Int. Immunol. 8:603–611.

Nakamura, H., Nakamura, K., and Yodoi, J. 1997. Redox regulation of cellular activation. Annu. Rev. Immunol. 15:351–369.

Nakamura, H., Vaage, J., Valen, G., Padilla, C.A., Bjornstedt, M., and Holmgren, A. 1998. Measurements of plasma glutaredoxin and thioredoxin in healthy volunteers and during open-heart surgery. Free. Radical Biol. Med. 24:1176–1186.

Nimata, M., Kishimito, C., Shioji, K., Ishizaki, K., Kitaguchi, S., Hashimoto, T., Nagata, N., and Kawai, C. 2003. Upregulation of redox-regulating protein, thioredoxin, in endomyocardial biopsy samples of patients with myocarditis and cardiomyopathies. Mol. Cell. Biochem. 248:193–196.

Nishida, K., Miyazawa, Y., Hatano, M., Suzuki, K., Hirose, A., Fukushima, R., Okinaga, K. J. 1998 Reperfusion induces sublethal endothelial injury. Surg Res 79:85–90.

Nishida, M., Maruyama, Y., Tanaka, R., Kontani, K., Nagao, T., and Kurose, H.G. 2000. Alpha(i) and G alpha(o) are target proteins of reactive oxygen species. Nature 408:492–495.

Ohira, A., Honda, O., Gauntt, C.D., Yamamoto, M., Hori, K., Masutani, H., Yodoi, J., and Honda, Y. 1994. Oxidative stress induces adult T cell leukemia derived factor/thioredoxin in the rat retina. Lab. Invest. 70:279–285.

Petrosillo, G., Di Venosa, N., Ruggiero, F.M., Pistolese, M., D'Agostino, D., Tiravanti, E., Fiore, T., and Paradies, G. 2005. Mitochondrial dysfunction associated with cardiac ischemia/reperfusion can be attenuated by oxygen tension control. Role of oxygen-free radicals and cardiolipin. Biochim. Biophys. Acta 1710:78–86.

Piccoli, C., D'Aprile, A., Ripoli, M., Scrima, R., Lecce, L., Boffoli, D., Tabilio, A., and Capitanio, N. 2007. Bone-marrow derived hematopoietic stem/progenitor cells express multiple isoforms of NADPH oxidase and produce constitutively reactive oxygen species. Biochem. Biophys. Res. Commun. 353:965–972.

Pimentel, D.R., Adachi, T., Idom, Y., Heibeck, T., Jiang, B., Lee, Y., Melendez, J.A., Cohen, R.A., and Colucci, W.S. 2006. Strain-stimulated hypertrophy in cardiac myocytes is mediated by reactive oxygen species-dependent Ras S-glutathiolation. J. Mol. Cell. Cardiol. 41:613–622.

Powis, G., and Monfort, W.R. 2001. Properties and biological activities of thioredoxins. Annu. Rev. Pharmacol. Toxicol. 41:261–295.

Proctor, P.H., Kirpatrick, D.S., Morehead, L.A., Ginness, J.E., Hilton, J.G., and Hokanson, J.A. 1979. Role of active oxygen species in ocular and neurological diseases. Abstract, Conference on Active Oxygen and Medicine, Honolulu, Hawaii, March 3–4.

Sachi, Y., Hirota, K., Masutani, H., Toda, K., and Takahashi, O. 1995. Induction of ADF/TRX by oxidative stress in keratinocytes and lymphoid cells. Immunol. Lett. 44:189–193.

Sano, H., Sata, T., Nanri, H., Ikeda, M., and Shigematsu, A. 2002. Thioredoxin is associated with endotoxin tolerance in mice. Crit. Care Med. 30:190–194.

Sato, M., Cordis, G.A., Maulik, N., and Das, D.K. 2000. SAPKs regulation of ischemic preconditioning. Am. J. Physiol. 279:H901–H907.

Shinmura, K., Xuan, Y.T., Tang, X.L., Kodani, E., Han, H., Zhu, Y., and Bolli, R. 2002. Inducible nitric oxide synthase modulates cyclooxygenase-2 activity in the heart of conscious rabbits during the late phase of ischemic preconditioning. Circ. Res. 90:602–608.

Shioji, K., Kishimoto, C., Nakamura, H., Masutani, H., Yuan, Z., Oka, S., and Yodoi, J. 2002. Overexpression of thioredoxin-1 in transgenic mice attenuates adriamycin-induced cardiotoxicity. Circulation 106:1403–1409.

Shioji, K., Nakamura, H., Masutani, H., and Yodoi, J. 2003. Redox regulation by thioredoxin in cardiovascular diseases. Antioxid. Redox Signal. 5:795–802.

Skibska, B., Jozefowicz-Okonkwo, G., and Goraca, A. 2006. Protective effects of early administration of alpha-lipoic acid against lipopolysaccharide-induced plasma lipid peroxidation. Pharmacol. Rep. 58:399–404.

Takagi, S., Bhat, G.B., Hummel, B.C., and Walfish, P.G. 1989. Thioredoxin and glutaredoxin enhance the binding of L-triiodothyronine to its hepatic nuclear receptors. Biochem. Cell Biol. 67:477–480.

Takagi, Y., Mitusi, A., Nishiyama, A., Nozaki, K., and Sono, H. 1999. Overexpression of thioredoxin in transgenic mice attenuates focal ischemic brain damage. Proc. Natl. Acad. Sci. USA 96:4131–4136.

Takimoto, E., and Kass, D.A. 2007. Role of oxidative stress in cardiac hypertrophy and remodeling. Hypertension 49:241–248.

Tanaka, T., Nishiyama, Y., Okada, K., Hirota, K., Matsui, M., Yodoi, J., Hiai, H., and Toyokuni, S. 1997. Induction and nuclear translocation of thioredoxin by oxidative damage in the mouse kidney: independence of tubular necrosis and sulfhydryl depletion. Lab. Invest. 77:145–155.

Tanguy, S., Boucher, F.R., Malfroy, B., and de Leiris, J.G. 1996. Free radicals in reperfusion-induced arrhythmias: study with EUK 8, a novel nonprotein catalytic antioxidant. Free. Radical Biol. Med. 21:945–954.

Tanito, M., Nakamura, H., Kwon, Y.W., Teratani, A., Masutani, H., Shioji, K., Kishimoto, C., Ohira, A., Horie, R., and Yodoi, J. 2004. Enhanced oxidative stress and impaired thioredoxin expression in spontaneously hypertensive rats. Antioxid. Redox Signal. 6:89–97.

Tao, L., Jiao, X., Gao, E., Lau, W.B., Yuan, Y., Lopez, B., Christopher, T., Ramachandra, Rao, S.P., Williams, W., Southan, G., Sharma, K., Koch, W., and Ma, X.L. 2006a. Nitrative inactivation of thioredoxin-1 and its role in postischemic myocardial apoptosis. Circulation 114:1395–1402.

Tao, L., Gao, E., Hu, A., Coletti, C., Wang, Y., Christopher, T.A., Lopez, B.L., Koch, W., and Ma, X.L. 2006b. Thioredoxin reduces post-ischemic myocardial apoptosis by reducing oxidative/nitrative stress. Br. J. Pharmacol, 149:311–318.

Tortolani, A.J., Powell, S.R., Misik, V., Weglicki, W.B., Pogo, G.J., and Kramer, J.H. 1993. Detection of alkoxyl and carbon-centered free radicals in coronary sinus blood from patients undergoing elective cardioplegia. Free Radical Biol. Med. 14:421–426.

Tosaki, A., Bagchi, D., Pali, T., Cordis, G.A., and Das, D.K. 1993. Comparison of ESR and HPLC methods for the detection of OH radicals in ischemic reperfused hearts. Biochem. Pharmacol. 45:961–969.

Tsutsui, H., Ide, T., and Kinugawa, S. 2006. Mitochondrial oxidative stress, DNA damage, and heart failure. Antioxid. Redox Signal. 8:1737–1744.

T., Chang, V.W.S., Engelman, R.M., Maulik, N., Ho, Y.S., and Das, D.K. 2003a. Thioredoxin redox signaling in the ischemic heart: an insight with transgenic mice overexpressing Trx1. J. Mol. Cell. Cardiol. 35:695–704.

Turoczi, T., Jun, L., Cordis, G., Morris. J.E., Maulik, N., Stevens, R.G., and Das, D.K. 2003b. HFE mutation and dietary iron content interact to increase ischemia/reperfusion injury of the heart in mice. Circ. Res. 92:1240–1246.

Valen, G., Semb, A.G., and Vaage, J. 1993. Inhibition of lipoxygenase and cyclooxygenase augments cardiac injury by H2O2. Free Radical Biol. Med. 15:27–35.

Valen, G., Starkopf, J., Takeshima, S., Kullisaar, T., Vihalemm, T., Kengsepp, A.T., Lowbeer, C., Vaage, J., and Zilmer, M., 1998. Preconditioning with hydrogen peroxide (H2O2) or ischemia in H2O2-induced cardiac dysfunction. Free Radical Res. 29:235–245.

Washburn, M.P., and Wells, W.W. 1999. Identification of the dehydroascorbic acid reductase and thioltransferase (glutaredoxin) activities of bovine erythrocyte glutathione peroxidase. Biochem. Biophys. Res. Commun. 257:567–571.

Yamagata, K., Tagami, M., Ikeda, K., Yamori, Y., and Nara, Y. 2000. Altered gene expressions during hypoxia and reoxygenation in cortical neurons isolated from stroke-prone spontaneously hypertensive rats. Neurosci. Lett. 284:131–134.

Yamamoto, M., Yang, G., Hong, C., Liu, J., Holle. E., Yu, X., Wagner, T., Vatner, S.F., and Sadoshima, J. 2003. Inhibition of endogenous thioredoxin in the heart increases oxidative stress and cardiac hypertrophy. J. Clin. Invest.112:1395–1406.

Yang, J., Marden, J.J., Fan, C., Sanlioglu, S., Weiss, R.M., and Ritchie, T.C. 2003. Genetic redox preconditioning differentially modulates AP-1 and NF kappa B responses following cardiac ischemia/reperfusion injury and protects against necrosis and apoptosis. Mol. Ther. J. Am. Soc. Gene Ther. 7:341–353.

Yokomise, H., Fukuse, T., Hirata, T., Olikubo, K., and Go, T. 1996. Effect of recombinant human adult T cell leukomia-derived factor on rat lung reperfusion injury. Respiration 61:99–104.

Yoshioka, J., Schreiter, E.R., and Lee, R.T. 2006. Role of thioredoxin in cell growth through interactions with signaling molecules. Antioxid. Redox Signal. 8:2143–2151.

Yoshitake, S., Nanri, H., Fernando, M.R., and Minakami, S. 1994. Possible differences in the regenerative roles played by thioltransferase and thioredoxin for oxidatively damaged proteins. J. Biochem. (Tokyo) 116:42–46.

Zhang, P., Liu, B., Kang, S.W., Seo, M.S., Rhee, S.G., and Obeid, L.M. 1997. Thioredoxin peroxidase is a novel inhibitor of apoptosis with a mechanism distinct from that of Bcl-2. J. Biol. Chem. 272:30615–30618.

Zhang, S., Yang, J.H., Yu, F., Zhao, J., Jiang, P., Chang, L., Tang, C., and Xu, J. 2006. Protective role of 3-nitro-N-methyl-salicylamide on isolated rat heart during 4 hours of cold storage and reperfusion. Transplant. Proc. 38:1247–1252.

Zhou, J., Damdimopoulos, A.E., Spyrou, G., and Brune, B. 2007. Thioredoxin 1 and thioredoxin 2 have opposed regulatory functions on hypoxia inducible factor-1alpha. J. Biol. Chem. 282:7482–7490.

Zorov, D.B., Juhaszova, M., and Sollott, S.J. 2006. Mitochondrial ROS-induced ROS release: an update and review. Biochim. Biophys. Acta 1757:509–517.

Chapter 8
Relationship Between Redox Regulation and β-Adrenergic Responses in the Heart

Belma Turan

Abstract Catecholamines have physiologically important effects on the performance of the heart through the activation of adrenergic receptors. In general, it is known that sympathetic nervous system activation modulates the signaling pathway that controls excitation–contraction coupling (ECC) in the heart. Coordinated myocyte handling of Ca^{2+} is essential for efficient ECC in the heart. A growing body of knowledge on cardiac β-adrenergic receptor (β-AR) signal transduction demonstrates that the agonist-bound β-AR selectively interacts with the stimulatory G protein (G_S), which activates adenylyl cyclase (AC), catalyzing cAMP formation. Subsequently, activation of cAMP-dependent protein kinase A (PKA) leads to phosphorylation of regulatory proteins involved in cardiac ECC and energy metabolism. Published data have shown that the altered cardiac responses in pathological conditions are closely related to the function of the β-AR system. From the current literature it is clear that the β-AR system and its importance in regulating cardiac function under both physiological and pathophysiological situations has attracted the attention of many investigators. Ca^{2+} functions as a critical second messenger in mediating fast intracellular responses in all tissues through signaling proteins to coordinate cell function with different intracellular mechanisms. In addition, the identification of oxidatively sensitive proteins that modulate intracellular signaling mechanisms and the associated generation of reactive oxygen species (ROS) are critical to understanding how cells respond to oxidative stress. Therefore, any disturbance in the intracellular ionic homeostasis due to the excess ROS, was shown to result in the occurrence of impaired cardiac contractile activity. Since β-ARs and AC are known to participate in the regulation of cardiac function, it is possible that the β-AR-linked signal transduction pathway is also affected by ROS.

Introduction

Cardiovascular disease is the leading cause of death of most humans. However, differences in age, heart size, physiological status, and other factors confound

A. K. Srivastava, M. B. Anand-Srivastava (eds.) *Advances in Biochemistry in Health and Disease.* Signal Transduction in the Cardiovascular System in Health and Disease,

comparisons leading to variable and conflicting conclusions (see review by Chu et al. 2005). Myocellular Ca^{2+} cycling is a primary determinant of normal cardiac contractile function and any abnormality in this process is likely associated with cardiac dysfunction and heart disease (Marks 2001; Chu et al. 2005; Wold et al. 2006). Furthermore, it is well known that functional changes of heart have been associated with altered expression and/or activity of Ca^{2+} handling proteins (Choi et al. 2002; Eisner et al. 1998; Wehrens and Marks 2003; Yaras et al. 2005).

Contraction and relaxation of heart muscle cells are regulated by Ca^{2+} cycling between the cytoplasm and sarcoplasmic reticulum (SR). Therefore, myocyte mishandling of Ca^{2+} is a central cause of both contractile dysfunction and arrhythmias under pathological conditions (Pogwizd et al. 2001). β-Adrenergic stimulation is a major physiological mechanism to meet increases in circulatory demand, acting through positive inotropic and lusitropic effects. β-Adrenergic signaling and myocellular Ca^{2+} homeostasis are importantly linked because β-adrenergic receptor (β-AR) stimulation regulates the activity and expression of Ca^{2+} handling proteins. Although the β-adrenergic signal transduction pathway in the heart is known to participate in regulating cardiac performance by identifying, amplifying, and transmitting the catecholamine-initiated signals, this system is impaired under both chronic and acute pathological conditions (see review by Dhalla et al. 1997).

A large number of studies have demonstrated the role of reactive oxygen species (ROS) in the pathogenesis of the cardiovascular diseases. The risk factors for cardiovascular disease can widely depend on many factors such as contents of daily diets and/or environmental conditions of the individuals. Most recently, these factors became important in the early prevention of cardiovascular diseases. Oxidative stress, the imbalance between ROS production and breakdown by endogenous antioxidants, has been implicated in the onset and progression of cardiovascular diseases. Antioxidant therapy has shown promise in preventing the development of several different heart diseases. Thus, this article will attempt to explain the relationship between redox regulation and β-adrenergic responses in heart under oxidative stress.

Heart and β-Adrenergic System

It is well documented that coordinated myocyte handling of Ca^{2+} is essential for efficient excitation–contraction coupling in the heart. Since cardiac pump is able to alter its function in response to any requirement in the body and the regulation of contractile function of individual myocytes obtained by modulation of intracellular Ca^{2+} signaling, the characteristics of regulation induced by adrenergic stimulations are very important for maintaining the normal heart function in humans.

The sympathetic nervous system plays a central role in regulating heart function and response to most types of stress through β-AR stimulation. Binding of β-adrenergic agonists to receptors in the heart activates adenylyl cyclase (AC) via a stimulatory G protein. On the other hand, it is known that the altered responses of the diseased heart to catecholamines are associated with defects in one or more components of this signal transduction system which include β-ARs, stimulatory and inhibitory guanine nucleotide-binding proteins (G_s- and G_i-proteins), and AC (Das et al. 1997). Published data have shown that the altered response of β-adrenoceptor-linked pathway in pathological conditions is closely related to the expression and activity of β-adrenergic receptor kinase (βARK) (Iaccarino et al. 1999). On the other hand, it is recognized that the β-adrenergic system in chronic pathological conditions is desensitized (Bristow et al. 1982) which is associated with downregulation of β-ARs (Bohm 1995; Kaura et al. 1996), increased activity and expression of βARK (Bohm and Lohse 1994), reduced efficiency of β-ARs coupled to G-proteins (Kaura et al. 1996), decreased G_s-protein content and/or increased G_i-protein content in the myocardium (Kaura et al. 1996). Moreover, the β-adrenergic system is generally sensitized under acute pathological conditions, which is characterized by increased density of β-ARs (Strasser et al. 1990) and impaired G_i-protein function (Hammond et al. 1993).

From the current literature it is clear that the β-AR system has attracted the attention of many investigators to establish its importance in regulating cardiac function under both physiological and pathophysiological situations (Homcy et al. 1991). Several studies have been carried out in the human heart, indicating some important features of ARs (Brodde et al. 1992; Bristow et al. 1982). β-AR stimulation is a physiological mechanism to cause increases in circulation through improving cardiac performance. β-AR-mediated activation of AC, and subsequent elevation of intracellular cAMP, stimulates PKA that phosphorylates G_s-protein-coupled receptors. This kinase, in advance, provides a negative feedback regulation of β-AR activity (Hausdorff et al. 1990).

The sympathetic nervous system is central for the neurohumoral regulation of the cardiovascular system and largely involved in many cardiovascular diseases. During the 1980s, the classification of β-ARs included two subtypes (β_1 and β_2) (Lands et al. 1967). Thus, it is now known that three different subtypes, β_1-, β_2-, and β_3-AR subtypes, could at least participate in the regulation of cardiovascular function (see review by Rozec and Gauthier 2006). Cardiac electrophysiological effects of β_3-AR stimulation have not been investigated extensively yet.

In human ventricle, the activation of G_i proteins did not appear to cause an inhibition of AC, but rather an activation of ROS such as nitric oxide (NO) pathway, which is expressed both in ventricular cardiomyocytes and in endothelial cells (Schulz et al. 2006). It has been demonstrated that the NO production produces an activation of soluble guanylyl cyclase leading to an increase in intracellular cyclic guanosine monophosphate (cGMP) in rat and canine cardiomyocytes which causes significant decrease in L-type Ca^{2+}-current through activating cGMP-dependent protein kinase or downregulates the contractile response of myofilament proteins independently of changes in Ca^{2+}-transients (Zhang et al. 2005, Cheng et al. 2001). Alternativley, it will be appropriate to give a general definition as any oxidant, similar to NO, might regulate cardiac function in a cGMP-dependent and/or -independent manner through different intracellular pathways. Consistent with these ideas, we found that selenite, an oxyanion that reacts strongly with protein thiols (Björnstedt et al. 1995),

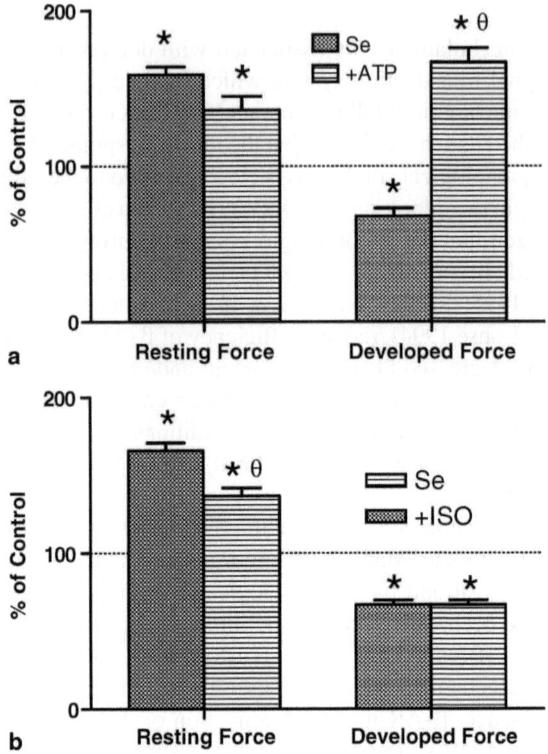

Fig. 8.1 Effects of ATP (0.1 mM) (a) and ISO (0.1 μM) (b) pretreatments (10 min) on selenite-induced (1 mM) contractures in papillary muscle preparations. The muscle strips were stimulated with rectangular electrical pulses at a rate of 0.1 Hz and resting and developed forces were recorded isometrically. The mean values (±SEM) represent the values at 25 min following selenite exposures. The values with selenite only exposure (Se) without and with ATP or ISO pretreatments (+) are given as % of the control values (without Se exposure). * $p < 0.05$ with respect to control, $^{\Theta}$ $p < 0.05$ with respect to selenite only exposure. (Adapted from Ugur and Turan, 2001.)

caused significant inhibition in both Na^+- and L-type Ca^{2+}-currents of rat ventricular cardiomyocytes when it is applied extracellularly with a concentration of 1 mM during a 5-min exposure and then a complete contracture in heart preparations (Ugur et al. 2002; Ugur and Turan 2001; Turan et al. 1996). These effects were reversed by the disulfide reducing agent dithiothreitol (DTT). Furthermore, we investigated the effects of two types of agonists, ATP and isoproterenol (ISO), on the selenite-induced contracture in the heart preparations (Figure 8.1). Our results demonstrated that ATP, the naturally occurring ligand for all P2-type receptors, seemed to delay the contracture when it was used as preexposure for 10 min (Ugur and Turan 2001). At the same time, we have shown that ISO had no effect on the selenite-induced contracture in these heart preparations. Different types of agonist having different effects in the same heart preparations provide reliable information indicating that different signaling pathways in the cell could have possible cross-talk under pathophysiological conditions. This hypothesis is further supported by published data on related subjects. For example, it is now well documented that ATP is an agonist for both P2Y (G-protein-coupled) and P2X (ligand-gated ion channels) types of receptors and P2Y1,2,6 are known to be expressed in adult rat cardiomyocytes (Webb et al. 1996). In addition, P2Y receptors are reported to stimulate cAMP production in cardiac myocytes (Puceat et al. 1998), which is also a case of β-ARs.

On the other hand, it has been shown that a variety of cardiac ion channels are regulated by the signaling cascade initiated by the activation of β_1-ARs (Hartzell 1988) and this type of regulation also involves the production of cAMP and subsequent phosphorylation by PKA. Recalling that the above signaling pathways together with the pathway on tyrosine kinases can modulate the activity of cardiac ion channels (Zhou et al. 1997), it is clear that there might be a cross-talk between tyrosine kinases, β_1-AR signaling, and regulation of cardiac function. Consistent with this idea, Hool et al. (1998) demonstrated that a tyrosine kinase inhibitor could increase the sensivity of cardiac ion channels to β_1-AR stimulation. Results of several other studies on this subject have suggested that the redox state of the cell is an important factor affecting β-adrenergic regulation of cardiac function (Sims and Harvey 2004; Turan et al. 1996).

Redox Regulation in the Heart

Contraction is initiated when a small amount of Ca^{2+} entering the cell following membrane depolarization, triggers larger release of Ca^{2+} from SR. On the other hand, it is not yet established that any alteration of Ca^{2+} signaling could be a main source of cardiac dysfunction such as cardiomyopathy (Choi et al. 2002). Ca^{2+} functions as a critical second messenger in mediating fast intracellular responses in all tissues through signaling proteins to coordinate cell function with different intracellular mechanisms. The identification of oxidatively sensitive proteins that modulate intracellular signaling mechanisms and the associated generation of ROS are critical to understanding how cells respond to oxidative stress. In another respect, oxidant-induced functional losses of the Ca^{2+} regulatory proteins contribute to the downregulation of cellular metabolism and the associated generation of ROS (see review by Wold et al. 2005). Reduction in the rate of ROS generation, in turn, will minimize protein oxidation and facilitate intracellular repair and degradative systems that function to eliminate damaged proteins (see review by Bigelow and Squier 2005). Shortly, oxidative stress, the imbalance between ROS production and breakdown by endogenous antioxidants, has been implicated in the onset and progression of cardiovascular diseases. Thus, any kind of antioxidant therapy has shown to be promising in preventing the development of heart complications (Quin et al. 2006; Nishizawa et al. 2004; Ye et al. 2003).

We and other researchers have demonstrated previously that the oxidation of cellular sulfhydryls can have both deleterious (Eley et al. 1991; Turan et al. 1996,

1997; Ayaz et al. 2005) and potentially beneficial (Tatsumi and Fliss 1994; Ayaz et al. 2004) effects in cardiac myocytes and endothelial cells. In an early study, Kaul et al. (1993) have pointed out the roles of ROS and their products in damaging and later causing dysfunction in the heart.

A large number of studies have led to the conclusion that ROS, acting as signal transducers, can regulate cellular function in a variety of organs including heart tissues (Das et al. 2004; Nishizawa et al. 2004). ARs are known to function as sensors for the neurotransmitter released from sympathetic nerve endings (Stiles and Lefkowitz 1984). The regulatory mechanisms related to the AR system have been studied extensively in the heart (Homcy et al. 1991). Alteration in the signal transduction pathway has been detected in many pathological conditions of the heart (Feldman 1993; Brodde et al. 1995). Dysfunction of cardiac contractility has been linked to the excess release and/or excess production of ROS which are accumulated in the myocardium. Thus, disturbances in the intracellular ionic homeostasis due to the excess ROS, have been shown to result in the occurrence of intracellular Ca^{2+} overload and the impairment of cardiac contractile force development (Eley et al. 1991; Turan et al. 1996). Since β-ARs and AC are known to participate in the regulation of cardiac function (Dhalla et al. 1982), it is possible that the β-AR-linked signal transduction pathway is also affected by ROS. It has been pointed out that β-AR-linked signal transduction mechanisms are impaired under pathological conditions of the heart (Persad et al. 1997). Although ROS, which are formed in the heart under pathologies, have been shown to alter characteristics of cardiac β-ARs and AC, the role of ROS in these pathologies is still under investigation. It is generally accepted that β-AR stimulation of heart muscle cells results in an increase in L-type Ca^{2+}-channel current. Binding of β-AR agonists to receptors in the heart activates AC via G_s. The resultant production of second messenger cAMP activates PKA and the kinase phosphorylates cellular proteins including the α_{1c} subunit of this type of Ca^{2+} channels (Gao et al. 1997). Therefore, it is highly likely that any type of involvement of oxidant stress in the heart will cause a defect in the β-AR system and/or β-AR-linked signal transduction mechanism. Although selenium compounds are known to play an antioxidant role in mammals, their effects are likely to be concentration and tissue dependent, and the effects of sodium selenite on heart function have yet to be characterized in detail. Experimental chronic selenium toxicity in animals affects the major organs; more particularly, sodium selenite has been shown to cause cellular dysfunction in tissues via forming covalent linkages with intracellular proteins (Lin-Shiau et al. 1989; Turan et al. 1996; Ugur et al. 2002). On the other hand, most pathology associated with selenium deficiency appears to be directly attributable to increased free-radical damage in tissues (Bettger 1993). It is almost clear that while low concentrations of selenium are essential for the synthesis of selenocysteine-containing enzymes (Koller and Exon 1986), an excess of selenium is also toxic to both animals and man and many selenium compounds are potent prooxidants and/or oxidants. So, the effects of selenium on functions of the muscle are controversial yet.

Considering the importance of sodium selenite as a supplement in deficiency and its possible specific effects in protein thiols, we investigated and compared the effects of dietary selenium on the mechanical activity and β-adrenergic responses of rat heart. Weanling rats of both sexes were fed a selenium-deficient (Se−) diet, a selenium-excess (Se+) diet, or a standard diet. The hearts of both the deficient group and the excess group rats revealed some significant alterations in contractile performance with increased heart rate and coronary perfusion pressure. Furthermore, the maximal increases in the peak contractile force of papillary muscle obtained with 100 nM isoproterenol (ISO) were 26% in the deficient group and 34% in the excess group when expressed as a percentage of the control (Turan et al. 1999). As can be seen from Figure 8.2, the maximal response was obtained with 1 μM ISO in the control group. However, the responses were somewhat different for the deficient and excess groups. The highest concentration of ISO was applied alone, with no cumulative doses to avoid desensitization. The mean values of EC_{50} of both the deficient and the excess groups calculated from the dose–response curves were not significantly different from the control group. On the other hand, the maximum responses to ISO in these two groups were significantly less than the controls (Figure 8.2).

To further investigate the effects of both the deficiency and excess of selenium on β-adrenergic responses in rat myocardium, in a later study, we focused on investigating molecular mechanisms that might underlie the changes in responsiveness of cardiac cells to β-adrenergic stimulation. We have shown that the amplitudes of L-type Ca^{2+}-currents recorded in isolated cardiomyocytes from rats fed with the deficient or the excess diet were not significantly different from those of the controls (Figure 8.3a) (Sayar et al. 2000). The average current–voltage relationships for peak I_{CaL} are given in Figure 8.3b. It can be seen that both deficiency and excess of selenium in the diet caused a significant shift in the peak amplitude of I_{CaL}. In both experimental groups, threshold potentials for activations were significantly lower than the controls, and potentials for maximum activations were higher than the controls (Figure 8.3c). Furthermore, two inactivation time constants (slow and fast) for these currents (estimated from the current induced by the voltage pulse from −70 to 0 mV for 200 ms) of these groups were significantly shorter than the control group (Sayar et al. 2000). In addition, the maximum current responses to ISO, when compared to the control cells, were significantly less for both experimental groups with no changes in EC_{50} values (Figure 8.4). These data are further supported by our previous observation related to the depressed maximum responses to ISO in the electrically stimulated papillary muscle preparations although the basal contractile activity of these preparations did not differ from the corresponding control (Turan et al. 1999). This observation could suggest that both Se deficiency and Se excess in rats did not interfere directly with the contractile machinery itself, but instead could affect the signal transduction pathway of the β-AR system including G-proteins, AC, PKA, and also Ca^{2+}-currents present in cardiac tissues and thus could depress the ISO responses.

To provide further explanation for the underlying mechanisms on the depressed β-adrenergic responses in both papillary muscle contractile activity and L-type Ca^{2+}-currents, we investigated the state of coupling between β-AR, AC, and the density of β-ARs in the hearts of these Se groups. Basal AC activity of the membranes from the deficient group was lower than the control group, but the activity

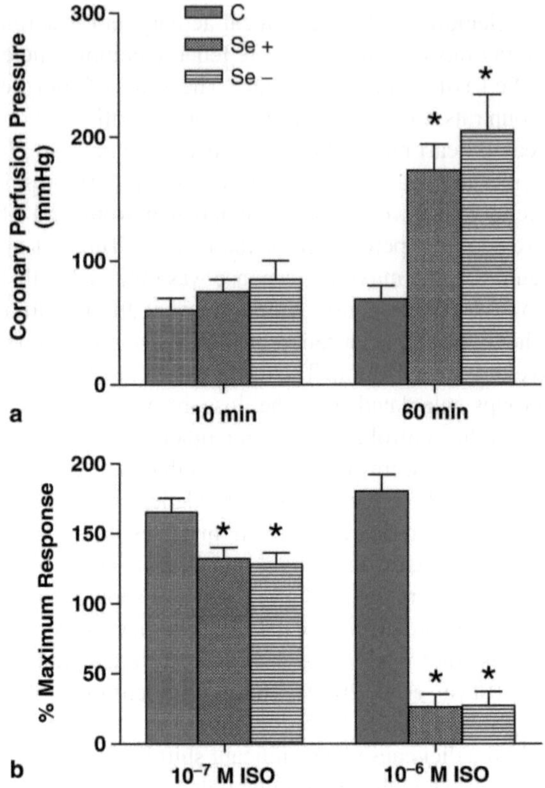

Fig. 8.2 Effects of dietary selenium on mechanical cardiac function. (a) The coronary perfusion pressures measured during the first 10 min were similar in the selenium-deficient (Se−), selenium-excess (Se+), and control group rat hearts in Langendorff system measurements. Later (during 60 min) these values increased significantly ($p < 0.05$) with respect to the control while the heart rates of both experimental groups increased from 200 ± 40 beats/min to 500 ± 40 beats/min. (b) Isoproterenol (ISO) responses in the papillary muscle preparations recorded under isometric conditions from Se− and Se+ group rats. The mean values of EC_{50} of both the deficient and the excess groups calculated from the dose–response curves were not significantly different from the control group. On the other hand, the maximum responses to ISO in these two groups were significantly less than the controls. (Adapted from Turan et al. 1999.)

of the excess group was not significantly different from the control. Likewise, the ISO- or forskolin-induced increase in AC activity was less in the deficient group but did not differ in the excess group with respect to the control (Sayar et al. 2000). Additionally, saturation binding experiments were performed in these membranes to determine total numbers of β-ARs. We also measured the efficiency of receptor-G-protein coupling in the same groups. Total numbers of β-ARs in cardiac tissues and basal or stimulated AC activities were less only for the deficient group compared to the control. However, in agonist competition experiments, GppNH induced a significant shift in agonist IC_{50} in the excess group. Furthermore, we measured G_s content of the membranes and found no significant differences between both

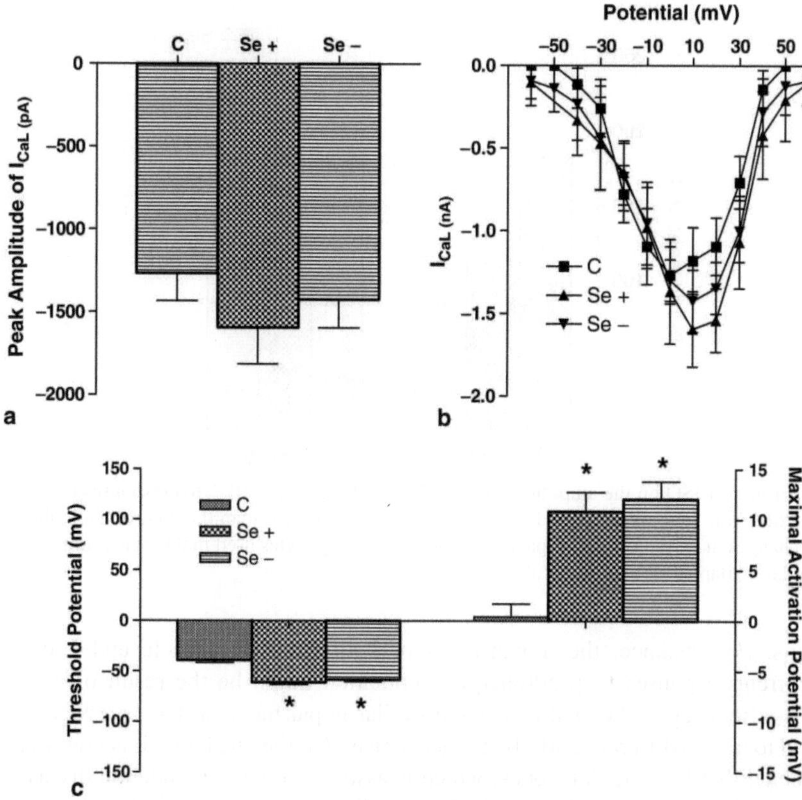

Fig. 8.3 Effects of dietary selenium on β-adrenergic responses in rat heart. (a) L-type Ca^{2+} currents (I_{CaL}) recorded from ventricular myocytes with depolarization from −70 mV to 0 mV, for 200 ms. Mean (±SEM) values of peak amplitudes of I_{CaL} in both experimental and control groups. The cell capacitances of these three groups of cardiomyocytes were similar. (b) Average current–voltage relationships for peak I_{CaL} (measured as the difference between the peak Ca^{2+} current and the end of 200-ms depolarization). The maximums of I_{CaL} of both experimental groups were shifted to the right with respect to the control. (c) The threshold potentials were significantly more negative and activation potentials were more positive in both experimental groups with respect to the control. (Adapted from Sayar et al. 2000.)

Se groups compared to the control. When we interpreted our electrophysiological results with biochemical data under known hypothesis related to the β-adrenergic signal transduction pathway and redox regulation of cells in these groups, it is seen that mostly they fit very well in our experimental animal models which have oxidant stress in their hearts (Turan et al. 1999). That is, in cardiac cells, β-adrenergic agonists increase cAMP production through stimulation of AC via G_s. cAMP which in turn activates PKA that modulates Ca^{2+} channels. Ca^{2+} channels and PKA can be the points, distal to the AC (Strasser et al. 1990).

The results of our studies indicated some common but also some different types of alteration in the signaling pathway of β-ARs in both Se-deficient and Se-excess

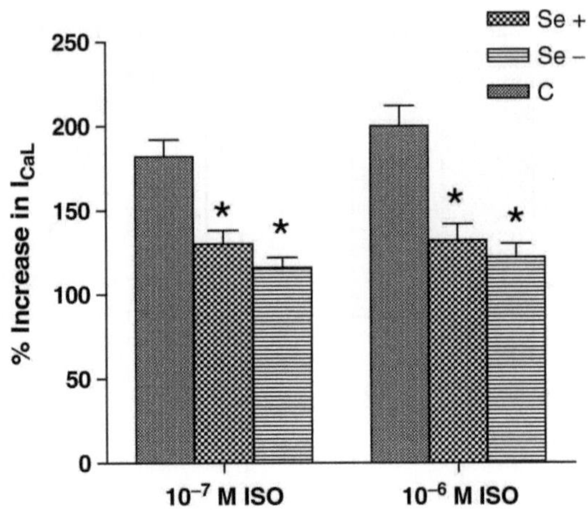

Fig. 8.4 Effects of ISO on the amplitude of I_{CaL}. The ISO (10^{-7} and 10^{-6} M) responses of peak I_{CaL} recorded from both experimental groups were significantly less than the control while there were no changes in EC_{50}. The cell capacitances of all cardiomyocytes used in these measurements were similar. (Adapted from Sayar et al. 2000.)

rat hearts. For instance, the impairments of both papillary muscle and L-type Ca^{2+}-current responses to β-adrenergic stimulation might be the result of a defect in β-adrenoceptor-AC pathway while similar impairments in the excess group appeared to have a different underlying mechanism. On the one hand, it is now well documented that both Se deficiency, which is associated with an endemic disease, known as Keshan disease (Li et al. 1985), as well as other type of cardiac diseases (Salonen et al. 1985), and Se toxicity, which affects the function of several major organs including heart via mainly reacting intracellularly with sulfhydryl groups of the proteins (Björnstedt et al. 1995; Handel et al. 1995), can cause an imbalance between prooxidants and antioxidants and consequently marked alterations in the redox state of the cells. On the other hand, altered redox potentials can also profoundly affect cell function by modifying the structure and function of proteins, in a common way of sulfhydryl modification of amino acid side chains (redox-mediated transduction mechanisms) (Oudit et al. 2001). Once oxidative stress occurs during any type of overproduction of ROS or deficits in antioxidant defense mechanisms, there is a shift in cell redox state, reflected by a decrease in the level of glutathione (GSH). Supporting this hypothesis, a group of researchers showed the important roles of cellular GSH level and its redox state in the regulation of heart function in diseased states, particularly related to coordinated activation of phosphatidylinositol 3-kinase (PI 3-kinase) and mitogen-activated protein kinase (MAPK) (Xu et al. 2002; Rozanski et al. 1998). Furthermore, two different groups have shown that selenium is a potent stimulator of tyrosyl phosphorylation and activator of MAPK and S6 kinase via affecting the overall phosphorylation state of the cell (Stapleton et al. 1997).

Overall, selenium is recognized as both an essential mammalian nutrient to maintain normal cardiac function and a hazardous element. Therefore, estimation of therapeutic and toxic doses of the element are very critical. Keeping in mind this important point, recently we have shown that sodium selenite administration (5 μmol/kg body weight/day) for 4 weeks to normal rats caused a slight but significant increase in blood glucose level, a significant decrease in plasma insulin level, and significant alterations in ionic mechanisms of the cardiomyocytes together with a significant shift in cell redox status and antioxidant defense system, even if there were no apparent deleterious changes in the heart function (Ayaz et al. 2005) and there were beneficial effects on both altered mechanical and electrical cardiac activities of diabetic rats (Ayaz et al. 2004). From these points, it can be concluded that there is a very delicate border between antioxidant/prooxidant levels of the supplements, which have important roles in the redox status of cardiomyocytes, when they are used as supplements in normal human diets.

Regulation of the Heart Under Oxidative Stress by Antioxidants

Oxidant-induced functional losses of Ca^{2+} regulatory mechanisms contribute to heart dysfunction in many diseases. The mechanisms of oxidatively sensitive intracellular targets that modulate Ca^{2+} homeostasis and the associated generation of ROS are critical to understanding how cells respond to oxidative stress. In this latter respect, oxidant-induced functional losses of Ca^{2+} regulatory proteins contribute to the downregulation of cellular metabolism and the associated generation of ROS (see recent reviews by Bigelow and Squier 2005, Endoh 2006). Reduction in the rate of ROS generation, in turn, will minimize protein oxidation and facilitate intracellular repair and degradative systems that function to eliminate damaged proteins (Bigelow and Squier 2005). Generally, it is now accepted that the general mechanisms underlying oxidant-dependent changes in cardiomyocyte function are likely to be relevant to diverse pathologies in the heart.

Cells and tissues are routinely subjected to sublethal doses of various oxidants, either exogenously or endogenously. An intense interest in research on antioxidants has been stimulated by the discovery of a new regulatory function of ROS, that is, as intracellular second messengers (Cruzado et al. 2005; see reviews by Das et al. 2004, and Gopalakrisna and Jaken 2000). There is considerable evidence that the resulting oxidative stress increases activity of signaling pathways including growth factor signal transduction, adrenergic signal transduction pathway. Many proteins are regulated by a redox environment through the reversible oxidation of their cysteine residues. The amino acid cysteine combines catalytical activity with an extensive redox chemistry and unique metal binding properties. Biologically, selenium can act both as antioxidants and as pro-oxidants (Ayaz et al. 2004, 2005: Turan et al. 2005). It is well documented that both selenium deficiency and selenium excess elicit oxidative stress (Turan et al. 1996, 1997, 1999; Sayar et al. 2000; Ugur et al. 2002). Oxidant exposure also increases the activity of a variety of protein kinases (Schieke et al. 1999). Moreover, oxidative stress has been shown to increase the activity of transcription factors such as AP-1 and NF-κB (Sun and Oberley 1996). Thus, both the oxidant-induced increase in kinase activity and redox regulation are important mechanisms for oxidant-induced changes in cellular responses to changes in redox status.

A consequence of the cellular damaging effect of oxidants is that antioxidants could have protective properties in the cells. Indeed, antioxidant micronutrients are one of the body's primary defenses against oxidants (Ayaz et al. 2004). Various chemical forms of selenium and vitamin E have been found to have beneficial effects in the heart under different pathological conditions (Ayaz and Turan 2006; Turan et al. 2005; Overcast et al. 2001). Similar to many other types of cardiovascular complications, chronic diabetes has been shown to be associated with a primary defect in the contractile function of the heart (Fein et al. 1980). Recent experimental data suggest that components of glucose mechanism are involved in reduced GSH homeostasis (Le et al. 1995). Once oxidative stress occurs during diabetes, which is proposed to result from deficits in antioxidant defense mechanisms, a marked shift in cell redox state appears which is reflected by a decrease in the level of reduced GSH. Additionally, Xu et al. (2002) and Li et al. (2003) reported that *in vitro* treatment of cardiomyocytes from diabetic rats with GSH upregulated inhibited transient K^+-currents and insulin-signaling cascade could regulate GSH level in cardiomyocytes by a coordinated activation of PI 3-kinase and MAPK. On the other hand, it has been shown that selenium compounds and also other antioxidants can stimulate glucose transport and have many beneficial effects on several different altered parameters of diabetic animals (Da Ros et al. 2004; Ye et al. 2003; Coppey et al. 2001; Gocmen et al. 2000). Since selenium compounds can restore some metabolic parameters of diabetic heart, we were tempted to investigate whether these beneficial effects extend to the diabetic rat cardiac function and cell redox status. We treated streptozotocin-induced diabetic rats with sodium selenite for 4 weeks and demonstrated that prolongation in both action potential duration and twitch duration of papillary muscle as well as the diminished amplitudes of the K^+ currents were reversed significantly (Ayaz et al. 2004). We also showed significant augmentation in the levels of soluble sulfhydryls, which were significantly lower in diabetics compared to the controls, with sodium selenite treatment. In experiments with the same setup, we have also demonstrated that sodium selenite treatment prevented diabetes-induced increases of intracellular Ca^{2+} and Zn^{2+} concentrations of rat heart (Ayaz and Turan 2006). In addition, this treatment markedly restored the increased levels of lipid peroxidation and NO products as well as decreased levels of metallothionein and reduced glutathione and depressed activities of superoxide dismutase, glutathione reductase, and glutathione peroxidase in the diabetic hearts (Table 8.1). Therefore, our data suggested that the beneficial effects of sodium selenite on the depressed heart function under diabetes appeared to be partially related to the restoration of the cell GSH redox cycle. Several previously published findings are also in line with our results. Epstein and colleagues showed that type 1 diabetic cardiomyopathy and ROS production could be prevented when the antioxidant metallothionein was overexpressed specifically in the heart (Ye et al. 2003).

One major intracellular target of oxidative stress is nuclear factor-κB (NF-κB), which can be activated by a variety of exogenous and endogenous stimuli, including

Table 8.1 Effects of selenium treatment on the biochemical parameters of rat hearts

	Untreated Control	Treated Control	Untreated Diabetic	Treated Diabetic
LPO (μM/g tissue)	0.49 ± 0.03	0.49 ± 0.07	0.82 ± 0.18*	0.48 ± 0.07
NOPs (μM/g tissue)	1.35 ± 0.19	1.29 ± 0.23	1.98 ± 0.34*	1.30 ± 0.25
SOD (mU/mg protein)	0.67 ± 0.04	0.70 ± 0.18	0.48 ± 0.06*	0.72 ± 0.15
GSH (μM/g tissue)	68.17 ± 0.39	59.67 ± 0.29	47.56 ± 1.19*	57.07 ± 4.09
GSSG (μM/g tissue)	3.17 ± 1.03	5.40 ± 0.85*	8.67 ± 1.68*	5.18 ± 1.37
GR (mU/mg protein)	2.58 ± 0.15	2.94 ± 0.16	1.88 ± 0.16*	2.9 ± 0.32
GSHPx (mU/mg protein)	1.25 ± 0.14	1.61 ± 0.20	2.22 ± 0.37*	1.34 ± 0.20
MT (μg/g tissue)	39.0 ± 1.8	37.7 ± 6.9	4.4 ± 4.0*	43.2 ± 7.0

Values are mean ± SEM. LPO: lipid peroxidation; NOPs: nitric oxide products; SOD: superoxide dismutase; GSH: reduced glutathione; GSSG: oxidized glutathione; GR: glutathione reductase; GSHPx: glutathione peroxidase; MT: metallothionein. * $p < 0.05$ versus untreated control group. (Adapted from Ayaz and Turan, 2006.)

ROS, hyperglycemia, and elevated free fatty acids (Mohamed et al. 1999) and alterations in NF-κB signaling are associated with a number of heart diseases. On the other hand, accumulation of ROS is considered an important mediator of ischemia/reperfusion-induced cardiac dysfunction that occurs following certain surgical procedures (Dhalla et al. 1999). In view of the importance of ROS in the activation of NF-κB, a redox-sensitive transcription factor for the regulation of different cellular processes (Valen et al. 2001), it has been suggested that ROS might enhance NF-κB activation by modifying the activity of one or more of the kinases responsible for NF-κB activation by changing the cellular redox state (Bowie and O'Neill 2000). Furthermore, the activation of NF-κB has been shown to be induced by pro-oxidants and several stimuli eliciting oxidative stress (Cargnoni et al. 2002). In fact, different antioxidants were found to inhibit this activation (Cargnoni et al. 2002). Thus, it would be interesting in explaning the critical roles of oxidative stress and activation of NF-κB in ischemia/reperfusion-mediated cardiac dysfunction. Furthermore, cell culture studies have revealed that sodium selenite inhibited the binding of NF-κB to nuclear responsive elements with a substantial increase in the activity of glutathione peroxidase and a significant inhibition in the activity of inducible nitric oxide (Kim and Stadtman 1997). Although the beneficial effects of selenium compounds have been demonstrated in ischemia/reperfusion-induced injury in the heart (Poltronieri et al. 1992), the mechanisms of such protection have not been elucidated. Therefore, we studied the effects of selenium in ischemia/reperfusion-induced hearts and showed the link with the level of phosphorylated ischemia/reperfusion-induced, as well as the subcellular distribution of NF-κB (Turan et al. 2005).

Sodium selenite, at nanomolar concentrations, improved the depressed cardiac performance of the isolated heart subjected to ischemia/reperfusion injury. As ischemia/reperfusion-mediated cardiac function has been reported to occur mainly due to the development of oxidative stress and intracellular Ca^{2+} overload (Bolli and Marban 1999), it is likely that selenium could protect the heart

Table 8.2 Effects of sodium selenite (Se) on mechanical performance of hearts under ischemia/reperfusion (I/R), or perfused with xanthine plus xanthine oxidase (X + XO), or hydrogen peroxide (H_2O_2), or subjected to Ca^{2+} paradox model

Groups	LVDP (% of preexposure)	LVEDP (mm Hg)
I/R	18.7 ± 2.7	85.3 ± 12.1
I/R + Se	$95.2 \pm 1.8^*$	$16.7 \pm 1.9^*$
X + XO	14.5 ± 3.7	91.7 ± 8.9
X + XO + Se	$32.3 \pm 1.9^*$	$58.4 \pm 6.2^*$
H_2O_2	11.6 ± 3.2	85.9 ± 10.2
H_2O_2 + Se	$38.2 \pm 5.2^*$	$62.2 \pm 8.2^*$
Ca^{2+} Paradox	10.2 ± 2.2	95.2 ± 12.7
Ca^{2+} Paradox + Se	$82.1 \pm 6.3^*$	$12.4 \pm 4.7^*$

The hearts were exposed to 75 nM sodium selenite (Se) 10 min before ischemia (or perfusions with X + XO or H_2O_2 or subjected to Ca^{2+} paradox) and 30 min during reperfusion (or 30 min perfusion with the Ca^{2+}-containing solution). LVDP: left ventricular developed pressure; LVEDP: left ventricular end developed pressure. Data represent means (±SEM). $^* p < 0.05$ versus I/R. (Adapted from Turan et al. 2005.)

against ischemia/reperfusion injury by attenuating the magnitude of both oxidative stress and Ca^{2+} overload. In this regard, we tested whether selenium has protective effects against depressed left ventricular developed pressure during both H_2O_2-perfused heart and hearts subjected to Ca^{2+} paradox. As can be seen from Table 8.2, the effects of selenium in preventing the ischemia/reperfusion-induced alterations in cardiac performance appear to be similar to those of H_2O_2-perfused hearts or hearts subjected to Ca^{2+} paradox. These effects of selenium regarding changes in different pathologic heart models seemed to be due to its antioxidant property. Sodium selenite treatment of the hearts partially, but significantly, attenuated the ischemia/reperfusion-induced changes in both MDA and reduced GSH/oxidized GSH levels, which are excellent markers of the redox state of the cells (Table 8.3).

Table 8.3 Effects of selenium treatment on ischemia/reperfused (I/R) cardiac tissue contents of malondialdehyde (MDA), reduced glutathione (GSH), oxidized glutathione (GSSG), and total and phospho nuclear factor kappa B (NF-κB)

Groups	Control	I/R	I/R + Se
MDA (nmol/g wet wt)	48.7 ± 2.7	$68.1 \pm 3.6^*$	$60.7 \pm 2.9^*$
GSH/GSSG ratio	6.1 ± 0.8	$2.7 \pm 1.2^*$	$4.2 \pm 1.9^{*\dagger}$
Total NF-κB in particulate (% of control)	100	$65 \pm 3^*$	$90 \pm 5^*$
Total NF-κB in cytosolic (% of control)	100.0	$45 \pm 3^*$	$70 \pm 5^*$
Phospho-NF-κB/total NF-κB (in homogenate)	0.0	$20 \pm 4^*$	3 ± 2

Values are mean (±SEM) in each group. Control, I/R, and I/R + Se groups represent 60 min perfused hearts, 30 min global ischemia plus 30 min reperfused hearts, and 10 min selenium (Se) perfusion prior to I plus 30 min global I plus 30 min R (with Se) hearts, respectively. $^* p < 0.05$ versus control. (Adapted from Turan et al. 2005.)

Since oxidative stress has been reported to produce subcellular redistribution and activation of NF-κB, we measured total NF-κB content in cytosolic and particulate fractions of the hearts subjected to ischemia/reperfusion. In another set of experiments, the activation of NF-κB protein was studied by measuring the total NF-κB and phospho-NF-κB in the heart homogenate. Elevated ratios of NF-κB in particulate versus cytosolic fraction and of phospho-NF-κB and total NF-κB in the hearts subjected to ischemia/reperfusion were reduced by selenium. As oxidative stress has been reported to produce subcellular redistribution and activation of NF-κB and antioxidants have been shown to prevent this alteration (Cargnoni et al. 2002), it is possible that our observation on selenium-induced NF-κB translocation might be due to its antioxidant action. It should be noted that since selenium treatment could reduce the tumor necrosis factor-α (TNF-α)-mediated activation of NF-κB (Kim and Stadtman 1997), it suggests a possible involvement of a reduction in the formation of TNF-α in the hearts subjected to ischemia/reperfusion by selenium.

Conclusion

Several lines of evidence demonstrate the roles of regulation of β-adrenergic signaling in β-adrenergic responses in the heart. Furthermore, redox modification of this link contributes new events due to redox-mediated new signaling pathways. Dysfunction of cardiac contractility has been linked to the excess amount of ROS in the myocardium. Therefore, any disturbance in the intracellular ionic homeostasis due to the excess ROS has been shown to result in impaired cardiac contractile activity. Since β-ARs and AC are known to participate in the regulation of cardiac function, it is possible that the β-AR-linked signal transduction pathway is also affected by ROS. Although ROS, which are formed in the heart under pathologies, have been shown to alter characteristics of cardiac β-ARs and AC, the role of ROS in these pathologies is still under investigation. Thus, antioxidants similar to β-blockers may counteract the effects of oxidants on β-adrenergic responses in the heart. Although studies with antioxidants firmly established that the protective effects on the depressed heart function under pathological conditions appeared to be partially related to the restoration of the cell GSH redox cycle, their relevance as well as their relation to β-adrenergic signaling pathway for human health have not been identified. Thus, redox regulation of cardiomyocytes may have an important role in the regulation of β-adrenergic signaling, and, therefore, in β-adrenergic responses of the heart, particularly under disease states. Further understanding the relationship between redox regulation and β-adrenergic responses in the heart may help to develop effective therapeutic agents to prevent heart dysfunctions.

Acknowledgments This work has been supported by grants from Ankara University (Projects Nos. 2001K-120-240, 2006-080-9233, 2003K-1201-9025) and projects of TUBITAK (Projects Nos. SBAG-PIA-10 and SBAG-3056).

References

Ayaz, M., and Turan, B. 2006. Selenium prevents diabetes-induced alterations in $[Zn^{2+}]_i$ and metallothionein level of rat heart via restoration of cell redox cycle. Am. J. Physiol. Heart Circ. Physiol. 290: H1071–H1080.

Ayaz, M., Ozdemir, S., Ugur, M., Vassort, G., and Turan, B. 2004. Effects of selenium on altered mechanical and electrical cardiac activities of diabetic rat. Arch. Biochem. Biophys. 426:83–90.

Ayaz, M., Ozdemir, S., Yaras, N., Vassort, G., and Turan, B. 2005. Selenium-induced alterations in ionic currents of rat cardiomyocytes. Biochem. Biophys. Res. Commun. 327:163–173.

Bettger, W. J. 1993. Zinc and selenium, site-specific versus general antioxidation. Can. J. Pharmacol. 71:721–724.

Bigelow, D. J., and Squier, T. C. 2005. Redox modulation of cellular signaling and metabolism through reversible oxidation of methionine sensors in calcium regulatory proteins. Biochim. Biophys. Acta 1703:121–134.

Björnstedt, M., Kumar, S., and Holgren, A. 1995. Selenite and selenodiglutathione. Reactions with thioredoxin systems. Methods Enzymol. 252:209–219.

Bohm, M. 1995. Alterations of beta-adrenoceptor-G-protein-regulated adenylyl cyclase in heart failure. Mol. Cell. Biochem. 147:147–160.

Bohm, M., and Lohse, M. J. 1994. Quantification of beta-adrenoceptors and beta-adrenoceptor kinase on protein and mRNA levels in heart failure. Eur. Heart J. 15:30–34.

Bolli, R., and Marban, E. 1999. Molecular and cellular mechanisms of myocardial stunning. Physiol. Rev. 79:609–634.

Bowie, A., and O'Neill, L. A. 2000. Oxidative stress and nuclear factor kappaB activation: a reassessment of the evidence in the light of recent discoveries. Biochem. Pharmacol. 59:13–23.

Bristow, M. R., Cubicciotti, R., Ginsburg, R., Stinson, E. B., and Johnson, C. 1982. Histamine-mediated adenylate cyclase stimulation in human myocardium. Mol. Pharmacol. 21:671–679.

Brodde, O.-E., Hillemann, S., Kunde, K., Vogesang, M., and Zerkowski, H.-R. 1992. Receptor systems affecting force of contraction in the human heart and their alterations in chronic heart failure. J. Heart Lung Transplant. 11:S164–S174.

Brodde, O.-E., Michel, M. C., and Zerkowski, H.-R. 1995. Signal transduction mechanisms controlling cardiac contractility and their alterations in chronic heart failure. Cardiovasc. Res. 30: 570–584.

Cargnoni, A., Ceconi, C., Gaia, G., Agnoletti, L., and Ferrari, R. 2002. Cellular thiols redox status: a switch for NF-κB activation during myocardial post-ischemic reperfusion. J. Mol. Cell. Cardiol. 34: 997–1005.

Cheng, H. J., Zhang, Z. S., Onishi, K., Ukai, T., Sane, D. C., and Cheng, C. P. 2001. Upregulation of functional beta(3)-adrenergic receptor in the failing canine myocardium. Circ. Res. 89:599–606.

Choi, K. M., Zhong; Y., Hoit, B. D., Grupp, I. L., Hahn, H., Dilly, K. W., Guatimosim, S., Lederer, W. J., and Matlib, M. A. 2002. Defective intracellular Ca(2+) signaling contributes to cardiomyopathy in Type 1 diabetic rats. Am. J. Physiol. Heart Circ. Physiol. 283:H1398–H1408.

Chu, S. H., Sutherland, K., Beck, J., Kowalski, J., Goldspink, P., and Schwertz, D. 2005. Sex differences in expression of calcium-handling proteins and beta-adrenergic receptors in rat heart ventricle. Life Sci. 76:2735–2749.

Coppey, L. J., Gellett, J. S., Davidson, E. P., Dunlap, J. A., Lund, D. D., and Yorek, M. A. 2001. Effect of antioxidant treatment of streptozotocin-induced diabetic rats on endoneurial blood flow, motor nerve conduction velocity, and vascular reactivity of epineurial arterioles of the sciatic nerve. Diabetes 50:1927–1937.

Cruzado, M. C., Risler, N. R., Miatello, R. M., Yao, G., Schiffrin, E. L., and Touyz, R. M. 2005. Vascular smooth muscle cell NAD(P)H oxidase activity during the development of hypertension: effect of angiotensin II and role of insulinlike growth factor-1 receptor transactivation. Am. J. Hypertens. 18:81–87.

Da Ros, R., Assaloni, R., and Ceriello, A. 2004. Antioxidant therapy in diabetic complications: what is new? Curr. Vasc. Pharmacol. 2:335–341.

Das, D. K., Maulik, N., and Engelman, R. M. 2004. Redox regulation of angiotensin II signaling in the heart. J. Cell. Mol. Med. 8(1):144–152.

Das, P. K., Temsah, R., Panagia, V., and Dhalla, N. S. 1997. Beta-adrenergic linked signal transduction mechanisms in developing and aging hearts. Heart Fail. Rev. 2:23–41.

Dhalla, N. S., Golfman, L., Takeda, N., and Nagano, M. 1999. Evidence for the role of oxidative stress in acute ischemic heart disease: a brief review. Can. J. Cardiol. 15:587–593.

Dhalla, N. S., Pierce, G. N., Panagia, V., Singal, P. K., and Beamish, R. E. 1982. Calcium movements in relation to heart function. Basic Res. Cardiol. 77:117–139.

Dhalla, N. S., Wang, X., Sethi, R., and Das, P. K. 1997. Beta-adrenergic linked signal transduction mechanisms in failing heart. Heart Fail. Rev. 2:55–65.

Eisner, D. A., Trafford, A. W., Diaz, M. E., Overend, C. L., and O'Neill, S. C. 1998. The control of Ca release from the cardiac sarcoplasmic reticulum: regulation versus autoregulation. Cardiovasc. Res. 38:589–604.

Eley, D. W., Eley, J. M., Korecky, B., Fliss, H., and Desilets, M. 1991. Calcium homeostasis in rabbit ventricular myocytes: disruption by hypochlorous acid and restoration by dithiothreitol. Circ. Res. 69: 1132–1138.

Endoh, M. 2006. Signal transduction and Ca^{2+} signaling in intact myocardium. J. Pharmacol. Sci. 100: 525–537.

Fein, F. S., Kornstein, L. B., Strobeck, J. E., Capasso, J. M., and Sonnenblick, E. H. 1980. Altered myocardial mechanics in diabetic rats. Circ. Res. 47:922–933.

Feldman, A. M. 1993. Classification of positive inotropic agents. J. Am. Coll. Cardiol. 22:1223–1227.

Gao, T., Yatani, A., Dell'Acqua, M. L., Sako, H., Green, S. A., Dascal, N., Scott, J. D., and Hosey, M. M. 1997. cAMP-dependent regulation of cardiac L-type Ca^{2+} channels requires membrane targeting of PKA and phosphorylation of channel subunits. Neuron 19:185–196.

Gocmen, C., Secilmis, A., Kumcu, E. K., Ertug, P. U., Onder, S., Dikmen, A., and Baysal, F. 2000. Effects of vitamin E and sodium selenate on neurogenic and endothelial relaxation of corpus cavernosum in the diabetic mouse. Eur. J. Pharmacol. 398:93–98.

Gopalakrishna, R., and Jaken, S. 2000. Protein kinase C signaling and oxidative stress. Free Radic. Biol. Med. 28:1349–1361.

Hammond, H. K., Roth, D. A., McKirnan, M. D., and Ping, P. 1993. Regional myocardial downregulation of the inhibitory guanosine triphosphate-binding protein (Gi alpha 2) and beta-adrenergic receptors in a porcine model of chronic episodic myocardial ischemia. J. Clin. Invest. 92:2644–2652.

Handel, M. L., Watts, C. K. W., DeFazio, A., and Grupp, G. 1995. Inhibition of AP-1 binding and transcription by gold and selenium involving conserved cysteine residues in Jun and Fos. Proc. Natl. Acad. Sci. USA 92:4497–4501.

Hartzell, H. C. 1988. Regulation of cardiac ion channels by catecholamines, acetylcholine and second messenger systems. Prog. Biophys. Mol. Biol. 52:165–247.

Hausdorff, W. P., Hnatowich, M., O'Dowd, B. F., Caron, M. G., and Lefkowitz, R. J. 1990. A mutation of the beta 2-adrenergic receptor impairs agonist activation of adenylyl cyclase without affecting high affinity agonist binding. Distinct molecular determinants of the receptor are involved in physical coupling to and functional activation of Gs. J. Biol. Chem. 265:1388–1393.

Homcy, C. J., Vatner, S. F., and Vatner, D. E. 1991. Beta-adrenergic receptor regulation in the heart in pathophysiologic states: abnormal adrenergic responsiveness in cardiac disease. Annu. Rev. Physiol. 53:137–159.

Hool, L. C., Middleton, L. M., and Harvey, R. D. 1998. Genistein increases the sensitivity of cardiac ion channels to β-adrenergic receptor stimulation. Circ. Res. 83:33–42.

Iaccarino, G., Lefkowitz, R. J., and Koch, W. J. 1999. Myocardial G protein-coupled receptor kinases: implications for heart failure therapy. Proc. Assoc. Am. Physicians 111:399–405.

Kaul, N., Siveski-Iliskovic, N., Hill, M., Slezak, J., and Singal, P. K. 1993. Free radicals and the heart. J. Pharmacol. Toxicol. Methods 30:55–67.

Kaura, D., Takeda, N., Sethi, R., Wang, X., Nagano, M., and Dhalla, N. S. 1996. Beta-adrenoceptor mediated signal transduction in congestive heart failure in cardiomyopathic (UM-X7.1) hamsters. Mol. Cell. Biochem. 157:191–196.

Kim, I. Y., and Stadtman, T. C. 1997. Inhibition of NF-κB DNA binding and nitric oxide induction in human T cells and lung adenocarcinoma cells by selenite treatment. Proc. Natl. Acad. Sci. USA 94: 12904–12907.

Koller, L. D., and Exon, J. H. 1986. The two faces of selenium—deficiency and toxicity—-are similar in animals and man. Can. J. Vet. Res. 50:297–306.

Lands, A. M., Arnold, A., McAuliff, J. P., Luduena, F. P., and Brown, T. G., Jr. 1967. Differentiation of receptor systems activated by sympathomimetic amines. Nature 214:597–598.

Le, C. T., Hollaar, L., Van der Valk, E. J., Franken, N. A., Van Ravels, F. J., Wondergem, J., and Van der Laarse, A. 1995. Protection of myocytes against free radical-induced damage by accelerated turnover of the glutathione redox cycle. Eur. Heart J. 16:553–562.

Li, G. S., Wang, F., Kang, D. R., and Li, C. 1985. Keshan disease: an endemic cardiomyopathy in China. Hum. Pathol. 16:602–609.

Li, S., Li, X., and Rozanski, G. J. 2003. Regulation of glutathione in cardiac myocytes. J. Mol. Cell. Cardiol. 35:1145–1152.

Lin-Shiau, S. Y., Liu, S. H., and Fu, W. M. 1989. Studies on the contracture of the mouse diaphragm induced by sodium selenite. Eur. J. Pharmacol. 167:137–146.

Marks, A. R. 2001. Ryanodine receptors/calcium release channels in heart failure and sudden cardiac death. J. Mol. Cell. Cardiol. 33:615–624.

Mohamed, A. K., Bierhaus, A., Schiekofer, S., Tritschler, H., Ziegler, R., and Nawroth, P. P. 1999. The role of oxidative stress and NF-κB activation in late diabetic complications. Biofactors 10: 157–167.

Nishizawa, T., Iwase, M., Kanazawa, H., Ichihara, S., Ichihara, G., Nagata, K., Obata, K., Kitaichi, K., Yokoi, T., Watanabe, M., Tsunematsu, T., Ishikawa, Y., Murohara, T., and Yokota, M. 2004. Serial alterations of beta-adrenergic signaling in dilated cardiomyopathic hamsters: possible role of myocardial oxidative stress. Circ. J. 68:1051–1060.

Oudit, G. Y., Kassiri, Z., Sah, R., Ramirez, R. J., Zobel, C., and Backx, P. H. 2001. The molecular physiology of the cardiac transient outward potassium current (I_{to}) in normal and diseased myocardium. J. Mol. Cell. Cardiol. 33:851–872.

Overcast, J. D., Ensley, A. E., Buccafusco, C. J., Cundy, C., Broadnax, R. A., He, S., Yoganathan, A. P., Pollock, S. H., Hartley, C. J., and May, S. W. 2001. Evaluation of cardiovascular parameters of a selenium-based antihypertensive using pulsed Doppler ultrasound. J. Cardiovasc. Pharmacol. 38: 337–346.

Persad, S., Takeda, S., Panagia, V., and Dhalla, N. S. 1997. β-adrenoceptor-linked signal transduction in ischemic-reperfused heart and scavenging of oxyradicals. J. Mol. Cell. Cardiol. 29:545–558.

Pogwizd, S. M., Schlotthauer, K., Li, L., Yuan, W., and Bers, D. M. 2001. Arrhythmogenesis and contractile dysfunction in heart failure: Roles of sodium-calcium exchange, inward rectifier potassium current, and residual beta-adrenergic responsiveness. Circ. Res. 88:1159–1167.

Poltronieri, R., Cevese, A., and Sbartani, A. 1992. Protective effect of selenium in cardiac ischemia and reperfusion. Cardioscience 3:155–160.

Puceat, M., Bony, C., Jaconi, M., and Vassort, G. 1998. Specific activation of adenylyl cyclase V by a purinergic agonist. FEBS Lett. 431:189–194.

Quin, F., Yan, C., Patel, R., Liu, W., and Dong, E. 2006. Vitamin C and E attenuate apoptosis, β-adrenergic receptor desensitization, and sarcoplasmic reticular Ca^{2+} ATPase downregulation after myocardial infarction. Free Radic. Biol. Med. 40:1827–1842.

Rozanski, G. J., Xu, Z., Zhang, K., and Patel, K. P. 1998. Altered K^+ current of ventricular myocytes in rats with chronic myocardial infarction. Am. J. Physiol. 274 (1 Pt 2):H259–H265.

Rozec, B., and Gauthier, C. 2006. β_3-adrenoceptors in the cardiovascular system: putative roles in human pathologies. Pharmacol. Ther. 111:652–673.

Salonen, J. T., Salonen, R., Penttila, I., Herranen, J., Jauhiainen, M., Kantola, S., Lappetelainen, R., Maenpaa, P., Alfthan, G., and Puska, P. 1985. Serum fatty acids, apolipoproteins, selenium and vitamin antioxidants and the risk of death from coronary artery disease. Am. J. Cardiol. 56:226–231.

Sayar, K., Ugur, M., Gurdal, H., Onaran, O., Hotomaroglu, O., and Turan, B. 2000. Dietary selenium and vitamin E intakes alter β-adrenergic response of L-type Ca-current and β-adrenoceptor-adenylate cyclase coupling in rat heart. J. Nutr. 130:733–740.

Schieke, S. M., Briviba, K., Klotz, L. O., and Sies, H. 1999. Activation of mitogen-activated protein kinases elicited by peroxynitrite: attenuation by selenite supplementation. FEBS Lett. 448:301–303.

Schulz, R., Rassaf, T., Massion, P. B., Kelm, M., and Balligand, J. L. 2006. Recent advances in the understanding of the role of nitric oxide in cardiovascular homeostasis. Pharmacol. Ther. 108:225–256.

Sims, C., and Harvey, R. D. 2004. Redox modulation of basal and beta-adrenergically stimulated cardiac L-type Ca^{2+} channel activity by phenylarsine oxide. Br. J. Pharmacol. 142:797–807.

Stapleton, S. R., Garlock, G., Foellmi-Adam, L., and Kletzien, R. F. 1997. Selenium: potent stimulator of tyrosyl phosphorylation and activator of MAP kinase. Biochim. Biophys. Acta 1355: 259–269.

Stiles, G. L., and Lefkowitz, R. J. 1984. Cardiac adrenergic receptors. Annu. Rev. Med. 35:149–164.

Strasser, R. H., Krimmer, J., Braun-Dullaeus, R., Marquetant, R., and Kubler, W. 1990. Dual sensitization of the adrenergic system in early myocardial ischemia: independent regulation of the beta-adrenergic receptors and the adenylyl cyclase. J. Mol. Cell. Cardiol. 22:1405–1423.

Sun, Y., and Oberley, L. W. 1996. Redox regulation of transcriptional activators. Free Radic. Biol. Med. 21:335–348.

Tatsumi, T., and Fliss, H. 1994. Hypochlorous acid mobilizes intracellular zinc in isolated rat heart myocytes. J. Mol. Cell. Cardiol. 26:471–479.

Turan, B., Desilets, M., Acan, L. N., Hotomaroglu, O., Vannier, C., and Vassort, G. 1996. Oxidative effects of selenite on rat ventricular contractility and Ca movements. Cardiovasc. Res. 32:351–361.

Turan, B., Fliss, H., and Desilets, M. 1997. Oxidants increase intracellular free Zn^{2+} concentration in rabbit ventricular myocytes. Am. J. Physiol. Heart Circ. Physiol. 272: H2095–H2106.

Turan, B., Hotomaroglu, O., Kilic, M., and Demirel-Yilmaz, E. 1999. Cardiac dysfunction induced by low and high diet antioxidant levels. Comparing selenium and vitamin E in rats. Regul. Pharmacol. Toxicol. 29:142–150.

Turan, B., Saini, H. K., Zhang, M., Prajapati, D., Elimban, V., and Dhalla, N. S. 2005. Selenium improves cardiac function by attenuating the activation of NF-κB due to ischemia-reperfusion injury. Antioxid. Redox Signal. 7:1388–1397.

Ugur, M., and Turan, B. 2001. Adenosine triphosphate alters the selenite-induced contracture and negative inotropic effect on cardiac muscle contractions. Biol. Trace Elem. Res. 79:235–245.

Ugur, M., Ayaz, M., Ozdemir, S., and Turan, B. 2002. Toxic concentrations of selenite shortens repolarization phase of action potential in rat papillary muscle. Biol. Trace Elem. Res. 89:227–238.

Valen, G., Yan, Z. Q., and Hansson, G. K. 2001. Nuclear factor kappa-B and the heart. J. Am. Coll. Cardiol. 38:307–314.

Webb, T. E., Boluyt, M. O., and Barnard, E. A. 1996. Molecular biology of P2Y purinoceptors: expression in rat heart. J. Auton. Pharmacol. 16:303–307.

Wehrens, X. H., and Marks, A. R. 2003. Altered function and regulation of cardiac ryanodine receptors in cardiac disease. Trends Biochem. Sci. 28:671–678.

Wold, L. E., Ceylan-Isik, A. F., and Ren, J. 2005. Oxidative stress and stress signaling: menace of diabetic cardiomyopathy. Acta Pharmacol. Sin. 26:908–917.

Wold, L. E., Ceylan-Isik, A. F., Fang, C. X., Yang, X., Li, S. Y., Sreejayan, N., Privratsky, J. R., and Ren, J. 2006. Metallothionein alleviates cardiac dysfunction in streptozotocin-induced diabetes: Role of Ca^{2+} cycling proteins, NADPH oxidase, poly(ADP-ribose) polymerase and myosin heavy chain isozyme. Free Radic. Biol. Med. 40:1419–1429.

Xu, Z., Patel, K. P., Lou, M. F., and Rozanski, G. J. 2002. Up-regulation of K^+ channels in diabetic rat ventricular myocytes by insulin and glutathione. Cardiovasc. Res. 53:80–88.

Yaras, N., Ozdemir, S., Ugur, M., Puralı, N., Gurdal, H., Lacampagne, A., Vassort, G., and Turan, B. 2005. Effect of diabetes on ryanodine receptor Ca release channel (RyR2) and Ca^{2+} homeostasis in rat heart. Diabetes 54:3082–3088.

Ye, G., Metreveli, N. S., Ren, J., and Epstein, P. N. 2003. Metallothionein prevents diabetes-induced deficits in cardiomyocytes by inhibiting reactive oxygen species production. Diabetes 52:777–783.

Zhang, M., Xu, Y. J., Saini, H. K., Turan, B., Liu, P. P., and Dhalla, N. S. 2005. TNF-alpha as a potential mediator of cardiac dysfunction due to intracellular Ca^{2+}-overload. Biochem. Biophys. Res. Commun. 327:57–63.

Zhou, Y. Y., Yao, J. A., and Tsen, G. N. 1997. Role of tyrosine kinase activity in cardiac slow delayed rectifier channel modulation by cell swelling. Pflügers Arch. 433:750–757.

Chapter 9
Role of Hyperglycemia and Redox-Induced Signaling in Vascular Complications of Diabetes

Ashok K. Srivastava and Madhu B. Anand-Srivastava

Abstract Vascular complications including impaired contractility and increased cell proliferation are the most common complications associated with diabetes. Chronic hyperglycemia appears to be an important contributing factor in this process. Various signaling pathways are implicated in diabetes/hyperglycemia-induced impaired vascular functions. Increased oxidative stress, nonenzymatic glycation, enhanced production of diacylglycerol, increased activity of protein kinase C, mitogen-activated protein kinases (MAPK), and other signaling pathways have been proposed to explain the adverse effects of hyperglycemia on vascular smooth muscle cells. Hyperglycemia-induced stimulation of the L-type Ca^{2+} channel via G-protein-coupled adenylyl cyclase/cAMP and phospholipase C/protein kinase C (PKC) pathways has also been shown. In addition, hyperglycemia has been reported to decrease the availability of nitric oxide and increase the formation of peroxynitrite which may contribute to all of the hemodynamic and physiological changes occurring in diabetes. G-protein/adenylyl cyclase signaling that plays an important role in the regulation of cardiovascular functions has also been reported to be impaired in diabetes and under hyperglycemic conditions. In this review, we have highlighted some key signaling pathways, including PKC, MAPK, and G-protein/adenylyl cyclase, which are altered in diabetes, and in response to hyperglycemia/oxidative stress, and discussed their contributions in the development of vascular complications.

Introduction

Cardiovascular disease represents a major secondary complication of diabetes and the evidence generated in recent years has suggested a strong correlation between hyperglycemia and vascular disease (Singh and Jialal 2006; Pyorala et al. 1987; Laakso 1999; Reaven et al. 2004). Although the precise mechanism by which hyperglycemia contributes to vascular disorders in diabetes remains poorly defined, a potential role of excessive generation of reactive oxygen species (ROS) in this process has been suggested (Chiasson et al. 2006) based on studies showing that

A. K. Srivastava, M. B. Anand-Srivastava (eds.) *Advances in Biochemistry in Health and Disease.* Signal Transduction in the Cardiovascular System in Health and Disease,

both chronic hyperglycemia as well as spiking postprandial hyperglycemia are associated with enhanced production of ROS (Monnier et al. 2006). Further support for a role of hyperglycemia-induced ROS generation as a mediator of diabetic complications has been provided by several studies in which use of antioxidants was shown to improve cardiovascular complications of diabetes (Davi et al. 1999; Singh and Jialal 2006; Chiasson et al. 2006). Thus, there is a lot of interest to define the mechanism by which hyperglycemia contributes to ROS generation.

Multiple mechanisms for hyperglycemia-induced ROS generation have been suggested, which include increased polyol pathway, enhanced production of advanced glycation end products (AGE), and increased levels of diacylglycerol (DAG) and PKC activities (Brownlee 2001; Koya and King 1998). PKC-dependent activation of NAD(P)H oxidase has been suggested as a major source of ROS generation in response to hyperglycemia (Singh and Jialal 2006; Lee et al. 1999).

Diabetes and hyperglycemia also induce an increase in expression levels of endothelin-1 (ET-1) in several tissues (Khan and Chakrabarti 2003; Farhangkhoee et al. 2006). Similarly, hyperglycemia also increases the formation of angiotensin II (Ang II) (Koka et al. 2006; Lavrentyev et al. 2007). Both ET-1 and Ang II also enhance NAD(P)H oxidase activity and ROS generation (Lee et al. 2003). Thus, diabetes-induced ET-1 and Ang II generation and associated increase in ROS and upregulation of several key signaling systems have been implicated in the pathogenesis of aberrant vascular function. The aim of this article is to provide an overview on some of the key signal transduction pathways, most notably PKC, mitogen-activated protein kinases (MAPK), and G-protein/adenylyl cyclase systems, which are regulated by hyperglycemia and ROS.

Hyperglycemia, ROS, and PKC

PKC is a serine/threonine protein kinase comprised of at least 11 isozymic forms (Nishizuka 1995; Liu and Heckman 1998). These isozymic forms have been classified as atypical, classical, and novel. Classical PKCs (α, βI, βII, and γ) are activated by Ca^{2+}, DAG, phosphatidylserine (PS), and the tumor promoter phorbol 12-myristate 13-acetate (PMA). Novel PKCs (δ, ε, η, μ, and θ) are activated by DAG, PS, and unsaturated fatty acids, while atypical PKCs (ζ, λ, and ι) are insensitive to DAG but are activated by PS and phosphatidylinositides (reviewed in Liu and Heckman 1998; Newton and Johnson 1998; Nakanishi et al. 1993). PKCs have been implicated in a wide variety of cellular responses, including growth, differentiation, gene expression, angiogenesis, contractility, and vesicle trafficking (Nishizuka 1995).

Both PKC-βII and PKC-δ isoforms have been shown to be activated in aortic and vascular smooth muscle cells (VSMC) from diabetic rats, as well as in response to hyperglycemia (Koya and King 1998; Craven and DeRubertis 1989; Inoguchi et al. 1992; Kunisaki et al. 1994; Lee et al. 1999). Treatment of VSMCs with 22 mM glucose for 3 days increased the levels of DAG as well as PKC-βII by 50 and 110%, respectively, compared to VSMC incubated with 5.5 mM glucose (Kunisaki et al. 1994). Under these conditions, no changes in PKC-α levels were detected (Kunisaki et al. 1994). Interestingly, α-tocopherol treatment of diabetic animals or incubation of VSMC with α-tocopherol prevented the DAG and PKC elevation due to diabetes or hyperglycemia (Kunisaki et al. 1994).

Hyperglycemia has also been found to activate total PKC activity in particulate fractions of VSMC, which was almost completely blocked by pretreatment of cells with the antioxidants probucol and D-α-tocopherol (Yasunari et al. 1999). The contribution of DAG–PKC that activates NADPH oxidase in the enhanced production of ROS has been reported in diabetic tissues (Inoguchi et al. 2003) and in cultured endothelial as well as in aortic VSMC exposed to high glucose (Inoguchi et al. 2000). In addition, expression of NADPH oxidase components was shown to be upregulated in vascular tissues and kidney from animal models of diabetes as well as in micro- and macrovascular tissues in diabetic patients and obese subjects (Inoguchi et al. 2000, 2003; Inoguchi and Nawata 2005). The increased levels of various vasoactive peptides including Ang II and endothelin in diabetes and under hyperglycemic conditions (Hargrove et al. 2000; Park et al. 2000; Sodhi et al. 2003) may contribute to the enhanced oxidative stress, because these peptides have been shown to increase oxidative stress by activating NADPH oxidase (Touyz et al. 2004). In this regard, we have recently reported that Ang II treatment of VSMC increased the production of O_2^- and the expression of Nox^4 and p^{47phox}—two subunits of NADPH oxidase (Li et al. 2007). Furthermore, AT_1 receptor blocker candesartan and ACE inhibitor quinapril have been shown to attenuate the enhanced expression of p^{47phox} in kidney from STZ-diabetic rats, implicating Ang II in diabetes-induced increased oxidative stress (Onozato et al. 2002). An increase of total cellular PKC in cultured human VSMC has also been demonstrated in response to high glucose (Williams et al. 1997). Results showing that an orally active inhibitor of PKC-β, LY33353, reversed some of the vascular abnormalities in diabetic rats have strengthened the proposed role of PKC as a mediator of diabetic vascular complications (Ishii et al. 1996).

Hyperglycemia, ROS, and MAPK

MAPK are serine/threonine protein kinases, which are activated in response to a variety of external stimuli, including growth factors, hormones, and stress. MAPK have been classified into several subfamilies: MAPK ERK 1/2 (extracellular signal-regulated kinases 1 and 2), $p38^{mapk}$, JNK/SAPK (c-Jun NH_2-terminal kinase/stress-activated protein kinase), ERK 3/4, ERK 5 (reviewed in Seger and Krebs 1995; Widmann et al. 1999). MAPK are activated by dual phosphorylation on both tyrosine and threonine residues by dual-specificity protein kinases known as MAPKK or MEK (*m*itogen *e*xtracellular signal-regulated kinase *k*inase) (Seger and Krebs 1995). The sequential upstream signaling molecules to MEK are Raf, serine/threonine kinase, and ras, a small GTP-binding protein (Widmann et al. 1999). MAPK phosphorylate downstream cytosolic and nuclear substrate/transcription factors, such as $p90^{rsk}$, and many transcription factors, such as c-Jun, ATF-2, Elk-1, CHOP, CREB, and MEF-2 (Widmann et al. 1999; Ip and Davis 1998; Force and Bonventre 1998; Davis 1993; Denton and Tavare 1995; Tan et al. 1996; Wang and Ron 1996; Gupta et al. 1995; Zinck et al. 1995). $P90^{rsk}$ phosphorylates ribosomal proteins and participates in protein synthesis (Frodin and Gammeltoft 1999), whereas the phosphorylation of transcription factors by MAPK leads to activation of several genes involved in growth and differentiation (Widmann et al. 1999). Thus, activation of the MAPK pathway can potentially result in increased growth, gene expression, and proliferation of VSMC in response to hyperglycemia. Hyperglycemia has been shown to activate several members of the MAPK family. For example, in porcine VSMC, hyperglycemia (25 mM glucose) markedly stimulated the activation state of ERK 1/2, JNK/SAPK as well as $p38^{mapk}$ (Natarajan et al. 1999). The impact of hyperglycemia on ERK and JNK/SAPK activation was detectable within 1 hour of treatment of VSMCs whereas at least 3-hour exposure with high glucose was required to elicit any stimulatory effect on $p38^{mapk}$ activity (Natarajan et al. 1999). A similar effect of hyperglycemia on ERK 1/2 and $p38^{mapk}$ activity in rat aortic VSMC (Igarashi et al. 1999) has been reported, where high glucose caused a 3- to 4-fold increase in $p38^{mapk}$ phosphorylation, as compared to cells treated with low glucose (Igarashi et al. 1999). A general inhibitor of PKC suppressed $p38^{mapk}$ activation, suggesting an involvement of PKC in the response (Igarashi et al. 1999). More recently, an increase in the total, as well as phosphorylated forms of ERK 1/2, JNK-1, and $p38^{mapk}$ were shown to be significantly elevated in aorta isolated from STZ-diabetic rats (Igarashi et al. 2007). In these studies, VSMCs cultured from diabetic aorta exhibited a heightened phosphorylation of ERK 1/2 and $p38^{mapk}$ in response to inflammatory cytokines, such as interleukin-1β and TNF-α (Igarashi et al. 2007).

In the last few years, many reports have documented ROS-induced activation of the MAPK pathway in different cell types, including Rat-2 fibroblasts (Esposito et al. 2003), rabbit renal proximal tubular cells (Zhuang and Schnellmann 2004), Chinese hamster ovary (CHO) cells (Mehdi et al. 2005), rat cardiomyocytes and heart fibroblasts (Purdom and Chen 2005) as well as VSMCs (Blanc et al. 2003, 2004; Tabet et al. 2005). Moreover, ROS generation was shown to be critical in the activation of MAPK by Ang II (Ushio-Fukai et al. 1998; Touyz et al. 2004) and ET-1 (Daou and Srivastava 2004; Touyz et al. 2004) in VSMCs.

Several lines of evidence have indicated that an aberrant activation of MAPK is often associated with vascular remodeling in cardiovascular diseases. For example, the enhanced activation of vascular MAPK has been demonstrated in different models of vascular disease (Xu et al. 1996; Touyz et al. 2002) and a heightened activation of MAPK has been implicated in vascular remodeling, which is a hallmark of vascular disease. MAPK are also activated by ROS, such as H_2O_2 or O_2^- in a variety of cell types, including VSMCs and cardiomyocytes (Azar et al. 2006, 2007). It may thus be suggested that diabetes/hyperglycemia-induced upregulation of Ang II/ET-1 as well as an enhanced generation of ROS, and resultant activation of PKC and MAPK pathways contributes to abnormal vascular functions due to heightened growth, migration, and proliferation of VSMC (Figure 9.1).

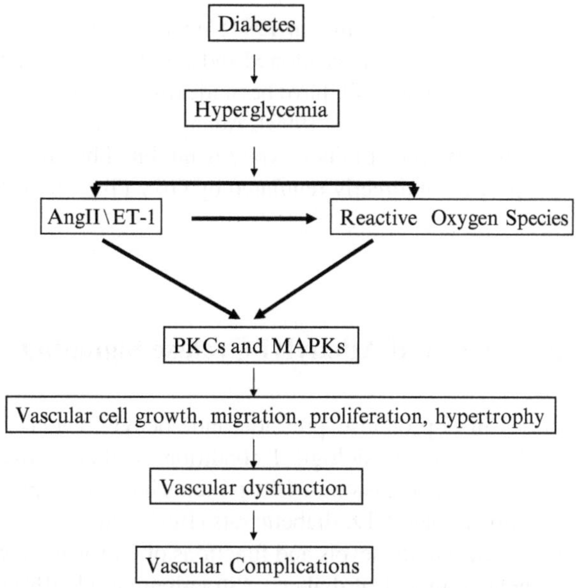

Fig. 9.1 Schematic model depicting the key events which mediate diabetes-associated vascular complications. Diabetes/hyperglycemia augments the levels of vasoactive peptides including Ang II/ET-1 that enhance the generation of reactive oxygen species (ROS). ROS-induced activation of growth-promoting signaling pathways, such as PKC and MAPK, contributes to aberrant vascular functions.

Hyperglycemia/Diabetes, ROS, and G-protein/Adenylyl Cyclase System

In addition to PKC and MAPK pathways, the adenylyl cyclase/cAMP system is also modified in diabetes and in response to hyperglycemia. The adenylyl cyclase/cAMP is one of the signal transduction systems implicated in the regulation of cardiovascular functions including arterial tone, reactivity, and cell proliferation. The hormone-sensitive adenylyl cyclase system is composed of three components: receptor, catalytic subunit, and G-proteins—stimulatory (Gs) and inhibitory (Gi)—which mediate the stimulatory and inhibitory responses of hormones on adenylyl cyclase respectively (Gilman 1984; Stryer and Bourne 1986; Spiegel 1987). G-proteins are heterotrimeric proteins composed of α, β, and γ subunits, and the specificity of G-proteins is attributed to α-subunits (Stryer and Bourne 1986). Four different forms of Gsα and three distinct forms of Giα—Giα-1, Giα-2, and Giα-3—have been identified. All three forms of Giα are implicated in adenylyl cyclase inhibition (Wong et al. 1992). Both the Gα and G$\beta\gamma$ mediate G-protein signaling. Five different β-ubunits of 35–36 kDa and seven γ-subunits of 8–10 kDa have been identified by molecular cloning (Strathmann et al. 1989; Cali et al. 1992). The effectors regulated by G$\beta\gamma$ include K^+ channels, phospholipase C-β, and adenylyl cyclase

(Strathmann et al. 1989; Tang and Gilman 1991; Wickman et al. 1994) Of the nine types of adenylyl cyclase that have been cloned and expressed (Ishikawa and Homcy 1997) only two types, types V and VI, have been identified in heart, aorta, and brain (Katsushika et al. 1992; Premont et al. 1992). Adenylyl cyclase types II and IV are activated by G$\beta\gamma$ in the presence of Gsα, type I is inhibited by G$\beta\gamma$, and types III, V, and VI do not appear to be directly regulated by G$\beta\gamma$ (Toro et al. 1987; Taussig et al. 1993).

Diabetes, G-proteins, and Adenylyl Cyclase Signaling

Several abnormalities in G-protein expression and adenylyl cyclase activity have been shown in various pathophysiological conditions including diabetes (Lynch et al. 1989). The decreased expression of Giα proteins has been reported in hepatocytes from human diabetics and STZ-diabetic rats (Bushfield et al. 1990; Caro et al. 1994), whereas an increase in the levels and functions of Giα was shown in diabetic adipocytes from a genetic model of diabetes (Strassheim et al. 1991). Livingstone et al. (1991) have shown a decreased expression of Giα proteins in platelets from diabetic subjects as compared to nondiabetic subjects. In addition, diabetic retina has been shown to exhibit decreased levels and functions of Giα (Hadjiconstantinou et al. 1988). Hattori et al. (2000) have also reported a similar decrease in Gi protein in aorta from long-term diabetic rats; however, these investigators did not examine adenylyl cyclase Gi-protein signaling in their studies. However, Weber and McLeod (1997) were unable to observe any changes in the levels of Giα proteins in aorta or caudal artery from 12- to 14-week STZ-diabetic rats as compared to control rats. Further support and involvement of Giα-2 protein in the pathogenesis of diabetes has been provided by studies showing that the overexpression of constitutively activated Giα-2 ameliorates STZ-induced diabetes in rats (Zheng et al. 1998). In addition, a complete knockout of the Giα-2 gene that has been reported to produce a metabolic state resembling type II diabetes suggests the relationship between the decreased levels of Giα protein and diabetes (Moxham and Malbon 1996). However, Hashim et al. (2002) have recently shown that the aorta from STZ-diabetic rats exhibited decreased expression of Giα-2 and Giα-3 but not of Gsα proteins. An unaltered expression of Gsα in hearts from STZ-induced diabetic rats has also been reported (Griffiths et al. 1990; Hashim et al. 2002). A relationship between the development of diabetes and Giα protein expression was also demonstrated (Hashim et al. 2002). The rats treated with STZ showed enhanced blood glucose levels within 2 days after injection with a concurrent decrease in the levels of Giα proteins, suggesting that the decrease in the levels of Giα proteins is associated with the development of diabetes. Subsequent increase in the levels of blood glucose through day 5 resulted in further decrease in the levels of Giα proteins, suggesting a close relationship between decreased levels of Giα proteins and severity of diabetes (Hashim et al. 2002). The decreased levels of Giα proteins were reflected in decreased Gi functions, i.e., the inhibitory effect of GTPγS on FSK-stimulated adenylyl cyclase activity was attenuated (Hashim et al. 2002). In addition, the inhibitory effect of Ang II, oxotremorine, and C-ANP_{4-23} on adenylyl cyclase activity was completely abolished (Hashim et al. 2002). On the other hand, the Gsα-mediated stimulatory effects of hormones as well as of FSK and NaF on adenylyl cyclase were augmented in STZ-aorta as compared to control aorta resulting in enhanced levels of cAMP, whereas the basal cAMP levels were reduced in diabetic aorta (Hashim et al. 2002).

Hyperglycemia, G-proteins, and Adenylyl Cyclase Signaling

We have further shown that aorta as well as A10 VSMC exposed to high glucose (26 mM) that simulate diabetic state also exhibited decreased levels of Giα-2 and Giα-3 proteins, whereas the levels of Gsα were not altered (Hashim et al. 2004). The decreased expression of Giα proteins was concentration and time dependent. A significant decrease was observed at 20 mM glucose; below that concentration, the levels of Giα proteins were not altered. However, VSMC exposed to higher concentration of glucose (52 mM) or treated for a longer period of time (96 hours) resulted in further decrease in the levels of Giα proteins, whereas the levels of Gsα were not altered under these conditions (Hashim et al. 2004). These data indicate a correlation between the levels of glucose (*in vivo* and *in vitro*) and decreased expression of Giα proteins and suggest that hyperglycemia may be a contributing factor in diabetes-induced decreased expression of Giα proteins. In addition, aortic VSMC from STZ-diabetic rats have also been shown to exhibit decreased expression of Giα proteins (Li et al. 2008), suggesting that aortic VSMC cultured from STZ-diabetic rats retained the diabetic phenotype.

Hyperglycemia was also shown to stimulate adenylyl cyclase activity in bovine aortic endothelial cells which causes an inhibition of glucose-6-phosphate dehydrogenase and thereby results in decreased levels of NADPH that may be responsible for hyperglycemia-induced apoptosis (Zhang et al. 2000). In addition, an increased stimulation of cAMP levels by OP-1206 alpha-CD, an analog of prostaglandin E_1 (PGE_1), was reported in sciatic nerve from STZ-diabetic rats that were shown to increase Na^+/K^+ ATPase activity (Yasuda et al. 1994). In support of these studies, Hashim et al. (2004) have recently reported an increased stimulation of adenylyl cyclase activity by isoproterenol and glucagon in A10 vascular smooth muscle cells exposed to high glucose (26 mM) as compared to the cells exposed to normal glucose (5.5 mM). These enhanced stimulations were shown to be attributed to the decreased expression of Giα-2 and Giα-3 levels and not to increased levels of Gsα proteins, because the levels of Gsα proteins were not altered in hyperglycemia cells. Taken together, it can be suggested that hyperglycemic may be a contributing factor in diabetes-induced decreased expression of Giα proteins. However, Mancusi et al. (1996) were unable to show any changes in Gi protein expression in human umbilical vein endothelial cells (HUVEC) exposed to high glucose for 15 days.

High-glucose-induced decreased expression of Giα proteins was shown to be reflected in decreased Gi functions (Hashim et al. 2006). For example, Ang II, oxotremorine (Oxo), and C-ANP$_{4–23}$ (a ring-deleted peptide of ANP) which inhibit adenylyl cyclase activity through Gi proteins (Tucek et al. 2001; Anand-Srivastava et al. 1987; Anand-Srivastava 1989) inhibited the enzyme activity by about 20, 40, and 25%, respectively, in control cells. However, the exposure of these cells to high glucose eliminated the inhibitory effect of Ang II and C-ANP$_{4–23}$, whereas Oxo-mediated inhibition was only diminished by 50%. In addition, GTPγS inhibited FSK-stimulated activity (receptor-independent functions of Giα) in a concentration-dependent manner in control cells, which was almost completely abolished in cells exposed to high glucose, suggesting a correlation between decreased levels of Giα proteins. Thus, it appears that about 50–60% decrease in Giα-2 and Giα-3 proteins by high glucose may be sufficient to inhibit Gi functions and to uncouple the hormone receptors from the adenylyl cyclase system, or alternatively, some other mechanisms at the receptor level, such as receptor downregulation, may also be responsible for a complete attenuation of inhibitory responses on adenylyl cyclase. In this context, acute hyperglycemia induced by STZ or alloxan has been shown to decrease the levels of vascular ANP-C, AT1, and arginine-vasopressin (AVP) receptors (Kook et al. 2002; Williams et al. 1992). Hyperglycemia has also been shown to impair voltage-gated K^+-channel current in rat small coronary VSMC (Liu et al. 2001). Since Giα proteins are implicated in the activation of K^+ channels, it may be possible that the impairment of K^+-channel activity is the result of decreased levels of Giα protein induced by high glucose. On the other hand, basal adenylyl cyclase activity was significantly decreased in A10 VSMC or aorta exposed to high glucose (Hashim et al. 2004). Since decreased cAMP levels have been shown to augment cell proliferation (Hayashi et al. 2000), it may be possible that the decreased basal adenylyl cyclase activity and thereby decreased cAMP levels induced by high glucose is a contributing factor in increased cell proliferation observed under hyperglycemic conditions and diabetes (Fujita et al. 2002).

Hyperglycemia, Oxidative/Nitrosative Stress, and G-protein/Adenylyl Cyclase Signaling

Hyperglycemia-induced enhanced oxidative stress has also been reported in cultured VSMC and different tissues from STZ-diabetic rats (Baynes and Thorpe 1999; Baynes 1991; Cai and Harrison 2000). In addition, the contribution of enhanced production of superoxide anion (O_2^-) in the decreased expression of Giα proteins has recently been reported in aortic VSMC from STZ-diabetic rats and A10 cells exposed to high glucose (Li et al. 2008). Antioxidants such as α-tocopherol and NAC—scavengers of O_2^-—and DPI (an inhibitor of NADPH oxidase) that restored the enhanced levels of O_2^- induced by hyperglycemia also restored the hyperglycemia-induced decreased expression of Giα-2 and Giα-3 to control levels (Li et al. 2008). These studies implicate NADPH oxidase/O_2^- in hyperglycemia-evoked decreased expression of Giα proteins. We also showed that antioxidants that restored the hyperglycemia-induced decreased expression of Giα proteins to control levels also restored to control levels the decreased Gi-mediated functions (receptor-dependent and -independent), as demonstrated by the restoration of decreased inhibition of adenylyl cyclase by Ang II, C-ANP$_{4–23}$, and oxotremorine to control levels (Li et al. 2008). In addition, GTPγS-mediated decreased inhibition of forskolin-stimulated adenylyl cyclase activity (receptor-independent functions of Giα proteins) as well as enhanced stimulation of adenylyl cyclase by GTPγS and stimulatory hormones such as isoproterenol and glucagon in hyperglycemic cells were also restored to control levels by DPI (Li et al. 2008).

Hyperglycemia-induced decreased expression of Giα proteins was shown to be attributed to the increased levels of peroxynitrite because scavengers of peroxynitrite—uric acid and MnTBAP—were able to restore the hyperglycemia-induced decreased expression of Giα proteins to control level (Li et al. 2008).The implication of peroxynitrite in NO-induced decreased expression of Giα proteins in aortic and A10 VSMC has recently been reported (Bassil and Anand-Srivastava 2006b). In addition, the treatment of VSMC with peroxynitrite has also been reported to result in decreased expression of Giα proteins (Bassil and Anand-Srivastava 2006a). There is accumulating evidence supporting the hypothesis that diabetes is associated with increased nitrosative stress and peroxynitrite formation

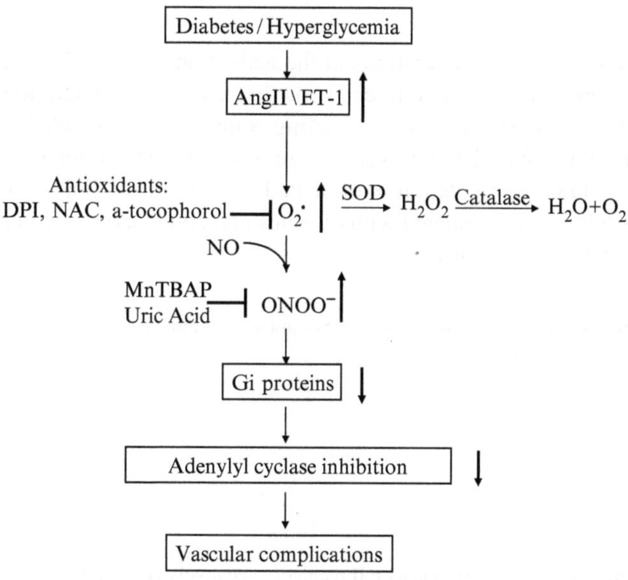

Fig. 9.2 Schematic diagram depicting the possible mechanisms by which hyperglycemia/diabetes decreases the expression of Giα proteins and adenylyl cyclase signaling. Diabetes/hyperglycemia augments the levels of vasoactive peptides including Ang II/ET-1 that enhance the oxidative stress by increasing the levels of superoxide anion (O_2^-) and peroxynitrite ($ONOO^-$). O_2^- and $ONOO^-$ decrease the levels of Gi proteins.The treatment with antioxidants and $ONOO^-$ scavengers reverses the hyperglycemia-induced decreased expression of Giα proteins and adenylyl cyclase signaling.

in several tissues both in experimental animals and in humans (Pacher et al. 2005). The increased level of nitrotyrosine, a relatively specific marker of peroxynitrite formation, has been shown in different tissues from STZ-diabetic rats and diabetic subjects (Pacher and Szabo 2006). For example, increased nitrotyrosine plasma levels were shown in type 2 diabetic patients (Cerillo et al. 2001) and iNOs-dependent peroxynitrite production was shown to be increased in platelets from diabetic individuals (Tannous et al. 1999). In addition, hyperglycemia has also been reported to induce increased nitrotyrosine formation in the artery wall of monkeys (Pennathur et al. 2001). It thus may be possible that hyperglycemia-induced enhanced levels of peroxynitrite, formed by the interaction of NO and O_2^-, may contribute to hyperglycemia-induced decreased expression of Giα proteins in VSMC. Taken together, diabetes/hyperglycemia was associated with a decreased expression of Giα proteins and associated adenylyl cyclase signaling which may be attributed to the augmented levels of vasoactive peptides such as Ang II and ET-1 that enhance the oxidative stress by increasing the levels of O_2^- and $ONOO^-$ (Figure 9.2). The ability of antioxidants to reverse the hyperglycemia-induced decreased expression of Giα proteins and adenylyl cyclase signaling to control levels suggests a key role of ROS generation in this process.

Conclusion

Diabetes is associated with alterations in the activation of several signaling events which play a key role in mediating cell growth, hypertrophy, migration, proliferation, and contractility. Many of these signaling components that include PKC, MAP kinase, G-proteins, and adenylyl cyclases are also activated in response to hyperglycemia and ROS. Dysregulation of these pathways has been linked to the cardiovascular complications associated with diabetes and could serve as potential targets in pharmacotherapy of the disease.

Acknowledgments The work in the authors laboratories are supported by grants from Canadian Institutes of Health Research.

References

Anand-Srivastava, M.B. 1989. Angiotensin II receptors negatively coupled to adenylate cyclase in rat myocardial sarcolemma. Involvement of inhibitory guanine nucleotide regulatory protein. Biochem. Pharmacol. 38:489–496.

Anand-Srivastava, M.B., A.K. Srivastava, and M. Cantin. 1987. Pertussis toxin attenuates atrial natriuretic factor-mediated inhibition of adenylate cyclase. Involvement of inhibitory guanine nucleotide regulatory protein. J. Biol. Chem. 262:4931–4934.

Azar, Z.M., M.Z. Mehdi, and A.K. Srivastava. 2006. Activation of insulin-like growth factor type-1 receptor is required for H_2O_2-induced PKB phosphorylation in vascular smooth muscle cells. Can. J. Physiol. Pharmacol. 84:777–786.

Azar, Z.M., M.Z. Mehdi, and A.K. Srivastava. 2007. Insulin-like growth factor type-1 receptor transactivation in vasoactive peptide and oxidant-induced signaling pathways in vascular smooth muscle cells. Can. J. Physiol. Pharmacol. 85:105–111.

Bassil, M., and M.B. Anand-Srivastava. 2006a. Nitric oxide modulates Gi-protein expression and adenylyl cyclase signaling in vascular smooth muscle cells. Free Radic. Biol. Med. 41:1162–1173.

Bassil, M., and M.B. Anand-Srivastava. 2006b. Peroxynitrite modulates the expression of Giα protein and adenylyl cyclase signaling in vascular smooth muscle cells. FASEB J. 20:A664.

Baynes, J.W. 1991. Role of oxidative stress in development in diabetes. Diabetes 40: 405–412.

Baynes, J.W., and S.R. Thorpe. 1999. Role of oxidative stress in diabetic complications: a new perspective on an old paradigm. Diabetes 48:1–9.

Blanc, A., N.R. Pandey, and A.K. Srivastava. 2003. Synchronous activation of ERK 1/2, $p38^{mapk}$ and PKB/Akt signaling by H_2O_2 in vascular smooth mucle cells: Potential involvement in vascular disease (review). Int. J. Mol. Med. 11:229–234.

Blanc, A., N.R. Pandey, and A.K. Srivastava. 2004. Distinct roles of Ca^{2+}, calmodulin, and protein kinase C in H_2O_2-induced activation of ERK1/2, p38 MAPK, and protein kinase B signaling in vascular smooth muscle cells. Antioxid. Redox Signal. 6:353–366.

Brownlee, M. 2001. Biochemistry and molecular cell biology of diabetic complications. Nature 414:813–820.

Bushfield, M., S.L. Griffiths, G.J. Murphy, N.J. Pyne, J.T. Knowler, G. Milligan, P.J. Parker, S. Mollner, and M.D. Houslay. 1990. Diabetes-induced alterations in the expression, functioning and phosphorylation state of the inhibitory guanine nucleotide regulatory protein Gi-2 in hepatocytes. Biochem. J. 271:365–372.

Cai, H., and D.G. Harrison. 2000. Endothelial dysfunction in cardiovascular diseases: the role of oxidative stress. Circ. Res. 87:840–844.

Cali, J.J., E.A. Balcueva, I. Rybalkin, and J.D. Robishaw. 1992. Selective tissue distribution of G protein gamma subunits, including a new form of the gamma subunits identified by cDNA cloning. J. Biol. Chem. 267:24023–24027.

Caro, J.F., M.S. Raju, M. Caro, C.J. Lynch, J. Poulos, J.H. Exton, and J.K. Thakkar. 1994. Guanine nucleotide binding regulatory proteins in liver from obese humans with and without type II diabetes: evidence for altered "cross-talk" between the insulin receptor and Gi-proteins. J. Cell Biochem. 54:309–319.

Ceriello, A., F. Mercuri, L. Quagliaro, R. Assaloni, E. Motz, L. Tonutti, and C. Taboga. 2001. Detection of nitrotyrosine in the diabetic plasma: evidence of oxidative stress. Diabetologia 44:834–848.

Chiasson, J.L., R. Rabasa-Lhoret, and A.K. Srivastava. 2006. Oxidative stress in the development of diabetes and its complication. In *Antioxidants and Cardiovascular Disease*. J.C. Tardif and M.G. Bourassa, editors. Springer, New York. 380–391.

Craven, P.A., and F.R. DeRubertis. 1989. Protein kinase C is activated in glomeruli from streptozotocin diabetic rats. Possible mediation by glucose. J. Clin. Invest. 83:1667–1675.

Daou, G.B., and A.K. Srivastava. 2004. Reactive oxygen species mediate endothelin-1-induced activation of ERK1/2, PKB, and Pyk2 signaling, as well as protein synthesis, in vascular smooth muscle cells. Free Radic. Biol. Med. 37:208–215.

Davi, G., G. Ciabattoni, A. Consoli, A. Mezzetti, A. Falco, S. Santarone, E. Pennese, E. Vitacolonna, T. Bucciarelli, F. Costantini, F. Capani, and C. Patrono. 1999. *In vivo* formation of 8-iso-prostaglandin f2alpha and platelet activation in diabetes mellitus: effects of improved metabolic control and vitamin E supplementation. Circulation 99:224–229.

Davis, R.J. 1993. The mitogen-activated protein kinase signal transduction pathway. J. Biol. Chem. 268:14553–14556.

Denton, R.M., and J.M. Tavare. 1995. Does mitogen-activated-protein kinase have a role in insulin action? The cases for and against. Eur. J. Biochem. 227:597–611.

Esposito, F., G. Chirico, G.N. Montesano, I. Posadas, R. Ammendola, T. Russo, G. Cirino, and F. Cimino. 2003. Protein kinase B activation by reactive oxygen species is independent of tyrosine kinase receptor phosphorylation and requires SRC activity. J. Biol. Chem. 278:20828–20834.

Farhangkhoee, H., Z.A. Khan, H. Kaur, X. Xin, S. Chen, and S. Chakrabarti. 2006. Vascular endothelial dysfunction in diabetic cardiomyopathy: pathogenesis and potential treatment targets. Pharmacol. Ther. 111:384–399.

Force, T., and J.V. Bonventre. 1998. Growth factors and mitogen-activated protein kinases. Hypertension 31:152–161.

Frodin, M., and S. Gammeltoft. 1999. Role and regulation of 90 kDa ribosomal S6 kinase (RSK) in signal transduction. Mol. Cell. Endocrinol. 151:65–77.

Fujita, N., Y. Furukawa, J. Du, N. Itabashi, G. Fujisawa, K. Okada, T. Saito, and S. Ishibashi. 2002. Hyperglycemia enhances VSMC proliferation with NF-kappaB activation by angiotensin II and E2F-1 augmentation by growth factors. Mol. Cell. Endocrinol. 192:75–84.

Gilman, A.G. 1984. G proteins and dual control of adenylate cyclase. Cell 36:577–579.

Griffiths, S.L., J.T. Knowler, and M.D. Houslay. 1990. Diabetes-induced changes in guanine-nucleotide-regulatory-protein mRNA detected using synthetic oligonucleotide probes. Eur. J. Biochem. 193:367–374.

Gupta, S., D. Campbell, B. Derijard, and R.J. Davis. 1995. Transcription factor ATF2 regulation by the JNK signal transduction pathway. Science 267:389–393.

Hadjiconstantinou, M., Z.X. Qu, S.E. Moroi-Fetters, and N.H. Neff. 1988. Apparent loss of Gi protein activity in the diabetic retina. Eur. J. Pharmacol. 149:193–194.

Hargrove, G.M., J. Dufresne, C. Whiteside, D.A. Muruve, and N.C. Wong. 2000. Diabetes mellitus increases endothelin-1 gene transcription in rat kidney. Kidney Int. 58:1534–1545.

Hashim, S., Y.Y. Liu, R. Wang, and M.B. Anand-Srivastava. 2002. Streptozotocin-induced diabetes impairs G-protein linked signal transduction in vascular smooth muscle. Mol. Cell. Biochem. 240:57–65.

Hashim, S., Y. Li, A. Nagakura, S. Takeo, and M.B. Anand-Srivastava. 2004. Modulation of G-protein expression and adenylyl cyclase signaling by high glucose in vascular smooth muscle. Cardiovasc. Res. 63:709–718.

Hashim, S., Y. Li, and M.B. Anand-Srivastava. 2006. G protein-linked cell signaling and cardiovascular functions in diabetes/hyperglycemia. Cell Biochem. Biophys. 44:51–64.

Hattori, Y., N. Matsuda, A. Sato, S. Watanuki, H. Tomioka, H. Kawasaki, and M. Kanno. 2000. Predominant contribution of the G protein-mediated mechanism to NaF-induced vascular contractions in diabetic rats: association with an increased level of G(qalpha) expression. J. Pharmacol. Exp. Ther. 292:761–768.

Hayashi, S., R. Morishita, H. Matsushita, H. Nakagami, Y. Taniyama, T. Nakamura, M. Aoki, K. Yamamoto, J. Higaki, and T. Ogihara. 2000. Cyclic AMP inhibited proliferation of human aortic vascular smooth muscle cells, accompanied by induction of p53 and p21. Hypertension 35:237–243.

Igarashi, M., H. Wakasaki, N. Takahara, H. Ishii, Z.Y. Jiang, T. Yamauchi, K. Kuboki, M. Meier, C.J. Rhodes, and G.L. King. 1999. Glucose or diabetes activates p38 mitogen-activated protein kinase via different pathways. J. Clin. Invest. 103:185–195.

Igarashi, M., A. Hirata, H. Yamaguchi, N. Sugae, Y. Kadomoto-Antsuki, H. Nozaki, Y. Jimbu, and M. Tominaga. 2007. Characterization of activation of MAP kinase superfamily in vasculature from diabetic rats. J. Atheroscler. Thromb. 14:235–244.

Inoguchi, T., and H. Nawata. 2005. NAD(P)H oxidase activation: a potential target mechanism for diabetic vascular complications, progressive beta-cell dysfunction and metabolic syndrome. Curr. Drug Targets. 6:495–501.

Inoguchi, T., R. Battan, E. Handler, J.R. Sportsman, W. Heath, and G.L. King. 1992. Preferential elevation of protein kinase C isoform beta II and diacylglycerol levels in the aorta and heart of diabetic rats: differential reversibility to glycemic control by islet cell transplantation. Proc. Natl. Acad. Sci. USA 89:11059–11063.

Inoguchi, T., P. Li, F. Umeda, H.Y. Yu, M. Kakimoto, M. Imamura, T. Aoki, T. Etoh, T. Hashimoto, M. Naruse, H. Sano, H. Utsumi, and H. Nawata. 2000. High glucose level and free fatty acid stimulate reactive oxygen species production through protein kinase C-dependent activation of NAD(P)H oxidase in cultured vascular cells. Diabetes 49:1939–1945.

Inoguchi, T., T. Sonta, H. Tsubouchi, T. Etoh, M. Kakimoto, N. Sonoda, N. Sato, N. Sekiguchi, K. Kobayashi, H. Sumimoto, H. Utsumi, and H. Nawata. 2003. Protein kinase C-dependent increase in reactive oxygen species (ROS) production in vascular tissues of diabetes: role of vascular NAD(P)H oxidase. J. Am. Soc. Nephrol. 14:S227–S232.

Ip, Y.T., and R.J. Davis. 1998. Signal transduction by the c-Jun N-terminal kinase (JNK)—from inflammation to development. Curr. Opin. Cell Biol. 10:205–219.

Ishii, H., M.R. Jirousek, D. Koya, C. Takagi, P. Xia, A. Clermont, S.E. Bursell, T.S. Kern, L.M. Ballas, W.F. Heath, L.E. Stramm, E.P. Feener, and G.L. King. 1996. Amelioration of vascular dysfunctions in diabetic rats by an oral PKC beta inhibitor. Science 272:728–731.

Ishikawa, Y., and C.J. Homcy. 1997. The adenylyl cyclases as integrators of transmembrane signal transduction. Circ. Res. 80:297–304.

Katsushika, S., L. Chen, J. Kawabe, R. Nilakantan, N.J. Halnon, C.J. Homcy, and Y. Ishikawa. 1992. Cloning and characterization of a sixth adenylyl cyclase isoform: types V and VI constitute a subgroup within the mammalian adenylyl cyclase family. Proc. Natl. Acad. Sci. USA 89:8774–8778.

Khan, Z.A., and S. Chakrabarti. 2003. Endothelins in chronic diabetic complications. Can. J. Physiol. Pharmacol. 81:622–634.

Koka, V., W. Wang, X.R. Huang, S. Kim-Mitsuyama, L.D. Truong, and H.Y. Lan. 2006. Advanced glycation end products activate a chymase-dependent angiotensin II-generating pathway in diabetic complications. Circulation 113:1353–1360.

Kook, H., J. Lee, S.W. Kim, and Y.H. Baik. 2002. Augmented natriuretic peptide-induced guanylyl cyclase activity and vasodilation in experimental hyperglycemic rats. Jpn. J. Pharmacol. 88:167–173.

Koya, D., and G.L. King. 1998. Protein kinase C activation and the development of diabetic complications. Diabetes 47:859–866.

Kunisaki, M., S.E. Bursell, F. Umeda, H. Nawata, and G.L. King. 1994. Normalization of diacylglycerol-protein kinase C activation by vitamin E in aorta of diabetic rats and cultured rat smooth muscle cells exposed to elevated glucose levels. Diabetes 43:1372–1377.

Laakso, M. 1999. Hyperglycemia and cardiovascular disease in type 2 diabetes. Diabetes 48:937–942.

Lavrentyev, E.N., A.M. Estes, and K.U. Malik. 2007. Mechanism of high glucose induced angiotensin II production in rat vascular smooth muscle cells. Circ. Res. 101:455–464.

Lee, H.S., S.M. Son, Y.K. Kim, K.W. Hong, and C.D. Kim. 2003. NAD(P)H oxidase participates in the signaling events in high glucose-induced proliferation of vascular smooth muscle cells. Life Sci. 72:2719–2730.

Lee, I.K., D. Koya, H. Ishi, H. Kanoh, and G.L. King. 1999. d-Alpha-tocopherol prevents the hyperglycemia induced activation of diacylglycerol (DAG)-protein kinase C (PKC) pathway in vascular smooth muscle cell by an increase of DAG kinase activity. Diabetes Res. Clin. Pract. 45:183–190.

Li, Y., G. Lappas, and M.B. Anand-Srivastava. 2007.Role of oxidative stress in angiotensin II-induced enhanced expression of Gi(alpha) proteins and adenylyl cyclase signaling in A10 vascular smooth muscle cells. Am. J. Physiol. Heart Circ. Physiol. 292:H1922–H1930.

Li, Y., M. Descorbeth, and M.B. Anand-Srivastava. 2008 Role of oxidative stress in high glucose-induced decreased expression of Giα proteins and adenylyl cyclase signaling in vascular smooth muscle cells. Am. J. Physiol. Heart Circ. Physiol. 294:H2845–2854.

Liu, W.S., and C.A. Heckman. 1998. The sevenfold way of PKC regulation. Cell Signal. 10:529–542.

Liu, Y., K. Terata, N.J. Rusch, and D.D. Gutterman. 2001. High glucose impairs voltage-gated K(+) channel current in rat small coronary arteries. Circ. Res. 89:146–152.

Livingstone, C., A.R. McLellan, M.A. McGregor, A. Wilson, J.M. Connell, M. Small, G. Milligan, K.R. Paterson, and M.D. Houslay. 1991. Altered G-protein expression and adenylate cyclase activity in platelets of non-insulin-dependent diabetic (NIDDM) male subjects. Biochim. Biophys. Acta 1096:127–133.

Mancusi, G., C. Hutter, S. Baumgartner-Parzer, K. Schmidt, W. Schutz, and V. Sexl. 1996. High-glucose incubation of human umbilical-vein endothelial cells does not alter expression and function either of G-protein alpha-subunits or of endothelial NO synthase. Biochem. J. 315(Pt 1):281–287.

Mehdi, M.Z., N.R. Pandey, S.K. Pandey, and A.K. Srivastava. 2005. H_2O_2-induced phosphorylation of ERK1/2 and PKB requires tyrosine kinase activity of insulin receptor and c-Src. Antioxid. Redox Signal. 7:1014–1020.

Monnier, L., E. Mas, C. Ginet, F. Michel, L. Villon, J.P. Cristol, and C. Colette. 2006. Activation of oxidative stress by acute glucose fluctuations compared with sustained chronic hyperglycemia in patients with type 2 diabetes. JAMA 295:1681–1687.

Moxham, C.M., and C.C. Malbon. 1996. Insulin action impaired by deficiency of the G-protein subunit G ialpha2. Nature 379:840–844.

Nakanishi, H., K.A. Brewer, and J.H. Exton. 1993. Activation of the zeta isozyme of protein kinase C by phosphatidylinositol 3,4,5-trisphosphate. J. Biol. Chem. 268:13–16.

Natarajan, R., S. Scott, W. Bai, K.K. Yerneni, and J. Nadler. 1999. Angiotensin II signaling in vascular smooth muscle cells under high glucose conditions. Hypertension 33:378–384.

Newton, A.C., and J.E. Johnson. 1998. Protein kinase C: a paradigm for regulation of protein function by two membrane-targeting modules. Biochim. Biophys. Acta 1376:155–172.

Nishizuka, Y. 1995. Protein kinase C and lipid signaling for sustained cellular responses. FASEB J. 9:484–496.

Onozato, M.L., A. Tojo, A. Goto, T. Fujita, and C.S. Wilcox. 2002. Oxidative stress and nitric oxide synthase in rat diabetic nephropathy: effects of ACEI and ARB. Kidney Int. 61:186–194.

Pacher, P., and C. Szabo. 2006. Role of peroxynitrite in the pathogenesis of cardiovascular complications of diabetes. Curr. Opin. Pharmacol. 6:136–141.

Pacher, P., I.G. Obrosova, J.G. Mabley, and C. Szabo. 2005. Role of nitrosative stress and peroxynitrite in the pathogenesis of diabetic complications. Emerging new therapeutical strategies. Curr. Med. Chem. 12:267–275.

Park, J.Y., N. Takahara, A. Gabriele, E. Chou, K. Naruse, K. Suzuma, T. Yamauchi, S.W. Ha, M. Meier, C.J. Rhodes, and G.L. King. 2000. Induction of endothelin-1 expression by glucose: an effect of protein kinase C activation. Diabetes 49:1239–1248.

Pennathur, S., J.D. Wagner, C. Leeuwenburgh, K.N. Litwak, and J.W. Heinecke. 2001. A hydroxyl radical-like species oxidizes cynomolgus monkey artery wall proteins in early diabetic vascular disease. J. Clin. Invest. 107:853–860.

Premont, R.T., J. Chen, H.W. Ma, M. Ponnapalli, and R. Iyengar. 1992. Two members of a widely expressed subfamily of hormone-stimulated adenylyl cyclases. Proc. Natl. Acad. Sci. USA 89:9809–9813.

Purdom, S., and Q.M. Chen. 2005. Epidermal growth factor receptor-dependent and – independent pathways in hydrogen peroxide-induced mitogen-activated protein kinase activation in cardiomyocytes and heart fibroblasts. J. Pharmacol. Exp. Ther. 312:1179–1186.

Pyorala, K., M. Laakso, and M. Uusitupa. 1987. Diabetes and atherosclerosis: an epidemiologic view. Diabetes Metab. Rev. 3:463–524.

Reaven, G., F. Abbasi, and T. McLaughlin. 2004. Obesity, insulin resistance, and cardiovascular disease. Recent Prog. Horm. Res. 59:207–223.

Seger, R., and E.G. Krebs. 1995. The MAPK signaling cascade. FASEB J. 9:726–735.

Singh, U., and I. Jialal. 2006. Oxidative stress and atherosclerosis. Pathophysiology 13:129–142.

Sodhi, C.P., Y.S. Kanwar, and A. Sahai. 2003. Hypoxia and high glucose upregulate AT1 receptor expression and potentiate ANG II-induced proliferation in VSM cells. Am. J. Physiol. Heart Circ. Physiol. 284:H846–H852.

Spiegel, A.M. 1987. Signal transduction by guanine nucleotide binding proteins. Mol. Cell. Endocrinol. 49:1–16.

Strassheim, D., T. Palmer, and M.D. Houslay. 1991. Genetically acquired diabetes: adipocyte guanine nucleotide regulatory protein expression and adenylate cyclase regulation. Biochim. Biophys. Acta 1096:121–126.

Strathmann, M., T.M. Wilkie, and M.I. Simon. 1989. Diversity of the G-protein family: sequences from five additional alpha subunits in the mouse. Proc. Natl. Acad. Sci. USA 86:7407–7409.

Stryer, L., and H.R. Bourne. 1986. G proteins: a family of signal transducers. Annu. Rev. Cell Biol. 2:391–419.

Tabet, F., E.L. Schiffrin, and R.M. Touyz. 2005. Mitogen-activated protein kinase activation by hydrogen peroxide is mediated through tyrosine kinase-dependent, protein kinase C-independent pathways in vascular smooth muscle cells: upregulation in spontaneously hypertensive rats. J. Hypertens. 23:2005–2012.

Tan, Y., J. Rouse, A. Zhang, S. Cariati, P. Cohen, and M.J. Comb. 1996. FGF and stress regulate CREB and ATF-1 via a pathway involving p38 MAP kinase and MAPKAP kinase-2. EMBO J. 15:4629–4642.

Tang, W.J., and A.G. Gilman. 1991. Type-specific regulation of adenylyl cyclase by G protein beta gamma subunits. Science 254:1500–1503.

Tannous, M., R.A. Rabini, A. Vignini, N. Moretti, P. Fumelli, B. Zielinski, L. Mazzanti, and B. Mutus.1999. Evidence for iNOS-dependent peroxynitrite production in diabetic platelets. Diabetologia 42:539–544.

Taussig, R., L.M. Quarmby, and A.G. Gilman. 1993. Regulation of purified type I and type II adenylylcyclases by G protein beta gamma subunits. J. Biol. Chem. 268:9–12.

Toro, M.J., E. Montoya, and L. Birnbaumer. 1987. Inhibitory regulation of adenylyl cyclases. Evidence inconsistent with beta gamma-complexes of Gi proteins mediating hormonal effects by interfering with activation of Gs. Mol. Endocrinol. 1:669–676.

Touyz, R.M., C. Deschepper, J.B. Park, G. He, X. Chen, M.F. Neves, A. Virdis, and E.L. Schiffrin. 2002. Inhibition of mitogen-activated protein/extracellular signal-regulated kinase improves endothelial function and attenuates Ang II-induced contractility of mesenteric resistance arteries from spontaneously hypertensive rats 1. J. Hypertens. 20:1127–1134.

Touyz, R.M., G. Yao, E. Viel, F. Amiri, and E.L. Schiffrin. 2004. Angiotensin II and endothelin-1 regulate MAP kinases through different redox-dependent mechanisms in human vascular smooth muscle cells. J. Hypertens. 22:1141–1149.

Tucek, S., P. Michal, and V. Vlachova. 2001. Dual effects of muscarinic M2 receptors on the synthesis of cyclic AMP in CHO cells: background and model. Life Sci. 68:2501–2510.

Ushio-Fukai, M., R.W. Alexander, M. Akers, and K.K. Griendling. 1998. p38 mitogen-activated protein kinase is a critical component of the redox-sensitive signaling pathways activated by angiotensin II. Role in vascular smooth muscle cell hypertrophy. J. Biol. Chem. 273:15022–15029.

Wang, X.Z., and D. Ron. 1996. Stress-induced phosphorylation and activation of the transcription factor CHOP (GADD153) by p38 MAP kinase. Science 272:1347–1349.

Weber, L.P., and K.M. Macleod. 1997. Influence of streptozotocin diabetes on the alpha-1 adrenoceptor and associated G proteins in rat arteries. J. Pharmacol. Exp. Ther. 283:1469–1478.

Wickman, K.D., J.A. Iniguez-Lluhl, P.A. Davenport, R. Taussig, G.B. Krapivinsky, M.E. Linder, A.G. Gilman, and D.E. Clapham. 1994. Recombinant G-protein beta gamma-subunits activate the muscarinic-gated atrial potassium channel. Nature 368:255–257.

Widmann, C., S. Gibson, M.B. Jarpe, and G.L. Johnson. 1999. Mitogen-activated protein kinase: conservation of a three-kinase module from yeast to human. Physiol. Rev. 79:143–180.

Williams, B., P. Tsai, and R.W. Schrier. 1992. Glucose-induced downregulation of angiotensin II and arginine vasopressin receptors in cultured rat aortic vascular smooth muscle cells. Role of protein kinase C. J. Clin. Invest. 90:1992–1999.

Williams, B., B. Gallacher, H. Patel, and C. Orme. 1997. Glucose-induced protein kinase C activation regulates vascular permeability factor mRNA expression and peptide production by human vascular smooth muscle cells *in vitro*. Diabetes 46:1497–1503.

Wong, Y.H., B.R. Conklin, and H.R. Bourne. 1992. Gz-mediated hormonal inhibition of cyclic AMP accumulation. Science 255:339–342.

Xu, Q., Y. Liu, M. Gorospe, R. Udelsman, and N.J. Holbrook. 1996. Acute hypertension activates mitogen-activated protein kinases in arterial wall 2. J. Clin. Invest. 97:508–514.

Yasuda, H., K. Maeda, M. Sonobe, T. Kawabata, M. Terada, T. Hisanaga, Y. Taniguchi, R. Kikkawa, and Y. Shigeta. 1994. Metabolic effect of PGE1 analogue 01206.alpha CD on nerve Na(+)-K(+)-ATPase activity of rats with streptozocin-induced diabetes is mediated via cAMP: possible role of cAMP in diabetic neuropathy. Prostaglandins 47:367–378.

Yasunari, K., M. Kohno, H. Kano, K. Yokokawa, M. Minami, and J. Yoshikawa. 1999. Antioxidants improve impaired insulin-mediated glucose uptake and prevent migration and proliferation of cultured rabbit coronary smooth muscle cells induced by high glucose. Circulation 99:1370–1378.

Zhang, Z., K. Apse, J. Pang, and R.C. Stanton. 2000. High glucose inhibits glucose-6-phosphate dehydrogenase via cAMP in aortic endothelial cells. J. Biol. Chem. 275:40042–40047.

Zheng, X.L., J. Guo, H. Wang, and C.C. Malbon. 1998. Expression of constitutively activated Gialpha2 *in vivo* ameliorates streptozotocin-induced diabetes. J. Biol. Chem. 273:23649–23651.

Zhuang, S., and R.G. Schnellmann. 2004. H_2O_2-induced transactivation of EGF receptor requires Src and mediates ERK1/2, but not Akt, activation in renal cells. Am. J. Physiol. Renal Physiol. 286:F858–F865.

Zinck, R., M.A. Cahill, M. Kracht, C. Sachsenmaier, R.A. Hipskind, and A. Nordheim. 1995. Protein synthesis inhibitors reveal differential regulation of mitogen-activated protein kinase and stress-activated protein kinase pathways that converge on Elk-1. Mol. Cell Biol. 15:4930–4938.

Chapter 10
Methylglyoxal and Insulin Resistance

Lingyun Wu

Summary

Insulin resistance leads to a number of pathophysiological situations such as type 2 diabetes mellitus and hypertension. All of these clinical manifestations are associated with increased cardiovascular risks. Currently, the causes and mechanisms of insulin resistance remain poorly understood. Methylglyoxal (MG) is a highly reactive metabolite of sugar. Increased MG levels and MG-related advanced glycation endproducts have been reported in different insulin resistance states. MG increases oxidative stress and activates inflammatory genes including NF-κB p65. The correlation between enhanced MG formation and development of insulin resistance as well as the possible underlying mechanisms are discussed and future research directions in this regard foreseen.

Introduction

Insulin Resistance

Insulin resistance is a state in which increased concentrations of insulin are required to produce a given biological response (Rett 1999). Abnormalities of this action of insulin can lead to a number of important clinical and pathophysiological outcomes. Insulin resistance syndrome has been used as a common denominator to cluster high blood pressure, glucose intolerance, hyperinsulinemia, high triglyceride, low high-density lipoproteins (HDL)-cholesterol, and both central and overall obesity (Rett 1999). Other terms used to describe this cluster include "metabolic syndrome" and "syndrome X" (Rett 1999). Type 2 diabetes mellitus (T2DM) is characterized by varying degrees of impaired insulin secretion and insulin resistance (Martin et al. 1992). It is generally agreed that insulin resistance is a key pathogenetic factor for T2DM although the underlying mechanisms are not clear. It has also been noticed

A. K. Srivastava, M. B. Anand-Srivastava (eds.) *Advances in Biochemistry in Health and Disease.* Signal Transduction in the Cardiovascular System in Health and Disease,

that about 85% of T2DM patients (Stumvoll and Haring 2001) are obese. Interestingly, epidemiological, clinical, and experimental studies have now indicated that essential hypertension is an insulin-resistant state even in the absence of T2DM or obesity (Rocchini 1995). In addition to its effect on glucose metabolism, insulin is also a potent vasodilator. Some evidence showed that insulin-mediated vasodilation was impaired in insulin resistance syndrome (Heise et al. 1998), which might contribute to vascular dysfunction in insulin resistance syndrome.

Causes and Mechanisms for Insulin Resistance

Excess body fat is generally accepted as the strongest determinant of insulin resistance. Several factors have been identified through which obesity is thought to cause insulin resistance. These factors include free fatty acids (FFAs) (Fisher and Kahn 2003), tumor necrosis factor-α (TNF-α) (Galgani and Diaz 2000), and resistin (Steppan et al. 2001). In obese individuals, release of FFAs into plasma is proportional to adipose tissue mass. It has been suggested that FFAs inhibit muscle glycogen synthesis (Stumvoll and Haring 2001), which might cause insulin resistance. Interestingly, increased FFA oxidation accelerates endogenous glucose production by stimulation of gluconeogenesis (Randle et al. 1994). Obese humans express 2.5-fold more TNF-α in fat tissue than do lean control individuals (Hotamisligil et al. 1997) and expression level of this cytokine is correlated with degree of insulin resistance. Resistin is a new protein added to a panoply of factors involved in the cross-talk between adipose tissue and insulin-sensitive tissues (Steppan et al. 2001). At present, however, convincing evidence for an important role of TNF-α or resistin in development of insulin resistance associated with human obesity is still scarce (Hube and Hauner 1999). Although different knockout mouse models such as those lacking insulin receptor or insulin receptor substrates, or those lacking PI-3-kinase have been used in the studies of insulin resistance (Kadowaki 2000; Kulkarni et al. 1999), whether or how insulin resistance is determined genetically is still open for debate.

Currently, the cause and mechanisms of insulin resistance remain poorly understood. Both genetic and environmental factors may be involved (Alzaid 1996). Searching for the causes and mechanisms of insulin resistance is crucial for prevention and treatment of different insulin resistance states including T2DM and hypertension.

Methylglyoxal (MG) and Insulin Resistance

MG Formation and Degradation

MG Formation

The main source of MG in mammals is anaerobic glycolysis, i.e., nonenzymatic and enzymatic elimination of phosphate from glyceraldehyde-3-phosphate and

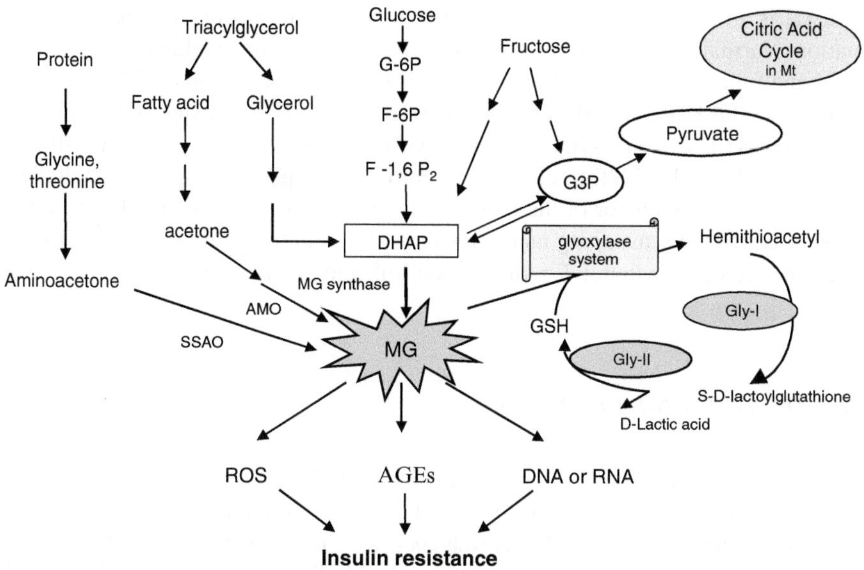

Fig. 10.1 Cellular formation and metabolism of methylglyoxal (MG). AGEs: advanced glycoxidation endproducts; DHAP: dihydroxyacetone phosphate; G3P: glyceraldehyde 3-phosphate; F-6P: fructose 6-phosphate; F-1, $6P_2$: fructose 1,6-bisphosphate; Gly-I & II: glyoxalase I & II; SSAO: semicarbazide-sensitive amine oxidase; AMO: amine oxidase.

dihydroxyacetone phosphate (Figure 10.1). MG synthase catalyzes dihydroxyacetone to form MG and this enzyme has been purified from goat liver (Ray and Ray 1981). Other sources of MG in mammalian metabolism are acetone catalyzed by acetol monooxygenase (AMO) (Casazza et al. 1984) and aminoacetone catalyzed by semicarbazide-sensitive amine oxidase (SSAO) (Yu 1998) (Figure 10.1). SSAO is found in high amounts within vascular smooth muscle cells (VSMC) and in human plasma (Ekblom 1998). It has been suggested that plasma SSAO might be secreted from VSMC (Blaschko 1974).

MG Degradation

MG is metabolized by the glyoxalase system. This system is present in the cytosol of all mammalian cells, expressed during embryogenesis and through maturation, adult life, and senescence. The glyoxalase system is comprised of two enzymes, glyoxalase-I and glyoxalase-II, and one cofactor, glutathione (GSH) (Chang and Wu 2006) (Figure 10.1). The main substrates for this system are MG and glyoxal formed from lipid peroxidation and glycation reactions (Thornalley 1996). MG is first converted irreversibly by glyoxalase-I to S-D-lactoylglutathione in the presence of reduced GSH as the cofactor, and then to D-lactate by glyoxalase-II. MG can also be converted to AGE precursor acetol catalyzed by aldose reductase, which has a 40-fold lower enzyme activity than glyoxalase-I (Thornalley 1996).

GSH is an important cofactor for metabolizing MG, thus decreasing AGE formation. Normal metabolism of MG is largely dependent on availability of cellular GSH and function of the GSH pathway. This pathway is composed of GSH, GSH-peroxidase (GSH-Px), GSH-reductase (GSH-Red), GSH-S-transferase (GST), and L-γ-glutamyl-L-cysteine synthase (GCS). GSH is used to form a conjugate with its targets, catalyzed by GST, to eliminate harmful electrophilic compounds including *o*-hydroquinone precursor of the redox cycler aminochrome (Kalapos 1999). GSH is also essential in scavenging many free radicals, including hydrogen peroxide, organic peroxide, and vitamin E radicals (Juurlink and Paterson 1998).

MG-Induced Protein Glycation

MG is very electrophilic and tends to interact readily with certain arginine or lysine residues in proteins, leading to glycation of proteins and therefore yield of irreversible advanced glycated endproducts (AGEs). In comparison with glucose-induced glycation that undergoes a reversible Maillard reaction with N-terminal amino groups or lysyl chains to form fructosamine, an early glycation adduct (Kalapos 1999), MG induces glycation through an oxidation process and generates irreversible end glycation adducts. MG is the most reactive AGE precursor. The irreversible reaction of MG with arginine residues of protein forms argpyrimidine (Ahmed et al. 2002), while reaction of MG with lysine residues of protein forms $N\varepsilon$-carboxymethyl-lysine (CML) and $N\varepsilon$-carboxyethyl-lysine (CEL). Subsequently, these proteins crosslink to form MG-derived lysine dimers, imidazolysin (MOLD) (Ahmed et al. 2002). Antibodies raised against some of these epitopes, such as mAB3C and mAb6B for argpyrimidine, have helped to identify these specific dicarbonyl modified proteins both in arterial tissues and in plasma (Uchida et al. 1997). Using immunofluorescence staining with a specific monoclonal antibodies against MG-induced CEL or CML, MG-induced glycation of lysine residue of proteins are also successfully identified in different tissues (Wang et al. 2004, 2005, 2006). It is worth noting that this glycation is selective depending on structural configuration and/or physical locations of the modified proteins that govern which residue(s) is glycated. In support of this scheme, recent studies on human peripheral blood lymphocytes or on plasma proteins from different animals (rat, mouse, and monkey) show that only a limited number of proteins undergo glycation (Jana et al. 2002).

Formation of AGEs in many cases results in reduced or abolished function of enzymes, receptors, carriers and structural proteins. For example, MG induces glycation of arginine residue of glutathione reductase, thus inactivating this enzyme and decreasing its scavenging ability for free radicals (Vander Jagt et al. 1997). MG and related AGEs such as CEL have been recognized as indicators of carbonyl overload *in vivo* (Baynes and Thorpe 1999), and are correlated with age (Li et al. 1996). AGEs can affect cellular metabolism in several ways. One is by alteration of cellular protein functions such as glycosylation of GSH reductase, inactivating this enzyme

(Blakytny and Harding 1992). Another is by interaction with the receptor of AGE (RAGE), which is present in endothelial cells, VSMC and mononuclear phagocytes (Kranzhofer et al. 1999). In addition to modifying free amino of proteins, MG also tends to interact with guanyl residues of nucleic acids (Thornalley 1996). Reaction of MG with guanyl residues in DNA and RNA results in transcriptional changes. MG may also directly affect contractile functions and proliferation of VSMC. However, study in this area is limited.

Increased MG Level in Insulin Resistance Syndromes

As mentioned earlier, MG is formed during various metabolic processes including metabolism of acetone from lipolysis and metabolism of threonine from protein catabolism (Yu 1998). Increased MG formation may occur due to an increased availability of its precursors such as increased plasma glucose, or administration of ethanol or threonine (Ma et al. 1989). Interestingly, there is a linkage between most obese subjects and their life style with over-consumption of food containing carbohydrates and fat and/or overdrinking beverages containing high levels of ethanol. This may not come as a surprise since MG is ubiquitous in beverages or coffee. Accumulation of glucose, fat or ethanol *in vivo* provides available precursors for MG generation and eventually causes an overproduction of MG, which may contribute to development of insulin resistance. On the other hand, MG levels are dependent on the balance between its synthesis and catabolism. When formation of MG exceeds its degradation, accumulation of MG occurs even with normal glucose levels or other MG precursors (Figure 10.1). By the same token, accumulation of MG may occur when degradation ability of the glyoxalase system is suppressed even with a normal cellular flux of glucose. An increased cellular MG concentration has been observed after application of glyoxalase-I inhibitors. Using Northern blot analysis, our previous study from human astroglial cells has demonstrated significant decrease in activity of glyoxalase-I and increase in AGE formation, which resulted from downregulation of glyoxalase-I expression (Wu et al. 2001).

Increased MG Level in Diabetes

Generally, physiological concentration of plasma MG in rats is approximately 5 μM (Nagaraj et al. 2002). Plasma level of MG increased from 1.4 μM in healthy humans to 3.6 μM in patients with type 1 diabetes mellitus (T1DM) (McLellan et al. 1994). Cultured cells may produce more MG since MG concentration up to 310 μM was detected in cultured Chinese hamster ovary cells (Chaplen et al. 1998). On the other hand, lower concentrations of plasma MG have been reported. For example, McLellan et al. (1994) detected plasma MG level at 470.7 pmole/g (about 0.47 μM) in insulin-dependent patients versus 79.8 pmole/g in control subjects (McLellan et al. 1994). Low levels of plasma MG, less than 0.2 μM, were also reported in normal and diabetic patients (Beisswenger et al. 1999). The reason for the reported difference in plasma MG levels is not clear and it can be partially ascribed to the use of different MG measurement methods. Since MG is mainly formed during glycolysis, it is likely that intracellular MG concentration would be considerably higher than circulating level (Shinohara et al. 1998).

It has been reported that formation of MG in cultured human red blood cells or bovine endothelial cells is increased under hyperglycemic conditions or by the addition of precursors of MG including fructose, D-glyceraldehyde, dihydroxyacetone, acetone, and hydroxyacetone (Shinohara et al. 1998; H. Wang et al. 2006; Wu 2006). Obviously, hyperglycemia even for a short period of time is sufficient to increase MG concentrations *in vivo*. Cellular MG level is dependent on the balance between its synthesis and catabolism. It has been shown that serum concentration of MG increases 5- to 6-fold in patients with T1DM and 2- to 3-fold in patients with T2DM (McLellan et al. 1994). Significantly increased activities of glyoxalase-I in red blood cells were observed in both types of diabetes patients (McLellan et al. 1994). Increased circulating MG in these patients may be largely due to increased activities of plasma SSAO (Ekblom 1998) and plasma AMO (Ekblom 1998). Biochemical and clinical evidence suggests that increased formation of MG in diabetes mellitus is linked to the development of diabetic complications such as microvascular damage in the eyes or kidneys (retinopathy or nephropathy), but the exact role of this dicarbonyl in this process remains largely unknown (Wu and Juurlink 2002).

Increased MG in Hypertension

High MG levels may also play an important role in other insulin resistance states such as hypertension. For instance, characteristic protein modification induced by MG has been found in the aorta of stroke-prone spontaneously hypertensive rats (SHRsp) (Mizutani et al. 1999b). Vasdev et al. (1998b) reported that the systolic blood pressure was significantly increased 1 week after Wistar Kyoto (WKY) rats were treated with high fructose (4%) or MG (0.2–0.8%) in drinking water. In addition, increased aldehyde conjugates and microvascular damage were observed in the kidney but not in other tested tissues such as the heart, liver, and muscle (Vasdev et al. 1998b). Two mechanisms have been proposed for MG-induced hypertension. One is nephropathy due to a large amount of MG or fructose that flowed through the kidney (Vasdev et al. 1998a). The other is oxidative stress induced by extracellular MG (Vasdev et al. 1998b). Unfortunately, this study did not measure plasma or cellular levels of MG after MG oral uptake treatment. Artificially increasing extracellular MG level most likely simulates the condition of diabetes with increased plasma MG or MG precursors with developed microvascular damage, but it is unlikely that this maneuver represents what happens in essential hypertension. No difference has been reported in plasma glucose level among WKY rats, spontaneously hypertensive rats (SHR) and SHRsp at the age of 12 weeks, although a significant elevation in blood pressure had developed in both SHR and SHRsp (Mizutani et al. 1999a).

We observed a significant increase in blood pressure coinciding with elevated MG level in plasma and aorta of SHR in an age-dependent fashion compared to

Table 10.1 MG levels in plasma, aorta, and kidney from WKY rats and SHR at different ages

	Rats	5 weeks	8 weeks	13 weeks	20 weeks
BP (mm Hg)	WKY	123 ± 1	121 ± 1	121 ± 1	119 ± 1
	SHR	130 ± 2	168 ± 1**	204 ± 1**	202 ± 1**
MG levels in plasma	WKY	11.2 ± 0.4	9.1 ± 0.8	18.5 ± 2.7	14.2 ± 3.5
(nmol/ml)	SHR	11.3 ± 0.6	13.8 ± 0.7*	30.3 ± 2.0*	33.6 ± 2.1*
MG levels in aorta	WKY	0.93 ± 0.12	0.75 ± 0.09	0.83 ± 0.01	1.38 ± 0.25
(nmol/mg protein)	SHR	0.92 ± 0.12	1.06 ± 0.1*	1.2 ± 0.11*	2.57 ± 0.2*
MG levels in kidney	WKY	0.67 ± 0.15	0.79 ± 0.20	0.71 ± 0.06	0.22 ± 0.04
(nmol/mg protein)	SHR	0.67 ± 0.12	0.82 ± 0.10	0.91 ± 0.09*	0.30 ± 0.06*

* $p < 0.05$, ** $p < 0.001$ for SHR versus WKY rats of the same age group, $n = 3$–5 for each group. (Reproduced with permission from Wang et al., 2005. J. Hypertens. 23:1565–1573; and Wang et al., 2004. Kidney Int. 66:2315–2321.)

age-matched WKY rats (Table 10.1). No difference was observed in blood glucose levels between these two strains (Chen et al. 1994). More interestingly, there was a significant increase of MG level in plasma and aorta, but not in kidney or heart, in SHR at an early age of 8 weeks (Wang et al. 2004, 2005). Our findings suggest that, in addition to diabetes/hyperglycemic or hyperlipidemic conditions, accumulation of MG in blood vessel wall plays an important role in development of hypertension or its complications even in the absence of diabetes. Using immunofluorescence staining with specific monoclonal antibodies against MG-induced CEL and CML, we also showed a strong association of MG and its AGE products, CML and CEL, with hypertension in SHR. The blood pressure of SHR was not different from that of WKY rats at 5 weeks of age. From 8 weeks onwards, blood pressure of SHR was significantly elevated compared to age-matched WKY rats. Importantly, this increase was associated with a progressively increased MG and MG-induced CML and CEL in kidney and aorta of SHR, compared to age-matched WKY rats (Wang et al. 2004, 2005). Our study suggests that increased MG and related AGEs may be causative for hypertension development.

Normal catabolism of MG is largely dependent on availability of cellular GSH and activities of GSH-related enzymes, including GSH-Px and GSH-Red. Malfunction of the GSH system could potentially evoke cellular damage. For instance, treating normotensive rats with buthionine sulfoximine (BSO) for 2 weeks has been reported to decrease cellular GSH by 70% and to induce hypertension (Vaziri et al. 2000). In addition, reduction of intracellular GSH level by BSO resulted in sustained and reversible narrowing of rat femoral arteries (Zhou et al. 1996). We showed increased MG level with decreased GSH and lower activities of GSH-Red and GSH-Px in aortic VSMC from 13-week-old SHR compared to those from normotensive WKY rats (Wu and Juurlink 2002). The question that remains to be answered is whether increased cellular MG level is the key link between an impaired GSH pathway and essential hypertension. We recently observed elevated levels of MG in plasma or the aorta and MG-induced CEL and CML in the aorta of SHR as early as 8 weeks of age. These changes preceded decreased GSH level (Table 10.2) and

Table 10.2 GSH levels in plasma, aorta and kidney from WKY rats and SHR at different ages

	Rats	5 weeks	8 weeks	13 weeks	20 weeks
GSH levels in plasma	WKY	35.06 ± 0.6	32.96 ± 0.9	32.93 ± 1.3	
(nmol/ml)	SHY	35.18 ± 1.2	32.04 ± 0.3	30.86 ± 1	
GSH levels in aorta	WKY	0.66 ± 0.14	1.05 ± 0.14	2.72 ± 0.1*	
(nmol/mg protein)	SHR	0.96 ± 0.21	1.03 ± 0.1	1.80 ± 0.2	
GSH levels in kidney	WKY	95.8 ± 10	82.5 ± 6.2	75.3 ± 11.4	60.7 ± 3
(nmol/mg protein)	SHR	94.6 ± 6.1	76.5 ± 8.7	67.4 ± 11.1	50.1 ± 4.3*

*$p < 0.05$ for SHR versus WKY rats of the same age group, $n = 3$–5 for each group. (Reproduced with permission from Wang et al., 2005. J. Hypertens. 23:1565–1573; and Wang et al., 2004. Kidney Int. 66:2315–2321.)

suppressed activities of GSH-Px and GSH-Red in the aorta of SHR, which appeared at the age of 13 weeks. Delayed decrease of GSH in SHR indicates that the primary reason for early onset of MG increase could be an abnormally increased MG production rather than dysfunctional MG degradation.

Mechanisms for Role of MG in Development of Insulin Resistance

MG and ROS/RNS

Reactive oxygen species (ROS) have been linked to pathological processes of insulin resistance (Wu and de Champlain 1999). ROS including O_2^- and H_2O_2 are constantly generated under normal conditions as a consequence of aerobic metabolism. Increased production of O_2^- and H_2O_2 has been observed in different cardiovascular diseases such as hypertension (Wu and Juurlink 2001). Nitric oxide (NO), one of the reactive nitrogen species (RNS), is closely linked to the ROS family (Chang et al. 2005). NO plays a crucial role in regulation of vasomotor tone and blood flow. Endothelial cells (ECs) generate NO in response to different stimuli including acetylcholine (ACh) or insulin (Moncada et al. 1991). NO, in turn, diffuses to neighboring smooth muscle cells, thereby stimulating soluble guanylyl cyclase (sGC), leading to cGMP formation and vasorelaxation (Furchgott and Zawadzki 1980). This gasotransmitter is unique in that it rapidly reacts with O_2^- to form peroxynitrite ($ONOO^-$). $ONOO^-$ can cross cell membranes freely and is an extremely strong and reactive oxidant. Obviously, formation of $ONOO^-$ results in alteration of the biological function of NO. Reduction of NO with decreased expression of eNOS has been observed in animal models of hypertension (Govers and Rabelink 2001).

Recently, it was shown that MG induces H_2O_2 in neutrophils (Ward and McLeish 2004). In the presence of thrombin, MG induces platelet H_2O_2 accumulation and aggregation (Leoncini and Poggi 1996). Our previous studies also showed that MG increased oxidative stress in rat VSMC (Chang et al. 2005; Wu and Juurlink 2002). These observations have been obtained using the DCFH assay. This assay detects oxidation of $DCFH_2$ inside cells by H_2O_2 or $ONOO^-$ (Kooy et al. 1997) to form oxidized DCF in the cytosol. This is the reason why inhibitors of NOS have been used for differentiation of $ONOO^-$ from H_2O_2 when detecting oxidized-DCF formation (Kooy et al. 1997). However, it remains unclear which type of ROS is induced by MG in VSMC or ECs of blood vessels and whether MG induces $ONOO^-$ generation. Our most recent study showed that MG significantly increased $O_2{}^-$ production, which was markedly reduced by $O_2{}^-$ scavenger SOD or NAD(P)H oxidase inhibitor DPI in VSMC. In the presence of NOS inhibitor L-NAME, SOD, or DPI, MG-induced oxidized DCF is significantly inhibited. This provides evidence that MG can induce not only $O_2{}^-$ and H_2O_2 but also $ONOO^-$ in VSMC.

Moreover, ROS can induce cell proliferation. From this point of view, MG and oxidative stress are inextricably and insidiously linked. Considering the importance of ROS/RNS in cell signaling pathways (Maulik and Das 2002), especially in redox cell signaling, it is tempting to portray MG as a linker between glucose metabolism and cellular signal transductions, or a regulator in redox tone and redox cell signaling, rather than a simple toxic by-product. Overproduction of MG and MG-induced ROS will eventually impair vascular function and cause insulin resistance.

MG and Calcium Homeostasis

Extensive evidence from clinical and basic research has demonstrated enhanced vascular contractility or impaired endothelial-dependent vascular relaxation in insulin resistance syndromes (Girouard et al. 2003). Both extracellular calcium entry and intracellular calcium release contribute to changes in intracellular free calcium concentration in VSMC, thus contraction of these cells. Altered calcium homeostasis in VSMC and ECs in insulin resistance has also been reported (Pogan et al. 2001). Because of its electrophilic nature, MG could directly interact with free amino acid residues such as sulfhydryl groups of proteins, altering biological functions of VSMC or ECs in blood vessel walls and causing increased peripheral resistance. Chronic treatment of normal rats with oral MG elevated resting intracellular calcium level in platelets (Vasdev et al. 1998a). Whether calcium level in VSMC was also changed by this treatment was not studied. Another study showed that application of MG to isolated rat pancreatic beta-cells increased intracellular Ca^{2+} and depolarized membrane within 5 min in a reversible manner (Cook et al. 1998). Growing evidence shows that PKA- or PKG-induced phosphorylation of sulfhydryl groups of IP_3 receptors and L-type Ca^{2+} channels (Opie 1998) would decrease intracellular calcium release from IP_3 sensitive pool and inhibit calcium influx across cell membrane. Conversely, interaction of MG with sulfhydryl groups of IP_3 receptors or L-type Ca^{2+} channels might increase calcium release and/or influx, resulting in increased cellular Ca^{2+} and enhanced vascular contractility. This possibility has not been examined in VSMC.

Recently, the role of oxidative stress in modifying different signal transduction pathways has received much attention. Enhanced cellular ROS formation such as GSSG with tert-butyl hydroperoxide induced an increase in calcium-induced calcium release in hepatocytes (Elliott and Koliwad 1997). Treatment of endothelial cells with GSSG leads to increased IP_3 receptor activity and depleted IP_3-sensitive Ca^{2+} stores, attenuating intracellular Ca^{2+} response to agonist stimulation (Elliott and Koliwad 1995). Decreased cellular GSH level with increased GSSG would lead to decreased activity of the glyoxalase system and enhanced cellular MG, which might play a role in GSSG-induced alteration in intracellular Ca^{2+} homeostasis. It has been reported that $ONOO^-$ impairs the sarcoplasmic reticulum Ca^{2+} pump in pig coronary artery smooth muscle (Schmidt et al. 2004) and triggers apoptosis of cultured rat aortic VSMC (Li et al. 2003). Increased production of O_2^- in SHR reduces NO function and leads to altered blood vessel tone (Cuzzocrea et al. 2004). Formation of $ONOO^-$ affects endothelial integrity and results in endothelial cell dysfunction (Cuzzocrea et al. 2004). The *in vitro* study showed that MG could directly react with hydrogen peroxide to form several new compounds, including acetic acid that potently acetylated amino groups of guanosine and induced mutagenicity (Nukaya et al. 1993). The mutagenicity of MG was enhanced about 30 times in the presence of hydrogen peroxide (Ferro and Webb 1997). Interestingly, hydrogen peroxide is also formed as a side-product in deamination reaction mediated by all amine oxidases. For instance, when aminoacetone is deaminated by SSAO, MG is generated and hydrogen peroxide is also formed from this reaction (Ekblom 1998). How interaction of MG with ROS affects calcium handling by VSMC and vascular contractility in vascular tissues from insulin resistance such as diabetes and hypertension has been unknown.

MG and Vascular Proliferation

MG and NF-κB

In quiescent cells, NF-κB is held inactive within the cytoplasm by interaction with inhibitor κB (IκBα) or other IκB family members (Beg and Baldwin 1993). After stimulation with tumor necrosis factor-α (TNF-α), lipopolysaccharide (LPS), or IL-1, IκBα undergoes rapid phosphorylation on serine residues 32 and 36; this signals proteolytic degradation of IκBα in proteasomes (Brown et al. 1995). Dissociation of IκBα from NF-κB exposes the nuclear localization signal (Jans and Hubner 1996) on NF-κB, which directs translocation of transcription factor to the nucleus to promote expression of genes that mediate inflammatory reactions. Superoxide-induced activation of NF-κB is due to phosphorylation of IκBα (Ogata et al. 2000). H_2O_2-induced activation of NF-κB is not associated with IκBα degradation and may involve tyrosine phosphorylation since H_2O_2-induced activation of NF-κB cannot be blocked by IκBα inhibitor MG-115, but is potentiated by inhibition of tyrosine phosphatases (Canty et al. 1999). Activation of NF-κB and induction of the associated genes have been suggested to be responsible for inflammatory and proliferative vascular response (Hoare et al. 1999). Proinflammatory cytokines (TNF-α and IL-1) and platelet-derived growth factor (PDGF) activated NF-κB in diabetes and atherosclerosis (Marumo et al. 1997). Increased activation of NF-κB has been observed in macrophages from DOCA-salt hypertensive rats, a secondary hypertension model (Beswick et al. 2001). Our previous study demonstrated increased activation of NF-κB in kidney of stroke-prone SHR. This increased expression of NF-κB can be significantly inhibited by antioxidant sulforaphane, a compound extracted from broccoli (Wu et al. 2004). We also observed that treatment with MG activated NF-κB p65 in cultured VSMC from aorta (Wu and Juurlink 2002) and mesenteric artery (Wu 2005). MG-induced oxidative stress might play an important role in activation of NF-κB since H_2O_2 activates NF-κB p65 with degradation of IκBα in SHR VSMC (Wu and Juurlink 2002). Activation of NF-κB by O_2^- and H_2O_2 in human ECs (Canty et al. 1999; Ogata et al. 2000) and by $ONOO^-$ in ECs (Cooke and Davidge 2002) has also been observed.

Activation of NF-κB has been demonstrated to be essential for proliferation of VSMC (Cooke and Davidge 2002). Recent studies suggest that NF-κB may function more generally as a central regulator of stress responses. NF-κB blocks cell apoptosis in several cell types including hepatocyte (Li et al. 1999). It is quite intriguing that many signals that initiate apoptosis also activate NF-κB, which suppresses apoptosis. This seemingly contradictory phenomenon has manifested itself in the case of the TNF pathway since NF-κB induces a group of gene products that function cooperatively at the earliest checkpoint to suppress TNF-α-mediated apoptosis (Wang et al. 1996). It is likely that MG plays an important role in activation of NF-κB, contributing to vascular proliferation during development of insulin resistance.

MG and MAPK

Mitogen-activated protein kinases (MAPK), including ERK1/2, JNK, and p38, are a group of enzymes that phosphorylate serine/threonine residues of proteins (Force et al. 1996). Activator protein-1 (AP-1), a transcription factor that can be activated by MAPK, is also involved in a variety of cellular response (Shaulian and Karin 2001). Activation of ERK, JNK, p38, and AP-1 can induce cell proliferation or growth but JNK, p38, and AP-1 can also induce apoptosis (Kim and Iwao 2003). Increased activation of MAPK and AP-1 in aorta or kidney of angiotensin II-induced hypertensive rats or stroke-prone SHR has been reported (Kim and Iwao 2003). Some MAPK can be activated by tyrosine kinase-linked receptors (Hollenberg 1994) and others by G-protein-coupled receptors (Alberts et al. 1994). It has been shown that growth factors and phorbol myristate acetate strongly activate ERK1/2 and weakly activate JNK and p38 kinases (Cano et al. 1994). TNF-α is a strong stimulus for p38 and ROS are potent activators for JNK (Devary et al. 1992). PDGF and angiotensin II are powerful activators of ERK1/2 (Liao et al. 1996). In rat aortic VSMC, MG induced expression of heparin-binding epidermal growth factor (HB-EGF) (Che et al. 1997). A recent study reported that MG inhibited cellular response to human insulinlike growth factor-1 through an MEK/ERK-dependent pathway in a cultured human embryonic kidney cell line (HEK293) and mouse fibroblast cell line (NIH3T3) (Du et al. 2003).

Conclusions and Perspectives

Increased generation of MG and MG-related AGEs in diabetes and hypertension has been observed. These findings suggest an important role of MG in development of insulin resistance. Among putative mechanisms for MG-induced insulin resistance are MG-induced glycoxidation of critical proteins, overproduction of ROS, increased vascular tone, and vascular proliferation and remodeling. Role of MG in insulin resistance development and the underlying mechanisms should be continuously investigated. For example, correlation of MG metabolism with obesity has not been addressed either in animal models or in patients. The possibility that MG contributes to etiology of insulin resistance is appealing as it may lead to discovery of new mechanisms and methods for management and prevention of insulin resistance syndromes.

Acknowledgments The author thanks the Heart and Stroke Foundation of Canada and Canadian Institutes of Health Research for supporting this project.

References

Ahmed, N., Argirov, O.K., Minhas, H.S., Cordeiro, C.A., and Thornalley, P.J. 2002. Assay of advanced glycation endproducts (AGEs): surveying AGEs by chromatographic assay with derivatization by 6-aminoquinolyl-N-hydroxysuccinimidyl-carbamate and application to Nepsilon-carboxymethyl-lysine- and Nepsilon-(1-carboxyethyl)lysine-modified albumin. Biochem J 364:1–14.

Alberts, B., Bray, D., Lewis, J., Raff, M., Roberts, K., and Watson, J.D. 1994. *Molecular Biology of the Cell.* Garland Publishing, New York.

Alzaid, A.A. 1996. Insulin resistance in non-insulin-dependent diabetes mellitus. A review. Acta Diabetol 33:87–99.

Baynes, J.W., and Thorpe, S.R. 1999. Role of oxidative stress in diabetic complications: a new perspective on an old paradigm. Diabetes 48:1–9.

Beg, A.A., and Baldwin, A.S. Jr. 1993. The I kappa B proteins: multifunctional regulators of Rel/NF-kappa B transcription factors. Genes Dev 7:2064–2070.

Beisswenger, P.J., Howell, S.K., Touchette, A.D., Lal, S., and Szwergold, B.S. 1999. Metformin reduces systemic methylglyoxal levels in type 2 diabetes. Diabetes 48:198–202.

Beswick, R.A., Zhang, H., Marable, D., Catravas, J.D., Hill, W.D., and Webb, R.C. 2001. Long-term antioxidant administration attenuates mineralocorticoid hypertension and renal inflammatory response. Hypertension 37:781–786.

Blakytny, R., and Harding, J.J. 1992. Glycation (non-enzymic glycosylation) inactivates glutathione reductase. Biochem J 288(Pt 1):303–307.

Blaschko, H. 1974 The natural history of amine oxidases. Rev Physiol Biochem Pharmacol 70:83–148.

Brown, K., Gerstberger, S., Carlson, L., Franzoso, G., and Siebenlist, U. 1995. Control of I kappa B-alpha proteolysis by site-specific, signal-induced phosphorylation. Science 267:1485–1488.

Cano, E., Hazzalin, C.A., and Mahadevan, L.C. 1994. Anisomycin-activated protein kinases p45 and p55 but not mitogen-activated protein kinases ERK-1 and -2 are implicated in the induction of c-fos and c-jun. Mol Cell Biol 14:7352–7362.

Canty, T.G. Jr., Boyle, E.M. Jr., Farr, A., Morgan, E.N., Verrier, E.D., and Pohlman, T.H. 1999. Oxidative stress induces NF-kappaB nuclear translocation without degradation of IkappaBalpha. Circulation 100:II361–364.

Casazza, J.P., Felver, M.E., and Veech, R.L. 1984. The metabolism of acetone in rat. J Biol Chem 259:231–236.

Chang, T., and Wu, L. 2006. Methylglyoxal, oxidative stress, and hypertension. Can J Physiol Pharmacol 84:1229–1238.

Chang, T., Wang, R., and Wu, L. 2005. Methylglyoxal-induced nitric oxide and peroxynitrite production in vascular smooth muscle cells. Free Radic Biol Med 38:286–293.

Chaplen, F.W., Fahl, W.E., and Cameron, D.C. 1998. Evidence of high levels of methylglyoxal in cultured Chinese hamster ovary cells. Proc Natl Acad Sci USA 95:5533–5538.

Che, W., Asahi, M., Takahashi, M., Kaneto, H., Okado, A., Higashiyama, S., and Taniguchi, N. 1997. Selective induction of heparin-binding epidermal growth factor-like growth factor by methylglyoxal and 3-deoxyglucosone in rat aortic smooth muscle cells. The involvement of reactive oxygen species formation and a possible implication for atherogenesis in diabetes. J Biol Chem 272:18453–18459.

Chen, C., Hosokawa, H., Bumbalo, L.M., and Leahy, J.L. 1994. Mechanism of compensatory hyperinsulinemia in normoglycemic insulin-resistant spontaneously hypertensive rats. Augmented enzymatic activity of glucokinase in beta-cells. J Clin Invest 94:399–404.

Cook, L.J., Davies, J., Yates, A.P., Elliott, A.C., Lovell, J., Joule, J.A., Pemberton, P., Thornalley, P.J., and Best, L. 1998. Effects of methylglyoxal on rat pancreatic beta-cells. Biochem Pharmacol 55:1361–1367.

Cooke, C.L., and Davidge, S.T. 2002. Peroxynitrite increases iNOS through NF-kappaB and decreases prostacyclin synthase in endothelial cells. Am J Physiol Cell Physiol 282:C395–C402.

Cuzzocrea, S., Mazzon, E., Dugo, L., Di Paola, R., Caputi, A.P., and Salvemini, D. 2004. Superoxide: a key player in hypertension. Faseb J 18:94–101.

Devary, Y., Gottlieb, R.A., Smeal, T., and Karin, M. 1992. The mammalian ultraviolet response is triggered by activation of Src tyrosine kinases. Cell 71:1081–1091.

Du, J., Cai, S., Suzuki, H., Akhand, A.A., Ma, X., Takagi, Y., Miyata, T., Nakashima, I., and Nagase, F. 2003. Involvement of MEKK1/ERK/P21Waf1/Cip1 signal transduction pathway in inhibition of IGF-I-mediated cell growth response by methylglyoxal. J Cell Biochem 88:1235–1246.

Ekblom, J. 1998. Potential therapeutic value of drugs inhibiting semicarbazide-sensitive amine oxidase: vascular cytoprotection in diabetes mellitus. Pharmacol Res 37:87–92.

Elliott, S.J., and Koliwad, S.K. 1995. Oxidant stress and endothelial membrane transport. Free Radic Biol Med 19:649–658.

Elliott, S.J., and Koliwad, S.K. 1997. Redox control of ion channel activity in vascular endothelial cells by glutathione. Microcirculation 4:341–347.

Ferro, C.J., and Webb, D.J. 1997. Endothelial dysfunction and hypertension. Drugs 53(Suppl 1):30–41.

Fisher, S.J., and Kahn, C.R. 2003. Insulin signaling is required for insulin's direct and indirect action on hepatic glucose production. J Clin Invest 111:463–468.

Force, T., Pombo, C.M., Avruch, J.A., Bonventre, J.V., and Kyriakis, J.M. 1996. Stress-activated protein kinases in cardiovascular disease. Circ Res 78:947–953.

Furchgott, R.F., and Zawadzki, J.V. 1980. The obligatory role of endothelial cells in the relaxation of arterial smooth muscle by acetylcholine. Nature 288:373–376.

Galgani, J., and Diaz, E. 2000. [Obesity and fatty acids in the etiology of insulin resistance]. Rev Med Chil 128:1354–1360.

Girouard, H., Chulak, C., Wu, L., Lejossec, M., and de Champlain, J. 2003. N-acetylcysteine improves nitric oxide and alpha-adrenergic pathways in mesenteric beds of spontaneously hypertensive rats. Am J Hypertens 16:577–584.

Govers, R., and Rabelink, T.J. 2001. Cellular regulation of endothelial nitric oxide synthase. Am J Physiol Renal Physiol 280:F193–F206.

Heise, T., Magnusson, K., Heinemann, L., and Sawicki, P.T. 1998. Insulin resistance and the effect of insulin on blood pressure in essential hypertension. Hypertension 32:243–248.

Hoare, G.S., Marczin, N., Chester, A.H., and Yacoub, M.H. 1999. Role of oxidant stress in cytokine-induced activation of NF-kappaB in human aortic smooth muscle cells. Am J Physiol 277:H1975–H1984.

Hollenberg, M.D. 1994. Tyrosine kinase pathways and the regulation of smooth muscle contractility. Trends Pharmacol Sci 15:108–114.

Hotamisligil, G.S., Arner, P., Atkinson, R.L., and Spiegelman, B.M. 1997. Differential regulation of the p80 tumor necrosis factor receptor in human obesity and insulin resistance. Diabetes 46:451–455.

Hube, F., and Hauner, H. 1999. The role of TNF-alpha in human adipose tissue: prevention of weight gain at the expense of insulin resistance? Horm Metab Res 31:626–631.

Jana, C.K., Das, N., and Sohal, R.S. 2002. Specificity of age-related carbonylation of plasma proteins in the mouse and rat. Arch Biochem Biophys 397:433–439.

Jans, D.A., and Hubner, S. 1996. Regulation of protein transport to the nucleus: central role of phosphorylation. Physiol Rev 76:651–685.

Juurlink, B.H., and Paterson, P.G. 1998. Review of oxidative stress in brain and spinal cord injury: suggestions for pharmacological and nutritional management strategies. J Spinal Cord Med 21:309–334.

Kadowaki, T. 2000. Insights into insulin resistance and type 2 diabetes from knockout mouse models. J Clin Invest 106:459–465.

Kalapos, M.P. 1999. Methylglyoxal in living organisms: chemistry, biochemistry, toxicology and biological implications. Toxicol Lett 110:145–175.

Kim, S., and Iwao, H. 2003. Stress and vascular responses: mitogen-activated protein kinases and activator protein-1 as promising therapeutic targets of vascular remodeling. J Pharmacol Sci 91:77–181.

Kooy, N.W., Royall, J.A., and Ischiropoulos, H. 1997. Oxidation of 2′, 7′-dichlorofluorescein by peroxynitrite. Free Radic Res 27:245–254.

Kranzhofer, R., Schmidt, J., Pfeiffer, C.A., Hagl, S., Libby, P., and Kubler, W. 1999. Angiotensin induces inflammatory activation of human vascular smooth muscle cells. Arterioscler Thromb Vasc Biol 19:1623–1629.

Kulkarni, R.N., Bruning, J.C., Winnay, J.N., Postic, C., Magnuson, M.A., and Kahn, C.R. 1999. Tissue-specific knockout of the insulin receptor in pancreatic beta cells creates an insulin secretory defect similar to that in type 2 diabetes. Cell 96:329–339.

Leoncini, G., and Poggi, M. 1996. Effects of methylglyoxal on platelet hydrogen peroxide accumulation, aggregation and release reaction. Cell Biochem Funct 14:89–95.

Li, J., Su, J., Li, W., Liu, W., Altura, B.T., and Altura, B.M. 2003. Peroxynitrite induces apoptosis in canine cerebral vascular muscle cells: possible relation to neurodegenerative diseases and strokes. Neurosci Lett 350:173–177.

Li, Q., Van Antwerp, D., Mercurio, F., Lee, K.F., and Verma, I.M. 1999. Severe liver degeneration in mice lacking the IkappaB kinase 2 gene. Science 284:321–325.

Li, Y.M., Steffes, M., Donnelly, T., Liu, C., Fuh, H., Basgen, J., Bucala, R., and Vlassara, H. 1996. Prevention of cardiovascular and renal pathology of aging by the advanced glycation inhibitor aminoguanidine. Proc Natl Acad Sci USA 93:3902–3907.

Liao, D.F., Duff, J.L., Daum, G., Pelech, S.L., and Berk, B.C. 1996. Angiotensin II stimulates MAP kinase kinase kinase activity in vascular smooth muscle cells. Role of Raf. Circ Res 79:1007–1014.

Ma, X.L., Baraona, E., Hernandez-Munoz, R., and Lieber, C.S. 1989. High levels of acetaldehyde in nonalcoholic liver injury after threonine or ethanol administration. Hepatology 10:933–940.

Martin, B.C., Warram, J.H., Krolewski, A.S., Bergman, R.N., Soeldner, J.S., and Kahn, C.R. 1992. Role of glucose and insulin resistance in development of type 2 diabetes mellitus: results of a 25-year follow-up study. Lancet 340:925–929.

Marumo, T., Schini-Kerth, V.B., Fisslthaler, B., and Busse, R. 1997. Platelet-derived growth factor-stimulated superoxide anion production modulates activation of transcription factor NF-kappaB and expression of monocyte chemoattractant protein 1 in human aortic smooth muscle cells. Circulation 96:2361–2367.

Maulik, N., and Das, D.K. 2002. Redox signaling in vascular angiogenesis. Free Radic Biol Med 33:1047–1060.

McLellan, A.C., Thornalley, P.J., Benn, J., and Sonksen, P.H. 1994. Glyoxalase system in clinical diabetes mellitus and correlation with diabetic complications. Clin Sci (London) 87:21–29.

Mizutani, K., Ikeda, K., Kawai, Y., and Yamori, Y. 1999a. Biomechanical properties and chemical composition of the aorta in genetic hypertensive rats. J Hypertens 17:481–487.

Mizutani, K., Ikeda, K., Kawai, Y., and Yamori, Y. 1999b. Extract of wine phenolics improves aortic biomechanical properties in stroke-prone spontaneously hypertensive rats (SHRSP). J Nutr Sci Vitaminol (Tokyo) 45:95–106.

Moncada, S., Palmer, R.M., and Higgs, E.A. 1991. Nitric oxide: physiology, pathophysiology, and pharmacology. Pharmacol Rev 43:109–142.

Nagaraj, R.H., Sarkar, P., Mally, A., Biemel, K.M., Lederer, M.O., and Padayatti, P.S. 2002. Effect of pyridoxamine on chemical modification of proteins by carbonyls in diabetic rats: characterization of a major product from the reaction of pyridoxamine and methylglyoxal. Arch Biochem Biophys 402:110–119.

Nukaya, H., Inaoka, Y., Ishida, H., Tsuji, K., Suwa, Y., Wakabayashi, K., and Kosuge, T. 1993. Modification of the amino group of guanosine by methylglyoxal and other alpha-ketoaldehydes in the presence of hydrogen peroxide. Chem Pharm Bull (Tokyo) 41:649–653.

Ogata, N., Yamamoto, H., Kugiyama, K., Yasue, H., and Miyamoto, E. 2000. Involvement of protein kinase C in superoxide anion-induced activation of nuclear factor-kappa B in human endothelial cells. Cardiovasc Res 45:513–521.

Opie, L.H. 1998. *The Heart: Physiology, from Cell to Circulation*. Lippincott-Raven, Philadelphia.

Pogan, L., Garneau, L., Bissonnette, P., Wu, L., and Sauve, R. 2001. Abnormal Ca^{2+} signalling in vascular endothelial cells from spontaneously hypertensive rats: role of free radicals. J Hypertens 19:721–730.

Randle, P.J., Priestman, D.A., Mistry, S., and Halsall, A. 1994. Mechanisms modifying glucose oxidation in diabetes mellitus. Diabetologia 37(Suppl 2):S155–S161.

Ray, S., and Ray, M. 1981. Isolation of methylglyoxal synthase from goat liver. J Biol Chem 256:6230–6233.

Rett, K. 1999. The relation between insulin resistance and cardiovascular complications of the insulin resistance syndrome. Diabetes Obes Metab 1(Suppl 1):S8–S16.

Rocchini, A.P. 1995. Insulin resistance, obesity and hypertension. J Nutr 125:1718S–1724S.

Schmidt, T., Zaib, F., Samson, S.E., Kwan, C.Y., and Grover, A.K. 2004. Peroxynitrite resistance of sarco/endoplasmic reticulum Ca^{2+} pump in pig coronary artery endothelium and smooth muscle. Cell Calcium 36:77–82.

Shaulian, E., and Karin, M. 2001. AP-1 in cell proliferation and survival. Oncogene 20:2390–2400.

Shinohara, M., Thornalley, P.J., Giardino, I., Beisswenger, P., Thorpe, S.R., Onorato, J., and Brownlee, M. 1998. Overexpression of glyoxalase-I in bovine endothelial cells inhibits intracellular advanced glycation endproduct formation and prevents hyperglycemia-induced increases in macromolecular endocytosis. J Clin Invest 101:1142–1147.

Steppan, C.M., Bailey, S.T., Bhat, S., Brown, E.J., Banerjee, R.R., Wright, C.M., Patel, H.R., Ahima, R.S., and Lazar, M.A. 2001. The hormone resistin links obesity to diabetes. Nature 409:307–312.

Stumvoll, M., and Haring, H. 2001. Insulin resistance and insulin sensitizers. Horm Res 55(Suppl 2):3–13.

Thornalley, P.J. 1996. Pharmacology of methylglyoxal: formation, modification of proteins and nucleic acids, and enzymatic detoxification – a role in pathogenesis and antiproliferative chemotherapy. Gen Pharmacol 27:565–573.

Uchida, K., Khor, O.T., Oya, T., Osawa, T., Yasuda, Y., and Miyata, T. 1997. Protein modification by a Maillard reaction intermediate methylglyoxal. Immunochemical detection of fluorescent 5-methylimidazolone derivatives *in vivo*. FEBS Lett 410:313–318.

Vander Jagt, D.L., Hunsaker, L.A., Vander Jagt, T.J., Gomez, M.S., Gonzales, D.M., Deck, L.M., and Royer, R.E. 1997. Inactivation of glutathione reductase by 4-hydroxynonenal and other endogenous aldehydes. Biochem Pharmacol 53:1133–1140.

Vasdev, S., Ford, C.A., Longerich, L., Gadag, V., and Wadhawan, S. 1998a. Role of aldehydes in fructose induced hypertension. Mol Cell Biochem 181:1–9.

Vasdev, S., Ford, C.A., Longerich, L., Parai, S., Gadag, V., and Wadhawan, S. 1998b. Aldehyde induced hypertension in rats: prevention by N-acetyl cysteine. Artery 23:10–36.

Vaziri, N.D., Wang, X.Q., Oveisi, F., and Rad, B. 2000. Induction of oxidative stress by glutathione depletion causes severe hypertension in normal rats. Hypertension 36:142–146.

Wang, C.Y., Mayo, M.W., and Baldwin, A.S. Jr. 1996. TNF- and cancer therapy-induced apoptosis: potentiation by inhibition of NF-kappaB. Science 274:784–787.

Wang, H., Meng, Q.H., Chang, T., and Wu, L. 2006. Fructose-induced peroxynitrite production is mediated by methylglyoxal in vascular smooth muscle cells. Life Sci 79:2448–2454.

Wang, X., Desai, K., Clausen, J.T., and Wu, L. 2004. Increased methylglyoxal and advanced glycation end products in kidney from spontaneously hypertensive rats. Kidney Int 66:2315–2321.

Wang, X., Desai, K., Chang, T., and Wu, L. 2005. Vascular methylglyoxal metabolism and the development of hypertension. J Hypertens 23:1565–1573.

Wang, X., Desai, K., Juurlink, B.H., de Champlain, J., and Wu, L. 2006. Gender-related differences in advanced glycation endproducts, oxidative stress markers and nitric oxide synthases in rats. Kidney Int 69:281–287.

Ward, R.A., and McLeish, K.R. 2004. Methylglyoxal: a stimulus to neutrophil oxygen radical production in chronic renal failure? Nephrol Dial Transplant 19:1702–1707.

Wu, L. 2005. The pro-oxidant role of methylglyoxal in mesenteric artery smooth muscle cells. Can J Physiol Pharmacol 83:63–68.

Wu, L. 2006. Is methylglyoxal a causative factor for hypertension development? Can J Physiol Pharmacol 84:129–139.

Wu, L., and de Champlain, J. 1999. Effects of superoxide on signaling pathways in smooth muscle cells from rats. Hypertension 34:1247–1253.

Wu, L., and Juurlink, B.H. 2001. The impaired glutathione system and its up-regulation by sulforaphane in vascular smooth muscle cells from spontaneously hypertensive rats. J Hypertens 19:1819–1825.

Wu, L., and Juurlink, B.H. 2002. Increased methylglyoxal and oxidative stress in hypertensive rat vascular smooth muscle cells. Hypertension 39:809–814.

Wu, L., Eftekharpour, E., Davies, G.F., Roesler, W.J., and Juurlink, B.H. 2001. Troglitazone selectively inhibits glyoxalase I gene expression. Diabetologia 44:2004–2012.

Wu, L., Ashraf, M.H., Facci, M., Wang, R., Paterson, P.G., Ferrie, A., and Juurlink, B.H. 2004. Dietary approach to attenuate oxidative stress, hypertension, and inflammation in the cardiovascular system. Proc Natl Acad Sci USA 101:7094–7099.

Yu, P.H. 1998. Deamination of methylamine and angiopathy; toxicity of formaldehyde, oxidative stress and relevance to protein glycoxidation in diabetes. J Neural Transm Suppl 52:201–216.

Zhou, D., Mayberg, M.R., London, S., and Gajdusek, C. 1996. Reduction of intracellular glutathione levels produces sustained arterial narrowing. Neurosurgery 39:991–997.

Part III
Growth Factor Signaling

Chapter 11
Role of Growth Factor Receptor Transactivation in Vasoactive Peptide-Induced Signaling Pathways in Vascular Smooth Muscle Cells

Ali Bouallegue and Ashok K. Srivastava

Abstract Vasoactive peptides, such as endothelin-1 (ET-1) and angiotensin II (AngII), are believed to contribute to the pathogenesis of vascular abnormalities such as hypertension, atherosclerosis, hypertrophy, and restenosis. These peptides elicit their biological effects through the activation of transmembrane guanine nucleotide-binding protein-coupled receptors (GPCRs). GPCR activation results in the stimulation of several signaling pathways including mitogen-activated protein kinases (MAPKs), phosphatidylinositol 3-kinase (PI3-K), and protein kinase B. During the last several years, the concept of transactivation of growth factor receptors such as EGF-R, PDGF-R, and IGF-1R in triggering vasoactive peptide signaling has gained much recognition. Activation of these pathways is intimately linked to the regulation of cellular hypertrophy, growth, proliferation, migration, and cell survival. In this chapter, we provide an overview on the role of growth factor receptor transactivation in mediating AngII and ET-1-induced signaling events in vascular smooth muscle cells and their potential implication in vascular pathophysiology.

Introduction

Aberrant vascular smooth muscle cell (VSMC) growth, migration, and proliferation is believed to contribute to abnormal vascular functions associated with vascular diseases. It is widely recognized that vasoactive peptides such as angiotensin II (AngII) and endothelin-1 (ET-1), in addition to modulating the smooth muscle contractility, also act as growth factors and modulate migratory and proliferative responses in VSMC (Bouallegue et al. 2007a; Touyz and Schiffrin 2000). However, in contrast to receptors of the growth factors which possess an intrinsic protein tyrosine kinase (PTK) activity, the receptors for vasoactive peptides are coupled through heterotrimeric G-protein-coupled receptors (GPCR) (Bouallegue et al. 2007a). Nonetheless, the activation of vasoactive peptide receptors stimulates

A. K. Srivastava, M. B. Anand-Srivastava (eds.) *Advances in Biochemistry in Health and Disease.* Signal Transduction in the Cardiovascular System in Health and Disease,

intracellular signaling pathways similar to those activated by growth factor receptors (R) that possess intrinsic PTK activity, and GPCR have been shown to enhance the activity of many R-PTKs (Bouallegue et al. 2007a). The process through which ligands of GPCR activate R-PTK has been termed "transactivation." This process was first identified by Daub and colleagues who showed that ET-1, lysophosphatidic acid, and thrombin enhanced the tyrosine phosphorylation of the epidermal growth factor (EGF) receptor (EGF-R) (Daub et al. 1996). Since then, transactivation of other R-PTK such as insulinlike growth factor-1 receptor (IGF-1R), and platelet-derived growth factor receptor (PDGF-R) has been documented in response to several GPCR agonists in a variety of cell types (Hua et al. 2003; Kodama et al. 2002). Although earlier work focused on the transactivation of EGF-R, recent studies have shown that transactivation of other R-PTK also participates in transducing vasoactive peptide-induced signaling events. This review will examine the role and mechanism of growth factor receptor transactivation in triggering vasoactive peptide signaling in VSMC with a focus on the extracellular signal regulated kinase 1/2 (ERK1/2) and protein kinase B (PKB) signaling pathways.

ERK1/2 Pathway

In VSMC, both AngII and ET-1 have been shown to activate the MAPK pathway (Daou and Srivastava 2004; Liao et al. 1996). ERK1/2 is an important member of the mitogen-activated protein kinase (MAPK) family of serine/threonine protein kinases which are classically associated with cell growth, proliferation, differentiation, and death (Kyosseva 2004; Pearson et al. 2001). MAPK are activated in response to a variety of external stimuli such as growth factors, hormones, stress, and vasoactive peptides. ERK1/2 is activated by sequential activation of several upstream signaling components in this cascade (Figure 11.1). The first component in this cascade is Ras, a member of small GTP-binding protein family. Ras cycles between an active GTP-bound conformation and an inactive GDP-bound form (Kyriakis and Avruch 2001). Once activated, Ras binds to membrane and recruits Raf, which in turn phosphorylates MEK in specific serine/threonine residues. MEK is a dual-specificity protein kinase and phosphorylates ERK1/2 in threonine (Thr) and tyrosine (Tyr) residues. Activation of ERK1/2 leads to the phosphorylation of downstream cytosolic regulatory proteins, such as $p90^{rsk}$ which phosphorylates ribosomal proteins and participates in protein synthesis (Frodin and Gammeltoft 1999). Also, ERK1/2 and other MAPK family members are translocated from the cytosol to the nucleus (Lenormand et al. 1993; Widmann et al. 1999) and phosphorylate many transcription factors which lead to activation of genes involved in growth and differentiation (Chen et al. 1992; Widmann et al. 1999) (Figure 11.1).

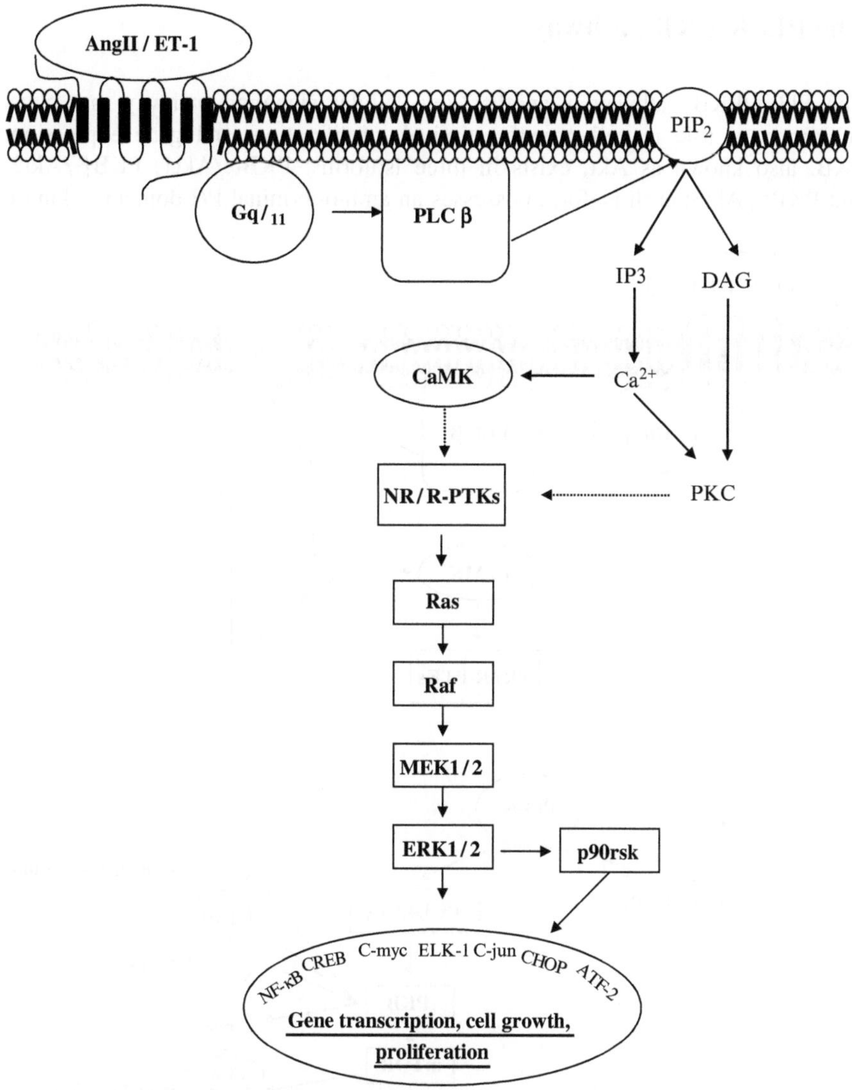

Fig. 11.1 Activation of MAPK pathway by AngII and ET-1 in VSMC. Stimulation of AngII and ET-1 receptors through $Gq/_{11}$ activation enhances the activity of PLCβ. Activated PLCβ converts PIP_2 to IP_3 and diacylglycerol (DAG). IP_3 elevates the concentration of intracellular calcium and DAG activates PKC. PKC and/or Ca^{2+}/calmodulin (CaM)-dependent protein kinases (CaMK) activate nonreceptor (NR) and/or receptor (R) protein tyrosine kinases. Activation of these components signals the stimulation of Ras/Raf/MEK/ERK1/2 and p70 s6k. ERK1/2 and p70 s6k are translocated to nucleus and regulate nuclear events by activating transcription factors through phosphorylation.

The PI3-K/PKB Pathway

The PI3-K/PKB pathway is another key cellular signaling pathway that plays an important role in cell growth, survival, proliferation, and gene expression. PKB, also known as Akt, exists in three isoforms: PKBα/Akt1, PKBβ/Akt2, and PKBγ/Akt3; each isoform possesses an amino-terminal PH domain, a kinase

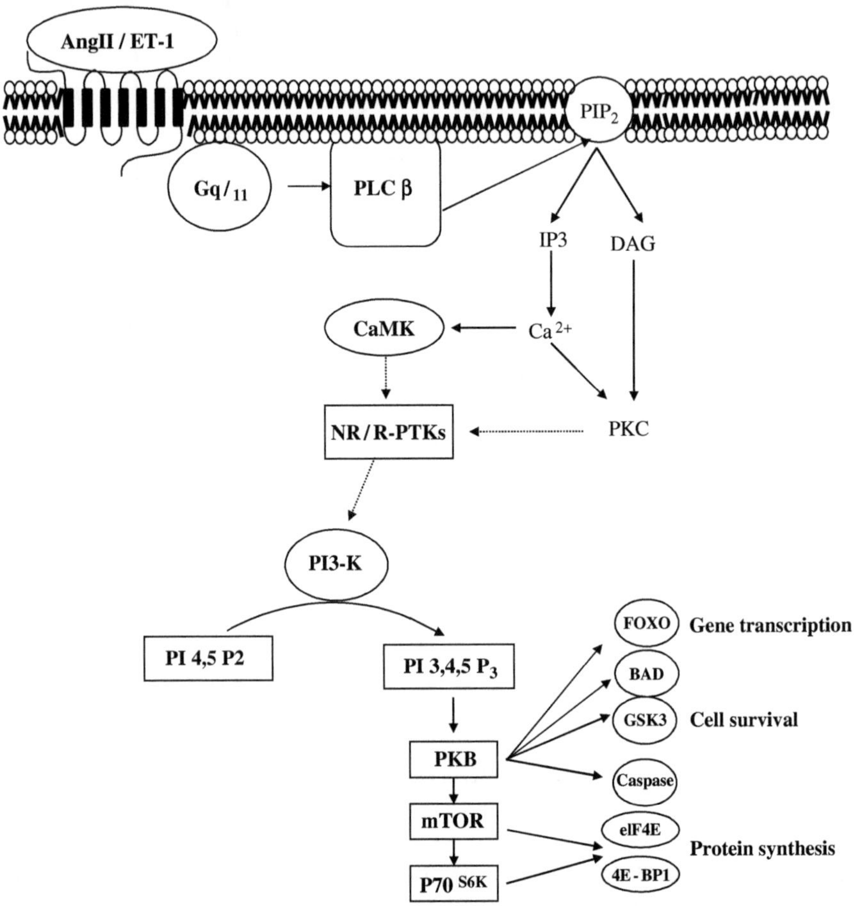

Fig. 11.2 AngII- and ET-1-induced PI3-K/PKB signaling pathway. Stimulation of AngII and ET-1 receptors through $Gq/_{11}$ activation enhances the activity of PLCβ. Activated PLCβ converts PIP_2 to IP_3 and diacylglycerol (DAG). IP_3 elevates the concentration of intracellular calcium and DAG activates PKC. PKC and/or Ca^{2+}/calmodulin (CaM)-dependent protein kinases (CaMK) activate nonreceptor (NR) and/or receptor (R) protein tyrosine kinases. Activation of these components signals the stimulation of PI3-K. The activation of PI3-K leads to the production of PI(3,4,5) P_3 from PI (3,4) P_2, which results in the activation of PKB. PKB has several effectors such as forkhead box transcription factors (FOXO), Bcl-X_L antagonist causing cell death (BAD), glycogen synthase kinase 3 (GSK-3), caspases, mammalian target of rapamycin (mTOR), and 70-kDa ribosomal protein S6 kinase (P^{70}S6K). These signaling intermediates participate in a variety of cellular processes including protein synthesis, cell growth, survival, and gene transcription.

domain, and a carboxy-terminal regulatory domain. PKB is activated in response to insulin (Farese et al. 2005; van der Heide et al. 2005), AngII (Saward and Zahradka 1997), ET-1 (Daou and Srivastava 2004), and many other growth factors (Markadieu et al. 2005; Zaka et al. 2005) in a PI3-K-dependent manner (Figure 11.2).

PI3-K is a heterodimeric lipid kinase, composed of an 85-kDa (p85) regulatory subunit and a 110-kDa (p110) catalytic subunit. The p85 subunit contains the src homology-2 (SH-2) domain and interacts with phosphorylated tyrosine residues on receptor or other docking proteins, leading to stimulation of catalytic activity of its 110-kDa catalytic subunit. p110 catalyzes the phosphorylation of phosphatidylinositol (PI), PI 4-phosphate, and PI 4,5-biphosphate in the 3′ position of the inositol ring. This reaction promotes the generation of PI 3′-P, PI 3,4-biphosphate (PI-3, 4-P_2), and PI-3,4,5-triphosphate (PIP_3) (Bouallegue et al. 2007a; Leevers et al. 1999) (Figure 11.2).

Binding of PIP_3 to PKB recruits it to the plasma membrane for phosphorylation by phosphoinositide-dependent kinases 1 (PDK-1) and 2 (PDK-2). PDK-1 phosphorylates PKB at Thr^{308} in the catalytic domain while putative PDK-2 phosphorylates it at Ser^{473} in the C-terminal regulatory domain of PKB (reviewed in Whiteman et al. 2002)). Activated PKB phosphorylates several downstream substrates, such as glycogen synthase kinase-3β (GSK-3β), Forkhead transcription factor (FKHR, also termed FOXO), Bcl-2/Bcl-X_L antagonist, causing cell death (BAD), Caspase 9, mammalian target of rapamycin (mTOR), nuclear factor-κB (NF-κB), and endothelial nitric oxide synthase (eNOS) (reviewed in Song et al. 2005; Whiteman et al. 2002). The phosphorylated form of these substrates regulates diverse cellular functions, such as glucose transport, cell growth, gene expression, cell survival and death as well as protein synthesis (Saward and Zahradka 1997) (reviewed in Datta et al. 1999) (Figure 11.2).

It has been shown that vasoactive peptides enhance the phosphorylation of PKB in many cell types. For example, ET-1 has been shown to increase PKB activation in cardiomyocytes (Pham et al., 2001), myofibroblasts (Shi-Wen et al. 2004), human umbilical vein endothelial cells (Dong et al. 2005) as well as in A-10 VSMC (Bouallegue et al. 2007b; Daou and Srivastava 2004). Similar to ET-1, AngII also induces PKB activation in mesangial cells (Gorin et al. 2003), epidermal cells (Kippenberger et al. 2005), brain neurons (Yang et al. 2002), and VSMC (Takahashi et al. 1999). Recent evidence has suggested that ET-1 and AngII receptor activation couples with transactivation of R-PTK and thereby mediates the ERK1/2 and PKB signaling events. The main growth factor receptors that have been implicated in triggering ET-1- and AngII-induced activation of ERK1/2 and PKB are EGF-R, IGF-1R, and PDGF-R.

EGF-R Transactivation in AngII and ET-1 Signaling

The EGF-R is an R-PTK that is ubiquitously expressed in a variety of cell types, with the most abundant expression in epithelial cells and many cancer cells

(Carpenter 2000; Mendelson 2000; Prenzel et al. 2001). It contains an extracellular ligand binding domain, a single transmembrane domain, and a cytoplasmic tyrosine kinase autophosphorylation and regulatory domain (reviewed in Normanno et al. 2005). EGF-R undergoes dimerization to induce autophosphorylation of tyrosine residues in its tyrosine kinase domain in response to ligand binding (Mendelson 2000; Schlessinger 2002). Dimerization activates the intrinsic PTK activity of the intracellular domain leading to the phosphorylation of several key tyrosine residues. Phosphorylated tyrosines serve as docking sites for binding with Src homology 2 (SH-2) domain-containing signaling proteins which trigger downstream events. The phosphorylation of EGF-R on Tyr^{1068} recruits the adaptor protein Grb2, leading to the activation of the Ras/ERK1/2 pathway.

AngII has been shown to induce the tyrosine phosphorylation of EGF-R in many cell types, including VSMC (Eguchi et al. 1999a,b), cardiofibroblasts (Moriguchi et al. 1999), intestinal epithelial cells (Chiu et al. 2005), preglomerular VSMC (Andresen et al. 2003), C9 cells (Shah and Catt 2002), and pancreatic stellate cells (Hama et al. 2004). This increase in EGF-R phosphorylation or transactivation is believed to be caused by the binding of AngII to AT_1 receptors (Bokemeyer et al. 2000). Use of specific pharmacological inhibitors of EGF-R PTK such as AG1478 and PD153035 has suggested a role of EGF-R transactivation in AngII-induced growth, hypertrophy, and proliferation in VSMC (Bokemeyer et al. 2000; Eguchi et al. 1999a; Ohtsu et al. 2006b) and in other cell types (Chiu et al. 2005; Hama et al. 2004). AngII-induced activation of many key signaling events including ERK1/2 and PKB was also found to be dependent on EGF-R transactivation. For example, AG1478 attenuated activation of ERK1/2 and p38MAPK induced by AngII (Eguchi et al. 2001). Also, in endothelial and pancreatic cells, a requirement of EGF-R transactivation for ERK1/2 activation induced by AngII has been demonstrated using a similar approach (Hama et al. 2004; Li et al. 1998). Furthermore, in rat aorta VSMC (Li and Malik 2005) and in intestinal epithelial IEC-18 cells, AngII-stimulated PKB phosphorylation was associated with EGF-R transactivation (Chiu et al. 2005).

Similar to AngII, ET-1 has also been shown to transactivate EGF-R by tyrosine phosphorylation in many cell types (Iwasaki et al. 1998; Kodama et al. 2002), and a critical role of this transactivation in mediating ET-1-induced ERK1/2 activation has been suggested (Iwasaki et al. 1998). An increase in EGF-R phosphorylation and ERK1/2 activation was also reported in freshly isolated rat aorta in response to ET-1 stimulation, and PD153035, another EGF-R PTK inhibitor, blocked ERK1/2 phosphorylation in these studies (Kodama et al. 2002). An involvement of ET-1-induced EGF-R transactivation has been implicated in protein and DNA synthesis and c-Fox gene transcription in VSMC (Iwasaki et al. 1998, 1999). In contrast to the results of Iwasaki et al. showing that ET-1-induced protein synthesis could be blocked by AG1478, *in vivo* studies have failed to show any inhibitory effect of this inhibitor on total protein synthesis in small mesenteric arteries (Beaucage and Moreau 2004). However, these investigators did not examine the effect of AG1478 on EGF-R phosphorylation. A requirement of EGF-R transactivation in ET-1-induced vascular contraction has also been shown in rabbit basilary artery rings (Kawanabe et al. 2004) and in mouse aortic ring segments (Flamant et al. 2003). This contractile response was dependent on ERK1/2 activation because pharmacological blockade of this pathway inhibited ET-1-induced contraction in basilary artery rings (Kawanabe et al. 2004). Although several studies in VSMC have suggested an involvement of ET_A receptor in ET-1-induced EGF-R transactivation, recently, by using either the full-length ET_B receptor or one N-terminally truncated, a role of ET_B receptor in this process has also been suggested (Grantcharova et al. 2006).

It thus appears that EGF-R transactivation plays a role not only in growth-promoting/hypertrophic responses of AngII and ET-1, but also in mediating the contractile events induced by these two vasoactive peptides. Further support for a potential role of EGF-R transactivation in cardiovascular pathophysiology is provided by the finding that the use of antisense oligonucleotides of EGF-R was shown to attenuate AngII-induced cardiac hypertrophy in spontaneously hypertensive rat model (Kagiyama et al. 2002, 2003).

IGF-1R Transactivation in AngII and ET-1 Signaling

The IGF-1R is also an R-PTK that shares structural and functional homology with the insulin receptor. The mature receptor is a tetramer consisting of two extracellular α-chains and two intracellular β-chains (Hernandez-Sanchez et al. 1997). The β-chains include an intracellular tyrosine kinase domain that is believed to be essential for most of the receptor's biologic effects (Delafontaine 1995). Binding of IGF-1 or insulin (at very high concentrations) induces the activation of the PTK domain of IGF-1Rβ subunit which in turn activates the autophosphorylation of the receptor (reviewed in Adams et al. 2000). One of the earliest steps in signal transduction initiated by the IGF-1R is the phosphorylation of adaptor/docking proteins such as insulin receptor substrate (IRS-1 or IRS-2) and Shc (LeRoith et al. 1995; White 1998). IRS-1, an important substrate for both the insulin and the IGF-1 receptor, contains multiple tyrosine phosphorylation sites that recognize and bind to SH2 domain-containing signaling molecules, such as Grb2, Nck, the p85 subunit of PI3-K, and the SH2 domain-containing tyrosine phosphatase-2 (SHP-2) (LeRoith et al. 1995). Of these, the binding of Grb2/Sos to tyrosine-phosphorylated IRS-1 activates Ras, which then stimulates the Raf-1/MAPK cascade (Rozakis-Adcock et al. 1992). Shc can also interact directly with IGF-1R (Giorgetti et al. 1994). After tyrosine phosphorylation of Shc, it recruits the Grb2/Sos complex and activates the Ras/Raf-1/MEK/ERK pathway (Rozakis-Adcock et al. 1992). The activated IGF-1R also triggers the activation of PI3-K and its downstream targets PKB and p70s6k (Oldham and Hafen 2003; Zheng and Clemmons 1998).

In addition to the well-studied role of EGF-R transactivation in mediating AngII- and ET-1-induced signaling, there are several reports suggesting a role of IGF-1R in triggering the responses of AngII. In fact, a stimulatory effect of AngII on the tyrosine phosphorylation of IGF-1R β-subunit as well as on IRS-1 was reported in rat aortic VSMC a decade ago (Du et al. 1996). However, a requirement of IGF-1R activation in AngII-induced downstream signaling has been demonstrated relatively recently (Touyz et al. 2003). It was found that in VSMC isolated from mesenteric vascular bed, AG1024, a selective pharmacological inhibitor of IGF-1R-PTK activity, partially reduced AngII-induced phosphorylation of ERK1/2, p38MAPK, and ERK5 (Cruzado et al. 2005; Touyz et al. 2003). Zahradka and colleagues also reported an enhanced tyrosine phosphorylation of the β-subunit of IGF-1R in response to AngII treatment (Zahradka et al. 2004). However, in contrast to the studies of Cruzado et al. and Touyz et al., pharmacological blockade of IGF-1R failed to attenuate AngII-induced ERK1/2 activation in these studies whereas it inhibited AngII-induced phosphorylation of the p85 subunit of PI3-K and p70s6k was significantly attenuated by AG1024 (Zahradka et al. 2004).

It was recently demonstrated that AngII transactivates both EGF-R and IGF-1R which participate in the activation of hypoxia-inducible factor-1 (HIF-1) in VSMC (Lauzier et al. 2007). It thus appears that IGF-1R transactivation may also be essential in triggering AngII- and ET-1 induced signaling events in a tissue/cell-specific or signaling component-specific fashion. In fact, recent evidence indicating that transactivation of EGF-R mediates the responses of IGF-1R in some cell types suggests the existence of a cross-talk between IGF-R and EGF-R transactivation (Meng et al. 2007). Moreover, studies showing that dominant negative or antisense oligonucleotides of IGF-1R are able to attenuate neointima formation in an injured carotid artery model in rats (Lim et al. 2004) and reduce AT1 receptor expression and function in spontaneously hypertensive rats (Nguyen et al. 2006) support a potential pathogenic role of upregulated IGF-1R signaling in vascular disease.

PDGF-R Transactivation in AngII and ET-1 Signaling

The PDGF-R is also an R-PTK expressed in many cell types. It exists in two isoforms: PDGF-Rα and PDGF-Rβ that display affinity for the different isoforms of the PDGF family—PDGF-AA, PDGF-AB, PDGF-BB, PDGF-CC, and PDGF-DD (reviewed in Board and Jayson 2005). Tissue culture and *in vivo* mouse model studies have suggested that PDGF-Rα and PDGF-Rβ activate distinct signaling pathways. Downstream targets of PDGF-Rβ include ERK1/2, PKB, and small G proteins including Rho and Rac-1, which ultimately mediate PDGF-induced responses such as cell cycle progression, migration, and survival (reviewed in Rosenkranz and Kazlauskas 1999). Similar to EGF, binding of PDGF to its receptor on the cell surface induces its dimerization and autophosphorylation of the tyrosine kinase domain, which in turn recruits and activates SH2 domain-containing proteins such as Grb2, Src, p85 subunit of PI3-K, and phospholipase C γ (PLCγ). Tyr^{751} in the kinase domain of PDGF-Rβ is the docking site for PI3-K. Tyr^{740} is also important for the activation of PI3-K by PDGF-Rβ, while PDGF-stimulated PLCγ signaling is dependent on the phosphorylation at the two sites Tyr^{1009} and Tyr^{1021} (reviewed in Board and Jayson 2005).

Several studies have shown that PDGF-R undergoes tyrosine phosphorylation in response to AngII *in vitro* in various cell types such as in human mesanglial cells

(Mondorf et al. 2000), and *in vivo* in VSMC (Heeneman et al. 2000). In the case of VSMC this phosphorylation was sensitive to inhibition by PDGF-R-PTK inhibitor AG1296 as well as antioxidants such as *N*-acetylcysteine and Tiron (Heeneman et al. 2000). In contrast to these studies, however, Bokemeyer et al. were unable to detect any increase in PDGF-R phosphorylation in AngII-treated VSMC isolated from adult rat thoracic aorta (Bokemeyer et al. 2000). Furthermore, in these studies, AngII-induced ERK1/2 phosphorylation was insensitive to inhibition by AG1296 whereas AG1478 almost completely blocked the ERK1/2 phosphorylation. It should be noted that AngII was recently found to induce the tyrosine phosphorylation of an ectodomain-truncated PDGF-R having a molecular size of 70 kDa (Gao et al. 2006). Upon phosphorylation, this truncated PDGF-R interacts with the p85 subunit of PI3-K, and mediates downstream AngII signaling; however, its tyrosine phosphorylation induced by AngII is insensitive to inhibition by AG1296 (Gao et al. 2006). Further studies, in intact mesenteric arteries, that utilized another pharmacological blocker of PDGF-R-PTK, RPR 101511A, as well as PDGF-R antibodies provided evidence to support an involvement of PDGF-R activation in AngII-, as well as pressure-induced ERK1/2 phosphorylation (Eskildsen-Helmond and Mulvany 2003). Additional studies showing that *in vivo* infusion of AngII in Sprague-Dawley rats resulted in an increase in PDGF-R beta tyrosine phosphorylation and PKB phosphorylation in isolated mesenteric arteries (Kelly et al. 2004) substantiated a role of PDGF-R in AngII-induced responses. Notably, Kelly et al. also demonstrated that inhibition of PDGF-R signaling by using imatinib mesylate, a highly selective inhibitor of PDGF-R-PTK, significantly improved vascular remodelling in mesenteric arteries of AngII-treated animals (Kelly et al. 2004). An increase in PDGF-R phosphorylation has also been reported in balloon-injured arteries and in the aorta of stroke-prone spontaneously hypertensive rats (Abe et al. 1997; Eskildsen-Helmond and Mulvany 2003). The fact that ACE inhibitors or AngII, AT_1 receptor antagonist reduced the exaggerated PDGF-R phosphorylation and the intimal thickening in these models points toward a vasculotrophic effect of AngII-induced PDGF-R transactivation (Eskildsen-Helmond and Mulvany 2003; Kim et al. 2000). In contrast to the studies showing an important role of PDGF-R in mediating the responses of AngII, a similar role of PDGF-R transactivation in ET-1 action has not been reported as yet.

Mechanism of Receptor Transactivation by AngII and ET-1

Although the mechanism by which vasoactive peptides transactivate growth factor receptors is not fully understood, several possible mechanisms have been suggested. The first mechanism involves metalloproteinase-induced cleavage of proheparin-binding EGF (proHB-EGF) to HB-EGF which binds to the ectodomain of EGF-R, and activates downstream signaling events (Prenzel et al. 1999). Both ET-1- and AngII-induced EGF-R transactivation have been found to be sensitive to inhibition by a series of metalloproteinase inhibitors such as GM6001, doxycycline, and batimastat (Chansel et al. 2006; Prenzel et al. 1999). These inhibitors were also shown to block downstream signaling induced by ET-1 and AngII in VSMC (Chansel et al. 2006; Yang et al. 2005). Several matrix metalloproteinases have been identified and some of these have been implicated in cardiovascular pathophysiology. For example, an involvement of MMP2/9, a member of the MMP family, in the development of pressure-induced enhanced myogenic tone in mouse resistance artery has been reported (Lucchesi et al. 2004). A role of ADAM 17, also known as tumor necrosis factor (TNF) alpha-converting enzyme (TACE), in mediating AngII-induced

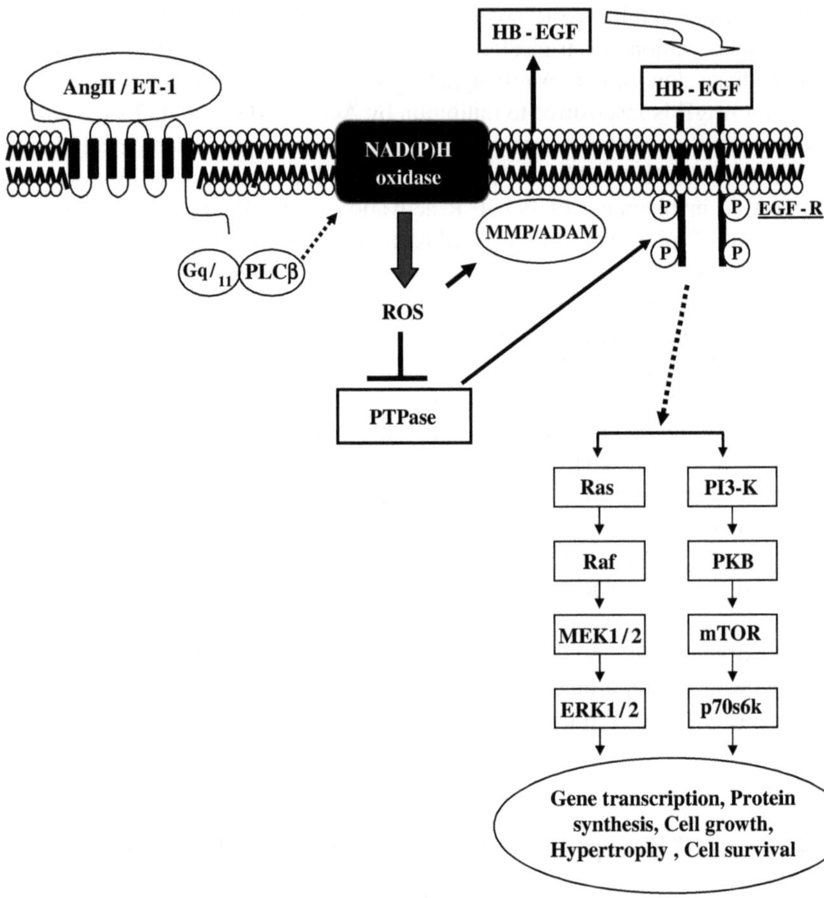

Fig. 11.3 Potential mechanism of EGF-R transactivation by AngII and ET-1. AngII and ET-1 receptor activation through PLCβ and/or other signaling pathways causes generation of reactive oxygen species (ROS) via NADPH oxidase activation. ROS-dependent or -independent matrix metalloproteinase such as MMP2/9 and/or ADAMs leads to transformation of proheparin-binding (HB) EGF to HB-EGF. Interaction of HB-EGF with EGF-R leads to activation of Ras/Raf/MEK and ERK1/2 as well as PI3-K/PKB cascade. ROS production also inhibits the PTPase which increases the EGF-R phosphorylation and thereby activates the Ras and PI3-K signaling cascade. These signaling intermediates participate in a variety of cellular processes including protein synthesis, cell growth, survival, and gene transcription.

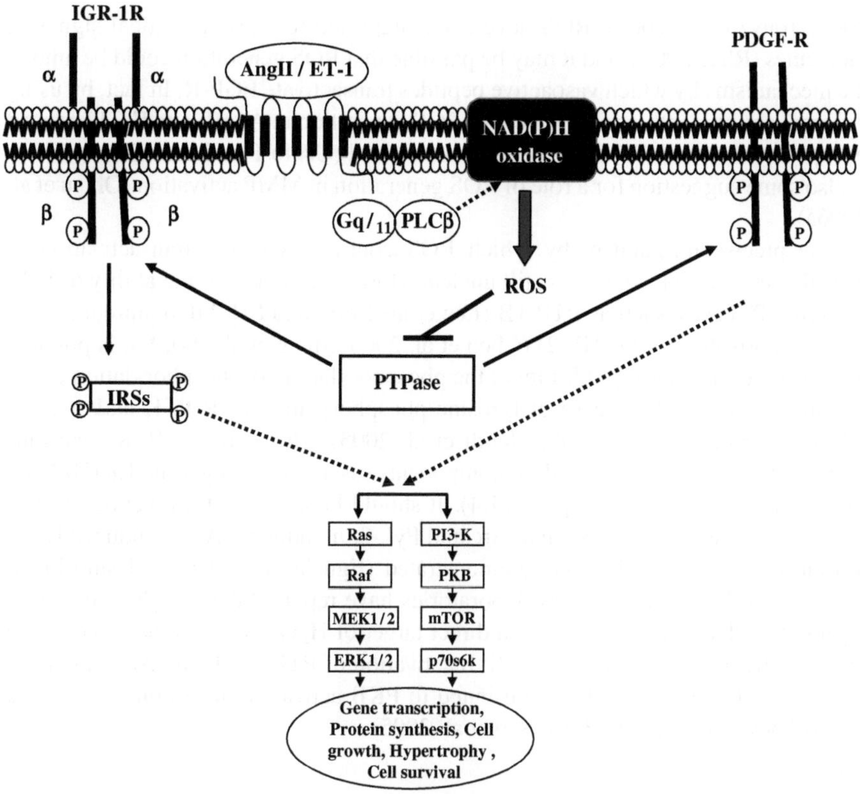

Fig. 11.4 Potential mechanism of PDGF-R and IGF-1R transactivation by AngII and ET-1. AngII and ET-1 receptor activation leads to $Gq/_{11}$-protein activation, which activates PLCβ and also leads to an increase in reactive oxygen species (ROS) generation, via NADPH oxidase activation. ROS inhibits PTPases which enhances the phosphorylation of PDGF-R and IGF-1R. The resultant activation of the PTK activity of these receptors phosphorylates several downstream substrates (e.g., IRSs, in case of IGF-1R) which promotes assembly of signaling components leading to the activation of Ras/Raf/MEK and ERK1/2 as well as PI3-K/PKB cascade. These signaling intermediates participate in a variety of cellular processes including protein synthesis, cell growth, survival, and gene transcription.

VSMC hypertrophy and EGF-R transactivation was recently demonstrated (Ohtsu et al. 2006b). Interestingly, another ADAM family member, ADAM 10, was shown to have no role in AngII-induced EGFR transactivation in this system (Ohtsu et al. 2006b), whereas a requirement of ADAM 12 in AngII-induced cardiac hypertrophy and HB-EGF release was demonstrated (Asakura et al. 2002) (Figure 11.3). It seems that the requirement of specific MMPs/ADAMs to release HB-EGF from pro-HB-EGF is dependent on cell type and physiological context. Although MMPs are crucial for EGF-R transactivation, a similar role of MMPs in PDGF-R or IGF-1R transactivation remains to be established.

In VSMC, AngII, ET-1, and other vasoactive peptides mediate their responses through the generation of reactive oxygen species (ROS) (Daou and Srivastava

2004; Touyz et al. 2004). ROS have been suggested to serve as critical signaling molecules (Rhee 2006), and it may be possible that ROS generation could be among the mechanisms by which vasoactive peptides transactivate EGF-R. In fact, by using different antioxidants, a requirement of ROS generation in AngII- and ET-1-induced phosphorylation of EGF-R has been demonstrated (Ushio-Fukai et al. 2001). There is also some suggestion for a role of ROS generation in MMP activation (Ohtsu et al. 2006a).

The precise mechanism by which ROS contributes to the transactivation of growth factor receptor-PTKs is still unclear. However, in view of the ability of ROS to inhibit PTPases, such as PTP-1B (Lee et al. 1998) and SH-2 domain-containing tyrosine phosphatase-2 (SHP-2) (Chen et al. 2006; Meng et al. 2002), it is possible that ROS can shift the equilibrium of the phosphorylation–dephosphorylation cycle, culminating in a net increase of tyrosine phosphorylation of R-PTK and/or other PTKs (Azar et al. 2006, 2007; Mehdi et al. 2005). The activated PTKs can thus promote the assembly of signaling components essential to trigger the ERK1/2 and PKB signaling pathways (Figure 11.4). It should be noted that a potential role of several nonreceptor PTKs, such as Src and Pyk-2, in inducing AngII-induced EGF-R transactivation has also been demonstrated (Eguchi et al. 1999b; Ushio-Fukai et al. 2001). In addition, several laboratories have reported that PTEN, which catalyzes PIP_3 dephosphorylation, is a direct target of H_2O_2 and becomes inactivated by H_2O_2 treatment (Lee et al. 2002). Inactivation of PTEN, which results in an increase of PIP_3 levels, has been attributed to PKB activation in response to several agents known to generate ROS (Seo et al. 2005).

Conclusion

There is an increasing body of evidence to suggest an important role of growth factor receptor transactivation in mediating vasoactive peptide-induced signaling pathways in VSMC. Among various growth factor receptors, the transactivation of EGF-R has been studied in great detail in response to AngII. However, transactivation of other growth factor receptors, such as PDGF-R and IGF-1R, in triggering the responses of both AngII and ET-1 has been recognized recently. Although the precise mechanism by which vasoactive peptides transactivate various growth factor receptors is unclear at present, a large body of evidence has suggested a role of metalloproteinase-induced generation of HB-EGF in the transactivation of EGF-R. Activation of vasoactive peptide receptors generates ROS, which, by their ability to oxidize the catalytic cysteine residues in many enzymes, can also contribute to growth factor receptor transactivation. PTPase are among this group of enzymes whose inhibition by ROS has been attributed to receptor transactivation and enhanced PTK activity of several receptor and nonreceptor PTK in response to vasoactive peptides. A potential role of ROS in MMP activation has also been suggested. Furthermore, studies showing that blockade of growth factor receptor transactivation by pharmacological or genetic approaches resulting in improved cardiovascular pathophysiology in animal models suggest that upregulation of growth factor R-PTK may be an important contributor in this process.

Acknowledgments The work in the authors' laboratory is supported by funding from the Heart and Stroke Foundation of Quebec and the Canadian Institutes of Health Research (CIHR) operating grant number 67037 to A.K.S. A.B. is a recipient of a CIHR/RX&D-Canadian Hypertension Society Doctoral Research Award.

References

Abe J, Deguchi J, Matsumoto T, Takuwa N, Noda M, Ohno M, Makuuchi M, Kurokawa K, Takuwa Y. 1997. Stimulated activation of platelet-derived growth factor receptor *in vivo* in balloon-injured arteries: a link between angiotensin II and intimal thickening. Circulation 96:1906–1913.

Adams TE, Epa VC, Garrett TP, Ward CW. 2000. Structure and function of the type 1 insulin-like growth factor receptor. Cell Mol Life Sci 57:1050–1093.

Andresen BT, Linnoila JJ, Jackson EK, Romero GG. 2003. Role of EGFR transactivation in angiotensin II signaling to extracellular regulated kinase in preglomerular smooth muscle cells. Hypertension 41:781–786.

Asakura M, Kitakaze M, Takashima S, Liao Y, Ishikura F, Yoshinaka T, Ohmoto H, Node K, Yoshino K, Ishiguro H, Asanuma H, Sanada S, Matsumura Y, Takeda H, Beppu S, Tada M, Hori M, Higashiyama S. 2002. Cardiac hypertrophy is inhibited by antagonism of ADAM12 processing of HB-EGF: metalloproteinase inhibitors as a new therapy. Nat Med 8:35–40.

Azar ZM, Mehdi MZ, Srivastava AK. 2006. Activation of insulin-like growth factor type-1 receptor is required for H2O2-induced PKB phosphorylation in vascular smooth muscle cells. Can J Physiol Pharmacol 84:777–786.

Azar ZM, Mehdi MZ, Srivastava AK. 2007. Insulin-like growth factor type-1 receptor transactivation in vasoactive peptide and oxidant-induced signaling pathways in vascular smooth muscle cells. Can J Physiol Pharmacol 85:105–111.

Beaucage P, Moreau P. 2004. EGF receptor transactivation in angiotensin II and endothelin control of vascular protein synthesis *in vivo*. J Cardiovasc Pharmacol 44 Suppl 1:S20–S23.

Board R, Jayson GC. 2005. Platelet-derived growth factor receptor (PDGFR): a target for anti-cancer therapeutics. Drug Resist Updates 8:75–83.

Bokemeyer D, Schmitz U, Kramer HJ. 2000. Angiotensin II-induced growth of vascular smooth muscle cells requires an Src-dependent activation of the epidermal growth factor receptor. Kidney Int 58:549–558.

Bouallegue A, Daou GB, Srivastava AK. 2007a. Endothelin-1-induced signaling pathways in vascular smooth muscle cells. Curr Vasc Pharmacol 5:45–52.

Bouallegue A, Daou GB, Srivastava AK. 2007b. Nitric oxide attenuates endothelin-1-induced activation of ERK1/2, PKB, and Pyk2 in vascular smooth muscle cells by a cGMP-dependent pathway. Am J Physiol Heart Circ Physiol 293:H2072–H2079.

Carpenter G. 2000. The EGF receptor: a nexus for trafficking and signaling. Bioessays 22:697–707.

Chansel D, Ciroldi M, Vandermeersch S, Jackson LF, Gomez AM, Henrion D, Lee DC, Coffman TM, Richard S, Dussaule JC, Tharaux PL. 2006. Heparin binding EGF is necessary for vasospastic response to endothelin. FASEB J 20:1936–1938.

Chen CH, Cheng TH, Lin H, Shih NL, Chen YL, Chen YS, Cheng CF, Lian WS, Meng TC, Chiu WT, Chen JJ. 2006. Reactive oxygen species generation is involved in epidermal growth factor receptor transactivation through the transient oxidization of Src homology 2-containing tyrosine phosphatase in endothelin-1 signaling pathway in rat cardiac fibroblasts. Mol Pharmacol 69:1347–1355.

Chen RH, Sarnecki C, Blenis J. 1992. Nuclear localization and regulation of erk- and rsk-encoded protein kinases. Mol Cell Biol 12:915–927.

Chiu T, Santiskulvong C, Rozengurt E. 2005. EGF receptor transactivation mediates ANG II-stimulated mitogenesis in intestinal epithelial cells through the PI3-kinase/Akt/mTOR/p70S6K1 signaling pathway. Am J Physiol Gastrointest Liver Physiol 288:G182–G194.

Cruzado MC, Risler NR, Miatello RM, Yao G, Schiffrin EL, Touyz RM. 2005. Vascular smooth muscle cell NAD(P)H oxidase activity during the development of hypertension: Effect of angiotensin II and role of insulinlike growth factor-1 receptor transactivation. Am J Hypertens 18:81–87.

Daou GB, Srivastava AK. 2004. Reactive oxygen species mediate endothelin-1-induced activation of ERK1/2, PKB, and Pyk2 signaling, as well as protein synthesis, in vascular smooth muscle cells. Free Radic Biol Med 37:208–215.

Datta SR, Brunet A, Greenberg ME. 1999. Cellular survival: a play in three Akts. Genes Dev 13:2905–2927.

Daub H, Weiss FU, Wallasch C, Ullrich A. 1996. Role of transactivation of the EGF receptor in signalling by G-protein-coupled receptors. Nature 379:557–560.

Delafontaine P. 1995. Insulin-like growth factor I and its binding proteins in the cardiovascular system. Cardiovasc Res 30:825–834.

Dong F, Zhang X, Wold LE, Ren Q, Zhang Z, Ren J. 2005. Endothelin-1 enhances oxidative stress, cell proliferation and reduces apoptosis in human umbilical vein endothelial cells: role of ETB receptor, NADPH oxidase and caveolin-1. Br J Pharmacol 145:323–333.

Du J, Sperling LS, Marrero MB, Phillips L, Delafontaine P. 1996. G-protein and tyrosine kinase receptor cross-talk in rat aortic smooth muscle cells: thrombin- and angiotensin II-induced tyrosine phosphorylation of insulin receptor substrate-1 and insulin-like growth factor 1 receptor. Biochem Biophys Res Commun 218:934–939.

Eguchi S, Iwasaki H, Hirata Y, Frank GD, Motley ED, Yamakawa T, Numaguchi K, Inagami T. 1999a. Epidermal growth factor receptor is indispensable for c-Fos expression and protein synthesis by angiotensin II. Eur J Pharmacol 376:203–206.

Eguchi S, Iwasaki H, Inagami T, Numaguchi K, Yamakawa T, Motley ED, Owada KM, Marumo F, Hirata Y. 1999b. Involvement of PYK2 in angiotensin II signaling of vascular smooth muscle cells. Hypertension 33:201–206.

Eguchi S, Dempsey PJ, Frank GD, Motley ED, Inagami T. 2001. Activation of MAPKs by angiotensin II in vascular smooth muscle cells. Metalloprotease-dependent EGF receptor activation is required for activation of ERK and p38 MAPK but not for JNK. J Biol Chem 276:7957–7962.

Eskildsen-Helmond YE, Mulvany MJ. 2003. Pressure-induced activation of extracellular signal-regulated kinase 1/2 in small arteries. Hypertension 41:891–897.

Farese RV, Sajan MP, Standaert ML. 2005. Insulin-sensitive protein kinases (atypical protein kinase C and protein kinase B/Akt): actions and defects in obesity and type II diabetes. Exp Biol Med (Maywood) 230:593–605.

Flamant M, Tharaux PL, Placier S, Henrion D, Coffman T, Chatziantoniou C, Dussaule JC. 2003. Epidermal growth factor receptor trans-activation mediates the tonic and fibrogenic effects of endothelin in the aortic wall of transgenic mice. FASEB J 17:327–329.

Frodin M, Gammeltoft S. 1999. Role and regulation of 90 kDa ribosomal S6 kinase (RSK) in signal transduction. Mol Cell Endocrinol 151:65–77.

Gao BB, Hansen H, Chen HC, Feener EP. 2006. Angiotensin II stimulates phosphorylation of an ectodomain-truncated platelet-derived growth factor receptor-beta and its binding to class IA PI3K in vascular smooth muscle cells. Biochem J 397:337–344.

Giorgetti S, Pelicci PG, Pelicci G, Van Obberghen E. 1994. Involvement of Src-homology/collagen (SHC) proteins in signaling through the insulin receptor and the insulin-like-growth-factor-I-receptor 1. Eur J Biochem 223:195–202.

Gorin Y, Ricono JM, Kim NH, Bhandari B, Choudhury GG, Abboud HE. 2003. Nox4 mediates angiotensin II-induced activation of Akt/protein kinase B in mesangial cells. Am J Physiol Renal Physiol 285:F219–F229.

Grantcharova E, Reusch HP, Grossmann S, Eichhorst J, Krell HW, Beyermann M, Rosenthal W, Oksche A. 2006. N-terminal proteolysis of the endothelin B receptor abolishes its ability to induce EGF receptor transactivation and contractile protein expression in vascular smooth muscle cells. Arterioscler Thromb Vasc Biol 26:1288–1296.

Hama K, Ohnishi H, Yasuda H, Ueda N, Mashima H, Satoh Y, Hanatsuka K, Kita H, Ohashi A, Tamada K, Sugano K. 2004. Angiotensin II stimulates DNA synthesis of rat pancreatic stellate cells by activating ERK through EGF receptor transactivation. Biochem Biophys Res Commun 315:905–911.

Heeneman S, Haendeler J, Saito Y, Ishida M, Berk BC. 2000. Angiotensin II induces transactivation of two different populations of the platelet-derived growth factor beta receptor. Key role for the p66 adaptor protein Shc. J Biol Chem 275:15926–15932.

Hernandez-Sanchez C, Werner H, Roberts CT Jr, Woo EJ, Hum DW, Rosenthal SM, LeRoith D. 1997. Differential regulation of insulin-like growth factor-I (IGF-I) receptor gene expression by IGF-I and basic fibroblastic growth factor. J Biol Chem 272:4663–4670.

Hua H, Munk S, Whiteside CI. 2003. Endothelin-1 activates mesangial cell ERK1/2 via EGF-receptor transactivation and caveolin-1 interaction. Am J Physiol Renal Physiol 284:F303–F312.

Iwasaki H, Eguchi S, Marumo F, Hirata Y. 1998. Endothelin-1 stimulates DNA synthesis of vascular smooth-muscle cells through transactivation of epidermal growth factor receptor. J Cardiovasc Pharmacol 31 Suppl 1:S182–S184.

Iwasaki H, Eguchi S, Ueno H, Marumo F, Hirata Y. 1999. Endothelin-mediated vascular growth requires p42/p44 mitogen-activated protein kinase and p70 S6 kinase cascades via transactivation of epidermal growth factor receptor1. Endocrinology 140:4659–4668.

Kagiyama S, Eguchi S, Frank GD, Inagami T, Zhang YC, Phillips MI. 2002. Angiotensin II-induced cardiac hypertrophy and hypertension are attenuated by epidermal growth factor receptor antisense. Circulation 106:909–912.

Kagiyama S, Qian K, Kagiyama T, Phillips MI. 2003. Antisense to epidermal growth factor receptor prevents the development of left ventricular hypertrophy. Hypertension 41:824–829.

Kawanabe Y, Masaki T, Hashimoto N. 2004. Involvement of epidermal growth factor receptor-protein tyrosine kinase transactivation in endothelin-1-induced vascular contraction. J Neurosurg 100:1066–1071.

Kelly DJ, Cox AJ, Gow RM, Zhang Y, Kemp BE, Gilbert RE. 2004. Platelet-derived growth factor receptor transactivation mediates the trophic effects of angiotensin II *in vivo*. Hypertension 44:195–202.

Kim S, Zhan Y, Izumi Y, Yasumoto H, Yano M, Iwao H. 2000. *In vivo* activation of rat aortic platelet-derived growth factor and epidermal growth factor receptors by angiotensin II and hypertension. Arterioscler Thromb Vasc Biol 20:2539–2545.

Kippenberger S, Loitsch S, Guschel M, Muller J, Knies Y, Kaufmann R, Bernd A. 2005. Mechanical stretch stimulates protein kinase B/Akt phosphorylation in epidermal cells via angiotensin II type 1 receptor and epidermal growth factor receptor. J Biol Chem 280:3060–3067.

Kodama H, Fukuda K, Takahashi T, Sano M, Kato T, Tahara S, Hakuno D, Sato T, Manabe T, Konishi F, Ogawa S. 2002. Role of EGF receptor and Pyk2 in endothelin-1-induced ERK activation in rat cardiomyocytes. J Mol Cell Cardiol. 34:139–150.

Kyosseva SV. 2004. Mitogen-activated protein kinase signaling. Int Rev Neurobiol 59:201–220.

Kyriakis JM, Avruch J. 2001. Mammalian mitogen-activated protein kinase signal transduction pathways activated by stress and inflammation. Physiol Rev 81:807–869.

Lauzier MC, Page EL, Michaud MD, Richard DE. 2007. Differential regulation of hypoxia-inducible factor-1 (HIF-1) through receptor tyrosine kinase transactivation in vascular smooth muscle cells. Endocrinology 148:4023–4031.

Lee SR, Kwon KS, Kim SR, Rhee SG. 1998. Reversible inactivation of protein-tyrosine phosphatase 1B in A431 cells stimulated with epidermal growth factor. J Biol Chem 273:15366–15372.

Lee SR, Yang KS, Kwon J, Lee C, Jeong W, Rhee SG. 2002. Reversible inactivation of the tumor suppressor PTEN by H2O2. J Biol Chem 277:20336–20342.

Leevers SJ, Vanhaesebroeck B, Waterfield MD. 1999. Signalling through phosphoinositide 3-kinases: the lipids take centre stage. Curr Opin Cell Biol 11:219–225.

Lenormand P, Sardet C, Pages G, L'Allemain G, Brunet A, Pouyssegur J. 1993. Growth factors induce nuclear translocation of MAP kinases (p42mapk and p44mapk) but not of their activator MAP kinase kinase (p45mapkk) in fibroblasts. J Cell Biol 122:1079–1088.

LeRoith D, Werner H, Beitner-Johnson D, Roberts CT Jr. 1995. Molecular and cellular aspects of the insulin-like growth factor I receptor. Endocr Rev 16:143–163.

Li F, Malik KU. 2005. Angiotensin II-induced Akt activation through the epidermal growth factor receptor in vascular smooth muscle cells is mediated by phospholipid metabolites derived by activation of phospholipase D. J Pharmacol Exp Ther 312:1043–1054.

Li X, Lee JW, Graves LM, Earp HS. 1998. Angiotensin II stimulates ERK via two pathways in epithelial cells: protein kinase C suppresses a G-protein coupled receptor-EGF receptor transactivation pathway. EMBO J 17:2574–2583.

Liao DF, Duff JL, Daum G, Pelech SL, Berk BC. 1996. Angiotensin II stimulates MAP kinase kinase kinase activity in vascular smooth muscle cells: Role of Raf. Circ Res 79:1007–1014.

Lim HJ, Park HY, Ko YG, Lee SH, Cho SY, Lee EJ, Jameson JL, Jang Y. 2004. Dominant negative insulin-like growth factor-1 receptor inhibits neointimal formation through suppression of vascular smooth muscle cell migration and proliferation, and induction of apoptosis 1. Biochem Biophys Res Commun 325:1106–1114.

Lucchesi PA, Sabri A, Belmadani S, Matrougui K. 2004. Involvement of metalloproteinases 2/9 in epidermal growth factor receptor transactivation in pressure-induced myogenic tone in mouse mesenteric resistance arteries. Circulation 110:3587–3593.

Markadieu N, Crutzen R, Blero D, Erneux C, Beauwens R. 2005. Hydrogen peroxide and epidermal growth factor activate phosphatidylinositol 3-kinase and increase sodium transport in A6 cell monolayers. Am J Physiol Renal Physiol 288:F1201–F1212.

Mehdi MZ, Pandey NR, Pandey SK, Srivastava AK. 2005. H2O2-induced phosphorylation of ERK1/2 and PKB requires tyrosine kinase activity of insulin receptor and c-Src. Antioxid Redox Signal 7:1014–1020.

Mendelson J. 2000. Blockade of receptors for growth factors: an anticancer therapy – the fourth annual Joseph H Burchenal American Association of Cancer Research Clinical Research Award Lecture. Clin Cancer Res 6:747–753.

Meng D, Shi X, Jiang BH, Fang J. 2007. Insulin-like growth factor-I (IGF-I) induces epidermal growth factor receptor transactivation and cell proliferation through reactive oxygen species. Free Radic Biol Med 42:1651–1660.

Meng TC, Fukada T, Tonks NK. 2002. Reversible oxidation and inactivation of protein tyrosine phosphatases *in vivo*. Mol Cell 9:387–399.

Mondorf UF, Geiger H, Herrero M, Zeuzem S, Piiper A. 2000. Involvement of the platelet-derived growth factor receptor in angiotensin II-induced activation of extracellular regulated kinases 1 and 2 in human mesangial cells 1. FEBS Lett 472:129–132.

Moriguchi Y, Matsubara H, Mori Y, Murasawa S, Masaki H, Maruyama K, Tsutsumi Y, Shibasaki Y, Tanaka Y, Nakajima T, Oda K, Iwasaka T. 1999. Angiotensin II-induced transactivation of epidermal growth factor receptor regulates fibronectin and transforming growth factor-beta synthesis via transcriptional and posttranscriptional mechanisms. Circ Res 84:1073–1084.

Nguyen TT, Cao N, Short JL, White PJ. 2006. Intravenous insulin-like growth factor-I receptor antisense treatment reduces angiotensin receptor expression and function in spontaneously hypertensive rats. J Pharmacol Exp Ther 318:1171–1177.

Normanno N, Bianco C, Strizzi L, Mancino M, Maiello MR, De Luca A, Caponigro F, Salomon DS. 2005. The ErbB receptors and their ligands in cancer: an overview. Curr Drug Targets 6:243–257.

Ohtsu H, Dempsey PJ, Eguchi S. 2006a. ADAMs as mediators of EGF receptor transactivation by G protein-coupled receptors. Am J Physiol Cell Physiol 291:C1–C10.

Ohtsu H, Dempsey PJ, Frank GD, Brailoiu E, Higuchi S, Suzuki H, Nakashima H, Eguchi K, Eguchi S. 2006b. ADAM17 mediates epidermal growth factor receptor transactivation and vascular smooth muscle cell hypertrophy induced by angiotensin II. Arterioscler Thromb Vasc Biol 26:e133–e137.

Oldham S, Hafen E. 2003. Insulin/IGF and target of rapamycin signaling: a TOR de force in growth control. Trends Cell Biol 13:79–85.

Pearson G, Robinson F, Beers GT, Xu BE, Karandikar M, Berman K, Cobb MH. 2001. Mitogen-activated protein (MAP) kinase pathways: regulation and physiological functions. Endocr Rev 22:153–183.

Pham FH, Cole SM, Clerk A. 2001. Regulation of cardiac myocyte protein synthesis through phosphatidylinositol 3′ kinase and protein kinase B. Adv Enzyme Regul 41:73–86.

Prenzel N, Zwick E, Daub H, Leserer M, Abraham R, Wallasch C, Ullrich A. 1999. EGF receptor transactivation by G-protein-coupled receptors requires metalloproteinase cleavage of proHB-EGF. Nature 402:884–888.

Prenzel N, Fischer OM, Streit S, Hart S, Ullrich A. 2001. The epidermal growth factor receptor family as a central element for cellular signal transduction and diversification. Endocr Relat Cancer 8:11–31.

Rhee SG. 2006. Cell signaling. H_2O_2, a necessary evil for cell signaling. Science 312:1882–1883.

Rosenkranz S, Kazlauskas A. 1999. Evidence for distinct signaling properties and biological responses induced by the PDGF receptor alpha and beta subtypes. Growth Factors 16:201–216.

Rozakis-Adcock M, McGlade J, Mbamalu G, Pelicci G, Daly R, Li W, Batzer A, Thomas S, Brugge J, Pelicci PG. 1992. Association of the Shc and Grb2/Sem5 SH2-containing proteins is implicated in activation of the Ras pathway by tyrosine kinases. Nature 360:689–692.

Saward L, Zahradka P. 1997. Angiotensin II activates phosphatidylinositol 3-kinase in vascular smooth muscle cells. Circ Res 81:249–257.

Schlessinger J. 2002. Ligand-induced, receptor-mediated dimerization and activation of EGF receptor. Cell 110:669–672.

Seo JH, Ahn Y, Lee SR, Yeol YC, Chung HK. 2005. The major target of the endogenously generated reactive oxygen species in response to insulin stimulation is phosphatase and tensin homolog and not phosphoinositide-3 kinase (PI-3 kinase) in the PI-3 kinase/Akt pathway. Mol Biol Cell 16:348–357.

Shah BH, Catt KJ. 2002. Calcium-independent activation of extracellularly regulated kinases 1 and 2 by angiotensin II in hepatic C9 cells: roles of protein kinase Cdelta, Src/proline-rich tyrosine kinase 2, and epidermal growth receptor trans-activation. Mol Pharmacol 61:343–351.

Shi-Wen X, Chen Y, Denton CP, Eastwood M, Renzoni EA, Bou-Gharios G, Pearson JD, Dashwood M, du Bois RM, Black CM, Leask A, Abraham DJ. 2004. Endothelin-1 promotes myofibroblast induction through the ETA receptor via a rac/phosphoinositide 3-kinase/Akt-dependent pathway and is essential for the enhanced contractile phenotype of fibrotic fibroblasts. Mol Biol Cell 15:2707–2719.

Song G, Ouyang G, Bao S. 2005. The activation of Akt/PKB signaling pathway and cell survival. J Cell Mol Med 9:59–71.

Takahashi T, Taniguchi T, Konishi H, Kikkawa U, Ishikawa Y, Yokoyama M. 1999. Activation of Akt/protein kinase B after stimulation with angiotensin II in vascular smooth muscle cells. Am J Physiol 276:H1927–H1934.

Touyz RM, Schiffrin EL. 2000. Signal transduction mechanisms mediating the physiological and pathophysiological actions of angiotensin II in vascular smooth muscle cells. Pharmacol Rev 52:639–672.

Touyz RM, Cruzado M, Tabet F, Yao G, Salomon S, Schiffrin EL. 2003. Redox-dependent MAP kinase signaling by Ang II in vascular smooth muscle cells: role of receptor tyrosine kinase transactivation. Can J Physiol Pharmacol 81:159–167.

Touyz RM, Yao G, Viel E, Amiri F, Schiffrin EL. 2004. Angiotensin II and endothelin-1 regulate MAP kinases through different redox-dependent mechanisms in human vascular smooth muscle cells. J Hypertens 22:1141–1149.

Ushio-Fukai M, Griendling KK, Becker PL, Hilenski L, Halleran S, Alexander RW. 2001. Epidermal growth factor receptor transactivation by angiotensin II requires reactive oxygen species in vascular smooth muscle cells. Arterioscler Thromb Vasc Biol 21:489–495.

van der Heide LP, Kamal A, Artola A, Gispen WH, Ramakers GM. 2005. Insulin modulates hippocampal activity-dependent synaptic plasticity in a N-methyl-d-aspartate receptor and phosphatidyl-inositol-3-kinase-dependent manner. J Neurochem 94:1158–1166.

White MF. 1998. The IRS-signalling system: a network of docking proteins that mediate insulin action. Mol Cell Biochem 182:3–11.

Whiteman EL, Cho H, Birnbaum MJ. 2002. Role of Akt/protein kinase B in metabolism. Trends Endocrinol Metab 13:444–451.

Widmann C, Gibson S, Jarpe MB, Johnson GL. 1999. Mitogen-activated protein kinase: conservation of a three-kinase module from yeast to human. Physiol Rev 79:143–180.

Yang H, Shaw G, Raizada MK. 2002. ANG II stimulation of neuritogenesis involves protein kinase B in brain neurons. Am J Physiol Regul Integr Comp Physiol 283:R107–R114.

Yang X, Zhu MJ, Sreejayan N, Ren J, Du M. 2005. Angiotensin II promotes smooth muscle cell proliferation and migration through release of heparin-binding epidermal growth factor and activation of EGF-receptor pathway. Mol Cells 20:263–270.

Zahradka P, Litchie B, Storie B, Helwer G. 2004. Transactivation of the insulin-like growth factor-I receptor by angiotensin II mediates downstream signaling from the angiotensin II type 1 receptor to phosphatidylinositol 3-kinase. Endocrinology 145:2978–2987.

Zaka M, Rafi MA, Rao HZ, Luzi P, Wenger DA. 2005. Insulin-like growth factor-1 provides protection against psychosine-induced apoptosis in cultured mouse oligodendrocyte progenitor cells using primarily the PI3K/Akt pathway. Mol Cell Neurosci 30:398–407.

Zheng B, Clemmons DR. 1998. Blocking ligand occupancy of the alphaVbeta3 integrin inhibits insulin-like growth factor I signaling in vascular smooth muscle cells. Proc Natl Acad Sci USA 95:11217–11222.

Chapter 12
Identifying the Cellular and Molecular Events Associated with the Divergent Phenotypes of Cardiac Hypertrophy

Angelino Calderone

Abstract Following the imposition of a hemodynamic load, the primary adaptive response of the heart involves an increase in size of individual cardiac myocytes in the absence of cell division. This process is characterized as cardiac hypertrophy and the pattern of individual myocyte growth is directly influenced by the nature of the hemodynamic load. Cardiac hypertrophy can be classified as either physiological or pathological and their disparate phenotypes are related in part to a divergent pattern of peptide growth factor expression. G_q-mediated recruitment of Ca^{2+}- and PKC-dependent signaling pathways may primarily be implicated in the progression of pathological hypertrophy characterized by a concentric pattern of remodeling. By contrast, recruitment of the phosphatidylinositol 3-kinase isoform p110α may selectively participate in the physiological growth of cardiac myocytes. However, there exists evidence to suggest that Ca^{2+}-dependent pathways may also play a supporting role in physiological cardiac hypertrophy. Thus, this review will provide a comprehensive analysis of the morphological, cellular, and molecular phenotypes of physiological and pathological cardiac hypertrophy and explore the relative contribution of G_q- and phosphatidylinositol 3-kinase-dependent pathways.

Introduction

The Frank-Starling mechanism represents an inherent feature of the heart to normalize cardiac output in response to acute changes in pressure or volume. However, sustained or progressive pressure- or volume-overload on the heart induces a response of cardiac hypertrophy that constitutes the basic adaptive mechanism for preserving cardiac function. Since the adult myocyte is terminally differentiated and no longer possesses the capacity to proliferate, cardiac hypertrophy constitutes an increase in cell size without a concomitant increase in cell number. The distinguishing features of hypertrophic growth consist of an overall increase in myocyte RNA and protein synthesis, an increase in contractile protein content, and the synthesis of new sarcomeres. However, depending on the nature of the initiating

A. K. Srivastava, M. B. Anand-Srivastava (eds.) *Advances in Biochemistry in Health and Disease*. Signal Transduction in the Cardiovascular System in Health and Disease,

mechanical stimuli, there are important qualitative differences in the phenotype of the hypertrophic myocyte. These phenotypic differences have been delineated at the morphological, cellular, and molecular level. Despite these observations, a primordial issue that remains unresolved is the identity of the signaling pathway(s) that translates a mechanical stimulus to a specific hypertrophic phenotype. This review will attempt to address this issue by describing the qualitative differences between pathological and physiological cardiac hypertrophy and focus primarily on the role of G_q-dependent signaling pathways and phosphatidylinositol 3-kinase in the divergent hypertrophic phenotypes.

Cellular and Morphological Adaptations of Pathological and Physiological Cardiac Hypertrophy

When the primary stimulus to hypertrophy is either aortic stenosis or hypertension, the resultant pressure overload will lead to an increase in systolic wall stress (Grossman et al. 1975). Elevated systolic wall stress has been defined as the primary signal promoting a *parallel* pattern of sarcomere assembly leading to an increase in myocyte cell width (Figure 12.1) (Morkin 1970; Grossman et al. 1975; Marino et al. 1985; Anversa et al. 1986). Morphologically, this cellular adaptation causes significant ventricular wall thickening and a reduction in chamber radius resulting in a "concentric" pattern of left ventricular hypertrophy (Figure 12.1). Grossman and colleagues observed that systolic wall stress was within normal limits in patients with chronic pressure-overload, supporting the thesis that concentric hypertrophy represents an adaptive mechanism to normalize elevated systolic wall stress (Grossman et al. 1975). It was further postulated that the normalization of systolic wall stress represents a negative feedback mechanism inhibiting further sarcomere synthesis (Braunwald 1992).

When the primary stimulus to hypertrophy is either mitral or aortic regurgitation, or secondary to arteriovenous shunting, the resultant volume overload will lead to an increase in diastolic wall stress (Grossman et al. 1975; Gerdes et al. 1988). Elevated diastolic wall stress represents the primary signal promoting an *in series* pattern of sarcomere assembly leading to an increase in myocyte cell length (Figure 12.1) (Grossman et al. 1975; Marino et al. 1985; Anversa et al. 1986; Braunwald 1992). Morphologically, the elongation of cardiac myocytes leads to chamber enlargement and an "eccentric" pattern of cardiac hypertrophy (Figure 12.1). However, the increase in chamber radius that accompanies eccentric remodeling will lead to a modest elevation of systolic wall stress thereby initiating the assembly of newly synthesized sarcomeres in a *parallel* fashion leading to wall thickening (Braunwald 1992). Thus, eccentric remodeling of the heart is associated with a modest increase in ventricular wall thickness that attempts to normalize elevated systolic wall stress. By contrast, the eccentric hypertrophic response is unable to normalize elevated diastolic wall stress (Grossman et al. 1975; Braunwald 1992).

When the primary stimulus to hypertrophy is exercise, the intermittent diastolic load imposed on the heart leads to an "eccentric" pattern of cardiac hypertrophy with enhanced cardiac function (Anversa et al. 1983, 1986; Moore et al. 1993; Jin

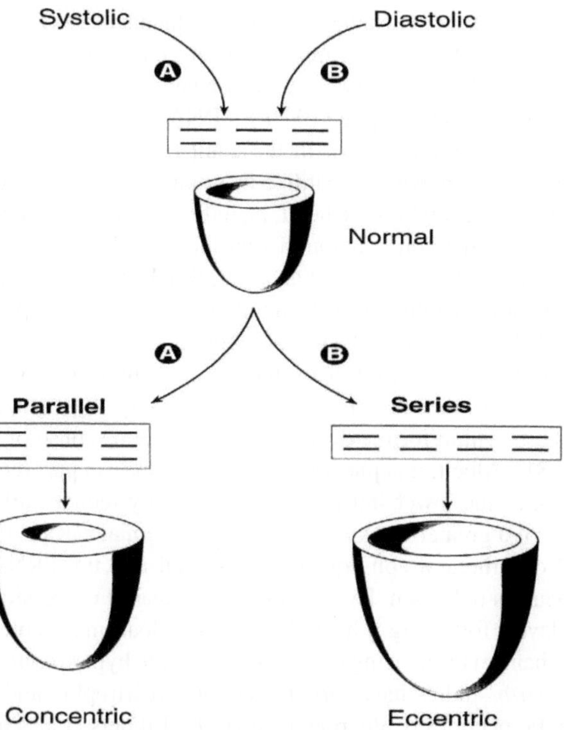

Fig. 12.1 Concentric and eccentric cardiac remodeling. In response to a systolic load, newly formed sarcomeres will be assembled in a *parallel* fashion leading to an increase in myocyte cell width. This mode of sarcomere assembly will result in a concentric pattern of cardiac hypertrophy characterized by an increase in wall thickness and reduction in chamber volume. In contrast, in response to a diastolic load, newly formed sarcomeres will be assembled in an *in series* pattern leading to an increase in myocyte cell length. This mode of sarcomere assembly will promote an eccentric pattern of cardiac hypertrophy characterized primarily by an increase in chamber volume. However, a modest increase in wall thickness will also occur because of the secondary increase in systolic wall stress associated with eccentric remodeling.

et al. 2000). Furthermore, despite similar morphological phenotypes between physiological and pathological eccentric cardiac hypertrophy, important qualitative differences exist at the cellular and molecular levels that distinguish these two forms of hypertrophy.

Molecular Phenotypes of Concentric and Eccentric Cardiac Hypertrophy

Concentric Hypertrophy

During the maturation process of the heart, the proliferative capacity of cardiac myocytes is lost, and important qualitative changes in the expression of cardiac-specific genes are evident (Zak 1974; Nadal-Ginard and Mahdavi 1989). Depending on the species, there is the selective downregulation of genes encoding fetal contractile proteins (e.g., skeletal α-actin and β-myosin heavy chain) and the decreased expression of atrial natriuretic peptide (ANP) mRNA in the ventricular myocardium (Nadal-Ginard and Mahdavi 1989). In the adult rat, aortic constriction (pressure overload) promotes an increase in cardiac mass and the induction of a program of genes reminiscent of the embryonic heart, including the reexpression of cardiac fetal contractile genes and the upregulation of ventricular ANP mRNA (Mercadier et al. 1981; Izumo et al. 1988; Schiaffino et al. 1989; Black et al. 1991). Coincident with these changes is downregulation of the adult-specific gene sarcoplasmic reticulum Ca^{2+}-ATPase (SERCA2) (de la Bastie et al. 1990).

In the pressure-overloaded rat heart, the α- → β-myosin heavy chain isoform switching leads to a diminished myosin ATPase activity (V3 myosin isoform; ββ) and is in part responsible for the decreased contractile response of the heart (Mercadier et al. 1981). Albeit, this phenotypic alteration may in part represent an adaptive response since the lower intrinsic ATPase activity reduces utilization of ATP in the presence of a greater work load and oxygen demand (Katz 1990). Presently, the functional or structural consequence of skeletal α-actin mRNA expression in the adult rat heart is unknown. By contrast, the increased expression of ventricular ANP mRNA levels following a hemodynamic overload may represent a counter-regulatory mechanism influencing the magnitude of the hypertrophic response. Both *in vivo* and *in vitro* studies have reported an antihypertrophic and antifibrotic action of natriuretic peptides (Calderone et al. 1998; Ellmers et al. 2002). Lastly, the downregulation of SERCA2 mRNA levels is in part responsible for abnormal sarcoplasmic reticulum Ca^{2+} handling and subsequent impairment of myocardial relaxation observed in concentric hypertrophy of both experimental animal models and patients with cardiac disease (de la Bastie et al. 1990; Masashi et al. 1994).

Eccentric Hypertrophy (Pathological Stimulus)

In the ventricular myocardium of rats with volume overload due to an arteriovenous fistula, increased ANP mRNA expression has been documented (Calderone et al. 1995). Therefore, despite the apparent differences in the morphological and cellular phenotypes of volume and pressure overload, the reexpression of ANP mRNA represents a conserved molecular event of pathological cardiac hypertrophy. By contrast, during the early phase of left ventricular hypertrophy secondary to an arteriovenous fistula and despite a similar degree of left ventricular hypertrophy as that observed in response to suprarenal aortic constriction, volume-overload-induced cardiac hypertrophy was not associated with either the increased expression of β-myosin heavy chain, skeletal α-actin or the reciprocal downregulation of SERCA2 (Calderone et al. 1995). These data demonstrate the development of volume- and pressure-overload cardiac hypertrophy in response to a pathological stimulus is associated with distinct molecular phenotypes. Second, ANP, skeletal α-actin, and β-myosin heavy chain have been generally viewed as a conserved feature of pressure-induced cardiac hypertrophy. However, the discordant pattern of fetal cardiac gene expression observed in volume-induced cardiac hypertrophy suggests that at least at the mRNA level, these genes can be independently regulated in response to different mechanical stimuli. Thus, the increased expression of ANP mRNA in response to divergent pathological stimuli suggests that its reexpression represents a secondary event dependent on the hypertrophic growth of cardiac myocytes. By contrast, the regulation of skeletal α-actin, β-myosin heavy chain, and SERCA2 mRNA appear to be less conserved events of the hypertrophic process and more stimulus specific (pressure versus volume).

Eccentric Hypertrophy (Physiological)

An intermittent diastolic load associated with exercise leads to an eccentric pattern of cardiac hypertrophy with enhanced ventricular function (Anversa et al. 1983, 1986; Moore et al. 1993; Jin et al. 2000). The preponderance of studies have reported either no change or decreased expression of ventricular ANP mRNA in response to a physiological stimulus (Jin et al. 2000; Calderone et al. 2001; Diffee et al. 2003). Thus, the pattern of ventricular ANP mRNA expression represents a seminal feature distinguishing physiological and pathological eccentric cardiac hypertrophy. The underlying mechanism attributed to the latter paradigm may be related in part to the intensity and duration of the diastolic load in response to a pathological versus a physiological stimulus. Furthermore, swimming-induced physiological ventricular hypertrophy was not associated with either the upregulation of cardiac contractile fetal genes or the downregulation of SERCA2 thereby further reaffirming the concept that the pattern of gene expression is stimulus specific (McMullen et al. 2003).

Peptide Growth Factor Expression in Response to Cardiac Hypertrophy

Pathological Cardiac Hypertrophy

The increased expression of peptide growth factors in the adult heart in response to a hemodynamic load or following an ischemic insult has suggested that these proteins may initiate and/or modulate ventricular remodeling. In the surviving myocardium of the infarcted rat heart, transforming growth factor-β_1 (TGF-β_1) and insulinlike growth factor-1 (IGF-1) mRNA levels were increased and TGF-β_1, TGF-β_3, and IGF-1 mRNA levels were upregulated in the hypertrophied left ventricle secondary to aortic constriction (Thompson et al. 1988; Calderone et al. 1995; Matthews et al. 1999). Further analysis revealed that the increased expression of peptide growth factors in response to pressure overload occurred exclusively in cardiac myocytes (Thompson et al. 1988; Calderone et al. 1995; Matthews et al. 1999). Analogous to that observed secondary to aortic constriction, TGF-β_1 mRNA levels were upregulated in the hearts exposed to a chronic volume overload and increased expression was detected in cardiac myocytes (Calderone et al. 1995). By contrast, TGF-β_3 and IGF mRNA levels were unchanged in the myocyte fraction of rats secondary to an arteriovenous fistula (Calderone et al. 1995). However, in the porcine heart, IGF-1 mRNA levels were increased in myocytes following an arteriovenous fistula and in the hypertrophied heart of patients with aortic regurgitation (Serneri et al. 1999; Modesti et al. 2000). Thus, regardless of the species, the cardiac myocyte appears to play a primary role in detecting the mechanical stimuli. Stimulus-specific expression was also reported for ET-1, as a selective upregulation of prepro-ET-1 mRNA in cardiac myocytes was observed following aortic constriction in the pig, whereas mRNA levels remained unchanged secondary to arteriovenous fistula (Modesti et al. 2000). Likewise, increased prepro-ET-1 mRNA levels were reported in the hypertrophied heart of patients with aortic stenosis whereas no change was detected in the hypertrophied heart of patients with aortic regurgitation (Serneri et al. 1999). Lastly, several studies have identified a hypertrophic role of AII in response to pathological stimuli and further demonstrated that cardiac myocytes contain the requisite machinery (e.g., angiotensinogen, renin, ACE) to synthesize the peptide (Ruzicka et al. 1995; Malhotra et al. 1999; Lijnen and Petrov 1999). In general, these data demonstrate a discordant pattern of regulation among a panel of peptide growth factors between pressure- and volume-overload induced left ventricular hypertrophy and highlight a stimulus-specific regulation. However, it would appear that the conserved expression of at least TGF-β_1, IGF-1, and AII, regardless of the pathological stimulus, suggests that their induction may represent an integral and/or adaptive response required to initiate and/or sustain the hypertrophic growth of individual myocytes.

Physiological Cardiac Hypertrophy

In response to swimming, myocardial IGF-1 expression was reported significantly increased (Scheinowitz et al. 2003). Likewise, in trained endurance athletes documented with eccentric physiological cardiac hypertrophy and enhanced ventricular function, increased cardiac IGF-1 synthesis was detected (Serneri et al. 2001). These data suggest that IGF-1 may participate in the development of physiological cardiac hypertrophy. However, the overexpression of IGF-1 receptor in the mouse heart does not faithfully reproduce the established phenotype of eccentric physiological hypertrophy, as an increase in ventricular wall thickness and concomitant reduction of chamber volume were reported (McMullen et al. 2004). Second, consistent with the morphological pattern of remodeling, cardiac myocyte length was unchanged whereas myocyte width was increased, indicative of a *parallel* pattern of sarcomere assembly. Lastly, a panel of cardiac fetal genes was induced in hearts overexpressing the IGF-1 receptor. The authors argue that enhanced systolic function and the lack of fibrosis in transgenic mice overexpressing the IGF-1 receptor in the heart represent more accurate markers of physiological hypertrophy rather than cardiac fetal gene reexpression. However, the authors failed to consider that increased systolic function in their transgenic model may be directly related to the positive inotropic action of IGF-1 (Redaelli et al. 1998; Kinugawa et al. 1999; von Lewinski et al. 2003). By contrast, in the study by Delaughter et al. (1999), a physiological hypertrophic response was initially observed in the mouse heart persistently expressing IGF-1. However, the latter hypertrophy progressed to a pathological condition characterized by depressed ventricular function, extensive fibrosis, and morphologically resembled concentric remodeling (Delaughter et al. 1999). Thus, it would appear that the chronic and sustained activation of the IGF-1 axis leads to the recruitment of signaling events coupled to pathological remodeling. By contrast, the tightly controlled regulation of IGF-1 synthesis by an intermittent diastolic load that accompanies physical training would appear to avert the pathological effects of the peptide growth factor. Despite these findings, the documented increase of cardiac IGF-1 in response to both physiological and pathological stimuli suggests that the peptide growth factor represents a conserved event of hypertrophy.

ET-1 synthesis was not detected in the heart of either endurance trained athletes or the hypertrophic rat heart following a swimming regiment (Iemitsu et al. 2001; Serneri et al. 2001). Furthermore, despite the apparent contribution of angiotensin II (AII) in pathological cardiac hypertrophy, the administration of an AT1 receptor antagonist failed to inhibit exercise-induced cardiac hypertrophy in the rat (Geenen et al. 1996). By contrast, TGF-β_1 mRNA levels were increased in the hypertrophic heart of voluntary exercised female rats whereas no change in TGF-β_3 mRNA expression was observed (Calderone et al. 2001). Likewise, TGF-β_3 mRNA was not induced in the hypertrophied heart of rats following an exercise regimen of swimming (McMullen et al. 2003). Thus, similar to IGF-1, TGF-β_1 mRNA induction in response to both pathological and physiological stimuli appears to represent a conserved event of the hypertrophic response.

The Contribution of Peptide Growth Factors to the Divergent Phenotypes of Pathological and Physiological Cardiac Hypertrophy

The Role of ET-1 and AII

In vitro studies have demonstrated that the exposure of cardiac myocytes to either ET-1 or AII promoted hypertrophy, the reexpression of cardiac fetal genes, and the downregulation of SERCA2 mRNA (Shubeita et al. 1990; Sadoshima and Izumo 1993; Calderone et al. 2000; Strait et al. 2001). In the pressure-overloaded heart, myocardial ET-1 expression was increased and the pharmacological antagonism of the ET_a receptor attenuated the hypertrophic response, thereby reaffirming the role of the peptide in concentric remodeling (Ito et al. 1994). By contrast, increased cardiac ET-1 expression was not observed during the progression of pathological eccentric cardiac hypertrophy, and its lack of induction was consistent with the absence of either skeletal α-actin or β-myosin heavy chain mRNA reexpression (Calderone et al. 1995; Serneri et al. 1999; Modesti et al. 2000). Consistent in part with the *in vitro* findings, the *in vivo* infusion of AII, a transgenic approach to increase the production of AII in the heart, or the cardiac-specific overexpression of the AT1 receptor promoted cardiac hypertrophy (Mazzolai et al. 1998; Paradis et al. 2000; Schultz et al. 2002; Kee et al. 2006). Albeit, employing a pharmacological approach, a putative role of AII in concentric remodeling remains ambiguous (Mohabir et al. 1994; Rockman et al. 1994; Weinberg et al. 1997). Despite the equivocal role in pressure-overloaded rat hearts, AII contributed to the hypertrophic response in volume-overloaded hearts. Following the creation of an arteriovenous fistula in the rat, plasma and cardiac AII levels were increased at 7 days and either enalapril (e.g., ACE inhibitor) or losartan (AII receptor blocker) therapy attenuated the hypertrophic response (Ruzicka et al. 1994, 1995). Consequently, the AII-mediated hypertrophic response secondary to an arteriovenous fistula can be segregated from the recapitulation cardiac fetal contractile genes documented *in vitro*. The latter paradigm was likewise observed following the infusion of a subpressor dose of AII in mice, as the modest hypertrophic response and concomitant reexpression of ANP mRNA was not associated with the reexpression of β-myosin heavy chain mRNA (Schultz et al. 2002). Collectively, these data suggest that in response to a pathological stimulus, AII may preferentially contribute to the growth of individual cardiac myocytes and the reexpression of ventricular ANP mRNA, rather than the recapitulation of cardiac fetal contractile genes. Lastly, neither AII nor ET-1 was implicated in the growth response of cardiac myocytes in response to a physiological stimulus (Geenen et al. 1996; Iemitsu et al. 2001; Serneri et al. 2001).

The Role of IGF-1 and TGF-β_1

There is increasing evidence to suggest that IGF-1 plays an integral role in physiological cardiac hypertrophy. However, its conserved upregulation in both physiological and pathological stimuli suggests that IGF-1 may be required to initiate and/or sustain the hypertrophic growth of individual myocytes and preserve cardiac function. The latter premise is supported by the well-established physiological effects of IGF-1 which include the stimulation of cardiac myocyte hypertrophy, protection against apoptosis, and enhancing contractility (Li et al. 1997; Delaughter et al. 1999; Redaelli et al. 1998; Kinugawa et al. 1999; Huang et al. 2002; von Lewinski et al. 2003; McMullen et al. 2004). With regard to TGF-β_1, *in vitro* studies have revealed that the treatment of neonatal rat ventricular myocytes promoted hypertrophy and the reexpression of a panel of fetal cardiac genes and AII stimulation of cardiac myocytes increased peptide growth factor expression (Parker et al. 1990;

Sadoshima and Izumo 1993). These data support the premise that TGF-β_1 was a candidate mediator for the cardiac remodeling initiated by AII and may play an integral role in concentric cardiac remodeling. However, secondary to an arteriovenous fistula, eccentric cardiac remodeling was likewise associated with increased cardiac AII and TGF-β_1 expression. Based on the latter observations, neither AII nor TGF-β_1 can be directly implicated in either the *in series* or *parallel* pattern of sarcomere assembly. In addition, these data have demonstrated that the upregulation of AII and TGF-β_1 in the volume-overloaded heart were not sufficient to promote the reexpression of cardiac contractile fetal genes and reciprocal downregulation of SERCA2 mRNA. Nonetheless, the conserved increased expression of TGF-β_1 to disparate pathological stimuli suggests that the peptide growth factor may contribute preferentially to the hypertrophic growth of individual cardiac myocytes in response to AII. Support for this premise was demonstrated in TGF-β_1-deficient transgenic mice, as both AII-mediated cardiac hypertrophy and ANP mRNA expression were inhibited (Schultz et al. 2002). Lastly, the increased expression of TGF-β_1 mRNA levels in the heart of voluntary exercised female rats further supports the premise that the peptide growth factor may contribute to the physiological growth of cardiac myocytes (Calderone et al. 2001).

A Synopsis of the Cellular and Molecular Events of Cardiac Hypertrophy and Potential Contribution of Peptide Growth Factors

A review of the literature has clearly demonstrated that pathological and physiological cardiac hypertrophies are associated with distinct morphological, cellular, and molecular patterns of adaptation. Furthermore, additional differences including the pattern of peptide growth factor expression were prevalent between pathological eccentric and concentric cardiac hypertrophy. Collectively, these findings have provided the framework to generate the following hypothetical models to address the establishment of the divergent phenotypes of hypertrophy. First, the disparate morphological and cellular phenotypes associated with volume- and pressure-overload induced pathological cardiac hypertrophy suggest that recruitment of distinct primordial signaling events was implicated in the assembly of sarcomeres in either an *in series* or *parallel* fashion (Figure 12.2). Second, ANP mRNA induction represents a conserved event of pathological hypertrophy, regardless of the initiating mechanical stimulus (Figure 12.2). Thus, ANP mRNA expression in response to chronic volume and pressure overload most likely occurred via the recruitment of a common signaling event. Third, the conserved upregulation of AII, TGF-β_1, and IGF-1 in pathological cardiac hypertrophy may collectively facilitate the hypertrophic growth response of cardiac myocytes (Figure 12.2). However, the increased expression of these latter stimuli in response to both pressure and volume overload suggests that these peptide growth factors do not directly contribute to the distinct pattern of sarcomere assembly associated with these disparate forms of hypertrophy. Fourth, AII,

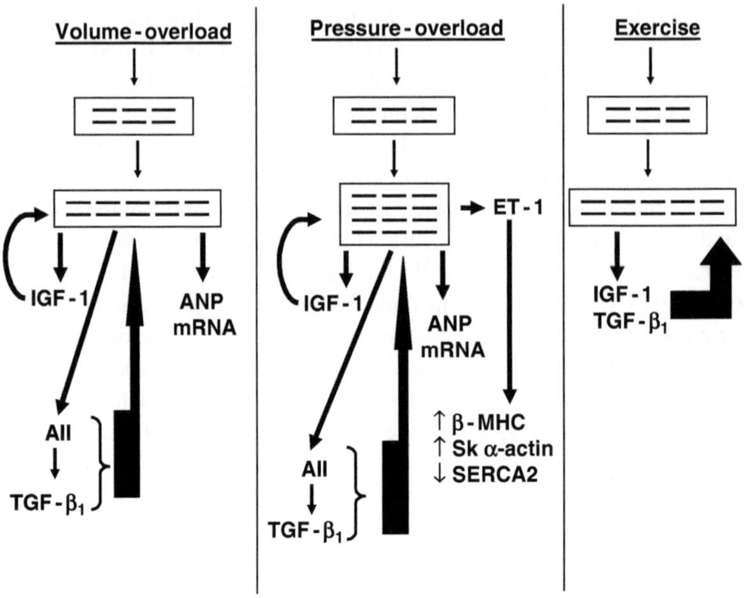

Fig. 12.2 Peptide growth factor expression and potential role in pathological and physiological cardiac hypertrophy. In volume- and pressure-overload induced cardiac hypertrophy, the induction of ANP mRNA represents a conserved event. Furthermore, the increased expression of IGF-1, TGF-β_1, and AII in both pathological models supports their involvement in the hypertrophic growth of cardiac myocytes. The selective upregulation of ET-1 in pressure-overloaded hearts may represent an integral event promoting the recapitulation of cardiac contractile fetal genes (β-MHC: β-myosin heavy chain; Sk α-actin: skeletal α-actin) and the downregulation of SERCA2. By contrast, exercise-induced cardiac hypertrophy was not associated with the recapitulation of cardiac fetal genes and neither AII nor ET-1 was implicated in physiological growth. However, the upregulation of both IGF-1 and TGF-β_1 in response to exercise further supports their conserved role in the hypertrophic growth of cardiac myocytes, regardless of the initiating stimulus.

acting directly or indirectly via TGF-β_1, may further promote the reexpression of ventricular ANP mRNA in response to a pathological stimulus (Figure 12.2). Lastly, the upregulation of cardiac contractile fetal genes and downregulation of SERCA2 represents a stimulus-specific response to pressure overload (Figure 12.2). Consequently, the latter pattern of gene expression reflects either the recruitment of an additional signaling event and/or stimulus-specific induction of the peptide growth factor ET-1 in the pressure-overloaded heart (Figure 12.2). The lack of ANP mRNA expression in response to exercise demonstrates that induction of the cardiac fetal gene can be segregated from a physiological hypertrophic growth of individual myocytes (Figure 12.2). The disparate regulation of ANP mRNA expression between eccentric physiological and pathological cardiac hypertrophy could be related at least in part to the intensity and duration of the diastolic load imposed on the heart and/or the selective increase of AII in chronically volume-overloaded hearts (Figure 12.2). Alternatively, exercise training may repress the recruitment of signaling event(s) coupled to a pathological pattern of molecular remodeling. Lastly, the upregulation of cardiac IGF-1 and TGF-β_1 in both concentric and eccentric remodeling may proceed via a common pathway and further supports the thesis that these peptides may preferentially facilitate a physiological adaptive growth response of cardiac myocytes, regardless of the nature (e.g., pathological versus physiological) of the initiating stimulus.

Signaling Events Coupled to the Divergent Phenotypes of Cardiac Hypertrophy; G_q- and Phosphatidylinositol 3-Kinase-Dependent Pathways

G_q- and phosphatidylinositol 3-kinase-dependent signaling pathways have been linked to the hypertrophic growth of cardiac myocytes. Data highlighting their involvement were reported in numerous *in vivo* and *in vitro* studies employing both pharmacological and transgenic approaches. The following section will attempt to summarize their relative role in eccentric and concentric cardiac hypertrophy.

The Role of PKC and Calcium in G_q-Dependent Induction of Concentric Cardiac Hypertrophy

The cardiac-specific overexpression of G_q induced a hypertrophic response associated with the upregulation of a panel of cardiac fetal genes in the absence of a hemodynamic overload (Adams et al. 1998). Furthermore, the magnitude of cardiac-specific G_q overexpression either led to a predominant hypertrophic response or progressed to heart failure characterized by marked cardiac myocyte apoptosis (Adams et al. 1998). The downstream signaling events implicated in G_q-dependent cardiac hypertrophy include protein kinase C (PKC) and Ca^{2+}. In the myocardium, the Ca^{2+}/DAG-dependent PKCα is the most abundant isoform and a partial inhibition of phorbol ester-mediated growth was observed in neonatal rat cardiac myocytes transfected with an adenovirus containing a dominant negative PKCα construct (Vijayan et al. 2004). By contrast, the hypertrophic response secondary to transverse aortic constriction was unaffected in PKCα-deficient mice suggesting that this isoform does not play an obligatory role in pressure-overload induced cardiac myocyte growth (Braz et al. 2004). In contrast to PKCα, a hypertrophic role *in vivo* has been described for the Ca^{2+}-insensitive DAG-dependent novel isoform PKCε. The cardiac-specific overexpression of PKCε resulted in increased wall thickness and the recapitulation of the cardiac fetal contractile genes β-myosin heavy chain and skeletal α-actin—characteristic features of concentric hypertrophy (Takeishi et al. 2000). Furthermore, PKCε-overexpressing mice did not progress to overt heart failure but were associated with normal left ventricular function (Takeishi et al. 2000).

A primary target of intracellular Ca^{2+} is calmodulin and the Ca^{2+}/calmodulin complex in turn promotes the concomitant activation of the calmodulin-dependent serine/threonine kinase CaMKII and the serine/threonine phosphatase calcineurin (Colomer and Means 2000; Obata et al. 2005). Their respective role in cardiac hypertrophy was reported *in vitro* as the pharmacological inhibition of either calmodulin with W7, CaMKII with KN-62, or calcineurin with cyclosporine attenuated cardiac myocyte hypertrophy following the activation of G_q-coupled receptors (Molkentin et al. 1998; Calderone et al. 2000; Zhu et al. 2000). Employing a transgenic approach, the overexpression of CaMKII δ_B isoform in the myocardium was highly concentrated to the nucleus and associated with the reexpression of cardiac fetal genes and hypertrophy (Zhang et al. 2002). Furthermore, a transition to a dilated cardiomyopathy with ventricular dysfunction was observed in CaMKII δ_B-overexpressing mice (Zhang et al. 2002). A hypertrophic role was also demonstrated for the CaMKII δ_C isoform as either a low or high copy of expression in the heart was associated with the recapitulation of the panel of cardiac fetal genes and the reciprocal downregulation of adult-specific genes (Zhang et al. 2003). Furthermore, in mice with a medium to high copy number of the transgene, survival was severely compromised and associated with marked ventricular contractile dysfunction. These data have demonstrated that the overexpression of CaMKII isoforms led to a concentric pattern of cardiac remodeling and can further initiate a transition from compensated to decompensated cardiac hypertrophy and subsequent heart failure.

α-Myosin heavy chain-driven expression of a constitutively active form of the calcineurin catalytic subunit led to an increase in ventricular wall thickness and the reexpression of cardiac fetal genes—a pattern consistent with concentric cardiac remodeling (Molkentin et al. 1998). The hypertrophic action of calcineurin occurred via the recruitment of the transcriptional factor NFAT (Molkentin et al. 1998). Furthermore, in constitutively active calcineurin-expressing transgenic mice, ventricular dilation was observed with increasing age and reactive fibrosis was evident (Molkentin et al. 1998). These data support the premise that the increased expression and activity of calcineurin reported in the heart secondary aortic banding in the rat may play an important role in the development of concentric cardiac hypertrophy (Lim et al. 2000). Thus, G_q-dependent recruitment of calcineurin, CaMKII, and PKC signaling events may act in a coordinated fashion to facilitate concentric cardiac hypertrophy. However, sustained calcineurin as well as CaMKII activity secondary to abnormal intracellular calcium homeostasis may represent precipitating events facilitating the transition from hypertrophy to decompensated hypertrophy.

Ca^{2+} – Dependent Pathways and Physiological Cardiac Hypertrophy

Elevated CaMKII activity was reported in the heart following voluntary running, albeit its direct role in physiological hypertrophy remains undefined (Konhilas et al. 2004). By contrast, indirect evidence of a role for calcineurin in physiological cardiac hypertrophy was demonstrated by Eto et al. (2000) as cyclosporine administration to voluntary exercised rats attenuated the increase in heart size. In contrast, cyclosporine treatment failed to inhibit the hypertrophic response in rats exposed to a swimming regimen (Hainsey et al. 2002). An alternative approach to assess the role of calcineurin in physiological cardiac hypertrophy involves the cardiac-specific overexpression of myocyte-enriched calcineurin interacting protein (MCIP)—endogenous inhibitors of calcineurin (Rothermel et al. 2000, 2001). The inhibitory action of MCIP1 was confirmed *in vivo* as its forced expression completely inhibited the hypertrophic response and recapitulation of cardiac fetal genes in transgenic mice expressing constitutively active calcineurin in the heart (Rothermel et al. 2001). In response to voluntary exercise, the cardiac-specific overexpression of MCIP1 in mice attenuated but did not abolish the hypertrophic response (Rothermel et al. 2001). These data suggest that calcineurin may contribute to the hypertrophic growth of cardiac myocytes in mice following voluntary exercise, but at least an additional pathway is required to initiate the full response. Furthermore, calcineurin-mediated growth of individual cardiac myocytes following voluntary exercise can be segregated from the recapitulation of cardiac fetal genes. Thus, it would appear that calcineurin may directly contribute to the growth of cardiac myocytes in both concentric and physiological eccentric hypertrophy but not to the disparate pattern of sarcomere assembly.

Phosphatidylinositol_3-Kinase (PI3-K)-Dependent Pathways and Cardiac Hypertrophy

Increased PI3-K activity was demonstrated in the heart of pressure-overloaded mice and attributed to the selective recruitment of the class IB p110γ isoform (Naga Prasad et al. 2000). In contrast, activity of the class IA p110α isoform remained unchanged despite a sustained pressure overload (Naga Prasad et al. 2000). A transgenic approach was used to assess the role of the class IA PI3-K family as the cardiac-specific expression of a constitutively active p110α kinase led to an increase in heart size, normal contractile function and histology (e.g., absence of fibrosis) (Shioi et al. 2000). The hypertrophic response was attributed to an increase in individual cardiac myocyte cell length and a modest increase in cell width, thereby supporting a role for the p110α kinase in an *in series* pattern of sarcomere assembly. Consistent with these data, the cardiac-specific overexpression of a catalytically inactive p110α kinase (dominant negative) was associated with smaller hearts attributed to a decrease in myocyte cell size (Shioi et al. 2000). The latter findings were reaffirmed in transgenic mice lacking expression of the p110α regulatory subunits: p85α and p85β (Luo et al. 2005). Thus, it would appear that the class IA family of PI3-K participates in postnatal physiological growth of cardiac myocytes. Furthermore, the exposure of either dominant negative p110α or p85α/p85β-deficient mice to a swimming regimen was associated with a reduced hypertrophic response (McMullen et al. 2003; Luo et al. 2005). Additional *in vivo* evidence supporting a role of the class IA PI3-K family in physiological cardiac hypertrophy was provided by the study of Crackower et al. (2002) in a transgenic mouse lacking the tumor suppressor PTEN (phosphatase and tensin homologue on chromosome ten). Collectively, these data support the premise that the class IA PI3-K family represents an integral signaling event coupled to physiological cardiac hypertrophy and may directly or indirectly facilitate the *in series* pattern of sarcomere assembly.

Conclusion

Important qualitative differences between pathological and physiological cardiac hypertrophy have been reported at the morphological, cellular, and molecular level. These differences may be related in part to the stimulus-specific regulation of a panel of peptide growth factors. It would appear that the recapitulation of cardiac fetal genes is stimulus specific as ANP mRNA induction can be dissociated from β-myosin heavy chain and skeletal α-actin mRNA expression. Furthermore, since physiological cardiac hypertrophy can occur in the absence of ANP mRNA expression, these events are not intimately linked. Numerous studies have demonstrated that G_q-dependent generation of concentric cardiac hypertrophy may involve the concerted effort of PKC and Ca^{2+}-dependent signaling events. Further dissection of these latter pathways demonstrated that the chronic activation of Ca^{2+}-dependent signaling events leads to decompensated cardiac hypertrophy and subsequent heart failure. In contrast, several studies strongly suggest that the p110α subunit of the class IA PI3-K family represents an integral event in the induction of physiological cardiac hypertrophy. There also exists evidence to support a potential role of calcineurin in exercise-induced cardiac hypertrophy. It is tempting to speculate that the physiological activation of calcineurin in response to an intermittent physiological stimulus could contribute to the overall growth response of the cardiac myocyte in the absence of ANP mRNA expression, since these two events can be dissociated. However, it is important to note that since eccentric and concentric remodeling are morphologically distinct, calcineurin would not be directly implicated in sarcomere assembly. What remains is an important, albeit daunting challenge to utilize these findings as a platform to pharmacologically target signaling events in the hearts of patients that may lead to maladaptive remodeling and progression to heart failure. To achieve this goal, it is essential that the various hypertrophic signaling events identified in experimental models are likewise recruited in the hypertrophied human heart and exert a similar phenotype.

Acknowledgments This work was supported by the Heart and Stroke Foundation of Canada and Quebec, Canadian Institutes of Health Research, and "La Fondation de l'Institut de Cardiologie de Montréal." Dr. Calderone is a "Chercheur-Boursier National du Fonds de la recherche en santé du Québec." The author thanks France Thériault for secretarial help.

References

Adams, J.W., Sakata, Y., Davis, M.G., Sah, V., Wang, Y., Ligget, S.B., Chein, K.R., Brown, J.H., and Dorn, G.W., II. 1998. Enhanced Gαq signalling: A common pathway mediates cardiac hypertrophy and apoptotic heart failure. Proc. Natl. Acad. Sci. USA 95:10140–10145.

Anversa, P., Levicky, V., Beghi, C., McDonald, S.L., and Kikkawa, Y. 1983. Morphometry of exercise-induced right ventricular hypertrophy in the rat. Circ. Res. 52:57–64.

Anversa, P., Ricci, R., and Olivetti, G. 1986. Quantitative structural analysis of the myocardium during physiological growth and induced cardiac hypertrophy: a review. J. Am. Coll. Cardiol. 7:1140–1149.

Black, F.M., Packer, S.E., Parker, T.G., Michael, L.H., Roberts, R., Schwartz, R.J., and Schneider, M.D. 1991. The vascular smooth muscle α-actin gene is reactivated during cardiac hypertrophy provoked by load. J. Clin. Invest. 79:970–977.

Braunwald, E. 1992. *Heart Disease: A Textbook of Cardiovascular Medicine* (4th ed.). Saunders, Philadelphia.

Braz, J., Gregory, K., Pathak, A., Zhao, W., Sahin, B., Klevitsky, R., Kimball, T.F., Lorenz, J.N., Nairn, A.C., Liggett, S.B., Bodi, I., Wang, S., Schwartz, A., Lakatta, E.G., DePaoli-Roach, A.A., Robbins, J., Hewett, T.E., Bibb, J.A., Westfall, M.V., Kranias, E.G., and Molkentin, J.D. 2004. PKC-alpha regulates cardiac contractility and propensity toward heart failure. Nature Med. 10:248–254.

Calderone, A., Takahashi, N., Izzo, N.J., Thaik, C.M., and Colucci, W.S. 1995. Pressure- and volume-induced left ventricular hypertrophies are associated with distinct myocyte phenotypes and differential induction of peptide growth factor mRNAs. Circulation 92:2385–2390.

Calderone, A., Takahashi, N., Thaik, C., and Colucci, W.S. 1998. Atrial natriuretic peptide, nitric oxide and cGMP inhibit the growth-promoting effects of norepinephrine in cardiac myocytes and fibroblasts. J. Clin. Invest. 101:812–818.

Calderone, A., Abdelaziz, N., Colombo, F., Schreiber, K.L., and Rindt, H. 2000. A farnesyltransferase inhibitor attenuates cardiac myocyte hypertrophy and gene expression. J. Mol. Cell. Cardiol. 32:1127–1140.

Calderone, A., Murphy, R.J.L., Lavoie, J., Colombo, F., and Beliveau, L. 2001. TGF-β_1 and prepro-ANP mRNAs are differentially regulated in exercise-induced cardiac hypertrophy. J. Appl. Physiol. 91:771–776.

Colomer, J.M., and Means, A.R. 2000. Chronic elevation of calmodulin in the ventricles of transgenic mice increases the autonomous activity of calmodulin-dependent protein kinase II, which regulates atrial natriuretic factor gene expression. Mol. Endocrinol. 14:1125–1136.

Crackower, M.A., Oudit, G.Y., Kozieradzki, I., Sarao, R., Sun, H., Sasaki, T., Irie-Sasaki, J., Sah, R., Cheng, H.-Y.M., Rybin, V.O., Lembo, G., Fratta, L., Oliveira-dos-Santos, A.J., Benovic, J.L., Kahn, C.R., Izumo, S., Steinberg, S.F., Wymann, M.P., Backx, P.H., and Penninger, J.P. 2002. Regulation of myocardial contractility and cell size by distinct PI3K-PTEN signalling pathways. Cell 110:737–749.

de la Bastie, D., Levitsky, D., Rappaport, L., Mercadier, J.J., Marotte, F., Wisnewsky, C., Brovkovich, V., Schwartz, K., and Lompre, A.M. 1990. Function of the sarcoplasmic reticulum and expression of its Ca^{2+} ATPase gene in pressure overload-induced cardiac hypertrophy in the rat. Circ. Res. 66:554–565.

Delaughter, M.C., Taffet, G.E., Fiorotto, M.L., Entman, M.L., and Schwartz, R.L. 1999. Local insulin-like growth factor expression induces physiologic, then pathologic, cardiac hypertrophy in transgenic mice. FASEB J. 13:1923–1929.

Diffee, G.M., Seversen, E.A., Stein, T.D., and Johnson, J.A. 2003. Microarray expression analysis of effects of exercise training: increase in atrial MLC-1 in rat ventricles. Am. J. Physiol. 284:H830–H837.

Ellmers, L.J., Knowles, J.W., Kim, H.S., Smithies, O., Maeda, N., and Cameron, V.A. 2002. Ventricular expression of natriuretic peptides in Npr1(-/-) mice with cardiac hypertrophy and fibrosis. Am. J. Physiol. 283:H707–H714.

Eto, Y., Yonekura, K., Sonoda, M., Arai, N., Sata, M., Sugiura, S., Takenaka, K., Gualberto, A., Hixon, M.L., Wagner, M.W., and Aoyagi, T. 2000. Calcineurin is activated in rat hearts with physiological left ventricular hypertrophy induced by voluntary exercise training. Circulation 101:2134–2137.

Geenen, D.L., Malhotra, A., and Buttrick, P.M. 1996. Angiotensin receptor 1 blockade does not prevent physiological cardiac hypertrophy in the adult rat. Am. J. Physiol. 81:816–821.

Gerdes, A.M., Campbell, S.E., and Hilbelink, D.R. 1988. Structural remodelling of cardiac myocytes in rats with arteriovenous fistula. Lab. Invest. 59:857–861.

Grossman, W., Jones, D., and McLaurin, L.P. 1975. Wall stress and patterns of hypertrophy in the human left ventricle. J. Clin. Invest. 56:56–64.

Hainsey, T., Csiszar, A., Sun, S., and Edwards, J.G. 2002. Cyclosporin A does not block exercise-induced cardiac hypertrophy. Med Sci. Sports. Exerc. 34:1249–1254.

Huang, C.-H., Hao, L.-Y., and Buetow, D.E. 2002. Insulin-like growth factor-induced hypertrophy of cultured adult rat cardiomyocytes is L-type calcium-channel-dependent. Mol. Cell. Biochem. 231:51–59.

Iemitsu, M., Miyauchi, T., Maeda, S., Sakai, S., Kobayashi, T., Fujii, N., Miyazaki, H., Matsuda, M., and Yamaguchi, I. 2001. Physiological and pathological cardiac hypertrophy induce different molecular phenotypes in the rat. Am. J. Physiol. 281:R2029–R2036.

Ito, H., Hiroe, M., Hirata, Y., Fujisaki, H., Adachi, S., Akimoto, H., Ohta, Y., and Marumo, F. 1994. Endothelin ET_A receptor antagonist blocks cardiac hypertrophy provoked by hemodynamic overload. Circulation 89:2198–2203.

Izumo, S., Nadal-Ginard, B., and Mahdavi, V. 1988. Proto-oncogene induction and reprogramming of the cardiac gene expression produced by pressure-overload. Proc. Natl. Acad. Sci. USA 85:339–343.

Jin, H., Yang, R., Li, W., Ryan, A., Ogasawara, A.K., Peborgh, J.V., and Paoni, N.F. 2000. Effects of exercise training on cardiac function gene expression, and apoptosis in rats. Am. J. Physiol. 279:H2994–H3002.

Katz, A.M. 1990. Cardiomyopathy of overload: A major determinant of prognosis in congestive heart failure. N. Engl. J. Med. 322:100–110.

Kee, H.J., Sohn, I.S., Nam, K.I., Park, J.E., Qian, Y.R., Yin, Z., Ahn, Y., Jeong, M.H., Bang, Y.J., Kim, N., Kim, J.K., Epstein, J.A., and Kook, H. 2006. Inhibition of histone deacetylation blocks cardiac hypertrophy induced by angiotensin II and aortic banding. Circulation 113:51–59.

Kinugawa, S., Tsutsui, H., Ide, T., Nakamura, R., Arimura, K., Egashira, K., and Takeshita, A. 1999. Positive inotropic effect of insulin-like growth factor-1 on normal and failing cardiac myocytes. Cardiovasc. Res. 43:157–164.

Konhilas, J.P., Maass, A.H., Luckey, S.W., Stauffer, B.L., Olson, E.N., and Leinwand, L.A. 2004. Sex modifies exercise and cardiac adaptation in mice. Am. J. Physiol. 287:H2768–H2776.

Li, Q., Li, B., Wang, X., Leri, A., Jana, K., Lui, Y., Kajstura, J., Baserga, R., and Anversa, P. 1997. Overexpression of insulin-like growth factor-1 in mice protects from death after infarction, attenuating ventricular dilatation, wall stress, and cardiac hypertrophy. J. Clin. Invest. 100:1991–1999.

Lijnen, P., and Petrov, V. 1999. Renin-angiotensin system, hypertrophy and gene expression in cardiac myocytes. J. Mol. Cell. Cardiol. 31:949–970.

Lim, H.W., De Windt, L.J., Steiberg, L., Taigen, T., Witt, S.A., Kimball, T.R., and Molkentin, J.D. 2000. Calcineurin expression, activation, and function in cardiac pressure-overload hypertrophy. Circulation 101:2431–2437.

Luo, J., McMullen, J.R., Sobkiw, C.L., Zhang, L., Dorfman, A.L., Sherwood, M.C., Logsdon, M.N., Horner, J.W., DePinho, R.A., Izumo, S., and Cantley, L.C. 2005. Class 1_A phosphoinositide 3-kinase regulates heart size and physiological cardiac hypertrophy. Mol. Cell. Biol. 25:9491–9502.

Malhotra, R., Sadoshima, J., Brosius, F.C. 3rd, and Izumo, S. 1999. Mechanical stretch and angiotensin II differentially upregulate the renin-angiotensin system in cardiac myocytes *in vitro*. Circ. Res 85:137–146.

Marino, T.A., Kent, R.L., Uboh, C.E., Fernandez, E., Thompson, E.W., and Cooper, G., IV. 1985. Structural analysis of pressure versus volume overload hypertrophy of cat right ventricle. Am. J. Physiol. 249:H371–H379.

Masashi, A., Matsui, H., and Periasamy, M. 1994. Sarcoplasmic reticulum gene expression in cardiac hypertrophy and heart failure. Circ. Res. 64:555–564.

Matthews, K.G., Devlin, G.P., Conaglen, J.V., Stuart, S.P., Mervyn Aitken, W., and Bass, J.J. 1999. Changes in IGFs in cardiac tissue following myocardial infarction. J. Endocrinol. 163:433–445.

Mazzolai, L., Nussberger, J., Aubert, J.F., Brunner, D.B., Gabbiani, G., Bruner, H.R., and Pedrazzini, T. 1998. Blood pressure-independent cardiac hypertrophy induced by locally activated renin-angiotensin system. Hypertension 31:1324–1330.

McMullen, J., Shioi, T., Zhang, L., Tarnavaski, O., Sherwood, M.C., Kang, P.M., and Izumo, S. 2003. Phosphoinositide 3-kinase (p110α) plays a critical role for the induction of physiological, but not pathological cardiac hypertrophy. Proc. Natl. Acad. Sci. USA 100:12355–12360.

McMullen, J., Shioi, T., Huang, W.-Y., Zhang, L., Tarnavaski, O., Bisping, E., Schinke, M., Kong, S., Sherwood, M.C., Brown, J., Riggi, L., Kang, P.M., and Izumo, S. 2004. The insulin-like growth factor 1 receptor induces physiological heart growth via the phosphoinositide 3-kinase (p110α) pathway. J. Biol. Chem. 279:4782–4793.

Mercadier, J.J., Lompre, A.M., Wisnewsky, C., Samuel, J.L., Bercovici, J., Swynghedauw, B., and Schwartz, K. 1981. Myosin isoenzymatic changes in several models of cardiac hypertrophy. Circ. Res. 49:525–532.

Modesti, P.A., Vanni, S., Bertolozzi, I., Cecioni, I., Polidori, G., Paniccia, R., Bandinelli, B., Perna, A., Liguori, P., Boddi, M., Galanti, G., and Serneri, G.G.N. 2000. Early sequence of cardiac adaptations and growth factor formation in pressure- and volume-overload hypertrophy. Am. J. Physiol. 279:H976–H985.

Mohabir, R., Young, S.D., and Strosberg, A.M. 1994. Role of angiotensin in pressure overload-induced hypertrophy in rats: effects of angiotensin-converting enzyme inhibitors, an AT1 receptor antagonist, and surgical reversal. J. Cardiovasc. Pharmacol. 23:291–299.

Molkentin, J.D., Lu, J.-R., Antos, C.L., Robbins, J., Grant, S.R., and Olson, E.N. 1998. A calcineurin-dependent transcriptional pathway for cardiac hypertrophy. Cell 93:215–228.

Moore, R.L., Musch, T.I., Yelamarty, R.V., Scaduto, R.C., Jr., Semanchick, A.M., Elensky, M., and Cheung, J.Y. 1993. Chronic exercise alters contractility and morphology of isolated cardiac myocytes. Am. J. Physiol. 264: C1180–C1189.

Morkin, E. 1970. Postnatal muscle fiber assembly: localization of newly synthesized myofibrillar proteins. Science 167:1499–1501.

Nadal-Ginard, B., and Mahdavi, V. 1989. Molecular basis of cardiac performance: plasticity of the myocardium generated through protein isoform switches. J. Clin. Invest. 84:1693–1700.

Naga Prasad, S.V., Esposito, G., Mao, L., Koch, W.J., and Rockman, H.A. 2000. Gβγ-dependent phosphoinositide 3-kinase activation in hearts with *in vivo* pressure overload hypertrophy. J. Biol. Chem. 275:4693–4698.

Obata, K., Nagata, K., Iwase, M., Odashima, M., Nagasaka, T., Izawa, H., Murohara, T., Yamada, Y., and Yokota, M. 2005. Overexpression of calmodulin induces cardiac hypertrophy by a calcineurin-dependent pathway. Biochem. Biophys. Res. Commun. 338:1299–1305.

Paradis, P., Dali-Youcef, N., Paradis, F.W., Thibault, G., and Nemer, M. 2000. Overexpression of angiotensin II type 1 receptor in cardiomyocytes induced cardiac hypertrophy and remodelling. Proc. Natl. Acad. Sci. USA 97:931–936.

Parker, T.G., Packer, S.E., and Schneider, M.D. 1990. Peptide growth factors can provoke fetal contractile gene expression in rat cardiac myocytes. J. Clin. Invest. 85:507–514.

Redaelli, G., Malhotra, A., Li, B., Li, P., Sonneblick, E.H., Hofmann, P.A., and Anversa, P. 1998. Effects of constitutive overexpression of insulin-like growth factor-1 on the mechanical characteristics and molecular properties of ventricular myocytes. Circ. Res. 82:594–603.

Rockman, H.A., Wachhorst, S.P., Mao, L., and Ross, J., Jr. 1994. ANG II receptor blockade prevents ventricular hypertrophy and ANF gene expression with pressure overload in mice. Am. J. Physiol. 266:H2468–H2475.

Rothermel, B.A., Vega, R.B., Yang, J., Wu, H., Bassel-Duby, R., and Williams, R.S. 2000. A protein encoded within the down syndrome critical region is enriched in striated muscles and inhibits calcineurin signalling. J. Biol. Chem. 275:8719–8725.

Rothermel, B.A., McKinsey, T.A., Vega, R.B., Nicol, R.L., Mammen, P., Yang, J., Antos, C.L., Shelton, J.M., Bassel-Duby, R., Olson, E.N., and Williams, R.S. 2001. Myocyte-enriched calcineurin-interacting protein, MCIP1, inhibits cardiac hypertrophy *in vivo*. Proc. Natl. Acad. Sci. USA 98:3328–3333.

Ruzicka, M., Yuan, B., and Leenen, F.H. 1994. Effects of enalapril versus losartan on regression of volume overload-induced cardiac hypertrophy in rats. Circulation 90:484–491.

Ruzicka, M., Skarda, V., and Leenen, F.H.H. 1995. Effects of ACE inhibitors on circulating versus cardiac angiotensin II in volume overload induced cardiac hypertrophy in rats. Circulation 92:3568–3573.

Sadoshima, J., and Izumo, S. 1993. Molecular characterization of angiotensin II-induced hypertrophy of cardiac myocytes and hyperplasia of cardiac fibroblasts. Critical role of the AT1 receptor subtype. Circ. Res. 73:413–423.

Scheinowitz, M., Kessler-Icekson, G., Freimann, S., Zimmerman, R., Schaper, W., Golomb, E., Savion, N., and Eldar, M. 2003. Short- and long-term swimming exercise training increases myocardial insulin-like growth factor-1 gene expression. Growth Hormone IGF Res. 13:19–25.

Schiaffino, S., Samuel, J.L., Sassoon, D., Lompre, A.M., Garner, I., Marotte, F., Buckingham, M., Rappaport, L., and Schwartz, K. 1989. Nonsynchronous accumulation of α-skeletal actin and β-myosin heavy chain mRNA's during the early stages of pressure-overload induced cardiac hypertrophy demonstrated by in situ hybridization. Circ. Res. 64:937–948.

Schultz, J.E.J., Witt, S.A., Glascocj, B.J., Neiman, M.L., Reiser, P.J., Nix, S.L., Kimball, T.R., and Doetschman, T. 2002. TGF-β1 mediates the hypertrophic cardiomyocyte growth induced by angiotensin II. J. Clin. Invest. 109:787–796.

Serneri, G.G.N., Boddi, M., Modesti, P.A., Cecioni, I., Coppo, M., Padeletti, L., Michelucci, A., Colella, A., and Galanti, G. 2001. Increased sympathetic activity and insulin-like growth factor-1 formation are associated with physiological hypertrophy in athletes. Circ. Res. 89:977–982.

Serneri, G.G.N., Modesti, P.A., Boddi, M., Cecioni, I., Paniccia, R., Coppo, M., Simonetti, I., Vanni, S., Papa, L., Bandinelli, B., Migliorini, A., Modesti, A., Maccherini, M., Sani, G., and Toscano, M. 1999. Cardiac growth factors in human hypertrophy: relations with myocardial contractility and wall stress. Circ. Res. 85:57–67.

Shioi, T., Kang, P., Douglas, P.S., Hampe, J., Yballe, C.M., Lawitts, J., Cantley, L.C., and Izumo, S. 2000. The conserved phosphoinositide 3-kinase pathway determines heart size in mice. Cell 19:2537–2548.

Shubieta, H.E., McDonough, P.M., Harris, A.N., Knowlton, K.U., Glembotski, C.C., Brown, J.H., and Chien, K.R. 1990. Endothelin induction of inositol phosphate hydrolysis, sarcomere assembly, and cardiac gene expression in ventricular myocytes. A paracrine mechanism for myocardial cell hypertrophy. J.Biol. Chem. 265:20555–20562.

Strait, J.B., Martin, J.L., Bayer, A., Mestril, R., Eble, D.M., and Samarel, A.M. 2001. Role of protein kinase C-ε in hypertrophy of cultured neonatal rat ventricular myocytes. Am. J. Physiol. 280:H756–H766.

Takeishi, Y., Ping, P., Bolli, R., Kirkpatrick, D.L., Hoit, B.D., and Walsh, R.A. 2000. Transgenic overexpression of constitutively active protein kinase C ε causes concentric cardiac hypertrophy. Circ. Res. 86:1218–1223.

Thompson, N.L., Bazoberry, F., Speir, E.H., Casscells, W., Ferrans, V.J., Flamnder, K.C., Kondaiah, P., Geiser, A.G., and Sporn, M.B. 1988. Transforming growth factor beta-1 in acute myocardial infarction in rats. Growth Factors 1:91–99.

Vijayan, K., Szotek, E.L., Martin, J.L., and Samarel, A.M. 2004. Protein kinase C-α-induced hypertrophy of neonatal rat ventricular myocytes. Am. J. Physiol. 287:H2777–H2789.

von Lewinski, D., Voss, K., Hulsmann, S., Kogler, H., and Pieske, B. 2003. Insulin-like growth factor-1 exerts Ca^{2+}-dependent positive inotropic effects in failing human myocardium. Circ. Res. 92:169–176.

Weinberg, E.O., Lee, M.A., Weigner, M., Lindpianter, K., Bishop, S.P., Benedict, C.R., Douglas, P.S., Chafizadeh, E., and Lorell, B.H. 1997. Angiotensin AT1 receptor inhibition. Effects on hypertrophic remodelling and ACE expression in rats with pressure-overload hypertrophy due to ascending aortic stensois. Circulation 95:1592–1600.

Zak, R. 1974. Development and proliferative capacity of cardiac muscle cells. Circ. Res. 34–35 (Suppl. II):17–25.

Zhang, T., Johnson, E.N., Gu, Y., Morissette, M.R., Sah, V.P., Gigena, M.S., Belke, D.D., Dillman, W.H., Rogers, T.B., Schulman, H., Ross, J., Jr., and Brown, J.H. 2002. The cardiac-specific nuclear δ_B isoform of Ca^{2+}/calmodulin dependent protein kinase II induces hypertrophy and dilated cardiomyopathy associated with increased protein phosphatase 2A activity. J. Biol. Chem. 277:1261–1267.

Zhang, T., Maier, L.S., Dalton, N.D., Miyamoto, S., Ross, J., Jr., Bers, D.M., and Brown, J.H. 2003. The δ_c isoform of CaMKII is activated in cardiac hypertrophy and induces dilated cardiomyopathy and heart failure. Circ. Res. 92:912–919.

Zhu, W., Zou, Y., Shiojima, I., Kudoh, S., Aikawa, R., Hayashi, D., Mizukami, M., Toko, H., Shibasaki, F., Yazaki, Y., Nagai, R., and Komuro, I. 2000. Ca^{2+}/calmodulin-dependent kinase II and calcineurin play critical roles in endothelin-1-induced cardiomyocyte hypertrophy. J. Biol. Chem. 275:15239–15245.

Chapter 13
Regulatory Role of TGF-β in Cardiac Myofibroblast Function and Post-MI Cardiac Fibrosis: Key Roles of Smad7 and c-Ski

Ian M.C. Dixon, Baiqiu Wang, Kristen M. Bedosky, Ryan H. Cunnington, Sunil G. Rattan, and Ahmed Almaqrhi

Abstract The molecular pathways that couple increased hemodynamic load to cardiac hypertrophy, cardiac fibrosis, and heart failure are incompletely understood. Cardiac fibrosis is recognized as a major disease modifier and as such is important in the pathogenesis of heart failure of most etiologies. This review is focused on R-Smad signaling in cardiac myofibroblasts and their role in remodeling the extracellular matrix of the failing myocardium after myocardial infarction (MI). As major mediators of TGF-β_1 signaling in cardiac fibroblasts and myofibroblasts as well as myocytic cells, Smad proteins are emerging as an important postreceptor class in post-MI heart failure. How cytosolic c-Ski and C184M proteins influence R-Smads (and possibly Smad7 itself) in cardiac myofibroblasts, and how c-Ski expression influences cardiac myofibroblast function are largely unknown. We suggest that decreased Smad7 expression and altered intracellular c-Ski expression and/or distribution may contribute to chronic imbalance of R-Smad activation in classic TGF-β_1 signaling in the context of cardiac myofibroblast function. Thus, reduced and insufficient expression and activation of c-Ski and Smad7 may contribute to abnormal stimulation of collagen synthesis by these cells. Exploitation of Smad7 and c-Ski in the modulation of cardiac myofibroblast function may provide the experimental basis for the development of highly specific drugs for treating heart failure with attendant cardiac fibrosis. Future studies to identify and characterize Smad-associated factors responsible for induction of pathological cardiac hypertrophy are warranted.

Introduction

The molecular pathways that couple increased hemodynamic load to cardiac hypertrophy, cardiac fibrosis, and heart failure are multiplex and incompletely understood, despite intense recent investigation (Hunter and Chien 1999; MacLellan and Schneider 2000). Although TGF-β_1 initiates multiple modes of signal activation, including the recently discovered TβRII/TAK1 pathway (Watkins et al. 2006),

A. K. Srivastava, M. B. Anand-Srivastava (eds.) *Advances in Biochemistry in Health and Disease.* Signal Transduction in the Cardiovascular System in Health and Disease,

this review focuses on proteins that inhibit the canonical TGF-β_1/TβRI kinase-mediated phosphorylation of R-Smads in cardiac myofibroblasts and their role in remodeling the extracellular matrix (ECM or matrix) of the failing myocardium after myocardial infarction (MI).

TGF-β_1 is a well-known profibrotic stimulator of matrix secretion and deposition by primary fibroblasts and myofibroblasts (Butt et al. 1995) while inhibitory Smad7 (Smad7) is an endogenous inhibitor of TGF-β_1 signaling (Park 2005). Other repressors of TGF-β_1 functions are known to exist, and while the precise role of c-Ski is relatively undefined at present, it has been suggested that this protein fits this mode of cardiac fibroblast/myofibroblast regulation. For the past few years, the general premise of our investigations has focused on the hypothesis that a loss of balanced excitatory and inhibitory Smad signaling in post-myocardial infarction (post-MI) heart contributes to overt cardiac fibrosis and heart failure. In post-MI or otherwise diseased myocardium, the primary function of adult cardiac myofibroblasts is to govern net fibrillar collagen deposition and maintain the integrity of the cardiac matrix. Usually the appearance of the myofibroblastic phenotype heralds the presence of overt tissue fibrosis. As cardiac fibrosis is an acknowledged disease modifier of many etiologies of heart disease and thereby contributes to heart failure, and as R-Smads, e.g., Smad2, may activate collagen genes, work to characterize the control of R-Smad function particularly by endogenous inhibitory proteins is of considerable interest. Nevertheless, the mechanisms for endogenous inhibition of cardiac R-Smad signaling are not well understood, particularly with respect to the regulation of myofibroblast function or in the pathogenesis of heart failure. Our past work supports a link between Smad7 expression and altered myofibroblast function as well as the onset of overt cardiac fibrosis and heart failure. Far less clear, however, is the involvement of either c-Ski or of C184M, a novel cytosolic c-Ski

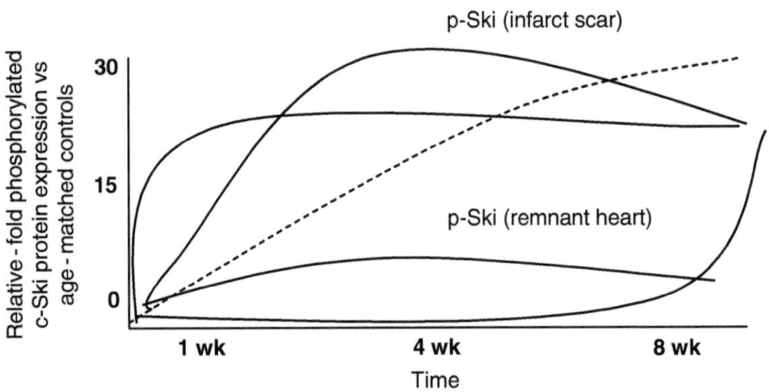

Fig. 13.1 Phosphorylated c-Ski expression in cytosolic fraction of cells in post-MI rat heart. In post-MI rat hearts with chronic large myocardial infarction, phosphorylated c-Ski expression was noted in the cytosol of cells populating the infarct scar (predominantly myofibroblasts) and the remnant heart (mixed cytosolic fraction from myocytes and nonmyocytes). Hearts were sampled at different times after surgical ligation of coronary artery occlusion and Western analysis was carried out to assess phosphorylated c-Ski expression. Trends in -fold protein expression (target band intensity corrected for loading) are shown in curves.

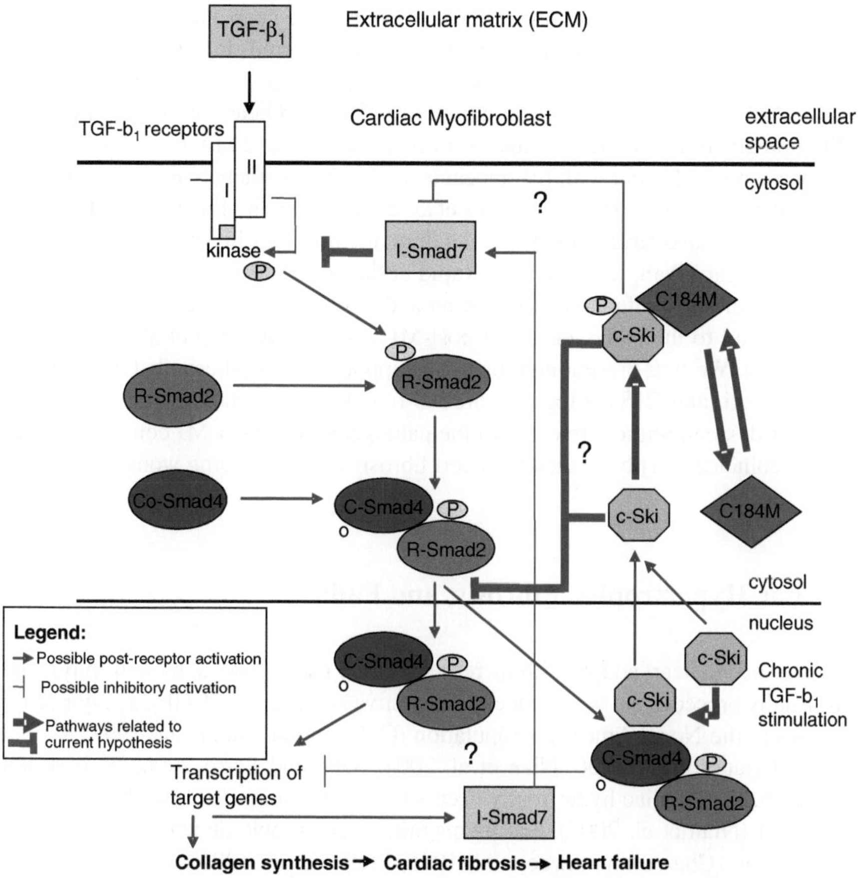

Fig. 13.2 R-Smad signaling and the putative roles of Smad7 and c-Ski. Schematic representation of integrated Smad signaling in cardiac myofibroblasts in response to TGF-β_1 stimulation. Receptor-activated Smad2 (R-Smad2) is phosphorylated by TGF-β_1; this potentiates enhanced synthesis of fibrillar collagens. Inhibition of R-Smads occurs via the expression and activation of Smad7. c-Ski is highly expressed in the post-MI heart, and we suggest that c-Ski, along with its novel binding partner C184M, may function like Smad7 and inhibit R-Smads in cardiac myofibroblasts. (See color plate.)

binding partner (Kokura et al. 2003). Investigation of the effects of c-Ski and/or C184M on Smad7 in modulation of cardiac matrix production, myofibroblast contraction and proliferation *in vitro* and in post-MI heart failure is ongoing. Our published data and very recent work have led us to suspect that an imbalance between R-Smads (both Smad2 and Smad3) versus Smad7 (and now c-Ski) may cue fibrosis and contribute to ongoing remodeling in post-MI congestive heart failure. The current review will sum the current state of knowledge of Smad7 and c-Ski/C184M as endogenous TGF-β_1 repressor proteins in cardiac myofibroblasts. An overview of classic Smad signaling is presented, as well as the highlights of our current hypotheses about putative inhibition of R-Smads by Smad7 and c-Ski (Figure 13.2). Further, a review of the pathophysiology of post-MI congestive heart failure as well as the biology of Smads and Smad-associated proteins is offered. Preliminary data from our laboratory are presented to indicate that phosphorylated c-Ski (p-Ski) is abnormally expressed within 48 hours of infarction and in overt heart failure (Figure 13.1). c-Ski upregulation may be coordinated to coincide with end of the acute inflammatory phase post-MI, and with the appearance of hypersynthetic cardiac myofibroblasts in the infarct zone (Frangogiannis et al. 2002). Data previously published from this laboratory also have demonstrated a dramatic drop-out in Smad7 expression in the infarct scar (Wang et al. 2002). Rapid collagen turnover by myofibroblasts is a hallmark of early infarct scar formation and in chronic remodeling of the matrix that is pivotal to the pathogenesis of post-MI heart failure (Hao et al. 2000; Wang et al. 2002). We will present arguments to support the hypothesis that endogenous Smad7, c-Ski, and C184M regulate normal function of cardiac myofibroblasts and that their dysregulation is relevant to the pathogenesis of post-MI congestive heart failure (coincident with the onset of overt fibrosis and/or ongoing wound healing).

Cardiac Hypertrophy, Fibrosis, and Failure

The overloaded heart adapts with increased muscle mass (cardiac hypertrophy), and this usually precedes the occurrence of congestive heart failure (CHF), a major cause of death in the North American population (Colucci and Braunwald 1997; Kannel 1997; Mettauer et al. 2006; Nian et al. 2004; Saha and Ferro 2006; Wollert and Drexler 2005). Cardiac hypertrophy occurs in compensation for loss of heart tissue due to MI (Nian et al. 2004), and its magnitude is variable depending on the size of infarction (Chareonthaitawee et al. 1995; Pfeffer et al. 1991). In the event of a large MI, the ventricular chamber may remodel by increasing in volume (Pfeffer et al.1995) and exhibiting severe hypertrophy which is associated with increased myocyte size and decreased intrinsic cardiac performance (Colucci and Braunwald 1997). Using an experimental model of post-MI heart failure we have observed (Hao et al. 1999; Ju et al. 1997; Liu et al. 1996; Makino et al. 1996) progressive cardiac dysfunction (prefailure, moderate and severe heart failure) with chronically elevated fibrillar collagen expression in remnant heart and infarct scar from animals with moderate heart failure (Dixon et al. 1990; Freed et al. 2005). In previous work, we have noted a profound decrease of Smad7 expression in moderate heart failure and a strong positive correlation between overdriven Smad7 expression and decreased fibrillar collagen synthesis in cardiac myofibroblasts (Wang et al. 2002). In the same series of published results, we found that Smad7 undergoes rapid activation, i.e., reciprocal translocation from *nuclei to cytosol* in COS-7 cells treated with TGF-β_1, and confers negative feedback for this ligand. Finally, our preliminary evidence points to a significant increase in p-Ski expression in myofibroblasts of the infarct scar at various times after MI versus remnant tissue and sham-operated heart (Figure 13.1). As cardiac fibrosis is associated with chronic wound healing and as this may initiate at the site of the infarction and subsequently "spill over" into myocardium which is remote to the infarct site, further investigation of the expression and function of Smad7 and c-Ski is of considerable interest. Collectively, our data provide the basis for the hypothesis that both Smad7 and c-Ski function as naturally occurring, i.e., endogenous, inhibitory proteins that may check otherwise unrestricted signaling of TGF-β_1 receptors in cardiac fibroblasts and myofibroblasts. Finally, as TGF-β_1 is clearly a pleuripotent ligand that exerts different effects on (i) metastatic cells and transformed cell lines versus non-cancerous primary cultured cells and (ii) cell types from different organs from the same species, the study of the specific regulation aspects of TGF-β_1 signaling, e.g., Smad7 and c-Ski effects in primary fibroblastic cells, is warranted.

TGF-β_1 is an Important Ligand in the Context of Heart Failure and in Cardiac Fibrosis

The myocardial ECM (mainly comprised of fibrillar collagens) is an organized network intimately associated with cardiac function, serving to direct, transmit, and distribute myocyte-generated contractile force (Caulfield and Borg 1979). Other functions include regulation of cell death, gene expression, and parenchymal cell differentiation (Birchmeier and Birchmeier 1993; Simon-Assmann et al. 1995). Nonetheless, elevated fibrillar collagen expression may be responsible for changing heart function in heart disease based on its adverse influence on myocardial stiffness (Jalil et al. 1989; Makino et al. 1996; Thiedemann et al. 1983). Pathological cardiac hypertrophy is associated with interstitial and perivascular fibrosis in remnant heart or as replacement fibrosis for necrosed muscle (Bartosova et al. 1969; Hao et al. 1999; Weber and Brilla 1991). In the latter, ongoing collagen remodeling may contribute to decompensated cardiac function in severe heart failure stage (Cleutjens et al. 1999; Ju et al. 1998).

TGF-β_1 mediates cell growth and differentiation, tissue wound repair, and extracellular matrix production (Brand and Schneider 1995, 1996; Kingsley 1994), including regulation of fibrillar collagens (Butt and Bishop 1997; Ignotz and Massague 1986; Inagaki et al. 1994; Powell et al. 1999; Roberts et al. 1990), and is expressed in the normal and hypertrophied myocardium (Brand and Schneider 1995; Dixon et al. 2000; Hao et al. 1999, 2000; Ohta et al. 1994; Sadoshima and Izumo 1993). Recent evidence generated with transgenic mice has provided data correlating TGF-β_1 ligand expression to cardiac fibrosis (Brooks and Conrad 2000), while another study indicates that this relationship exists in atria but not in ventricles in transgenic animals (Nakajima et al. 2000). The suggestion that postreceptor signal proteins (Smads) may potentiate cardiac myofibroblast function is reasonable in light of the evidence in the literature. TGF-β_1 signaling is well-known to be mediated by Smads and their transcriptional partners including coactivators and corepressors (Chen et al. 1998; Derynck et al. 1998; Massague 1998; Massague et al. 1997; Nakao et al. 1997b; Wrana 2000; Wrana and Pawson 1997; Zhou et al. 1998).

Chronic cardiac wound healing and the development of fibrosis in congestive heart failure is a complex process and may involve input from multiple factors (Hunter and Chien 1999; Makino et al. 1996). In this regard, TGF-β_1 is a known modulator of expression of fibrillar collagens (Butt et al. 1995; Ignotz and Massague 1986; Tomasek et al. 2002; Weber 1997b) and TGF-β_1 ligand overexpression is associated with matrix remodeling in post-MI heart failure (Hao et al. 2000; Sun and Weber 2000; Wang et al. 2002).

Fibroblasts and myofibroblasts are abundant in the heart (Weber and Brilla 1991), and wound healing/interstitial cardiac fibrosis is mediated by primarily the latter type (Eghbali et al. 1988; Peterson et al. 1999; Powell et al. 1999; Sun 1997). We have demonstrated the predominance of myofibroblasts in the infarct scar (Peterson et al. 1999) in post-MI rats. Myofibroblasts are highly synthetic phenotypic variants (Frangogiannis et al. 2002), expressing α-smooth muscle actin (α-SMA), vimentin, AT_1 receptors, TGF-β receptors, LIFR/gp-130, ACE, and fibrillar collagens (Hao et al. 2000; Hildebrand et al. 1994; Sun 1997; Sun et al. 1994; Sun and Weber 1994; Wang et al. 2002; Weber 1997b). This stable phenotype is induced by TGF-β_1 and *in vitro* culture seeding at low density (Dugina et al. 2001; Evans et al. 2003; Masur et al. 1996).

TGF-β_1 signaling in myofibroblasts of healthy hearts is the sum of different stimulatory R-Smads (e.g., Smad2 and Smad3) and inhibitory signals (e.g., Smad7 (Wrana 2000) and possibly c-Ski) which, when stimulated, undergo nucleocytoplasmic shuttling to carry out their specific function (Reguly and Wrana 2003). We speculate that the overactivation of R-Smads (i.e., phosphorylated Smad2 (p-Smad2)) in cardiac myofibroblasts may dominate Smad7 and/or c-Ski function in failing heart; this imbalance then contributes to increased matrix production in these cells. Smad proteins may subserve multiple systems in mammalian cells (Wrana 2000) and are linked to the regulation of collagen metabolism in cardiac fibroblasts (Hao et al. 2000; Wang et al. 2002), acting as a nexus for cellular control. Smad2 is a key stimulatory and receptor-regulated protein as it binds several known DNA-binding coactivators; this binding is stabilized by common Smad4 (co-Smad4) (Chen et al. 1996). Smad7 inhibits this system by binding and contributing to the degradation of the TβRI receptor, thereby eliminating the receptor kinase function and inhibiting R-Smad phosphorylation (Park 2005). Activation of Smad7 and its translocation from the nucleus to the cytosol is initiated by TGF-β_1 – its initial translocation to the cytosol is preceded by the binding of Smad ubiquitination regulatory factor 1/2 (Smurf 1/2 proteins), known to have E3 ubiquitin ligase activity (Kavsak et al. 2000; Murakami et al. 2003). Smad7 functions as an adapter protein for Smurf-dependent ubiquitination and degradation of target proteins (Blank et al. 1991). Thus, unlike the R-Smads, Smad7 is not activated by phosphorylation but rather recruitment of Smad7/Smurf complexes to TβRI results in degradation of the receptor and prevents phosphorylation of R-Smads, thereby inhibiting TGF-β_1 signaling (Kavsak et al. 2000; Murakami et al. 2003). Furthermore, Smad7 prevents the complexing of R-Smads and co-Smad4 (Park 2005), and thus prevents R-Smads from translocating to the nucleus. Smad7 also binds other proteins and some of these are known to control its degradation. For example, if Smad7 binds Jab1 in the nucleus, the subsequent translocation of the Smad7/Jab1 complex to the cytosol is associated with degradation of Smad7 and thus the TGF-β signal is derepressed (Kim et al. 2004).

The Biology of Smads and c-Ski in TGF-β_1 Signaling

TGF-β_1 ligand signaling from cell-surface receptors to the nucleus is transduced by Smads and their DNA-binding partners (Chen et al. 1998; Derynck et al. 1998; Massague 1998; Massague et al. 1997; Nakao et al. 1997b; Wrana 2000; Wrana and Pawson; 1997; Xu et al. 2000; Zhou et al. 1998). TGF-β_1 receptor type I and II are Ser/Thr kinase class proteins, and signal through receptor-regulated Smads (R-Smad2 and 3) by specific recognition and phosphorylation steps (Macias-Silva et al. 1996; Zhang et al. 1996). Smad access to Ser/Thr kinase receptors is regulated by SARA (Smad anchor for receptor activation) protein which binds unphosphorylated R-Smads (Tsukazaki et al. 1998). Activated R-Smads dissociate from SARA and complex with co-Smad 4 which then translocates to the nucleus (Lagna et al. 1996; Macias-Silva et al. 1996); an R- and co-Smad complex bound to nuclear DNA has the option to recruit not only coactivators (e.g., p300) (Nishihara et al. 1998) but also corepressors (e.g., c-Ski), respectively (Heldin et al. 1997; Massague and Wotton 2000; Wrana 2000; Wrana and Pawson 1997; Xu et al. 2000). R-Smads may be phosphorylated at different sites within the protein including a conserved SSXS carboxy-terminal region (Wrana 2000). R-Smad function closely depends on how much R-Smad is in the nucleus, and this activation is extensively regulated (Wrana 2000). Genes that are transcriptionally responsive to Smads contain Smad-binding elements (SBE) in their promoter regions (Zawel et al. 1998). Endogenous inhibitors of R-Smads exist as I-Smads (i.e., Smad7) and nuclear Smad corepressors, e.g., c-Ski, and they are known to competitively inhibit TβRI-mediated phosphorylation of R-Smads or directly downregulate R-Smad target gene transcription, respectively (Christian and Nakayama 1999; Hayashi et al. 1997; Imamura et al. 1997; Massague and Wotton 2000; Nakao et al. 1997a; Whitman 1997). In this regard, R-Smad activation has been linked to activation of collagen genes (Vindevoghel et al. 1998a,b). As discussed earlier, Smad7 has also been implicated in the direct degradation of TGF-β_1 receptors (Shi et al. 1995). I-Smad expression is stimulated by both TGF-β_1 and R-Smads, and a consensus SBE is known to exist in the Smad7 promoter region (Denissova et al. 2000; Ishida et al. 2000; Itoh et al. 1998; Nagarajan et al. 1999; Nakao et al. 1997a); together these data suggest an *autoinhibitory role* for I-Smads, wherein the R-Smad activation is balanced by that of I-Smad. Conversely, Smad6 function in mammalian systems is limited to BMP signaling (Ishida et al. 2000) and, unlike Smad7, may act in the nucleus by binding Hoxc8 (Villarreal et al. 1996); Smad6 function seems to be limited to cardiac development (Galvin et al. 2000). The mechanism for regulated I-Smad7 expression is unknown.

The c-Ski oncogene is a phosphoprotein originally isolated from chicken as the cellular homologue of v-ski carried by the Sloan-Kettering avian retrovirus

(Li et al. 1986). The c-Ski gene primary sequence is highly conserved among human, *Xenopus*, and mouse (Ludolph et al. 1995; Nomura et al. 1989; Sleeman and Laskey 1993), and is important in morphological transformation and growth of chicken embryo fibroblasts as well as for cellular differentiation (Ludolph et al. 1995; Reed et al. 2001). Transgenic mice expressing chicken v-Ski display elevated muscle mass due to hypertrophy of type II muscle fibers (Colmenares and Stavnezer 1989) and c-Ski "knock-out" mice exhibit multiple defects in the CNS and in skeletal muscle development (Sutrave et al. 1990). c-Ski appears to be a pleuripotent protein and its function may depend on the cell type. c-Ski may function in the cytosol (Kokura et al. 2003; Reed et al. 2001) and/or the nucleus—in this respect, early investigation is focused on c-Ski as a nuclear protein and it has been shown to bind specific DNA motifs for transcriptional repression mediated by Mad and various other hormones (Dahl et al. 1998a; Kokura et al. 2001; Nomura et al. 1999; Tokitou et al. 1999). c-Ski may also function as a coactivator in cells expressing NF1 (Tarapore et al. 1997). Although the control of c-Ski expression is unclear, the multiple roles of c-Ski in regulation of cellular function implicate its participation in both physiologic and pathophysiologic processes. c-Ski incorporates into DNA-binding complexes of Smad proteins in certain cells and may negatively regulate TGF-β_1-responsive transcriptional activation through recruitment of the HDAC complex (Dahl et al. 1998a). R-Smad2 and 3 interact with c-Ski in a ligand-dependent manner in some cell types (Xu et al. 2000). Ski interacting protein (SKIP) antagonizes Ski and may modulate its function (Dahl et al. 1998b). With recent emphasis on the likelihood of c-Ski's function(s) in the cytosol of cells, it has been suggested to bind to *cytosolic* Smads and this association may be facilitated by novel C184M protein (Kokura et al. 2003). Whether C184M is necessary for c-Ski suppression of R-Smad functions in cardiac myofibroblasts, or even whether it is sufficient to carry out a parallel inhibitory function is unknown. Further, despite early research that clearly demonstrates that c-Ski is a phosphoprotein, the physiological consequences of c-Ski phosphorylation are not well understood, nor is the role of c-Ski in myofibroblast function.

Despite the lack of a defined role of c-Ski in eukaryotic cells and particularly in cardiac myocytes and nonmyocytes, a number of recent studies have shed considerable light on its putative functions using mainly transformed or dedicated cell lines (Briones-Orta et al. 2006; Denissova and Liu 2004; Luo 2004; Macdonald et al. 2004; Marcelain and Hayman 2005; Suzuki et al. 2004; Ueki and Hayman 2003). While the pleuripotence of TGF-β (and that of c-Ski itself) is such that one should exercise some caution in extrapolating these findings to the physiology of primary cardiac cells, a discussion of these findings is likely important in seeking clues to the basis of cardiac fibrosis. For example, it is of considerable interest to consider whether c-Ski may inhibit not only R-Smad function and expression, but also that of I-Smads such as Smad7, and if so, which class is preferentially inhibited.

c-Ski and SnoN are ubiquitously expressed in eukaryotic tissues (Nomura et al. 1989); the status of their expression in cardiac tissues in the setting of heart failure is unknown. Of considerable interest is that the regulation of SnoN, but not c-Ski, is regulated by TGF-β_1, and thus the former repressor is thought to provide negative feedback (in a manner similar to that of Smad7) in certain cell types (Stroschein et al. 1999). Thus, the regulation of Smad7 and c-Ski is markedly distinct, and this difference may present an opportunity to exploit in studies of the control of TGF-β_1-mediated effects in hypertrophying cardiac tissues. Denissova and Liu have demonstrated that Ski binds Smad4 to then act as corepressors of Smad7 expression in various cell lines (Denissova and Liu 2004). Further, SnoN has been shown to confer similar inhibition of Smad7 expression in A549 carcinoma cells (Briones-Orta et al. 2006). However, whether this mechanism functions in cardiac myofibroblast cells is unknown. Given the widely accepted role of c-Ski as a TGF-β_1 signal repressor, it may preferentially interact with Smad3 and Smad4 to ablate R-Smad function in heart, as has been reported using COS-1 and Mv1Lu cells (Ueki and Hayman 2003). This hypothesis is supported by two other critical lines of evidence in the current literature. As phosphorylated R-Smads interact closely with co-Smad4 to deliver their profibrotic message to the nucleus and that c-Ski clearly binds Smad4 at specific motifs within its primary structure (Wu et al. 2002) and c-Ski also interferes with phosphorylation of both Smad2 and Smad3 by the activated TGF-β type 1 receptor (Thibault et al. 1995), the suggestion that inhibition of this specific signal may predominate over others is strongly supported. Nonetheless, the importance of TGF-β_1 as a modulator of cardiac hypertrophy underscores the need for further investigation to characterize the role of c-Ski in cardiac health and dysfunction.

Despite the lack of specific data to address c-Ski and SnoN function in fibroblasts and myofibroblasts, a relatively large body of data points to a growing consensus for the general function of these structurally related nuclear oncoproteins as corepressors for TGF-β_1 signaling (Akiyoshi et al. 1999; Cohen et al. 1999; Luo 2004; Luo et al. 1999; Suzuki et al. 2004). Again the versatility of these proteins is becoming clear with recent work showing that c-Ski itself may function within the nucleus (Vogel 1999; Xu et al. 2000) or in the cytosol (Medrano 2003) possibly in combination with C184M (Kokura et al. 2003); this may depend on cell type. More to the point, the current literature on cancer cell activation is characterized by intense interest in R-Smad and Smad7 signaling. *It has been often observed that phenomena that transpire in cancer cells or immortalized cell lines do not parallel intracellular Smad signaling in primary cells, and particularly in cardiac myofibroblasts.* For example, it has now been shown that in a human lung carcinoma cell line (A549) Ski and SnoN act with Smad4 to degrade Smad7 and thereby affect the cellular ratio of R-Smad/I-Smad7 proteins (Briones-Orta et al. 2006)—whether this occurs in cardiac myofibroblasts is unknown. Similarly, whether c-Ski inhibits R-Smads in cardiac myofibroblasts is also unclear. Emerging evidence from our laboratory has allowed us to form the hypothesis that c-Ski primarily serves to inhibit the R-Smad signal in cardiac fibroblasts and myofibroblasts in a manner parallel to Smad7. That is, c-Ski and Smad7 may be redundant in their function to suppress R-Smad signaling. Further, we suggest that a drop-out or significant loss of either gene's expression (previously found in Smad7 expression in post-MI hearts by members of our lab) may herald the onset of cardiac hypertrophy with attendant fibrosis and modulate collagen remodeling associated with heart failure in the chronic phase of MI. Control of Smad7 expression is important as ectopic overexpression of Smad7 is sufficient to ablate collagen synthesis (Wang et al. 2002), likely by inhibition of R-Smad2 phosphorylation in adult myofibroblasts. In post-MI hearts, the induction and activation of c-Ski and C184M may depend on the type and stage of failure, as does the activation of other effector proteins (Ju et al. 1998).

Cardiac Remodeling and the ECM in Remnant Heart and Scar

The cardiac ECM participates in active restoration of sarcomeric length, via release of stored potential energy in matrix proteins (Robinson et al. 1983, 1986), and overt fibrosis contributes to abnormal cardiac function by increasing myocardial stiffness. Collagen types I and III (fibrillar collagens) are the most abundant proteins in the matrix comprising nearly 95% of total collagen in the heart (Bashey et al. 1992; Robinson et al. 1983, 1986; Weber et al. 1989). The majority of DNA-synthesizing cells in the surviving myocardium and infarct scar of experimental animals are fibroblasts and myofibroblasts (Cleutjens et al. 1995; Peterson et al. 1999; van Krimpen et al. 1991). Altered matrix synthesis and deposition by these cells participates in the development of heart failure (Dixon et al. 1997, 2000; Hao et al. 1999; Ju et al. 1997, 1998; Pelouch et al. 1993). Furthermore, results of both clinical and experimental studies from this lab (Hao et al. 2000; Ju et al. 1997, 1998; Liu et al. 1996) and others (Weber 1997a,b; Weber and Brilla 1991) provide evidence of fibrosis in the left ventricle remote to the infarct site with increased myocardial stiffness and increased cross-linking of collagen fibrils (Dixon et al. 1998; Jalil et al. 1989; Makino et al. 1996; Thiedemann et al. 1983). Nonetheless, the regulation of collagen gene expression in post-MI heart tissues is not well understood.

Although scar formation is completed at 3 weeks post-MI (Fishbein et al. 1978), myofibroblasts, whose numbers dominate the infarct scar (Peterson et al. 1999; Sun 1997; Sun and Weber 1996), remain active even 8 weeks post-MI (Cleutjens et al. 1999; Ju et al. 1998; Peterson et al. 1999) whereas collagen content may continually increase over months (Fishbein et al. 1978; Jugdutt and Amy 1986). Limited fibrosis in the healing infarct scar may help to preserve ventricular function, as the new scar tissue selectively resists circumferential deformation (Holmes et al. 1997). *Thus, the scar is a distinctly anisotropic tissue with large collagen fibers oriented within 30 degrees of the local circumferential axis* (Holmes et al. 1997). The healed infarct scar is populated by myofibroblasts (Peterson et al. 1999; Willems et al. 1994) and these cells are the major source of R-Smad expression (Hao et al. 1999). These cells may persist in infarct scar in post-MI patients for many years, even decades, after the initial insult (Cleutjens et al. 1995, 1999; Willems et al. 1994). Myofibroblasts produce isometric tension within granulation tissue *in vivo* and in cultures (Serini and Gabbiani 1999). Tension is exerted at the level of focal adhesions (FAs), which connect cells to matrix (Dugina et al. 2001). It is suggested that this mechanical force is beneficial for cardiac pump function in post-MI heart (Holmes et al. 1997).

The involvement of the c-Ski and C184M in the development of overt cardiac fibrosis and attendant congestive heart failure is unclear. In future studies, we will determine whether R-Smad expression and function, which is tightly connected to the end-point of collagen synthesis and deposition, is inhibited by c-Ski and/or C184M in cardiac myofibroblasts from failing hearts. As we have recently demonstrated the endogenous antifibrotic effect of Smad7, we are well-positioned to compare those data to those arising from our studies in the immediate future. Our published and preliminary data suggest that I-Smad7 participates in inhibition of fibrillar collagen expression (Wang et al. 2002), likely by suppression of R-Smad2 phosphorylation in primary cultured myofibroblasts. Loss of Smad7 signal may be paralleled by a loss of c-Ski expression. Together, these events represent a harbinger of inappropriate myofibroblast function and ultimately in the development of cardiac fibrosis and dysfunction.

Synopsis

Cardiac fibrosis is recognized as a major disease modifier and as such is important in the pathogenesis of heart failure of most etiologies. As major mediators of TGF-β_1 signaling in cardiac fibroblasts and myofibroblasts as well as myocytic cells, Smad proteins are emerging as an important postreceptor class in post-MI heart failure. Precisely how cytosolic c-Ski and C184M proteins influence R-Smads (and possibly Smad7 itself) in cardiac myofibroblasts, and how c-Ski expression influences cardiac myofibroblast function are currently unknown. In combination with decreased Smad7 expression in post-MI hearts, altered c-Ski expression may contribute to chronic imbalance of R-Smad activation in classic TGF-β_1 signaling in the context of cardiac myofibroblast function. That is, we suggest that reduced and insufficient expression and activation of c-Ski and Smad7 may contribute to abnormal stimulation of collagen synthesis by these cells. Information concerning cardiac fibrosis and the role of c-Ski and C184M expression in cardiac fibrosis at different stages of cardiac hypertrophy and heart failure, as well as in fibroproliferative events in isolated myofibroblasts may rapidly become an area of intensive investigation. It seems likely that the exploitation of Smad7 and c-Ski in the modulation of cardiac myofibroblast function may provide the experimental basis for the development of highly specific drugs for treating heart failure with attendant cardiac fibrosis. Future studies along these lines will provide information regarding the identification and characterization of factors responsible for induction of pathological cardiac hypertrophy.

Acknowledgments We thank the Heart and Stroke Foundation of Manitoba for their support of this work on Smad7 and c-Ski in the context of cardiac fibrosis and cardiac myofibroblast function. I.M.C.D. is a scholar of the Myles Robinson Heart Health Fund. B.W. was supported by a postdoctoral fellowship of the Manitoba Health Research Council, and both K.M.B. and R.H.C. are supported with operating funds from the Canadian Institutes for Health Research. The authors would also like to thank Mr. Sunil Rattan and Mr. Stephen C. Jones for their technical support in completion of the experimental portion of these studies.

References

Akiyoshi, S., Inoue, H., Hanai, J., Kusanagi, K., Nemoto, N., Miyazono, K., and Kawabata, M. 1999. c-Ski acts as a transcriptional co-repressor in transforming growth factor-beta signaling through interaction with smads. J. Biol. Chem. 274:35269–35277.

Bartosova, D., Chvapil, M., Korecky, B., Poupa, O., Rakusan, K., Turek, Z., and Vizek, M. 1969. The growth of the muscular and collagenous parts of the rat heart in various forms of cardiomegaly. J. Physiol. (London) 200:285–295.

Bashey, R.I., Martinez Hernandez, A., and Jimenez, S.A. 1992. Isolation, characterization, and localization of cardiac collagen type VI. Associations with other extracellular matrix components. Circ. Res. 70:1006–1017.

Birchmeier, C., and Birchmeier, W. 1993. Molecular aspects of mesenchymal-epithelial interactions. Annu. Rev. Cell Biol. 9:511–540.

Blank, J.L., Ross, A.H., and Exton, J.H. 1991. Purification and characterization of two G-proteins that activate the beta 1 isozyme of phosphoinositide-specific phospholipase C. Identification as members of the Gq class. J. Biol. Chem. 266:18206–18216.

Brand, T., and Schneider, M.D. 1995. The TGF beta superfamily in myocardium: ligands, receptors, transduction, and function. J. Mol. Cell. Cardiol. 27:5–18.

Brand, T., and Schneider, M.D. 1996. Transforming growth factor-beta signal transduction. Circ. Res. 78:173–179.

Briones-Orta, M.A., Sosa-Garrocho, M., Moreno-Alvarez, P., Fonseca-Sanchez, M.A., and Macias-Silva, M. 2006. SnoN co-repressor binds and represses smad7 gene promoter. Biochem. Biophys. Res. Commun. 341:889–894.

Brooks, W.W., and Conrad, C.H. 2000. Myocardial fibrosis in transforming growth factor beta(1)heterozygous mice. J Mol. Cell. Cardiol. 32:187–195.

Butt, R.P., and Bishop, J.E. 1997. Mechanical load enhances the stimulatory effect of serum growth factors on cardiac fibroblast procollagen synthesis. J. Mol. Cell. Cardiol. 29:1141–1151.

Butt, R.P., Laurent, G.J., and Bishop, J.E. 1995. Collagen production and replication by cardiac fibroblasts is enhanced in response to diverse classes of growth factors. Eur. J. Cell Biol. 68:330–335.

Caulfield, J.B., and Borg, T.K. 1979. The collagen network of the heart. Lab. Invest. 40:364–372.

Chareonthaitawee, P., Christian, T.F., Hirose, K., Gibbons, R.J., and Rumberger, J.A. 1995. Relation of initial infarct size to extent of left ventricular remodeling in the year after acute myocardial infarction. J. Am. Coll. Cardiol. 25:567–573.

Chen, X., Rubock, M.J., and Whitman, M. 1996. A transcriptional partner for MAD proteins in TGF-beta signalling. Nature 383:691–696.

Chen, Y.G., Hata, A., Lo, R.S., Wotton, D., Shi, Y., Pavletich, N., and Massague, J. 1998. Determinants of specificity in TGF-beta signal transduction. Genes Dev. 12:2144–2152.

Christian, J.L., and Nakayama, T. 1999. Can't get no SMADisfaction: Smad proteins as positive and negative regulators of TGF-beta family signals. Bioessays 21:382–390.

Cleutjens, J.P., Verluyten, M.J., Smits, J.F., and Daemen, M.J. 1995. Collagen remodeling after myocardial infarction in the rat heart. Am. J. Pathol. 147:325–338

Cleutjens, J.P., Blankesteijn, W.M., Daemen, M.J., and Smits, J.F. 1999. The infarcted myocardium: simply dead tissue, or a lively target for therapeutic interventions. Cardiovasc. Res. 44:232–241.

Cohen, S.B., Zheng, G., Heyman, H.C., and Stavnezer, E. 1999. Heterodimers of the SnoN and Ski oncoproteins form preferentially over homodimers and are more potent transforming agents. Nucleic Acids Res. 27:1006–1014.

Colmenares, C., and Stavnezer, E. 1989. The ski oncogene induces muscle differentiation in quail embryo cells. Cell 59:293–303.

Colucci, W.S., and Braunwald, E. 1997. Pathophysiology of heart failure. In *Heart Failure*, E. Braunwald (ed.). Saunders, Philadelphia, 394–420.

Dahl, R., Kieslinger, M., Beug, H., and Hayman, M.J. 1998a. Transformation of hematopoietic cells by the Ski oncoprotein involves repression of retinoic acid receptor signaling. Proc. Natl. Acad. Sci. USA 95:11187–11192.

Dahl, R., Wani, B., and Hayman, M.J. 1998b. The Ski oncoprotein interacts with Skip, the human homolog of Drosophila Bx42. Oncogene 16:1579–1586.

Denissova, N.G., and Liu, F. 2004. Repression of endogenous Smad7 by Ski. J. Biol. Chem. 279:28143–28148.

Denissova, N.G., Pouponnot, C., Long, J., He, D., and Liu, F. 2000. Transforming growth factor beta-inducible independent binding of SMAD to the Smad7 promoter. Proc. Natl. Acad. Sci. USA 97:6397–6402.

Derynck, R., Zhang, Y., and Feng, X.H. 1998. Smads: transcriptional activators of TGF-beta responses. Cell 95:737–740.

Dixon, I.M.C., Lee, S.L., and Dhalla, N.S. 1990. Nitrendipine binding in congestive heart failure due to myocardial infarction. Circ. Res. 66:782–788.

Dixon, I.M.C., Reid, N.L., and Ju, H. 1997. Angiotensin II and TGF-b in the development of cardiac fibrosis, myocyte hypertrophy, and heart failure. Heart Failure Rev. 2:107–116.

Dixon, I.M.C., Ju, H., and Reid, N.L. 1998. The role of angiotensin II in post-translational regulation of fibrillar collagens in fibrosed and failing rat heart.471–498.

Dixon, I.M.C., Hao, J., Reid, N.L., and Roth, M. 2000. Effect of chronic AT1 receptor blockade on cardiac smad overexpression in hereditary cardiomyopathic hamsters. Cardiovasc. Res. 46:286–297.

Dugina, V., Fontao, L., Chaponnier, C., Vasiliev, J., and Gabbiani, G. 2001. Focal adhesion features during myofibroblastic differentiation are controlled by intracellular and extracellular factors. J. Cell Sci. 114:3285–3296.

Eghbali, M., Czaja, M.J., Zeydel, M., Weiner, F.R., Zern, M.A., Seifter, S., and Blumenfeld, O.O. 1988. Collagen chain mRNAs in isolated heart cells from young and adult rats. J. Mol. Cell. Cardiol. 20:267–276.

Evans, R.A., Tian, Y.C., Steadman, R., and Phillips, A.O. 2003. TGF-beta1-mediated fibroblast-myofibroblast terminal differentiation—the role of smad proteins. Exp. Cell Res. 282:90–100.

Fishbein, M.C., Maclean, D., and Maroko, P.R. 1978. Experimental myocardial infarction in the rat: qualitative and quantitative changes during pathologic evolution. Am. J. Pathol. 90:57–70.

Frangogiannis, N.G., Smith, C.W., and Entman, M.L. 2002. The inflammatory response in myocardial infarction. Cardiovasc. Res. 53:31–47.

Freed, D.H., Cunnington, R.H., Dangerfield, A.L., Sutton, J.S., and Dixon, I.M. 2005. Emerging evidence for the role of cardiotrophin-1 in cardiac repair in the infarcted heart. Cardiovasc. Res. 65:782–792.

Galvin, K.M., Donovan, M.J., Lynch, C.A., Meyer, R.I., Paul, R.J., Lorenz, J.N., Fairchild-Huntress, V., Dixon, K.L., Dunmore, J.H., Gimbrone, M.A., Falb, D., and Huszar, D. 2000. A role for smad6 in development and homeostasis of the cardiovascular system. Nat. Genet. 24:171–174.

Hao, J., Ju, H., Zhao, S., Junaid, A., Scammell-LaFleur, T., and Dixon, I.M. 1999. Elevation of expression of Smads 2, 3, and 4, decorin and TGF-beta in the chronic phase of myocardial infarct scar healing. J. Mol. Cell. Cardiol. 31:667–678.

Hao, J., Wang, B., Jones, S.C., Jassal, D.S., and Dixon, I.M.C. 2000. Interaction between angiotensin II and Smad proteins in fibroblasts in failing heart and *in vitro*. Am. J. Physiol. 279:H3020–H3030.

Hayashi, H., Abdollah, S., Qiu, Y., Cai, J., Xu, Y.Y., Grinnell, B.W., Richardson, M.A., Topper, J.N., Gimbrone, M.A., Jr., Wrana, J.L., and Falb, D. 1997. The MAD-related protein Smad7 associates with the TGFbeta receptor and functions as an antagonist of TGFbeta signaling. Cell 89:1165–1173.

Heldin, C.H., Miyazono, K., and ten Dijke, P. 1997. TGF-beta signalling from cell membrane to nucleus through SMAD proteins. Nature 390:465–471.

Hildebrand, A., Romaris, M., Rasmussen, L.M., Heinegard, D., Twardzik, D.R., Border, W.A., and Ruoslahti, E. 1994. Interaction of the small interstitial proteoglycans biglycan, decorin and fibromodulin with transforming growth factor beta. Biochem. J. 302:527–534.

Holmes, J.W., Nunez, J.A., and Covell, J.W. 1997. Functional implications of myocardial scar structure. Am. J. Physiol. 272:H2123–H2130.

Hunter, J.J., and Chien, K.R. 1999. Signaling pathways for cardiac hypertrophy and failure. N. Engl. J. Med. 341:1276–1283.

Ignotz, R.A., and Massague, J. 1986. Transforming growth factor-beta stimulates the expression of fibronectin and collagen and their incorporation into the extracellular matrix. J. Biol. Chem. 261:4337–4345.

Imamura, T., Takase, M., Nishihara, A., Oeda, E., Hanai, J., Kawabata, M., and Miyazono, K. 1997. Smad6 inhibits signalling by the TGF-beta superfamily. Nature 389:622–626.

Inagaki, Y., Truter, S., and Ramirez, F. 1994. Transforming growth factor-beta stimulates alpha 2(I) collagen gene expression through a cis-acting element that contains an Sp1-binding site. J. Biol. Chem. 269:14828–14834.

Ishida, W., Hamamoto, T., Kusanagi, K., Yagi, K., Kawabata, M., Takehara, K., Sampath, T.K., Kato, M., and Miyazono, K. 2000. Smad6 is a Smad1/5-induced smad inhibitor. Characterization of bone morphogenetic protein-responsive element in the mouse Smad6 promoter. J. Biol. Chem. 275:6075–6079.

Itoh, S., Landstrom, M., Hermansson, A., Itoh, F., Heldin, C.H., Heldin, N.E., and ten Dijke, P. 1998. Transforming growth factor beta1 induces nuclear export of inhibitory Smad7. J. Biol. Chem. 273:29195–29201.

Jalil, J.E., Doering, C.W., Janicki, J.S., Pick, R., Shroff, S.G., and Weber, K.T. 1989. Fibrillar collagen and myocardial stiffness in the intact hypertrophied rat left ventricle. Circ. Res. 64:1041–1050.

Ju, H., Zhao, S., Davinder, S.J., and Dixon, I.M. 1997. Effect of AT1 receptor blockade on cardiac collagen remodeling after myocardial infarction. Cardiovasc. Res. 35:223–232.

Ju, H., Zhao, S., Tappia, P.S., Panagia, V., and Dixon, I.M.C. 1998. Expression of Gqalpha and PLC-beta in scar and border tissue in heart failure due to myocardial infarction. Circulation 97:892–899.

Jugdutt, B.I., and Amy, R.W. 1986. Healing after myocardial infarction in the dog: changes in infarct hydroxyproline and topography. J. Am. Coll. Cardiol. 7:91–102.

Kannel, W.B. 1997. Epidemiology of Heart Failure in the United States 279–288.

Kavsak, P., Rasmussen, R.K., Causing, C.G., Bonni, S., Zhu, H., Thomsen, G.H., and Wrana, J.L. 2000. Smad7 binds to Smurf2 to form an E3 ubiquitin ligase that targets the TGF beta receptor for degradation. Mol. Cell. 6:1365–1375.

Kim, B.C., Lee, H.J., Park, S.H., Lee, S.R., Karpova, T.S., McNally, J.G., Felici, A., Lee, D.K., and Kim, S.J. 2004. Jab1/CSN5, a component of the COP9 signalosome, regulates transforming growth factor beta signaling by binding to Smad7 and promoting its degradation. Mol. Cell. Biol. 24:2251–2262.

Kingsley, D.M. 1994. The TGF-beta superfamily: new members, new receptors, and new genetic tests of function in different organisms. Genes Dev. 8:133–146.

Kokura, K., Kaul, S.C., Wadhwa, R., Nomura, T., Khan, M.M., Shinagawa, T., Yasukawa, T., Colmenares, C., and Ishii, S. 2001. The Ski protein family is required for MeCP2-mediated transcriptional repression. J. Biol. Chem. 276:34115–34121.

Kokura, K., Kim, H., Shinagawa, T., Khan, M.M., Nomura, T., and Ishii, S. 2003. The Ski-binding protein C184M negatively regulates tumor growth factor-beta signaling by sequestering the Smad proteins in the cytoplasm. J. Biol. Chem. 278:20133–20139.

Lagna, G., Hata, A., Hemmati-Brivanlou, A., and Massague, J. 1996. Partnership between DPC4 and SMAD proteins in TGF-beta signalling pathways. Nature 383:832–836.

Li, Y., Turck, C.M., Teumer, J.K., and Stavnezer, E. 1986. Unique sequence, ski, in Sloan-Kettering avian retroviruses with properties of a new cell-derived oncogene. J. Virol. 57:1065–1072.

Liu, K.-Z., Jackson, M., Sowa, M., Ju, H., Dixon, I.M.C., and Mantsch, H.H. 1996. Modification of the extracellular matrix following myocardial infarction monitored by FTIR spectroscopy. Biochim. Biophys. Acta 1315:73–77.

Ludolph, D.C., Neff, A.W., Parker, M.A., Mescher, A.L., Smith, R.C., and Malacinski, G.M. 1995. Cloning and expression of the axolotl proto-oncogene ski. Biochim. Biophys. Acta 1260:102–104.

Luo, K. 2004. Ski and SnoN: negative regulators of TGF-beta signaling. Curr. Opin. Genet. Dev. 14:65–70.

Luo, K., Stroschein, S.L., Wang, W., Chen, D., Martens, E., Zhou, S., and Zhou, Q. 1999. The Ski oncoprotein interacts with the Smad proteins to repress TGFbeta signaling. Genes Dev. 13:2196–2206.

Macdonald, M., Wan, Y., Wang, W., Roberts, E., Cheung, T.H., Erickson, R., Knuesel, M.T., and Liu, X. 2004. Control of cell cycle-dependent degradation of c-Ski proto-oncoprotein by Cdc34. Oncogene 23:5643–5653.

Macias-Silva, M., Abdollah, S., Hoodless, P.A., Pirone, R., Attisano, L., and Wrana, J.L. 1996. MADR2 is a substrate of the TGFbeta receptor and its phosphorylation is required for nuclear accumulation and signaling. Cell 87:1215–1224.

MacLellan, W.R., and Schneider, M.D. 2000. Genetic dissection of cardiac growth control pathways. Annu. Rev. Physiol. 62:289–319.

Makino, N., Hata, T., Sugano, M., Dixon, I.M.C., and Yanaga, T. 1996. Regression of hypertrophy after myocardial infarction is produced by the chronic blockade of angiotensin type 1 receptor in rats. J. Mol. Cell. Cardiol. 28:507–517.

Marcelain, K., and Hayman, M.J. 2005. The Ski oncoprotein is upregulated and localized at the centrosomes and mitotic spindle during mitosis. Oncogene 24:4321–4329.

Massague, J. 1998. TGF-beta signal transduction. Annu. Rev. Biochem. 67:753–791.

Massague, J., and Wotton, D. 2000. Transcriptional control by the TGF-beta/Smad signaling system. EMBO J.19:1745–1754.

Massague, J., Hata, A., and Liu, F. 1997. Tgf-beta signalling through the Smad pathway. Trends Cell Biol. 7:187–192.

Masur, S.K., Dewal, H.S., Dinh, T.T., Erenburg, I., and Petridou, S. 1996. Myofibroblasts differentiate from fibroblasts when plated at low density. Proc. Natl. Acad. Sci. USA 93:4219–4223.

Medrano, E.E. 2003. Repression of TGF-beta signaling by the oncogenic protein SKI in human melanomas: consequences for proliferation, survival, and metastasis. Oncogene 22:3123–3129.

Mettauer, B., Zoll, J., Garnier, A., and Ventura-Clapier, R. 2006. Heart failure: a model of cardiac and skeletal muscle energetic failure. Pflugers Arch. 452:653–666.

Murakami, G., Watabe, T., Takaoka, K., Miyazono, K., and Imamura, T. 2003. Cooperative inhibition of bone morphogenetic protein signaling by Smurf1 and inhibitory Smads. Mol. Biol. Cell. 14:2809–2817.

Nagarajan, R.P., Zhang, J., Li, W., and Chen, Y. 1999. Regulation of Smad7 promoter by direct association with Smad3 and Smad4. J. Biol. Chem. 274:33412–33418.

Nakajima, H., Nakajima, H.O., Salcher, O., Dittie, A.S., Dembowsky, K., Jing, S., and Field, L.J. 2000. Atrial but not ventricular fibrosis in mice expressing a mutant transforming growth factor-beta(1) transgene in the heart. Circ. Res.86:571–579.

Nakao, A., Afrakhte, M., Moren, A., Nakayama, T., Christian, J.L., Heuchel, R., Itoh, S., Kawabata, M., Heldin, N.E., Heldin, C.H., and ten Dijke, P. 1997a. Identification of Smad7, a TGFbeta-inducible antagonist of TGF-beta signalling. Nature 389:631–635.

Nakao, A., Roijer, E., Imamura, T., Souchelnytskyi, S., Stenman, G., Heldin, C.H., and ten Dijke, P. 1997b. Identification of Smad2, a human Mad-related protein in the transforming growth factor beta signaling pathway. J. Biol. Chem. 272:2896–2900.

Nian, M., Lee, P., Khaper, N., and Liu, P. 2004. Inflammatory cytokines and postmyocardial infarction remodeling. Circ. Res. 94:1543–1553.

Nishihara, A., Hanai, J.I., Okamoto, N., Yanagisawa, J., Kato, S., Miyazono, K., and Kawabata, M. 1998. Role of p300, a transcriptional coactivator, in signalling of TGF-beta. Genes Cells 3:613–623.

Nomura, N., Sasamoto, S., Ishii, S., Date, T., Matsui, M., and Ishizaki, R. 1989. Isolation of human cDNA clones of ski and the ski-related gene, sno. Nucleic Acids Res. 17:5489–5500.

Nomura, T., Khan, M.M., Kaul, S.C., Dong, H.D., Wadhwa, R., Colmenares, C., Kohno, I., and Ishii, S. 1999. Ski is a component of the histone deacetylase complex required for transcriptional repression by Mad and thyroid hormone receptor. Genes Dev. 13:412–423.

Ohta, K., Kim, S., Hamaguchi, A., Yukimura, T., Miura, K., Takaori, K., and Iwao, H. 1994. Role of angiotensin II in extracellular matrix and transforming growth factor-beta 1 expression in hypertensive rats. Eur. J. Pharmacol. 269:115–119.

Park, S.H. 2005. Fine tuning and cross-talking of TGF-beta signal by inhibitory Smads. J. Biochem. Mol. Biol. 38:9–16.

Pelouch, V., Dixon, I.M., Golfman, L., Beamish, R.E., and Dhalla, N.S. 1993. Role of extracellular matrix proteins in heart function. Mol. Cell. Biochem. 129:101–120.

Peterson, D.J., Ju, H., Hao, J., Panagia, M., Chapman, D.C., and Dixon, I.M.C. 1999. Expression of Gi-2 alpha and Gs alpha in myofibroblasts localized to the infarct scar in heart failure due to myocardial infarction. Cardiovasc. Res. 41:575–585.

Pfeffer, J.M., Pfeffer, M.A., Fletcher, P.J., and Braunwald, E. 1991. Progressive ventricular remodeling in rat with myocardial infarction. Am. J. Physiol. 260:H1406–H1414.

Pfeffer, J.M., Fischer, T.A., and Pfeffer, M.A. 1995. Angiotensin-converting enzyme inhibition and ventricular remodeling after myocardial infarction. Annu. Rev. Physiol. 57:805–826.

Powell, D.W., Mifflin, R.C., Valentich, J.D., Crowe, S.E., Saada, J.I., and West, A.B. 1999. Myofibroblasts. I. Paracrine cells important in health and disease. Am. J. Physiol. 277:C1–C9.

Reed, J.A., Bales, E., Xu, W., Okan, N.A., Bandyopadhyay, D., and Medrano, E.E. 2001. Cytoplasmic localization of the oncogenic protein Ski in human cutaneous melanomas *in vivo*: functional implications for transforming growth factor beta signaling. Cancer Res. 61:8074–8078.

Reguly, T., and Wrana, J.L. 2003. In or out? The dynamics of Smad nucleocytoplasmic shuttling. Trends Cell Biol. 13:216–220.

Roberts, A.B., Heine, U.I., Flanders, K.C., and Sporn, M.B. 1990. Transforming growth factor-beta. Major role in regulation of extracellular matrix. Ann. N.Y. Acad. Sci. 580:225–232.

Robinson, T.F., Cohen-Gould, L., and Factor, S.M. 1983. Skeletal framework of mammalian heart muscle. Arrangement of inter- and pericellular connective tissue structures. Lab. Invest. 49:482–498.

Robinson, T.F., Factor, S.M., and Sonnenblick, E.H. 1986. The heart as a suction pump. Sci. Am. 254:84–91.

Sadoshima, J., and Izumo, S. 1993. Molecular characterization of angiotensin II-induced hypertrophy of cardiac myocytes and hyperplasia of cardiac fibroblasts. Critical role of the AT1 receptor subtype. Circ. Res. 73:413–423.

Saha, M., and Ferro, A. 2006. Cardiac stem cell therapy: present and future. Br. J. Clin. Pharmacol. 61:727–729.

Serini, G., and Gabbiani, G. 1999. Mechanisms of myofibroblast activity and phenotypic modulation. Exp. Cell Res. 250:273–283.

Shi, B., Heavner, J.E., McMahon, K.K., and Spallholz, J.E. 1995. Dynamic changes in G alpha i-2 levels in rat hearts associated with impaired heart function after myocardial infarction. Am. J. Physiol. 269:H1073–H1079.

Simon-Assmann, P., Kedinger, M., De Arcangelis, A., Rousseau, V., and Simo, P. 1995. Extracellular matrix components in intestinal development. Experientia 51:883–900.

Sleeman, J.P., and Laskey, R.A. 1993. Xenopus c-ski contains a novel coiled-coil protein domain, and is maternally expressed during development. Oncogene 8:67–77.

Stroschein, S.L., Wang, W., Zhou, S., Zhou, Q., and Luo, K. 1999. Negative feedback regulation of TGF-beta signaling by the SnoN oncoprotein. Science 286:771–774.

Sun, Y. 1997. Local angiotensin II and myocardial fibrosis. In *Hypertension and the Heart*, A. Zanchetti (ed.). Plenum Press, New York, 55–61.

Sun, Y., and Weber, K.T. 1994. Angiotensin II receptor binding following myocardial infarction in the rat. Cardiovasc. Res. 28:1623–1628.

Sun, Y., and Weber, K.T. 1996. Angiotensin converting enzyme and myofibroblasts during tissue repair in the rat heart. J. Mol. Cell. Cardiol. 28:851–858.

Sun, Y., and Weber, K.T. 2000. Infarct scar: a dynamic tissue. Cardiovasc. Res. 46:250–256.

Sun, Y., Cleutjens, J.P., Diaz-Arias, A.A., and Weber, K.T. 1994. Cardiac angiotensin converting enzyme and myocardial fibrosis in the heart. Cardiovasc. Res. 28:1423–1432

Sutrave, P., Kelly, A.M., and Hughes, S.H. 1990. ski can cause selective growth of skeletal muscle in transgenic mice. Genes Dev. 4:1462–1472.

Suzuki, H., Yagi, K., Kondo, M., Kato, M., Miyazono, K., and Miyazawa, K. 2004. c-Ski inhibits the TGF-beta signaling pathway through stabilization of inactive Smad complexes on Smad-binding elements. Oncogene 23:5068–5076.

Tarapore, P., Richmond, C., Zheng, G., Cohen, S.B., Kelder, B., Kopchick, J., Kruse, U., Sippel, A.E., Colmenares, C., and Stavnezer, E. 1997. DNA binding and transcriptional activation by the Ski oncoprotein mediated by interaction with NFI. Nucleic Acids Res. 25:3895–3903.

Thibault, G., Arguin, C., and Garcia, R. 1995. Cardiac endothelin-1 content and receptor subtype in spontaneously hypertensive rats. J. Mol. Cell. Cardiol. 27:2327–2336.

Thiedemann, K.U., Holubarsch, C., Medugorac, I., and Jacob, R. 1983. Connective tissue content and myocardial stiffness in pressure overload hypertrophy. A combined study of morphologic, morphometric, biochemical, and mechanical parameters. Basic Res. Cardiol. 78:140–155.

Tokitou, F., Nomura, T., Khan, M.M., Kaul, S.C., Wadhwa, R., Yasukawa, T., Kohno, I., and Ishii, S. 1999. Viral ski inhibits retinoblastoma protein (Rb)-mediated transcriptional repression in a dominant negative fashion. J. Biol. Chem. 274:4485–4488.

Tomasek, J.J., Gabbiani, G., Hinz, B., Chaponnier, C., and Brown, R.A. 2002. Myofibroblasts and mechano-regulation of connective tissue remodelling. Nat. Rev. Mol. Cell Biol. 3:349–363.

Tsukazaki, T., Chiang, T.A., Davison, A.F., Attisano, L., and Wrana, J.L. 1998. SARA, a FYVE domain protein that recruits Smad2 to the TGFbeta receptor. Cell 95:779–791.

Ueki, N., and Hayman, M.J. 2003. Direct interaction of Ski with either Smad3 or Smad4 is necessary and sufficient for Ski-mediated repression of transforming growth factor-beta signaling. J. Biol. Chem. 278:32489–32492.

van Krimpen, C., Schoemaker, R.G., Cleutjens, J.P., Smits, J.F., Struyker-Boudier, H.A., Bosman, F.T., and Daemen, M.J. 1991. Angiotensin I converting enzyme inhibitors and cardiac remodeling. Basic Res. Cardiol. 86 (Suppl. 1):149–155.

Villarreal, F.J., Lee, A.A., Dillmann, W.H., and Giordano, F.J. 1996. Adenovirus-mediated overexpression of human transforming growth factor-beta 1 in rat cardiac fibroblasts, myocytes and smooth muscle cells. J. Mol. Cell. Cardiol. 28:735–742.

Vindevoghel, L., Kon, A., Lechleider, R.J., Uitto, J., Roberts, A.B., and Mauviel, A. 1998a. Smad-dependent transcriptional activation of human type VII collagen gene (COL7A1) promoter by transforming growth factor-beta. J. Biol. Chem. 273:13053–13057.

Vindevoghel, L., Lechleider, R.J., Kon, A., de Caestecker, M.P., Uitto, J., Roberts, A.B., and Mauviel, A. 1998b. SMAD3/4-dependent transcriptional activation of the human type VII collagen gene (COL7A1) promoter by transforming growth factor beta. Proc. Natl. Acad. Sci. USA 95:14769–14774.

Vogel, G. 1999. A new blocker for the TGF-beta pathway. Science 286:665.

Wang, B., Hao, J., Jones, S.C., Yee, M.S., Roth, J.C., and Dixon, I.M. 2002. Decreased Smad 7 expression contributes to cardiac fibrosis in the infarcted rat heart. Am. J. Physiol. Heart Circ. Physiol. 282:H1685–H1696.

Watkins, S.J., Jonker, L., and Arthur, H.M. 2006. A direct interaction between TGFbeta activated kinase 1 and the TGFbeta type II receptor: implications for TGFbeta signalling and cardiac hypertrophy. Cardiovasc. Res. 69:432–439.

Weber, K.T. 1997a. Extracellular matrix remodeling in heart failure: a role for de novo angiotensin II generation. Circulation 96:4065–4082.

Weber, K.T. 1997b. Fibrosis, a common pathway to organ failure: angiotensin II and tissue repair. Semin. Nephrol. 17:467–491.

Weber, K.T., and Brilla, C.G. 1991. Pathological hypertrophy and cardiac interstitium. Fibrosis and renin-angiotensin-aldosterone system. Circulation 83:1849–1865.

Weber, K.T., Jalil, J.E., Janicki, J.S., and Pick, R. 1989. Myocardial collagen remodeling in pressure overload hypertrophy. A case for interstitial heart disease. Am. J. Hypertens. 2:931–940.
Whitman, M. 1997. Signal transduction. Feedback from inhibitory SMADs. Nature 389:549–551.
Willems, I.E., Havenith, M.G., De Mey, J.G., and Daemen, M.J. 1994. The alpha-smooth muscle actin-positive cells in healing human myocardial scars. Am. J. Pathol. 145:868–875.
Wollert, K.C., and Drexler, H. 2005. Clinical applications of stem cells for the heart. Circ. Res. 96:151–163.
Wrana, J.L. 2000. Regulation of Smad activity. Cell 100:189–192.
Wrana, J., and Pawson, T. 1997. Signal transduction. Mad about SMADs. Nature 388:28–29.
Wu, J.W., Krawitz, A.R., Chai, J., Li, W., Zhang, F., Luo, K., and Shi, Y. 2002. Structural mechanism of Smad4 recognition by the nuclear oncoprotein Ski: insights on Ski-mediated repression of TGF-beta signaling. Cell 111:357–367.
Xu, W., Angelis, K., Danielpour, D., Haddad, M.M., Bischof, O., Campisi, J., Stavnezer, E., and Medrano, E.E. 2000. Ski acts as a co-repressor with Smad2 and Smad3 to regulate the response to type beta transforming growth factor. Proc. Natl. Acad. Sci. USA 97:5924–5929.
Zawel, L., Dai, J.L., Buckhaults, P., Zhou, S., Kinzler, K.W., Vogelstein, B., and Kern, S.E. 1998. Human Smad3 and Smad4 are sequence-specific transcription activators. Mol. Cell 1:611–617.
Zhang, Y., Feng, X., We, R., and Derynck, R. 1996. Receptor-associated Mad homologues synergize as effectors of the TGF-beta response. Nature 383:168–172.
Zhou, S., Zawel, L., Lengauer, C., Kinzler, K.W., and Vogelstein, B. 1998. Characterization of human FAST-1, a TGF beta and activin signal transducer. Mol. Cell 2:121–127.

Chapter 14
The Signaling Duel Between Virus and Host: Impact on Coxsackieviral Pathogenesis

Xiaoning Si, David Marchant, Decheng Yang, Bruce M. McManus, and Honglin Luo

Abstract Coxsackievirus B3 (CVB3) is the primary human pathogen of viral myocarditis, a disease which causes sudden, unexpected death of infants and youth. Myocarditis was originally considered predominantly an inflammatory disease, but subsequent studies have revealed that direct myocardial injury by CVB3 prior to host immune responses contributes significantly to the progression of myocarditis. Heart transplantation is the only definitive treatment for serious myocarditis; thus, development of therapeutic intervention based on the pathogenesis of CVB3 becomes a preferred approach. Studies by our laboratory and others have shown that CVB3 infection triggers apoptosis of host cells through cytochrome C release and activation of multiple caspases. Furthermore, CVB3 infection activates several intracellular signaling pathways, such as the ERK pathway, the p38 MAPK pathway, and the PI3K/Akt pathway, for its efficient replication and release of progeny virions in order to complete its life cycle. Major cellular protein degradation systems, such as the ubiquitin–proteasome system, are also exploited by CVB3. These findings on the critical roles of host factors in pathogenesis of CVB3 thus provide us cellular targets for development of possible therapeutic interventions at molecular levels to restrain CVB3 infection.

Viral Myocarditis

Myocarditis has become a major cause of sudden unexpected death in patients younger than 40 years of age (Woodruff 1980). Evidence from both animal studies and clinical studies suggests that viral myocarditis, which is caused by infectious agents like group B coxsackieviruses (CVBs), echovirus, and adenovirus, is important in the etiology of idiopathic dilated cardiomyopathy (DCM) (Kandolf et al. 1987; McManus et al. 1993). Because heart transplantation is the only definitive treatment after a serious myocarditis and subsequent DCM and heart failure, understanding of the pathogenesis and seeking alternative and intuitive therapeutics are imminently required to relieve the societal burden of heart failure.

A. K. Srivastava, M. B. Anand-Srivastava (eds.) *Advances in Biochemistry in Health and Disease.* Signal Transduction in the Cardiovascular System in Health and Disease,

Although enteroviral myocarditis was originally considered predominantly an inflammatory heart disease that is initiated by an influx of macrophages, natural killer cells, and humoral neutralizing antibody, followed by infiltration of antigen-specific T lymphocytes, results from our laboratory and others have shown that CVB type 3 (CVB3) directly injures cardiomyocytes prior to the host immune response, contributing significantly to the progression of myocardial injury (Chow et al. 1992; Hufnagel et al. 1995; McManus et al. 1993). In cultured cells, CVB3 infection leads to cytopathic effects (CPE) and apoptosis of infected cells (Carthy et al. 1998). Both cardiovirulent and noncardiovirulent CVB3 cause early and extensive cardiomyocyte necrosis in severe combined immune deficient mice as compared to wild-type counterparts (Chow et al. 1992; Hufnagel et al. 1995). Prominent cytopathic alterations observed in cells that acquire the viral genome further demonstrate the importance of direct virus-induced damage (Klingel et al. 1998). We (Taylor et al. 2000; Yang et al. 1999) and others (Liu et al. 2000; Metcalf et al. 2000; Opavsky et al. 2002; Peng et al. 2001; Yasukawa et al. 2003) have shown that early host responses to viral infection play a key role in determining the severity of myocarditis and progression of DCM through differential expression of genes. Such gene expression program is regulated by intracellular signaling pathways that are activated during viral infection. This review focuses primarily on the relative roles of intracellular signaling pathways in determining cellular and viral fate invoked during CVB3 infection. It is believed that a better understanding of the cross-talk between various signaling pathways will provide valuable insight into the mechanisms by which viral infection injures cardiac muscle. This will allow us to develop novel, multitarget therapeutic strategies.

CVB3 Genome Organization and Life Cycle

CVB3 is a nonenveloped, single-stranded, positive polarity, small RNA virus that belongs to the enterovirus genus in the Picornaviridae family. Its genome consists of an ~7.4-kb positive single-stranded RNA sequence that encodes both structural and nonstructural proteins and contains untranslated regions (UTRs) at the 5′ and 3′ termini. CVB3 is one of the most common human pathogens for myocarditis, pancreatitis, and hepatitis, mostly affecting infants and young children.

CVB3 has a relatively short life cycle (Figure 14.1), which typically results in apoptosis and necrosis of infected cells to release of progeny virus. CVB3 initiates infection by attaching to its receptors on target cells and delivering its RNA into the host cell. Known receptors for CVB3 include the coxsackievirus and adenovirus receptor (CAR) (Bergelson et al. 1997) and the decay accelerating factor (DAF) coreceptor (Bergelson et al. 1995). Recent studies have shown that CAR is the primary receptor for CVB3. Binding of the extracellular portion of the receptor is necessary for CVB3 internalization and infection (Dorner et al. 2004; Yanagawa et al. 2004). Soluble recombinant CAR abrogates CVB3-induced myocarditis and

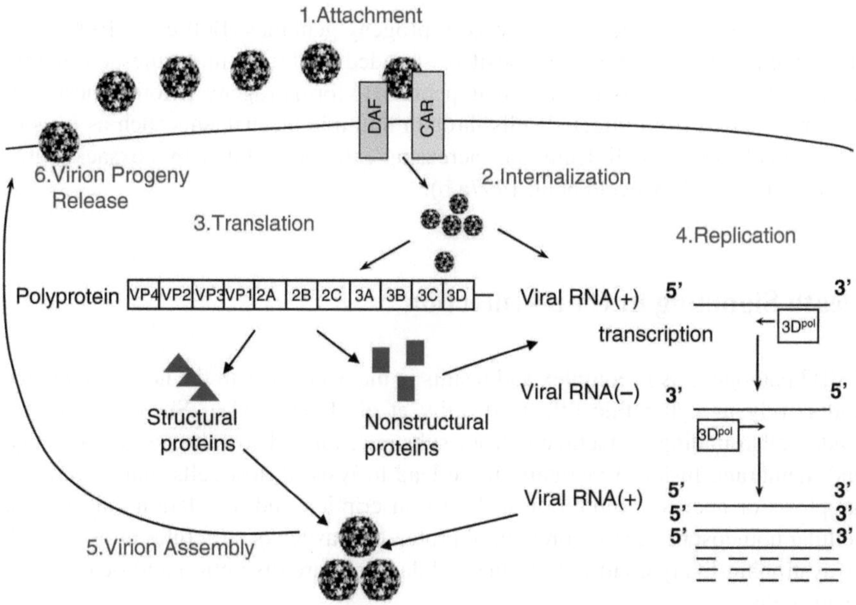

Fig. 14.1 The life cycle of coxsackievirus B3. CVB3 starts its life cycle by attaching to its receptor CAR and coreceptor DAF. Internalized virus releases its viral RNA, which can be used as the template for translation of polyprotein or transcription by RNA-dependent RNA polymerase 3D to replicate its genome. The polyprotein is self-cleaved by virus-encoded proteases to release structural proteins and nonstructural proteins. Later, structural proteins and viral RNA will assemble into progeny virions to be released from infected cell. Abbreviations: CVB3, coxsackievirus B3; DAF, decay accelerating factor; CAR, coxsackievirus and adenovirus receptor; $3D^{pol}$, RNA-dependent RNA polymerase.

pancreatitis in mice (Yanagawa et al. 2004). Furthermore, efficient CVB3 entry into host cells requires the coreceptor, DAF, which is a glycosyl-phosphatidylinositol (GPI)-anchored protein localized to both lipid rafts and caveolae (Nicholson-Weller and Wang 1994).

CVB3 viral RNA acts as a template for both the translation of the viral polyprotein and the replication of the viral genome. The polyprotein is sequentially cleaved into individual structural (Vp1–Vp4) and nonstructural proteins by viral proteases $2A^{pro}$ and $3C^{pro}$. Viral proteases also cleave host proteins to either shut-off host cell cap-dependent translation or alter the architecture of host cells (Badorff et al. 1999; Ehrenfeld 1982). For example, $2A^{pro}$ not only cleaves eukaryotic initiation factor 4-Gamma (eIF4γ) and poly(A) binding protein (PABP) (Kerekatte et al. 1999), proteins involved in host cap-dependent translation initiation, but it also cleaves structural proteins such as dystrophin (Badorff et al. 1999). Cleavage of dystrophin may eventually lead to dilated cardiomyopathy in patients infected with CVB3. The nonstructural viral protein CVB3 $3D^{pol}$ is an RNA-dependent RNA polymerase, which is required for transcription of the negative-strand viral RNA intermediate that serves as a template for synthesis of progeny genomes (Bell et al. 1999). The structural proteins associate with positive-stranded viral RNA molecules, covalently attached to VPg, a virus-encoded small peptide, to form progeny virions. These virions are released from infected cells through multiple mechanisms such as lysis of infected cells due to cell death and increasing cell permeability by coxsackievirus protein 2B (van Kuppeveld et al. 1997a,b).

Death Signaling in CVB3 Infection

CVB3 pathogenesis is complex and results in many changes to the host cell, known collectively as cytopathic effects (Carthy et al. 1998, 2003). Such changes include cell rounding, detachment from substrate, altered membrane permeability, and membrane fusion. Eventually these lead to lysis of host cells mainly through apoptosis or necrosis. Inhibition of host transcription and translation, and loss of cellular homeostasis due to direct viral protease cleavage of structural proteins such as dystrophin likely contribute to the cellular structure disruption and death of infected cells'.

Unlike necrosis that results from acute cellular injury, apoptosis is a type of programmed cell death (Green 2005; Yuan 2006). Apoptosis by itself is a deliberate process that results in death of a cell in a multicellular organism. It is a tightly regulated, highly ordered, and energy-consuming process. Apoptosis can occur when cells are exposed to stress such as starvation or damage to the cell's DNA due to toxicity, or exposure to ultraviolet or ionizing radiation. Programmed cell death also contributes to homeostasis and tissue development in multicellular organisms. In addition, apoptosis plays a major role in host defense responses since it occurs when a cell is damaged beyond repair or infected with a virus. The decisive signals for programmed cell death usually come from the damaged cell itself, or from surrounding tissue and immune cells. The apoptotic cell has its distinctive morphology under the microscope. The cell becomes round, and the plasma membrane of the apoptotic cell blebs followed by exposure of phosphatidylserine on the cell surface. The nuclear chromatin of apoptotic cells undergoes degradation and condensation, and the nuclear envelope becomes discontinuous with the DNA inside becoming fragmented.

Apoptosis can occur through death receptor-mediated pathways (Thorburn 2004). Binding of death-inducing ligands such as Fas-L, tumor necrosis factor-α (TNF-α), and other TNF-related apoptosis-inducing ligands to their specific receptors will induce the recruitment of adaptor proteins and procaspase-8 to form the death-inducing signaling complex. This leads to the proteolytic activation of caspase-8 followed by the activation of downstream caspases.

Biochemical execution of cell death depends on three major components: caspases, mitochondrial factors, and the Bcl-2 family of proteins (Green 2005; Yuan 2006). Caspases are a group of cysteine proteases acting as the central effectors of apoptosis. Caspases are normally suppressed by the inhibitor of apoptosis (IAP) in the healthy cell (Uren et al. 1996). Upon receiving apoptotic stimuli such as cellular stress or signals from the Fas or TNF pathways, a mitochondrial protein called second mitochondria-derived activator of caspases (SMAC) is released into the cytosol and in turn IAP activity is relieved (Du et al. 2000). The Bcl-2 family of proteins consists of proapoptotic proteins (such as BAX, BAK, BID, and BAD) and antiapoptotic proteins (such as Bcl-xL and Bcl-2) (Chao and Korsmeyer 1998). Downstream from the activaton of the Fas or TNF pathways, a balance between these pro- and antiapoptotic proteins is compromised.

Cytochrome *c* (Cyt *c*), the peripheral protein loosely associated with the inner membrane of mitochondria, is one of the most well-known factors involved in apoptosis (Green 2005). In healthy cells, Cyt *c* functions as an electron shuttle in the respiratory chain and its activity is necessary for life. Cyt *c* is released by the mitochondria as the consequence of elevated permeability of the outer membrane in responses to proapoptotic stimuli (Li et al. 1997). In the cytosol, Cyt *c* binds to the apoptosis-protease activating factor 1 (Apaf-1), which then recruits caspase-9 to form the apoptosome (Li et al. 1997). Caspase-9 in turn cleaves and activates executioner caspase-3, resulting in apoptotic cell death as described above. The whole process requires energy and relatively intact cell machinery.

Molecular and biochemical studies have also revealed that Bcl-2 and Bcl-xL can prevent apoptosis (Chao and Korsmeyer 1998). For example, overexpression of Bcl-2 and Bcl-xL in mitochondria can block Cyt *c* efflux and apoptosis (Kharbanda et al. 1997). Although Apaf-1 may not be the physiological target for Bcl-2 and Bcl-xL, the Bcl-2 homolog (BH) 4 domain of Bcl-2 and Bcl-xL can bind to the C-terminal part of Apaf-1, thus inhibiting the association of caspase-9 with Apaf-1 and preventing apoptosis of cells (Hu et al. 1998).

CVB3, like other picornaviruses, has a relatively short life cycle. Studies have shown that the virus interferes with cell death signaling pathways during different stages of its life cycle (Carthy et al. 1998), either by prolonging the survival of infected cells to evade host responses and facilitate replication in an early stage of its life cycle; or by inducing the death of host cells to facilitate progeny virus release at a later stage of its life cycle. Viral proteins, both structural and nonstructural proteins, contribute to this manipulation through interaction with host factors, cleavage of host proteins, or regulation of intracellular calcium homeostasis. CVB3 encodes several viral proteases such as $2A^{pro}$, $3C^{pro}$ and its precursor $3CD^{pro}$. These viral proteases are responsible for cleaving the structural and nonstructural viral proteins from the viral polyprotein precursor. In addition, a number of cellular proteins including eIF4γI and eIF4γII, PABP, and dystrophin are cleaved by these proteases (Badorff et al. 1999; Kerekatte et al. 1999). The eIF4γ, for instance, is necessary for host cap-dependent protein translation by priming the RNA and ribosome. The cleavage of eIF4γ by $2A^{pro}$ leads to the shutoff of host protein translation. CVB3, on the other hand, has an internal ribosome entry site (IRES), so the synthesis of viral proteins is unaffected by this cleavage and thus gives the virus a lead strategy for survival over the host cell. Viral proteases of CVB3 and other enteroviruses apparently cleave another key protein, PABP, to further down-regulate the cap-dependent translation of host proteins. It has been known that the

3′ poly(A) tail and the 5′ cap of the eukaryotic mRNA cooperate synergistically to enhance the translation of cellular proteins. PABP interacts with eIF4γ to circulate mRNA and facilitate the reattachment of the ribosome after a round of translation. Cleavage of the PABP by $3C^{pro}$ and $2A^{pro}$ (to a lesser extent) thus complements the effect of cleavage of eIF4γ to further shut off the synthesis of host proteins and benefit IRES-driven translation. In addition, such cleavage may also function to stimulate RNA packaging into infectious virion progeny, leading to late-stage CVB3 infection. However, we must remember that many cellular proteins also contain an IRES and undergo translation in a cap-independent fashion. These proteins include vascular endothelial growth factor, c-myc, Apaf-1, death-associated protein 5, X-linked inhibitor of apoptosis, and heat shock proteins. Some of these proteins, such as Apaf-1, can only be translated in a cap-independent manner, so the translation of these proteins could be a strategy for the host cell to fend off virus infection possibly through apoptosis. Transcriptional upregulation and cap-independent synthesis of heat shock proteins post-CVB3 infection, on the other hand, may reflect a defensive strategy deployed by host cells to initiate protective responses to cellular stress.

There is a delicate balance between cellular membrane permeability and intracellular calcium homeostasis during CVB3 infection. It has been well-documented that sustained elevation of calcium levels in the cytosol precedes Cyt *c* release from the mitochondria, and that the small amount of released Cyt *c* interacts with the inositol triphosphate receptor (IP3R) on the endoplasmic reticulum (ER) and prevents inhibition of ER calcium release. The overall increase of calcium leads to a massive release of Cyt *c* to maintain ER calcium release through interaction with the IP3Rs in a positive feedback loop, and to activate downstream caspases to execute apoptosis of damaged cells.

Viral protein 2B apparently plays a pivotal role in modulating calcium homeostasis in infected cells. This protein can modify the membrane permeability of the plasma membrane, mitochondrial membrane, Golgi apparatus, and ER (van Kuppeveld et al. 1997a,b). Damage of the ER and Golgi apparatus causes release of calcium into the cytosol and leads to elevated intracellular calcium concentrations (de Jong et al. 2006; van Kuppeveld et al. 1997a). In addition, 2B protein increases the influx of extracellular calcium due to modifying the permeability of the plasma membrane. Altogether, cytosolic calcium content will reach a pathological level and contribute to cellular damage at a comparably later time of infection, which may be beneficial for the release of viral progeny of CVB3. However, during early infection, it has been reported that CVB3 2B protein downregulates calcium signaling between intracellular calcium stores and the mitochondria by depleting the IP3-generating stimuli-sensitive calcium stores of both the ER and Golgi apparatus (Campanella et al. 2004; de Jong et al. 2006). The alteration of intracellular calcium signaling by 2B apparently decreases caspase activation and cell death of infected host cells during early infection, which allows CVB3 to complete its life cycle. The precursor protein 2BC of CVB4, a virus closely related to CVB3, can also associate with activated caspase-3 and suppress its activity (Salako et al. 2006). Thus, the antiapoptotic activity of the 2B protein most likely acts to delay apoptotic responses from the host cell antiviral defensive program and thereby provides the virus with enough time for genome replication and viral protein synthesis.

It has been shown that infection with CVB3 results in massive release of Cyt *c* from the mitochondria and caspase activation of host cells (Carthy et al. 2003; Cunningham et al. 2003). Caspase-2, -3, -6, -7, -8, and -9 are activated and cleaved into their characteristic large and small fragments during CVB3 infection, which further degrade other specific substrates: poly(ADP-ribose) polymerase, a nuclear protein involved in DNA repair, and DNA fragmentation factor, a cytoplasmic inhibitor of an endonuclease responsible for DNA fragmentation (Carthy et al. 1998). Because activated caspases can disrupt cellular architecture, host cytoskeleton, and membrane integrity, apoptosis is likely a viral mechanism that facilitates virus release and spread. Indeed, inhibition of caspase activation in CVB3-infected cells significantly decreases viral progeny release without an effect on virus genome and protein replication. It has been further demonstrated that overexpression of antiapoptotic proteins Bcl-2 and Bcl-xL prevented Cyt *c* release from the mitochondria to the cytosol following CVB3 infection in HeLa cells and HL-1 murine cardiomyocytes, thus delaying and decreasing caspase-like cleavage

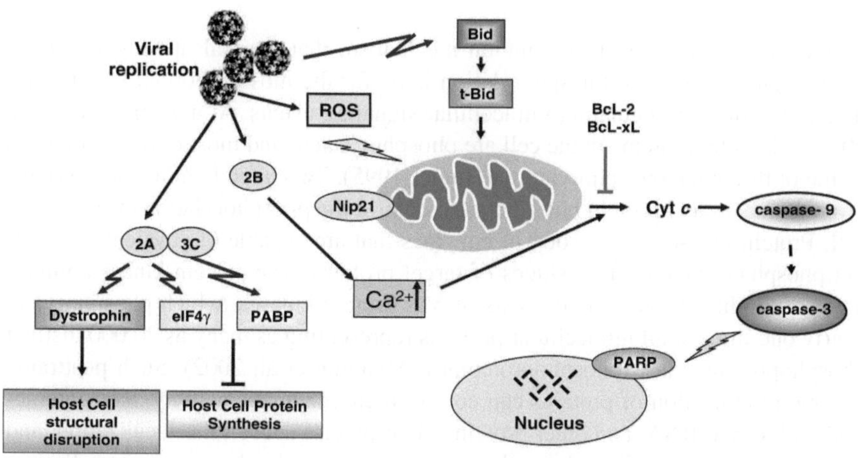

Fig. 14.2 CVB3-induced cell death events in infected cells. CVB3 infection increases mitochondrial membrane permeability, which leads to the release of Cyt *c* from mitochondria and subsequent activation of caspase family. This eventually results in apoptotic morphological changes of the infected cells. Increased Nip21 expression enhances CVB3-induced apoptosis of host cells via caspase-dependent mitochondria activation. CVB3 2B promotes Cyt *c* release via increasing intracellular calcium levels. Overexpression of antiapoptotic proteins Bcl-2 and Bcl-xL, blocks Cyt *c* release from mitochondria, indicating an important protective role of these proteins. Viral proteases 2A and 3C cleave eIF4γ and PABP, resulting in the blockage of host protein synthesis. 2A protease also cleaves dystrophin to disrupt structure of infected cells. The fate of infected host cells is finally determined by the balance between virus, on one side, trying to overcome the host cell defense machinery, and the host cells, on the other side, trying to eliminate the invading pathogen. Abbreviations: eIF4γ, eukaryotic initiation factor 4γ; PABP, protein poly(A) binding protein; Cyt *c*, cytochrome *c*.

activity (Carthy et al. 2003). Viral progeny release, which depends on the apoptosis of infected cells, is delayed in cells overexpressing Bcl-2 and Bcl-xL.

As well, infection of CVB3 in susceptible mice leads to altered expression of Nip21 and interferon -γ inducible protein 10 (IP10) in the heart that develops myocarditis (Yang et al. 1999; Zhang et al. 2005). Nip21 expression apparently enhances CVB3-induced apoptosis of host cells via caspase-dependent mitochondria activation (Zhang et al. 2002). IP10 expression also increases the apoptosis of infected cells through a p53-dependent pathway (Zhang et al. 2005). Overexpression of these genes inhibits the replication of CVB3 in infected cells (Zhang et al. 2002, 2005). Furthermore, the CVB3 structural protein VP2 specifically interacts with the proapoptotic protein, Siva, to induce apoptosis in a caspase-dependent pathway in infected tissues, such as pancreas and heart. Furthermore, expression of Siva is upregulated in infected mice (Henke et al. 2001).

The cell death signaling triggered by coxsackieviral infection is summarized in Figure 14.2.

Phosphorylation-Based Signaling Networks in CVB3 Replication

Protein phosphorylation is a common mechanism that the cell utilizes to control its biological processes. Phosphorylation is by far the most important posttranslational modification regulating intracellular signaling events. At a given time, up to 30% of the total proteins in the cell are phosphorylated, and most of this comprises serine or threonine phosphorylation (Hunter 1995). Two large families of enzymes, protein kinases and protein phosphatases, control the phosphorylation events in the cell. Protein kinases are a group of enzymes that are capable of covalently attaching phosphate moieties to residues of target proteins. The protein kinase complement of the human genome consists of 518 protein kinases, which phosphorylate nearly one-third of all intracellular proteins representing as many as 20,000 distinct phosphoprotein states (phosphoproteome) (Manning et al. 2002). Such posttranslational modification of proteins can control their enzymatic activity, their interaction with DNA, RNA, and other proteins, their precise subcellular localization, and their susceptibility to degradation by cellular proteases or the proteasome. Protein phosphatases, on the other hand, reverse this modification by removing the phosphate groups from target proteins. Phosphorylation on tyrosine residues occurs to a much lesser extent than that on serine or threonine residues, but it plays a key role in many cellular processes such as cell proliferation and differentiation, cell cycle control, cytoskeletal reorganization, cell migration, signal transduction, transcriptional regulation, ion channel activity, and immune responses (Hunter 1995; Tonks 2005). Therefore, excessive tyrosine phosphorylation often results in oncogenesis. The central role played by a constitutively activated protein tyrosine kinase (PTK) in oncogenesis was first illustrated by the discovery of the transforming v-src PTK (Hunter and Sefton 1980), and thereafter, by the identification of a large number of viral and cellular oncogenes that encode PTKs (Blume-Jensen and Hunter

2001). The observation that many PTKs are proto-oncogenes and growth factor receptors suggests that tyrosine phosphorylation is controlled by PTKs to activate transduction pathways. After growth factor stimulation or oncogenic transformation, the phosphotyrosine content of the cell can increase by as much as 100-fold to $\sim$1–2% of the total protein phosphorylation in the cell. In resting cells, protein tyrosine phosphatases (PTPs) maintain a very low basal level of tyrosine phosphorylation of PTKs and their substrates to ensure the proper function of the cell (Tonks 2005). In addition to negative regulatory roles, PTPs also play key roles as positive effectors of signal transduction pathways.

Increasing evidence supports the view that CVB3 usurps the host machinery, including intracellular signaling pathways, to facilitate its replication. However, the precise mechanisms utilized by viruses for temporal and spatial activation of the complicated signaling networks and relative roles of such activation remain to be investigated. Recently, studies by our group and others have indicated an important role of different signaling pathways in the life cycle of CVB3. Figure 14.3 illustrates the signaling events following CVB3 infection and their function roles. During CVB3 infection of host cells, several cellular proteins become tyrosine phosphorylated. For example, there is increased tyrosine phosphorylation of two proteins with molecular masses of 48 and 200 kDa in CVB3-infected HeLa and Vero cells (Huber et al. 1997, 1999). Subcellular fractionation experiments reveal that the 48-kDa CVB3-induced phosphoprotein is a soluble cytoplasmic protein, whereas the 200-kDa phosphoprotein is mostly associated with membrane structures. Interestingly, herbimycin A, an inhibitor of the src family kinases, reduces such tyrosine phosphorylation events and decreases the production of CVB3 progeny virions. This suggests that src family kinase (i.e., src, fyn, and yes) pathways are required for efficient replication of CVB3. Indeed, $p56^{lck}$, a member of the src family kinases, is required for viral replication in T cells and plays essential roles in CVB3 pathogenesis (Liu et al. 2000). Mice deficient in the $p56^{lck}$ gene are completely protected from CVB3-induced pancreatitis, hepatitis, myocarditis, and subsequent dilated cardiomyopathy. In addition, the CVB3 coreceptor DAF is associated with fyn, which is another member of the src family kinases (Shenoy-Scaria et al. 1992). The activation of fyn has been shown to be necessary for CVB3 entry through epithelial tight junctions in polarized cells (Coyne and Bergelson 2006). Furthermore, it was also observed that there is decreased phosphorylation of the src Tyr^{529} residue during CVB3 entry and replication in HeLa cells and HL-1 cardiomyocytes (unpublished results). Phosphorylation at this site often leads to reduced activity of src family kinases. This result together with previous findings further confirm that activation of src family kinase-mediated signaling pathways is beneficial for replication of CVB3. In addition, it has been found that the activity of glycogen synthase kinase 3β (GSK3β) is increased following CVB3 infection (Yuan et al. 2005). It was demonstrated that CVB3 infection likely stimulates the GSK3β activity via a tyrosine kinase-dependent mechanism, which contributes to CVB3-induced CPE and apoptosis through the deregulation of β-catenin.

Another finding is the biphasic phosphorylation of extracellular signal-regulated protein kinase (ERK) during CVB3 entry and replication (Cunningham et al. 2003;

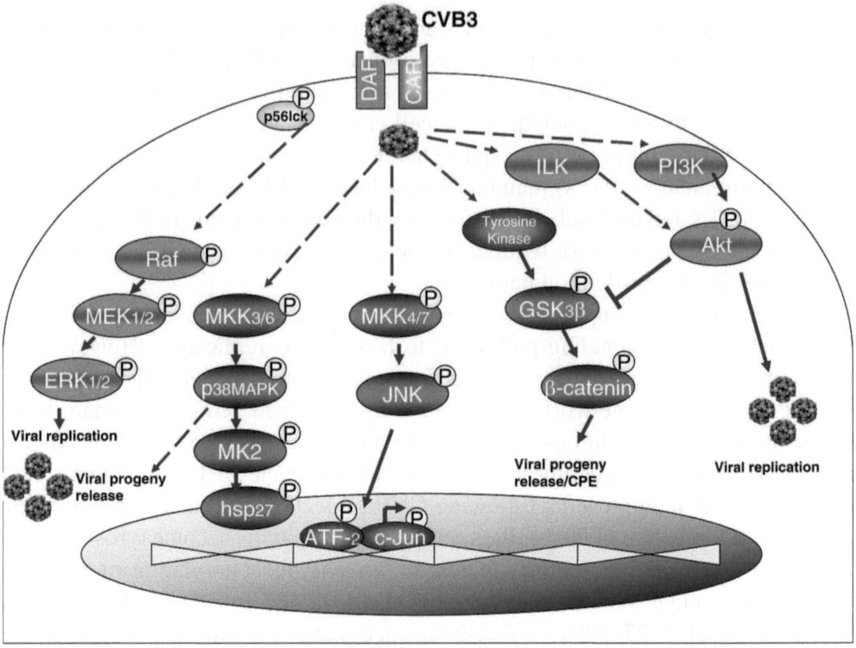

Fig. 14.3 CVB3-induced intracellular signaling networks in infected cells. CVB3 infection activates the src-family kinase and its downstream Raf/MEK/ERK signaling pathway, a necessary event for effective viral replication. CVB3 infection also leads to the activation of PI3K and ILK that subsequently activates Akt. Akt activation may play a role in CVB3-induced antiapoptotic effect during the early viral life cycle, and it is certainly beneficial for viral replication. In the late stage of CVB3 life cycle, stress-activated protein kinases including p38 MAPK and JNK are activated. Activation of p38 MAPK plays a role in CVB3 progeny virion release, while activation of JNK leads to the phosphorylation of transcription factors ATF-2 and c-Jun. In addition, GSK3β is activated through tyrosine phosphorylation, and such activation plays a role in CVB3-induced CPE and apoptosis for viral progeny release through dysregulation of β-catenin. Abbreviations: CVB3, coxsackievirus B3; DAF, decay accelerating factor; CAR, coxsackievirus and adenovirus receptor; MEK and MKK, mitogen-activated protein kinase kinase; ERK, extracellular signal-regulated kinase; MK2, mitogen-activated protein kinase-activated protein kinase 2; hsp27, heat shock protein 27; JNK, c-Jun NH_2-terminal kinase; ATF-2, activating transcription factor 2; GSK3β, glycogen synthase kinase 3β; PI3K, phosphatidylinositol 3-kinase; ILK, integrin-linked kinase.

Luo et al. 2002; Opavsky et al. 2002). ERK is one of the founding members of mitogen-activated protein kinases (MAPKs), which are evolutionarily conserved enzymes relaying extracellular signals from cell surface receptors to critical targets within cells, and thus regulating biological events such as cell proliferation, differentiation, and stress responses (Chang and Karin 2001; Johnson and Lapadat 2002). The MAPK pathway is part of a phosphorylation-based cascade composed of three tightly and sequentially activated kinases, including MAPK kinase kinases (MAPKKKs, MKKKs or MEKKs), MAPK kinases (MAPKKs, MKKs or MEKs), and MAPKs (Chang and Karin 2001; Johnson and Lapadat 2002). These modules may be activated by STE20 kinases or small GTP-binding proteins (GTPases). Many

MAPKs activate downstream effector kinases such as MAPK-activated protein kinases (MAPKAPKs or MKs), and are inactivated through a feedback loop by related MAPK phosphatases that belong to the family of dual-specificity protein tyrosine phosphatases. Studies from our laboratory and others have demonstrated that the early transient activation of ERK possibly results from the engagement of CVB3 with its native receptor CAR, and coreceptor DAF (Luo et al. 2002; Opavsky et al. 2002). Meanwhile, the late activation of ERK requires active replication of CVB3 and is necessary for the replication of CVB3. This activation is possibly mediated by cleavage of RasGAP by CVB3 viral proteases such as $3C^{pro}$. In cells, activities of small GTPases, such as Ras, are controlled by two groups of enzymes, guanine nuclear exchange factor (GEF) and GTPase-activating protein (GAP) (Overbeck et al. 1995; Quilliam et al. 1995; Tocque et al. 1997). GEF promotes the exchange of GDP-Ras to GTP-Ras and thus the activation of GTPases. GAP, on the contrary, stimulates the intrinsic GTP hydrolysis activity and returns the protein to its inactive GDP-bound state. Cleavage of RasGAP by viral proteases leads to accumulation of the active form of Ras (GTP-bound form of Ras), which in turn leads to activation of the well-documented Ras/Raf/MAPK pathway for ERK phosphorylation. In addition, src family kinases are likely involved in early activation of ERK because absence of $p56^{lck}$ in Jurkat cells reduces ERK phosphorylation following receptor engagement (Opavsky et al. 2002). The importance of the ERK pathway has been established since inhibition of MEKs, immediately upstream protein kinases of ERK, leads to reduced viral replication, decreased cleavage of host proteins, and suppression of host cell death (Luo et al. 2002).

Other protein kinases in the family of MAPKs include c-Jun NH_2-terminal kinases (JNK1, JNK2, and JNK3), p38 MAPKs (p38α, p38β, p38γ, and p38δ), and ERK5 (Chang and Karin 2001; Johnson and Lapadat 2002). Both JNKs and p38 MAPKs belong to stress-activated protein kinases (SAPKs) because they are activated in response to inhibition of protein synthesis, proinflammatory stimuli like TNFα and IL-1, ultraviolet radiation, growth hormones, ligands for G-protein-coupled receptors (GPCRs), and other environmental and cellular stresses such as osmotic shock and heat shock (Chang and Karin 2001; Johnson and Lapadat 2002). At least a dozen MEKKs are involved in the activation of JNKs and p38 MAPKs, but the physiological functions and specificities of these MEKKS are not well-characterized. It is possible that there is a redundancy in their roles to activate JNKs and p38 MAPKs in response to various stimuli and stresses, or there is a delicate balance of their activation to form a network or signalosome to relay the signal to specific JNKs, p38 MAPKs, and their effector kinases. The JNKs are necessary for the phosphorylation and activation of transcription factor c-Jun, which is part of the AP-1 transcription complex responsible for the expression of many cytokine genes in response to infection. The p38 MAPKs and their effector kinases are activated by inflammatory cytokines to regulate the expression of many cytokines and stabilization of mRNA transcripts of these cytokines, and in turn play an important role in activation of host immune responses (Ashwell 2006). Recent studies have shown that virus infections often lead to activation of JNKs and p38 MAPKs, and the activation of these SAPKs is required for efficient viral replication and viral progeny release (Hirasawa et al. 2003; Rahaus et al. 2005; Si et al. 2005a). However, the mechanisms leading to activation of these SAPKs and the specific roles of activation of specific SAPKs in virus replication differ among different viruses. We and others have investigated the relative roles of the SAPK pathways in CVB3 infection (Kim et al. 2004; Si et al. 2005a). Both the p38 MAPK and the JNK pathways are activated during the active replication steps of the CVB3 life cycle (Si et al. 2005a). It was initially reported that activation of JNK is required for the upregulation of cyr61 and induction of apoptosis of infected cells (Kim et al. 2004); however, the results suggest that JNK activation plays a role in host defense machinery through the phosphorylation and activation of activating transcription factor 2 (ATF-2), a component of the AP-1 transcription complex, rather than just inducing cell death (Si et al. 2005a). This was supported by earlier evidence that ATF-2 is dispensable for host antiviral defense against CVB3 infections (Reimold et al. 2001). Another player in host antiviral defense is the Janus kinase (JAK)-Signal Transducers and Activators of Transcription (STAT) pathway in CVB3 infection (Yasukawa et al. 2003). Furthermore, our results have demonstrated that activation of p38 MAPK is necessary for CVB3-induced caspase-3 activation and productive CVB3 viral progeny release (Si et al. 2005a). This caspase-3 activation also leads to prominent production of reactive oxygen species (ROS) in infected cells (Si et al. 2005b). Consequently, it is possible that activation of the p38 MAPK pathway, production of ROS, and subsequent activation of NFκB play a major role in the regulation of expression and secretion of proinflammatory cytokines in the progression of CVB3-induced pathogenesis.

As mentioned earlier in this review, CVB3 deploys antiapoptotic tactics to prevent host cell death during the early phase of infection. Such strategies provide the virus necessary time to complete its life cycle, which may include activation of prosurvival signaling pathways of host cells. One of the major intracellular prosurvival pathways is the phosphatidylinositol 3-kinase/Akt (PI3K/Akt) pathway (Hennessy et al. 2005). PI3K is a heterodimer of a regulatory subunit (p85) and a catalytic subunit (p110). The p85 binds and integrates signals from various cellular proteins such as receptor tyrosine kinases, src family kinases, protein kinase C, and small GTPases, and subsequently activates the p110 by recruiting it to the plasma membrane. Upon activation, PI3K phosphorylates phosphatidylinositol-4,5-biphosphate (PtdIns(4,5)P2, PIP2) to produce phosphatidylinositol-3,4,5-triphosphate (PIP3). PIP3 is a second messenger required for phosphorylation and activation of a downstream serine and threonine protein kinase, Akt, by 3′-phosphoinositide-dependent kinases (PDKs) such as PDK1 (Alessi et al. 1997) and integrin-linked kinase (ILK) (Delcommenne et al. 1998). Akt signaling inactivates many proapoptotic factors, which promotes the completion of the virus life cycle. A tumor suppressor, PTEN, is a physiological inhibitor of the PI3K/Akt pathway due to its unique 3′-phosphoinositide phosphatase activity to hydrolyze PIP3 to PIP2 (Maehama and Dixon 1998). The critical role of the PI3K/Akt pathway in CVB3 replication has recently been dissented (Esfandiarei et al. 2004). And it was found that CVB3 infection leads to the activation of the PI3K/Akt pathway, and this activation is mediated by upregulation of interferon-γ inducible GTPase (Zhang et al. 2003). Inhibition of this pathway using specific pharmacological inhibitors or dominant-negative mutants of Akt or ILK significantly impairs virus production yet increases apoptosis of infected cells (Esfandiarei et al. 2004, 2006). The role of ILK in the activation of Akt following CVB3 infection has also been investigated, and the results clearly demonstrate that kinase activity of ILK is required for efficient CVB3 replication in both HeLa cells and HL-1 murine cardiomyocytes (Esfandiarei et al. 2006).

The Ubiquitin–Proteasome System in CVB3 Infection

The ubiquitin–proteasome system (UPS) is a major intracellular proteolytic system for the removal of damaged or unwanted proteins in virtually all eukaryotic cells (Wolf and Hilt 2004). This ATP-dependent system begins with covalent attachment of ubiquitin to the ubiquitin-activating enzyme, the so-called E1. Then the ubiquitin moiety is transferred to the E2 ubiquitin-conjugating enzymes. Finally, the ubiquitin moiety is transferred to a substrate protein by the highly specific E3 ubiquitin ligases. Repeating of the above steps will result in poly-ubiquitination of substrate proteins that are targeted for degradation by the UPS. This system is involved in the degradation of mutant, damaged, and misfolded proteins due to environmental and cellular stresses, or degradation of other regulatory proteins such as cyclins (Glotzer et al. 1991), inhibitors of cyclin-dependent kinases (p21, p27) (Pagano et al. 1995), tumor suppressors (p53) (Scheffner et al. 1990), and inhibitor of NFκB (IκB) (Palombella et al. 1994). Thereby, the UPS is necessary for a variety of cellular mechanisms, such as control of cell-cycle progression, regulation of apoptosis, and stimulation of host immune responses (Roos-Mattjus and Sistonen 2004). Unlike poly-ubiquitination, which regulates substrate protein degradation, mono-ubiquitination of cellular proteins, such as histones, calmodulins, actin, proliferating cell nuclear antigen (PCNA), and receptor tyrosine kinases, plays much more diversified roles, including the regulation of chromatin remodeling, DNA repair, and endocytosis (Hicke 2001).

It has been shown that CVB3 infection results in downregulation of several host proteins such as cell-cycle protein cyclin D1, tumor suppressor p53, and transcription activator β-catenin in infected HeLa cells (Luo et al. 2003; Yuan et al. 2005). The downregulation of host proteins following CVB3 infection is possibly through the UPS because inhibitors of the 26 S proteasome reverse the degradation of proteins and reduce CVB3 replication in HeLa cells and HL-1 murine cardiomyocytes (Luo et al. 2003). It has also been shown that deregulation of the UPS by pyrrolidine dithiocarbamate (PDTC) (Si et al. 2005b) or curcumin (unpublished results) effectively reduces synthesis of CVB3 viral proteins and release of CVB3 progeny virions. However, there are no apparent proteasome activity changes in infected cells (Luo et al. 2003). It has been demonstrated that CVB3 infection results in increased poly-ubiquitination of cellular proteins and a decreased free ubiquitin pool in infected cells (unpublished results). Knockdown of ubiquitin expression by small interfering RNA (siRNA) also decreases CVB3 infection (unpublished results).

Furthermore, CVB3 likely utilizes the UPS to modify its viral proteins for its replication, as CVB3 RNA-dependent RNA polymerase 3D is either directly ubiquitinated or associated with ubiquitinated proteins in infected cells (unpublished results).

Conclusion

After decades of active research of viral infection of the heart, we have learned that the pathogenesis of viral myocarditis is not a one-sided attack on the myocardium by the virus, or clearance of infectious viral particles by the host immune system. Undoubtedly there is a signaling duel between virus and host, where the virus often prevails in the early stages of the disease and causes structural damage to the heart. Once the virus infects its host and cleaves host proteins by its proteases, the host cell is destined to undergo apoptotic and necrotic processes to limit virus replication despite that many prosurvival pathways, such as the ERK pathway and the PI3K/Akt pathway, are activated in CVB3 infection. Such irreversible cell death of infected cardiomyocytes due to direct viral injury and adaptive immune responses causes the loss of integrity of heart structure and impaired heart function. However, the window of opportunity for intervention may still exist if we can effectively block the necessary signaling pathways for CVB3 replication using small inhibitors to salvage the life and retain the function of infected cardiomyocytes, and thus delay the progression of myocarditis to DCM and heart failure. However, it is uncertain whether prolonging of the life of infected cells may provide a reservoir for the virus that facilitates chronic infection of the heart. Certainly the next step is to illustrate the complexity of the proapoptotic and prosurvival signaling networks of viral infections in heart diseases. Systems-like approaches, such as genomics, proteomics, infectomics, RNAi screens, phosphor-screens, and computational modeling, will be fundamental in deciphering key molecules and pathways in CVB3 pathogenesis, and in developing respective therapeutic regimes.

References

Alessi, D.R., Deak, M., Casamayor, A., Caudwell, F.B., Morrice, N., Norman, D.G., Gaffney, P., Reese, C.B., MacDougall, C.N., Harbison, D., et al. (1997). 3-Phosphoinositide-dependent protein kinase-1 (PDK1): structural and functional homology with the Drosophila DSTPK61 kinase. Curr Biol *7*, 776–789.

Ashwell, J.D. (2006). The many paths to p38 mitogen-activated protein kinase activation in the immune system. Nat Rev Immunol *6*, 532–540.

Badorff, C., Lee, G.H., Lamphear, B.J., Martone, M.E., Campbell, K.P., Rhoads, R.E., and Knowlton, K.U. (1999). Enteroviral protease 2A cleaves dystrophin: evidence of cytoskeletal disruption in an acquired cardiomyopathy. Nat Med *5*, 320–326.

Bell, Y.C., Semler, B.L., and Ehrenfeld, E. (1999). Requirements for RNA replication of a poliovirus replicon by coxsackievirus B3 RNA polymerase. J Virol *73*, 9413–9421.

Bergelson, J.M., Mohanty, J.G., Crowell, R.L., St John, N.F., Lublin, D.M., and Finberg, R.W. (1995). Coxsackievirus B3 adapted to growth in RD cells binds to decay-accelerating factor (CD55). J Virol *69*, 1903–1906.

Bergelson, J.M., Cunningham, J.A., Droguett, G., Kurt-Jones, E.A., Krithivas, A., Hong, J.S., Horwitz, M.S., Crowell, R.L., and Finberg, R.W. (1997). Isolation of a common receptor for coxsackie B viruses and adenoviruses 2 and 5. Science *275*, 1320–1323.

Blume-Jensen, P., and Hunter, T. (2001). Oncogenic kinase signalling. Nature *411*, 355–365.

Campanella, M., de Jong, A.S., Lanke, K.W., Melchers, W.J., Willems, P.H., Pinton, P., Rizzuto, R., and van Kuppeveld, F.J. (2004). The coxsackievirus 2B protein suppresses apoptotic host cell responses by manipulating intracellular Ca^{2+} homeostasis. J Biol Chem *279*, 18440–18450.

Carthy, C.M., Granville, D.J., Watson, K.A., Anderson, D.R., Wilson, J.E., Yang, D., Hunt, D.W., and McManus, B.M. (1998). Caspase activation and specific cleavage of substrates after coxsackievirus B3-induced cytopathic effect in HeLa cells. J Virol *72*, 7669–7675.

Carthy, C.M., Yanagawa, B., Luo, H., Granville, D.J., Yang, D., Cheung, P., Cheung, C., Esfandiarei, M., Rudin, C.M., Thompson, C.B., et al. (2003). Bcl-2 and Bcl-xL overexpression inhibits cytochrome c release, activation of multiple caspases, and virus release following coxsackievirus B3 infection. Virology *313*, 147–157.

Chang, L., and Karin, M. (2001). Mammalian MAP kinase signalling cascades. Nature *410*, 37–40.

Chao, D.T., and Korsmeyer, S.J. (1998). BCL-2 family: regulators of cell death. Annu Rev Immunol *16*, 395–419.

Chow, L.H., Beisel, K.W., and McManus, B.M. (1992). Enteroviral infection of mice with severe combined immunodeficiency. Evidence for direct viral pathogenesis of myocardial injury. Lab Invest *66*, 24–31.

Coyne, C.B., and Bergelson, J.M. (2006). Virus-induced Abl and Fyn kinase signals permit coxsackievirus entry through epithelial tight junctions. Cell *124*, 119–131.

Cunningham, K.A., Chapman, N.M., and Carson, S.D. (2003). Caspase-3 activation and ERK phosphorylation during CVB3 infection of cells: influence of the coxsackievirus and adenovirus receptor and engineered variants. Virus Res *92*, 179–186.

de Jong, A.S., Visch, H.J., de Mattia, F., van Dommelen, M.M., Swarts, H.G., Luyten, T., Callewaert, G., Melchers, W.J., Willems, P.H., and van Kuppeveld, F.J. (2006). The coxsackievirus 2B protein increases efflux of ions from the endoplasmic reticulum and Golgi, thereby inhibiting protein trafficking through the Golgi. J Biol Chem *281*, 14144–14150.

Delcommenne, M., Tan, C., Gray, V., Rue, L., Woodgett, J., and Dedhar, S. (1998). Phosphoinositide-3-OH kinase-dependent regulation of glycogen synthase kinase 3 and protein kinase B/AKT by the integrin-linked kinase. Proc Natl Acad Sci USA *95*, 11211–11216.

Dorner, A., Xiong, D., Couch, K., Yajima, T., and Knowlton, K.U. (2004). Alternatively spliced soluble coxsackie-adenovirus receptors inhibit coxsackievirus infection. J Biol Chem *279*, 18497–18503.

Du, C., Fang, M., Li, Y., Li, L., and Wang, X. (2000). Smac, a mitochondrial protein that promotes cytochrome c-dependent caspase activation by eliminating IAP inhibition. Cell *102*, 33–42.

Ehrenfeld, E. (1982). Poliovirus-induced inhibition of host-cell protein synthesis. Cell *28*, 435–436.

Esfandiarei, M., Luo, H., Yanagawa, B., Suarez, A., Dabiri, D., Zhang, J., and McManus, B.M. (2004). Protein kinase B/Akt regulates coxsackievirus B3 replication through a mechanism which is not caspase dependent. J Virol *78*, 4289–4298.

Esfandiarei, M., Suarez, A., Amaral, A., Si, X., Rahmani, M., Dedhar, S., and McManus, B.M. (2006). Novel role for integrin-linked kinase in modulation of coxsackievirus B3 replication and virus-induced cardiomyocyte injury. Circ Res *99*, 354–361.

Glotzer, M., Murray, A.W., and Kirschner, M.W. (1991). Cyclin is degraded by the ubiquitin pathway. Nature *349*, 132–138.

Green, D.R. (2005). Apoptotic pathways: ten minutes to dead. Cell *121*, 671–674.

Henke, A., Nestler, M., Strunze, S., Saluz, H.P., Hortschansky, P., Menzel, B., Martin, U., Zell, R., Stelzner, A., and Munder, T. (2001). The apoptotic capability of coxsackievirus B3 is influenced by the efficient interaction between the capsid protein VP2 and the proapoptotic host protein Siva. Virology *289*, 15–22.

Hennessy, B.T., Smith, D.L., Ram, P.T., Lu, Y., and Mills, G.B. (2005). Exploiting the PI3K/AKT pathway for cancer drug discovery. Nat Rev Drug Discov *4*, 988–1004.

Hicke, L. (2001). Protein regulation by monoubiquitin. Nat Rev Mol Cell Biol *2*, 195–201.

Hirasawa, K., Kim, A., Han, H.S., Han, J., Jun, H.S., and Yoon, J.W. (2003). Effect of p38 mitogen-activated protein kinase on the replication of encephalomyocarditis virus. J Virol *77*, 5649–5656.

Hu, Y., Benedict, M.A., Wu, D., Inohara, N., and Nunez, G. (1998). Bcl-XL interacts with Apaf-1 and inhibits Apaf-1-dependent caspase-9 activation. Proc Natl Acad Sci USA *95*, 4386–4391.

Huber, M., Selinka, H.C., and Kandolf, R. (1997). Tyrosine phosphorylation events during coxsackievirus B3 replication. J Virol *71*, 595–600.

Huber, M., Watson, K.A., Selinka, H.C., Carthy, C.M., Klingel, K., McManus, B.M., and Kandolf, R. (1999). Cleavage of RasGAP and phosphorylation of mitogen-activated protein kinase in the course of coxsackievirus B3 replication. J Virol *73*, 3587–3594.

Hufnagel, G., Chapman, N., and Tracy, S. (1995). A non-cardiovirulent strain of coxsackievirus B3 causes myocarditis in mice with severe combined immunodeficiency syndrome. Eur Heart J *16* Suppl O, 18–19.

Hunter, T. (1995). Protein kinases and phosphatases: the yin and yang of protein phosphorylation and signaling. Cell *80*, 225–236.

Hunter, T., and Sefton, B.M. (1980). Transforming gene product of Rous sarcoma virus phosphorylates tyrosine. Proc Natl Acad Sci USA *77*, 1311–1315.

Johnson, G.L., and Lapadat, R. (2002). Mitogen-activated protein kinase pathways mediated by ERK, JNK, and p38 protein kinases. Science *298*, 1911–1912.

Kandolf, R., Ameis, D., Kirschner, P., Canu, A., and Hofschneider, P.H. (1987). In situ detection of enteroviral genomes in myocardial cells by nucleic acid hybridization: an approach to the diagnosis of viral heart disease. Proc Natl Acad Sci USA *84*, 6272–6276.

Kerekatte, V., Keiper, B.D., Badorff, C., Cai, A., Knowlton, K.U., and Rhoads, R.E. (1999). Cleavage of poly(A)-binding protein by coxsackievirus 2A protease *in vitro* and *in vivo*: another mechanism for host protein synthesis shutoff? J Virol *73*, 709–717.

Kharbanda, S., Pandey, P., Schofield, L., Israels, S., Roncinske, R., Yoshida, K., Bharti, A., Yuan, Z.M., Saxena, S., Weichselbaum, R., et al. (1997). Role for Bcl-xL as an inhibitor of cytosolic cytochrome C accumulation in DNA damage-induced apoptosis. Proc Natl Acad Sci USA *94*, 6939–6942.

Kim, S.M., Park, J.H., Chung, S.K., Kim, J.Y., Hwang, H.Y., Chung, K.C., Jo, I., Park, S.I., and Nam, J.H. (2004). Coxsackievirus B3 infection induces cyr61 activation via JNK to mediate cell death. J Virol *78*, 13479–13488.

Klingel, K., Rieger, P., Mall, G., Selinka, H.C., Huber, M., and Kandolf, R. (1998). Visualization of enteroviral replication in myocardial tissue by ultrastructural in situ hybridization: identification of target cells and cytopathic effects. Lab Invest *78*, 1227–1237.

Li, P., Nijhawan, D., Budihardjo, I., Srinivasula, S.M., Ahmad, M., Alnemri, E.S., and Wang, X. (1997). Cytochrome c and dATP-dependent formation of Apaf-1/caspase-9 complex initiates an apoptotic protease cascade. Cell *91*, 479–489.

Liu, P., Aitken, K., Kong, Y.Y., Opavsky, M.A., Martino, T., Dawood, F., Wen, W.H., Kozieradzki, I., Bachmaier, K., Straus, D., et al. (2000). The tyrosine kinase p56lck is essential in coxsackievirus B3-mediated heart disease. Nat Med *6*, 429–434.

Luo, H., Yanagawa, B., Zhang, J., Luo, Z., Zhang, M., Esfandiarei, M., Carthy, C., Wilson, J.E., Yang, D., and McManus, B.M. (2002). Coxsackievirus B3 replication is reduced by inhibition of the extracellular signal-regulated kinase (ERK) signaling pathway. J Virol *76*, 3365–3373.

Luo, H., Zhang, J., Dastvan, F., Yanagawa, B., Reidy, M.A., Zhang, H.M., Yang, D., Wilson, J.E., and McManus, B.M. (2003). Ubiquitin-dependent proteolysis of cyclin D1 is associated with coxsackievirus-induced cell growth arrest. J Virol *77*, 1–9.

Maehama, T., and Dixon, J.E. (1998). The tumor suppressor, PTEN/MMAC1, dephosphorylates the lipid second messenger, phosphatidylinositol 3,4,5-trisphosphate. J Biol Chem *273*, 13375–13378.

Manning, G., Whyte, D.B., Martinez, R., Hunter, T., and Sudarsanam, S. (2002). The protein kinase complement of the human genome. Science *298*, 1912–1934.

McManus, B.M., Chow, L.H., Wilson, J.E., Anderson, D.R., Gulizia, J.M., Gauntt, C.J., Klingel, K.E., Beisel, K.W., and Kandolf, R. (1993). Direct myocardial injury by enterovirus: a central role in the evolution of murine myocarditis. Clin Immunol Immunopathol *68*, 159–169.

Metcalf, D., Di Rago, L., Mifsud, S., Hartley, L., and Alexander, W.S. (2000). The development of fatal myocarditis and polymyositis in mice heterozygous for IFN-gamma and lacking the SOCS-1 gene. Proc Natl Acad Sci USA *97*, 9174–9179.

Nicholson-Weller, A., and Wang, C.E. (1994). Structure and function of decay accelerating factor CD55. J Lab Clin Med *123*, 485–491.

Opavsky, M.A., Martino, T., Rabinovitch, M., Penninger, J., Richardson, C., Petric, M., Trinidad, C., Butcher, L., Chan, J., and Liu, P.P. (2002). Enhanced ERK-1/2 activation in mice susceptible to coxsackievirus-induced myocarditis. J Clin Invest *109*, 1561–1569.

Overbeck, A.F., Brtva, T.R., Cox, A.D., Graham, S.M., Huff, S.Y., Khosravi-Far, R., Quilliam, L.A., Solski, P.A., and Der, C.J. (1995). Guanine nucleotide exchange factors: activators of Ras superfamily proteins. Mol Reprod Dev *42*, 468–476.

Pagano, M., Tam, S.W., Theodoras, A.M., Beer-Romero, P., Del Sal, G., Chau, V., Yew, P.R., Draetta, G.F., and Rolfe, M. (1995). Role of the ubiquitin-proteasome pathway in regulating abundance of the cyclin-dependent kinase inhibitor p27. Science *269*, 682–685.

Palombella, V.J., Rando, O.J., Goldberg, A.L., and Maniatis, T. (1994). The ubiquitin-proteasome pathway is required for processing the NF-kappa B1 precursor protein and the activation of NF-kappa B. Cell *78*, 773–785.

Peng, T., Sadusky, T., Li, Y., Coulton, G.R., Zhang, H., and Archard, L.C. (2001). Altered expression of Bag-1 in coxsackievirus B3 infected mouse heart. Cardiovasc Res *50*, 46–55.

Quilliam, L.A., Khosravi-Far, R., Huff, S.Y., and Der, C.J. (1995). Guanine nucleotide exchange factors: activators of the Ras superfamily of proteins. Bioessays *17*, 395–404.

Rahaus, M., Desloges, N., and Wolff, M.H. (2005). ORF61 protein of Varicella-zoster virus influences JNK/SAPK and p38/MAPK phosphorylation. J Med Virol *76*, 424–433.

Reimold, A.M., Kim, J., Finberg, R., and Glimcher, L.H. (2001). Decreased immediate inflammatory gene induction in activating transcription factor-2 mutant mice. Int Immunol *13*, 241–248.

Roos-Mattjus, P., and Sistonen, L. (2004). The ubiquitin-proteasome pathway. Ann Med *36*, 285–295.

Salako, M.A., Carter, M.J., and Kass, G.E. (2006). Coxsackievirus protein 2BC blocks host cell apoptosis by inhibiting caspase-3. J Biol Chem *281*, 16296–16304.

Scheffner, M., Werness, B.A., Huibregtse, J.M., Levine, A.J., and Howley, P.M. (1990). The E6 oncoprotein encoded by human papillomavirus types 16 and 18 promotes the degradation of p53. Cell *63*, 1129–1136.

Shenoy-Scaria, A.M., Kwong, J., Fujita, T., Olszowy, M.W., Shaw, A.S., and Lublin, D.M. (1992). Signal transduction through decay-accelerating factor. Interaction of glycosyl-phosphatidylinositol anchor and protein tyrosine kinases p56lck and p59fyn 1. J Immunol *149*, 3535–3541.

Si, X., Luo, H., Morgan, A., Zhang, J., Wong, J., Yuan, J., Esfandiarei, M., Gao, G., Cheung, C., and McManus, B.M. (2005a). Stress-activated protein kinases are involved in coxsackievirus B3 viral progeny release. J Virol *79*, 13875–13881.

Si, X., McManus, B.M., Zhang, J., Yuan, J., Cheung, C., Esfandiarei, M., Suarez, A., Morgan, A., and Luo, H. (2005b). Pyrrolidine dithiocarbamate reduces coxsackievirus B3 replication through inhibition of the ubiquitin-proteasome pathway. J Virol *79*, 8014–8023.

Taylor, L.A., Carthy, C.M., Yang, D., Saad, K., Wong, D., Schreiner, G., Stanton, L.W., and McManus, B.M. (2000). Host gene regulation during coxsackievirus B3 infection in mice: assessment by microarrays. Circ Res *87*, 328–334.

Thorburn, A. (2004). Death receptor-induced cell killing. Cell Signal *16*, 139–144.

Tocque, B., Delumeau, I., Parker, F., Maurier, F., Multon, M.C., and Schweighoffer, F. (1997). Ras-GTPase activating protein (GAP): a putative effector for Ras. Cell Signal *9*, 153–158.

Tonks, N.K. (2005). Redox redux: revisiting PTPs and the control of cell signaling. Cell *121*, 667–670.

Uren, A.G., Pakusch, M., Hawkins, C.J., Puls, K.L., and Vaux, D.L. (1996). Cloning and expression of apoptosis inhibitory protein homologs that function to inhibit apoptosis and/or bind tumor necrosis factor receptor-associated factors. Proc Natl Acad Sci USA *93*, 4974–4978.

van Kuppeveld, F.J., Hoenderop, J.G., Smeets, R.L., Willems, P.H., Dijkman, H.B., Galama, J.M., and Melchers, W.J. (1997a). Coxsackievirus protein 2B modifies endoplasmic reticulum membrane and plasma membrane permeability and facilitates virus release. EMBO J *16*, 3519–3532.

van Kuppeveld, F.J., Melchers, W.J., Kirkegaard, K., and Doedens, J.R. (1997b). Structure-function analysis of coxsackie B3 virus protein 2B. Virology *227*, 111–118.

Wolf, D.H., and Hilt, W. (2004). The proteasome: a proteolytic nanomachine of cell regulation and waste disposal. Biochim Biophys Acta *1695*, 19–31.

Woodruff, J.F. (1980). Viral myocarditis. A review. Am J Pathol *101*, 425–484.

Yanagawa, B., Spiller, O.B., Proctor, D.G., Choy, J., Luo, H., Zhang, H.M., Suarez, A., Yang, D., and McManus, B.M. (2004). Soluble recombinant coxsackievirus and adenovirus receptor abrogates coxsackievirus b3-mediated pancreatitis and myocarditis in mice. J Infect Dis *189*, 1431–1439.

Yang, D., Yu, J., Luo, Z., Carthy, C.M., Wilson, J.E., Liu, Z., and McManus, B.M. (1999). Viral myocarditis: identification of five differentially expressed genes in coxsackievirus B3-infected mouse heart. Circ Res *84*, 704–712.

Yasukawa, H., Yajima, T., Duplain, H., Iwatate, M., Kido, M., Hoshijima, M., Weitzman, M.D., Nakamura, T., Woodard, S., Xiong, D., et al. (2003). The suppressor of cytokine signaling-1 (SOCS1) is a novel therapeutic target for enterovirus-induced cardiac injury. J Clin Invest *111*, 469–478.

Yuan, J. (2006). Divergence from a dedicated cellular suicide mechanism: exploring the evolution of cell death. Mol Cell *23*, 1–12.

Yuan, J., Zhang, J., Wong, B.W., Si, X., Wong, J., Yang, D., and Luo, H. (2005). Inhibition of glycogen synthase kinase 3beta suppresses coxsackievirus-induced cytopathic effect and apoptosis via stabilization of beta-catenin. Cell Death Differ *12*, 1097–1106.

Zhang, H.M., Yanagawa, B., Cheung, P., Luo, H., Yuan, J., Chau, D., Wang, A., Bohunek, L., Wilson, J.E., McManus, B.M., et al. (2002). Nip21 gene expression reduces coxsackievirus B3 replication by promoting apoptotic cell death via a mitochondria-dependent pathway. Circ Res *90*, 1251–1258.

Zhang, H.M., Yuan, J., Cheung, P., Luo, H., Yanagawa, B., Chau, D., Stephan-Tozy, N., Wong, B.W., Zhang, J., Wilson, J.E., et al. (2003). Overexpression of interferon-gamma-inducible GTPase inhibits coxsackievirus B3-induced apoptosis through the activation of the phosphatidylinositol 3-kinase/Akt pathway and inhibition of viral replication. J Biol Chem *278*, 33011–33019.

Zhang, H.M., Yuan, J., Cheung, P., Chau, D., Wong, B.W., McManus, B.M., and Yang, D. (2005). Gamma interferon-inducible protein 10 induces HeLa cell apoptosis through a p53-dependent pathway initiated by suppression of human papillomavirus type 18 E6 and E7 expression. Mol Cell Biol *25*, 6247–6258.

Chapter 15
Therapeutic Angiogenesis and Vasculogenesis for Ischemic Disease

Nilanjana Maulik

Abstract Despite significant advances in myocardial revascularization and reperfusion, coronary artery disease and subsequently myocardial infarction are the leading cause of morbidity and mortality in the United States. Thus, one of the main goals in the treatment of myocardial ischemia is the development of effective therapy for angiogenesis. The first evidence we found is the demonstration of alleviation of myocardial ischemia and increased number of collateral blood vessels in the early 1990s following intracoronary administration of basic fibroblast growth factor protein in dog. Multiple animal studies, including ours, and a small number of human studies have confirmed the concept of stimulation of collateral development by pharmacological and molecular means. This includes direct delivery of growth factors into the ischemic target tissues, or of genes that encode for synthesis of growth factors by target tissues. Both cell therapy and gene therapy have proven to be effective in promoting neovascularization in various animal models. Although cell therapy alone is proven to be beneficial, the combination of cell and gene therapy may enhance therapeutic neovascularization. Thus, clinically relevant, combined strategy could be an excellent strategy for treating patients with myocardial infarction.

Introduction

The discovery of the molecular mechanisms of physiological vasculogenesis and angiogenesis helped to recognize two classes of diseases: one where therapeutic angiogenesis can repair the tissue damages such as ischemic diseases and arteriosclerosis and the other where inhibition of pathological angiogenesis can cure the disease or delay its progression such as in retinopathies and tumor. Therapeutic angiogenesis in myocardial infarction (MI), in which we are interested, is an exciting new concept with significant clinical potential. Therapeutic angiogenesis is a strategy designed to increase the process of angiogenesis and enhance the reperfusion of the ischemic tissues. There are several approaches to induce angiogenesis, but proteins, genes, and cell therapy are the main ones that have been well studied.

A. K. Srivastava, M. B. Anand-Srivastava (eds.) *Advances in Biochemistry in Health and Disease.* Signal Transduction in the Cardiovascular System in Health and Disease,

Large MI in the left ventricle leads not only to expansion of the necrotic infarct zone, but also to compensate remodeling throughout the remainder of the left ventricle. Multiple animal studies, including ours, and a small number of human studies have confirmed the concept of stimulation of collateral development by pharmacological and molecular means. This includes direct delivery of growth factors into the ischemic target tissues, or of genes that encode for synthesis of growth factors by target tissues. Under ischemic conditions, coronary collateral vessels are formed by two main pathways: "angiogenesis," which is characterized by sprouting of new blood vessels from the existing one, and "arteriogenesis," which leads to the maturation of capillary blood vessels into mature arteriolar vessels. Vascular endothelial growth factor (VEGF) and basic fibroblast growth factor (bFGF), the main growth factors involved in the process of angiogenesis, appear to act by local upregulation of nitric oxide production and exhibit some synergism in their actions (Muorohara et al. 1998). Initiation of angiogenesis with intra-arterial injection of naked VEGF cDNA to relieve ischemia in atherosclerotic peripheral vascular disease was the first successful application of these research advances (Isner et al. 1996). Therapeutic angiogenesis aims at restoring perfusion to chronically ischemic myocardial territories by using growth factors or cells, without intervening on the epicardial coronary arteries. Therapeutic angiogenesis can be induced by different modalities including administration of angiogenic growth factors—protein, genes, stem cells, and most importantly endothelial progenitors derived from bone marrow or umbilical cord blood. All have shown promising results in inducing angiogenesis in experimental myocardial or limb ischemia as well as partial success in patients.

The question we are facing now is whether a single growth factor or a cocktail will be sufficient to produce its angiogenic effect. Significant studies are necessary to understand what cell type or types and origin are important with regard to achieving satisfactory improvement in myocardial angiogenesis and whether the observed improvement in blood supply will positively affect myocardial function and clinical outcome of these subjects.

Angiogenic Factors

VEGF and Its Receptors

The process of angiogenesis is regulated by the signals obtained from the transmembrane receptor tyrosine kinases (RTKs) and nonreceptor tyrosine kinases (Src family) of endothelial cells. Flk-1/KDR and Flt-1 are two such RTKs, which, together with their ligand VEGF, have been shown to control blood vessel development during embryogenesis (Fong et al. 1995; Takeshita et al. 1994). This receptor/ligand system has been shown to augment neovascularization (Isner et al. 1996; Asahara et al. 1998). VEGF is not only an endothelial cell-specific angiogenic factor but also a critical regulator of angiogenesis that stimulates proliferation, migration-and proteolytic activity of endothelial cells (Senger et al. 1990). Yet the signaling pathways that modulate the mitogenic effects of VEGF in vascular endothelial cells are still ill defined (Fong et al. 1995). A study has demonstrated VEGF localization and expression in the embryonic/fetal heart, and which remained high during the early postnatal period when capillary proliferation is high (Takeshita et al. 1994). It is now well established that alternate exon splicing of a single VEGF gene results in the generation of four different molecular species, having 121, 165, 189 and 206 amino acids following signal sequence cleavage. $VEGF_{165}$ is the predominant molecular species produced by a variety of normal and transformed cells. Transcripts encoding $VEGF_{121}$ and $VEGF_{189}$ are detected in the majority of cells and tissues expressing the VEGF gene (Ray et al. 2000). Among the mechanisms that have been proposed to participate in the regulation of VEGF gene expression, oxygen tension is a particularly important mediator, both *in vitro* and *in vivo*. Also, ischemia caused by occlusion of the left anterior descending coronary artery results in a dramatic increase in VEGF levels in the dog and rat myocardium, suggesting the possibility that VEGF may mediate the spontaneous revascularization that follows myocardial ischemia (Shizukuda et al. 1993; Sasaki et al. 2002).

VEGF is a potent vascular endothelial cell-specific mitogen that stimulates endothelial cell proliferation, microvascular permeability, vasodilatation, and angiogenesis (Ferrara and Davis-Smyth 1997; Neufeld et al. 1999). VEGF is the only growth factor proven to be specific and critical for blood vessel formation (Ferrara 1999). VEGF has also been shown to improve endothelial cell function and survival *in vitro* and vascular reactivity *in vivo* (Spyridopoulos et al. 1997). Various mechanisms have been shown to be involved in the regulation of VEGF expression. Oxygen tension appears to play a significant role, both *in vitro* and *in vivo*. VEGF mRNA expression is rapidly induced by exposure to low $P\ 0_2$ in a variety of cultured cells such as retinal pigmented epithelial cells (Shima et al. 1995), myoblast (Shweiki et al. 1992), cardiomyocyte (Banai et al. 1974), and tumor cells (Shweiki et al. 1992). Furthermore, occlusion of the left anterior descending coronary artery results in VEGF protein expression suggesting that VEGF is a mediator of the revascularization (spontaneous) that occurs in low oxygen conditions. There exist similarities between the mechanisms leading to hypoxic regulation of VEGF and erythropoietin (EPO). Cobalt chloride was found to induce both VEGF and EPO mRNA expression. Hypoxic induction of both genes was inhibited by carbon monoxide, suggesting the involvement of a heme protein in the process of sensing oxygen levels (Goldberg and Schneider 1994). EPO increases the oxygen-carrying capacity of the blood by stimulating red cell formation, while VEGF-induced angiogenesis allows the delivery of oxygen to ischemic tissues. Recent studies have demonstrated that nitric oxide (NO) plays an important role in many VEGF-induced actions. VEGF has been shown to induce the production of NO from rabbit, pig, bovine, and human vascular endothelial cells (Morbedelli et al. 1996; Van der Zee et al. 1997; Parenti et al. 1998). Inhibition of NO production by eNOS inhibitors significantly inhibited VEGF-induced mitogenic and angiogenic effects (Papapetropoulos et al. 1997). Endothelial NOS/NO has been implicated as one of the important mediators for

VEGF-induced hemodynamic changes and microvascular permeability. Although eNOS was originally described as a constitutive enzyme, more recent studies indicate that a variety of stimuli including hypoxia, shear stress, inflammatory cytokines, high glucose, and injury could modulate eNOS expression and activity (Sase and Michel 1997; Le Cras et al. 1996). *In vitro* experiments have shown that activation of KDR induces proliferation and migration of endothelial cells as well as expression of eNOS and iNOS (Waltenberger et al. 1994; Kroll and Waltenberger 1998). So far, the role of VEGF receptors in the release of NO is controversial. In human placental trophoblast it has been suggested that Flt-1 can induce the release of NO (Bussolati et al. 2001). The precise signaling pathways that mediate those responses have not been elucidated; however, phospholipase D and nitric oxide synthase have been identified as key enzymes activated by VEGF. These two enzymes have been shown to depend on protein kinase C (PKC) and its downstream mitogen-activated protein kinase (MAPK) activation (Wood et al. 1992; Seymour et al. 1996). Many studies found that the effects of preconditioning are mediated by PKC, MAPKs, and redox-regulated transcription factors such as NFκB and STAT 1 activation. Among 10 PKC isoforms, PKC α, δ, ε, and ζ isoforms are expressed in rat myocardium. Several groups have demonstrated that either PKC δ or ζ is required for the protective effect of preconditioning in rat myocardium. Inhibition of PKC activity with the chemical inhibitor calphostin prevents VEGF-induced vascular growth (Friedlander et al. 1992) but the role of specific isoenzymes in the processes that constitute angiogenesis is unknown, because the specificity of most PKC inhibitors for individual isoenzymes has not been established. Recently, peptide antagonist of several PKC isoforms has been synthesized by Mochly-Rosen and associates (Souroujon and Mochly 1998). Development of PKC translocation peptides is based on the hypothesis that binding of each PKC isoform to isoform-specific anchoring proteins termed receptors for activated C kinase (RACK) occurs after phospholipid-induced allosteric conformational changes that activate the enzyme and expose its RACK-binding domain (Mochly-Rosen 1995). This leads to PKC translocation and binding to isoform-specific RACKs at different subcellular sites, which is thought to determine the function of each isoform. The activation of Flk-1/KDR may be sufficient for embryonic angiogenesis, and activation of VEGF-1-mediated NO release may be required to promote the development of the intricate architecture of the vascular network in wound healing. Thus, NO seems to be a molecular switch for endothelial cell differentiation, and induction of eNOS and/or iNOS may have therapeutic activity in vascular insufficiencies. A final proof for the role of various VEGF receptors in NO release, however, remains to be demonstrated.

Fibroblast Growth Factors (FGFs)

FGFs are members of a family of polypeptides synthesized by a variety of cell types during the process of embryonic development and in adult tissues. These growth factors have been detected in normal and malignant cells and show a biological profile that includes mitogenic and angiogenic activity. FGF and FGF receptors (FGFR) play significant roles in many biological systems. The FGFRs are a family of transmembrane tyrosine kinases involved in signaling via interactions with the family of FGFs. The first agent to be used in an attempt to stimulate myocardial neovascularization was FGF-1 (Banai et al. 1991). In animal models, FGF has been shown to stimulate angiogenesis, granulation, tissue formation, epithelial growth, and wound tensile strength (Silverstein and Rifkin 1987) Recently, several animal studies have indicated that both basic fibroblast growth factor FGF and acidic fibroblast growth factor FGF enhance the regeneration of peripheral nerve (Aebischer et al. 1989). Both VEGF and FGF2 activate the ERK-1/2 pathway via the canonical Grb2-SOS-Ras pathway. Interactions between growth factors are not clearly understood. Various studies have shown that VEGF induces FGF2 expression in endothelial cells while FGF2 in turn can also induce VEGF expression (Seghezzi et al. 1998; Claffey et al. 2001). It is clear from the recent studies that neutralization of FGF2 activity inhibits VEGF-induced angiogenesis and vice versa (Tomanek et al. 2002). Overall, $VEGF_{165}$ and FGF2 share some similarities because both growth factors can stimulate the MAPK cascade and activation of these pathways is required for either factor's activity.

Revascularization/Angiogenic Strategies

In the past ten years, alternative revascularization/angiogenesis strategies have progressed from bench to bedside, focusing on the capillary sprouting and/or growth of new vessels to replace the old. However, most of the strategies involve the delivery of growth factors. As mentioned earlier, promotion of angiogenesis with intraarterial injection of naked VEGF cDNA to treat ischemia in atherosclerotic peripheral vascular disease was the first successful application of these research advances (Isner et al. 1996). Intracoronary injection of purified bFGF has also been used to reduce ischemic extent in coronary diseases.

Angiogenic Therapy

Angiogenic therapy is the stimulation of blood vessel growth in ischemic disease. Experimental research on angiogenesis for ischemia started in the 1960s (Svet-Moldavsky and Chimishkyan 1977). Yanagisawa-Miwa et al. (1992) infused bFGF into the coronary arteries of dogs and improved cardiac function after experimental MI. Takeshita et al. (1994) reported that VEGF infused into an iliac artery promoted the development of collateral vessels in an ischemic rabbit hind limb. Moreover, Morishita et al. (1999) reported that human HGF administered into an iliac artery improved collateral vessel formation using an ischemic rabbit hind limb. Shintani et al. (2001) found that mononuclear cells (MNCs), separated from autologus bone marrow, injected into muscle increased collateral vessel flow in a similar model. Angiogenic therapy for ischemic disease was first performed at Tufts University in the United States in 1994. Isner et al. (1996) reported that administration of a plasmid vector incorporating $VEGF_{165}$ into arteries supplying an ischemic limb of patients suffering from arteriosclerosis obliterans (ASO) increased collateralization.

Angiogenic therapy for ischemic disease has developed rapidly in the 10 years since its introduction. However, although *in vivo* experiments produced collateral vessels, some of the procedures used were not clinically appropriate. Recently, angiogenic therapy for patients with ischemic disease has been introduced. Angiogenic strategies for ischemic disease include: (a) angiogenic protein or gene administration, (b) bone marrow MNC implantation, (c) peripheral blood EPC implantation, and (d) hematogenous cytokine administration.

(a) Angiogenic therapy using angiogenic protein or gene administration for ischemic heart disease

Basic research using a porcine model of chronic ischemic heart showed that the injection of the $phVEGF_{165}$ gene (Vale et al. 1999b; Tio et al. 1999) or VEGF-2 gene (Vale et al. 1999a) into stunned (or hibernated) myocardium improved collateral vessel blood flow, and the injection of an adenovirus coded with the $VEGF_{121}$ gene (Mack et al. 1998; Lee et al. 2000) into stunned myocardium also increased collateral vessel flow and improved cardiac function. Although the administration of the FGF-2 protein into the coronary arteries improves myocardial ischemia in dogs (Lazarous et al. 1995; Rajanayagam et al. 2000) and pigs (Lopez et al. 1997), recombinant FGF-1 had no effect in a canine chronic myocardial infarction model (Banai et al. 1991). In the pig model, the injection of both DNA coded with FGF into stunned myocardium and adenovirus coded with FGF-5 into the coronary artery increased collateral blood flow (Giordano et al. 1996).

Animal experiments suggest that the administration of VEGF or FGF protein improves collateral blood flow in the ischemic myocardium. However, in clinical cases, these proteins had no effect (Henry et al. 2000; Simons et al. 2002). Esakof and colleagues administered a plasmid vector including the $phVEGF_{165}$ gene to patients with severe ischemic heart disease who were not candidates for PCI or CABG into their myocardium through a small left anterior thoracotomy under transesophageal echocardiographic guidance as a phase I clinical trial (Losordo et al. 1998; Esakof et al. 1999; Symes et al. 1999). This therapy improved patients' exercise tolerance and reduced their symptoms, demonstrating the safety and usefulness of this approach (Losordo et al. 2002). To minimize invasions, cardiac catheterization was used to administer the VEGF-2 DNA gene into stunned myocardium in patients with severe ischemic disease, and reduced the frequency of angina attacks compared with a placebo group (Vale et al. 2000). A few patients have received the bFGF gene for ischemic heart disease by injection into epicardial fat during CABG in areas where no graft could be placed. The results were not unequivocal since the clinical situation was complex and not easily categorized. Still, data suggested that angina symptoms improved and myocardial blood flow increased (Laham et al. 1999).

Further study to enhance effectiveness is needed, particularly to prolong its effect prior to its catabolism. Angiogenic gene therapy offers the additional benefit of avoiding the need for surgery in this group of very sick patients.

(b) Angiogenic therapy using bone marrow mononuclear cell (MNC) implantation for ischemic heart disease

A few vascular endothelial stem cells exist in the bone marrow. However, MNCs and hematogenous stem cells other than vascular endothelial stem cells can synthesize and release VEGF, bFGF, and angiopoietin-1 as factors, which induce the multiplication and maturation of endothelial cells (Brunner et al. 1993; Hamano et al. 2000). Furthermore, it has been reported that synthesis and release are enhanced under hypoxic conditions. Noishiki et al. (1996) reported that endothelialization of an artificial blood vessel occurred after implanting bone marrow MNCs in a canine abdominal aortic replacement model. Shi et al. (1998) reported that endothelial stem cells ($CD34^+$ cells), implanted from other animals, homogenously colonized endothelial flow surfaces of vascular prostheses of the descending thoracic aorta in dogs following whole body irradiation. It was suggested that bone marrow MNCs have the potential to differentiate into vascular endothelial cells. When using large animals, Kamihata et al. (2001) reported that after autologous bone marrow MNCs were introduced into stunned myocardium through the epicardium by the open chest method, collateral blood flow was significantly increased without arrhythmia.

Based on these studies, Matsubara and colleagues (Tateishi-Yuyama et al. 2002) reported a randomized double-blind control study of patients with severe ischemic limb disease (Fontaine grade III or IV) resistant to medical or surgical therapy. MNCs were separated and synthesized from autologous bone marrow and injected into the ischemic limb muscle, leading to improved ankle brachial pressure index (ABI), treadmill exercise tolerance, reduced symptoms, and collateral blood flow on angiography. Moreover, there were no complications, such as the change to cells other than endothelial cells, inflammation, edema, or new bone formation. This is the first clinical report on angiogenic therapy for ischemic disease using cell transplantation.

We have treated no-option severe ischemic limb ASO patients with autologous bone marrow MNC implantation with significant improvement in symptoms and in collateral blood flow as examined by angiography (Figure 15.1, adapted from Fukuda et al. 2004). Critical limb ischemia is an advanced form of peripheral artery disease, which is a general term used to describe those diseases affecting blood vessels of the arms and legs. Therefore, peripheral artery disease is strongly associated with coronary artery disease.

The initial clinical successes in conjunction with good results in basic animal studies warrant further investigation into the bone marrow MNC transplantation for clinical ischemic heart disease. Hamano et al. (2001) performed autologous

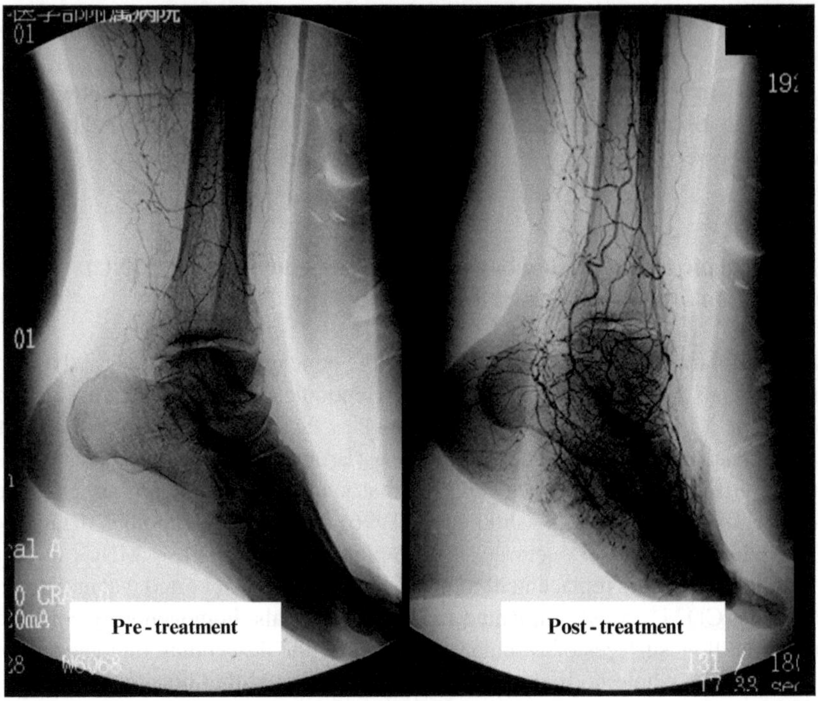

Fig. 15.1 Angiographies of a patient with arteriosclerosis obliterans who underwent bone marrow cell implantation. (Adapted from Mol. Cell. Biochem. 264:143–149, 2004.) (See color plate.)

bone marrow MNC transplantation to the domain inaccessible to surgical revascularization concomitantly with CABG. Although it is difficult to isolate the effect of angiogenic therapy because cell transplantation was performed simultaneously with CABG, left ventricular wall motion improved in distributions that received cell transplantation but no graft. There was no complication of angiogenic therapy in 18 months of follow-up. They performed autologous bone marrow MNC transplantation via small left thoracotomy in patients who previously underwent five PCIs and two CABGs, and reported decreased angina, improved left ventricular contraction, better exercise tolerance, and no arrhythmic complications.

At present, angiogenic therapy using bone marrow MNCs is performed worldwide in patients with treatment-resistant ischemic limbs. Moreover, angiogenic therapy using autologous bone marrow MNCs injected into stunned (hibernated) myocardium is performed in several institutions, and the number of trial centers is also increasing very fast.

(c) Angiogenic therapy using peripheral blood EPCs

The endothelial progenitor cell (EPC) of peripheral blood origin is the $CD34^+$, which differentiates into a vascular endothelial cell under the influence of VEGF, bFGF, and insulinlike growth factor (IGF) (Shi et al. 1998). However, since the number of peripheral blood $CD34^+$ EPCs is only 10% of that in bone marrow (Bender et al. 1991), angiogenic therapy is more effective when bone marrow MNCs are used rather than peripheral blood EPCs. Co-incubation of $CD34^+$ cells of low purity with $CD34^-$ cells reportedly increased the proliferation rate of EPCs by more than 10-fold than when the former were cultured alone.

(d) Angiogenic therapy using hematogenous cytokine

Takahashi et al. (1999) reported that granulocyte macrophage colony-stimulating factor (GM-CSF) is one of the hematogenous cytokines that cause EPCs to migrate bone marrow into the peripheral blood. Seiler et al. (2001) infused GM-CSF into the coronary arteries of 21 patients with ischemic heart disease who were not candidates for CABG due to severity of disease. The authors reported increased collateral blood flow in a randomized double-blind study. Another hematogenous cytokine, granulocyte colony-stimulating factor (G-CSF), was examined for its angiogenic potential, but no beneficial effect was achieved. GM-CSF has been shown to stimulate proliferation, differentiation, maturation, and survival of monocytes and macrophages. This potent cytokine was also found to reduce collagen synthesis (Tojo et al. 1999). Moreover, M-CSF is found to reduce apoptosis as reported by Fuller et al. (1993). Therefore, infusion of M-CSF might increase cell survival after MI in the heart and other organ.

(e) Angiogenic pretreatment with growth factors along with cell implantation more effective

The combination of myocardial cell transplantation and angiogenic gene transfer is a very fascinating way of inducing long-term angiogenesis (Figure 15.2). The potential importance of angiogenic pretreatment in avoiding an ischemic gap may affect cell survival followed by functional benefits of cellular cardiomyoplasty (Sakakibara et al. 2002). Sakakibara et al. administered bFGF- incorporating microspheres, 1 week prior to fetal cardiomyocyte transplantation in a rat coronary ligation model. The combination of therapeutic gene transfer strategies with myocardial cell transplantation may represent a promising strategy for patients with ischemic left ventricular dysfunction. Suzuki et al. (2001) demonstrated that the transplantation of control transfected skeletal myoblasts could reduce infarct size. Moreover, it has been shown that the transplantation of skeletal myoblasts expressing VEGF provides further advanced benefits in reducing infarct size and improves cardiac function. These effects correlated with the enhanced angiogenesis observed in comparison with control-myoblast transplanted hearts. Thus, transplantation of heart cells transfected with VEGF induced greater angiogenesis than transplantation of unmodified cells. In other study, it was also demonstrated that prior heat shock treatment of transplanted myoblasts enhances self-preservation systems

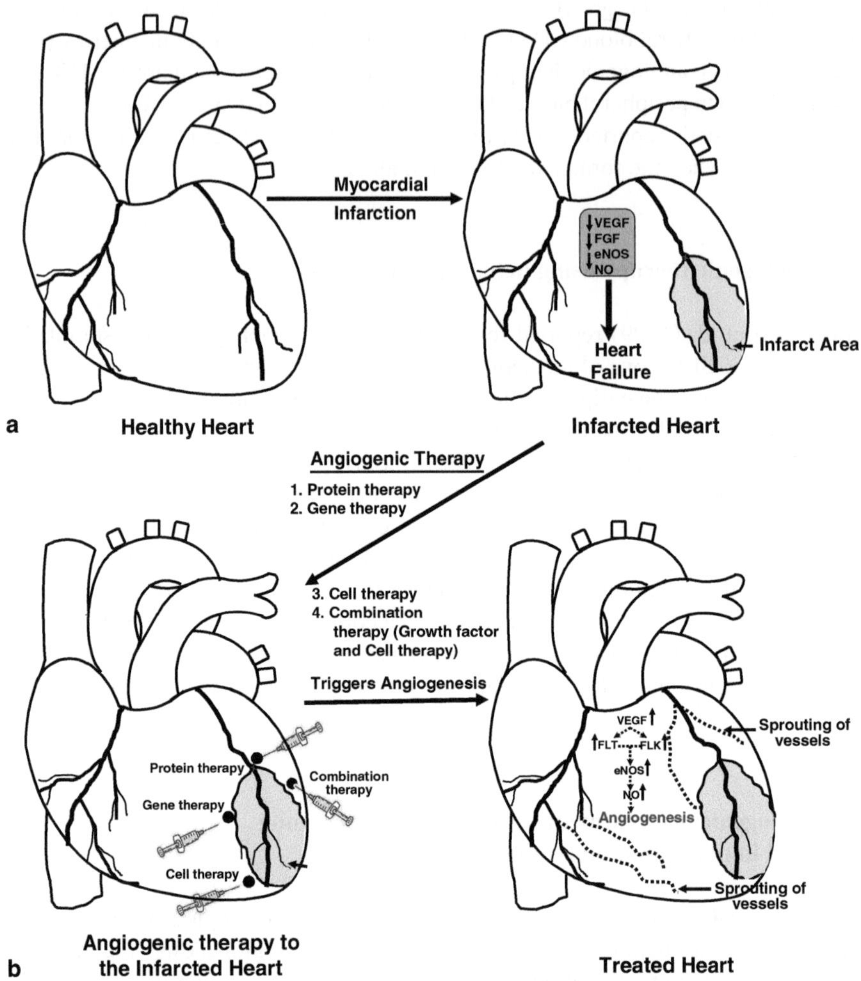

Fig. 15.2 (A) Healthy and infarcted heart. (B) Downregulation of angiogenic factors in the infarcted heart is rectified by various angiogenic therapies successfully leading to angiogenesis.

against environmental stress and this improves their survival after engraftment to the heart using beta-Gal-expressing skeletal myoblasts (Suzuki et al. 2000).

Conclusion

It is obvious that angiogenic therapy for ischemic heart disease using gene therapy, cell transplantation, and hematogenous cytokine administration will benefit increasing numbers of patients. However, long-term studies monitoring side effects to establish safety and effectiveness are required. Finally, a randomized double-blind study is needed to compare the efficacy of angiogenic therapy and conventional therapy with PCI and CABG. The angiogenic therapy is a promising medical development in the treatment of ischemia that will likely improve the quality of life in patients who have severe ischemic heart disease.

Acknowledgement This study was supported by National Institutes of Health Grant HL 56803 to the author, HL69910.

References

Aebischer, P., Salessiotis, A.N., and Winn, S.R. 1989. Basic fibroblast growth factor released from synthetic guidance channels facilitates peripheral nerve regeneration across long nerve gaps. J. Neurosci. Res. 23:282–289.

Asahara, T., Chen, D., Takahashi, T., Fujikawa, K., Kearney, M., Magner, M., Yancopoulos, G.D., and Isner, J.M. 1998. Tie-2 receptor ligands, angiopoietin-1 and angiopoietin-2, modulate VEGF-induced postnatal neovascularization. Circ. Res. 83:342–343.

Banai, S., Shweiki, D., Pinson, A., Chandra, M., Lazarovici, G., and Keshet, E. 1974. Upregulation of vascular endothelial growth factor expression induced by myocardial ischemia: implications for coronary angiogenesis. Cardiovasc. Res 28:1176–1179.

Banai, S., Jaklitsch, M.T., Casscells, W., Shou, M., Shrivastav, S., Correa, R., Epstein, S.E., and Unger, E.F. 1991. Effects of acidic fibroblast growth factor on normal and ischemic myocardium. Circ. Res. 69:76–85.

Bender, J.G., Unverzagt, K.L., Walker, D.E., Lee, W., Van Epps, D.E., Smith, D.H., Stewart, C.C., and To, L.B. 1991. Identification and comparison of CD34-positive cells and their subpopulations from normal peripheral blood and bone marrow using multicolor flow cytometry. Blood 77:2591–2596.

Brunner, G., Nguyen, H., Gabrilove, J., Rifkin, D.B., and Wilson, E.L. 1993. Basic fibroblast growth factor expression in human bone marrow and peripheral blood cells. Blood 81:631–638.

Bussolati, B., Dunk, C., Mason, M., Kontos, C.D., and Ahmed, A. 2001. Vascular endothelial growth factor receptor-1 modulates vascular endothelial growth factor mediated angiogenesis via nitric oxide. Am. J. Pathol. 159: 993–1008.

Claffey, K.P., Abrams, K., Shih, S.C., Brown, L.F., Mullen, A., and Keogh, M. 2001. Fibroblast growth factor 2 activation of stromal cell vascular endothelial growth factor expression and angiogenesis. Lab. Invest. 81:61–75.

Esakof, D.D., Maysky, M., Losordo, D.W., Vale, P.R., Lathi, K., Pastore, J.O., Symes, J.F., and Isner, J.M. 1999. Intraoperative multiplane transesophageal echocardiography for guiding direct myocardial gene transfer of vascular endothelial growth factor in patients with refractory angina pectoris. Hum. Gene Ther. 10:2307–2314.

Ferrara, N. 1999. Vascular endothelial growth factor: molecular and biological aspects. Curr. Top. Microbiol. Immunol. 237:1–30.

Ferrara, N., and Davis-Smyth, T. 1997. The biology of vascular endothelial growth factor. Endocr. Rev. 18:4–25.

Fong, G.H., Rossant, J., Gerstenstein, M., and Breitman, M.L. 1995. Role of the Flt-1 receptor tyrosine kinase in regulating the assembly of vascular endothelium. Nature 376:66–70.

Friedlander, M., Brooks, P.C., Shaffer, R.W., Kincaid, C.M., Varner, J.A., and Cheresh, D.A. 1992. Definition of two angiogenic pathways by distinct α v integrins. Science 270:607–614.

Fukuda, S., Yoshii, S., Kaga, S., Matsumoto, M., Kugiyama, K., and Maulik, N. 2004. Angiogenic strategy for human ischemic heart disease: Brief overview. Mol. Cell. Biochem. 264:143–149.

Fuller, K., Owens, J.M., Jagger, C.J., Wilson, A., Moss, R., and Chambers, T.J. 1993. Macrophage colony-stimulating factor stimulates survival and chemotactic behavior in isolated osteoclasts. J. Exp. Med. 178:1733–1744.

Giordano, F.J., Ping, P., McKirnan, M.D., Nozaki, S., DeMaria, A.N., Dillmann, W.H., Mathieu-Costello, O., and Hammond. H.K. 1996. Intracoronary gene transfer of fibroblast growth factor-5 increases blood flow and contractile function in an ischemic region of the heart. Nat. Med. 2:534–549.

Goldberg, M.A., and Schneider, T.J. 1994. Similarities between the oxygen-sensing mechanisms regulating the expression of vascular endothelial growth factor and erythropoietin. J. Biol. Chem. 269:4355–4361.

Hamano, K., Nishida, M., Hirata, K., Mikamo, A., Li, T.S., Harada, M., Miura, T., Matsuzaki, M., and Esato, K. 2001. Local implantation of autologous bone marrow cells for therapeutic angiogenesis in patients with ischemic heart disease: clinical trial and preliminary results. Jpn. Circ. J. 65:845–847.

Henry, T.D., Mckendall, G.R., Azrin, M.A., Lopez, J.J., Benza, R., Wiilerson, J.T., Giacomini, J., Olson, R., Bart, B.A., Roel, J.P., and Annex, B.H. 2000. VEGF in ischemia for vascular angiogenesis (VIVA) trial: one year follow up. Circulation 102:II-309.

Isner, J.M., Pieczek, A., Schainfed, R., Blair, R., Haley, L., Asahara, T., Rosenfield, K., Razvi, S., Walsh, K., and Symes, J. 1996. Clinical evidence of angiogenesis is after arterial gene transfer of ph VEGF165 in patient with ischemic limb. Lancet 34:370–374.

Kamihata, H., Matsubara, H., Nishiue, T., Fujiyama, S., Tsutsumi, Y., Ozono, R., Masaki, H., Mori, Y., Iba, O., Tateishi, E., Kosaki, A., Shintani, S., Murohara, T., Imaizumi, T., and Iwasaka, T. 2001. Implantation of bone marrow mononuclear cells into ischemic myocardium enhances collateral perfusion and regional function via side supply of angioblasts, angiogenic ligands, and cytokines. Circulation 4:1046–1052.

Kroll, J., and Waltenberger, J. 1998. VEGF-A induces expression of eNOS and iNOS in endothelial cells via VEGF receptor-2 (KDR). Biochem. Biophys. Res. Commun. 252:743–746.

Laham, R.J., Sellke, F.W., Edelman, E.R., Pearlman, J.D., Ware, J.A., Brown, D.L., Gold, J.P., and Simons, M. 1999. Local perivascular delivery of basic fibroblast growth factor in patients undergoing coronary bypass surgery: results of a phase I randomized, double-blind, placebo-controlled trial. Circulation 100:1865–1871.

Lazarous, D.F., Scheinowitz, M., Shou, M., Hodge, E., Rajanayagam, S., Hunsberger, S., Robison, W.G., Jr., Stiber, J.A., Correa, R., and Epstein, S.E. 1995. Effects of chronic systemic administration of basic fibroblast growth factor on collateral development in the canine heart. Circulation 91:145–153.

Le Cras, T.D., Xue, C., Rengasamy, A., and Johns, R.A. 1996. Chronic hypoxia up regulates endothelial and inducible NO synthase gene and protein expression in rat lung. Am. J. Physiol. 270:L164–L170.

Lee, L.Y., Patel, S.R., Hackett, N.R., Mack, C.A., Polce, D.R., El-Sawy, T., Hachamovitch, R., Zanzonico, P., Sanborn, T.A., Parikh, M., Isom, O.W., Crystal, R.G., and Rosengart, T.K. 2000. Focal angiogen therapy using intramyocardial delivery of an adenovirus vector coding for vascular endothelial growth factor 121. Ann. Thorac. Surg. 69:14–23.

Lopez, J.J., Edelman, E.R., Stamler, A., Hibberd, M.G., Prasad, P., Caputo, R.P., Carrozza, J.P., Douglas, P.S., Sellke, F.W., and Simons, M. 1997. Basic fibroblast growth factor in a porcine model of chronic myocardial ischemia: a comparison of angiographic, echocardiographic and coronary flow parameters. J. Pharmacol. Exp. Ther. 282:385–390.

Losordo, D.W., Vale, P.R., Symes, J.F., Dunnington, C.H., Esakof, D.D., Maysky, M., Ashare, A.B., Lathi, K., and Isner, J.M. 1998. Gene therapy for myocardial angiogenesis: initial clinical results with direct myocardial injection of phVEGF165 as sole therapy for myocardial ischemia. Circulation 98:2800–2804.

Losordo, D.W., Vale, P.R., Hendel, R.C., Milliken, C.E., Fortuin, F.D., Cummings, N., Schatz, R.A., Asahara, T., Isner, J.M., and Kuntz, R.E. 2002. Phase 1/2 placebo-controlled, double-blind, dose-escalating trial of myocardial vascular endothelial growth factor 2 gene transfer by catheter delivery in patients with chronic myocardial ischemia. Circulation 105:2012–2018.

Mack, C.A., Patel, S.R., Schwarz, E.A., Zanzonico, P., Hahn, R.T., Ilercil, A., Devereux, R.B., Goldsmith, S.J., Christian, T.F., Sanborn, T.A., Kovesdi, I., Hackett, N., Isom, O.W., Crystal, R.G., and Rosengart, T.K. 1998. Biologic bypass with the use of adenovirus-mediated gene transfer of the complementary deoxyribonucleic acid for vascular endothelial growth factor 121 improves myocardial perfusion and function in the ischemic porcine heart. J. Thorac. Cardiovasc. Surg. 115:168–176.

Meiler, C., Pohl, T., Wustmann, K., Hutter, D., Nicolet, P.A., Windecker, S., Eberli, F.R., and Meier, B. 2001. Promotion of collateral growth by granulocyte-macrophage colony-stimulating factor in patients with coronary artery disease: a randomized, double-blind, placebo-controlled study. Circulation 104:2012–2017.

Mochly-Rosen, D. 1995. Localization of protein kinases by anchoring proteins: A theme in signal transduction. Science 268:247–251.

Morbedelli, L., Chang, C.H., and Douglas, J.G. 1996. Nitric oxide mediates mitogenic effect of VEGF on coronary venular endothelium. Am. J. Physiol. 270:H411–H415.

Morishita, R., Nakamura, S., Hayashi, S., Taniyama, Y., Moriguchi, A., Nagano, T., Taiji, M., Noguchi, H., Takeshita, S., Matsumoto, K., Nakamura, T., Higaki, J., and Ogihara, T. 1999. Therapeutic angiogenesis induced by human recombinant hepatocyte growth factor in rabbit hind limb ischemia model as cytokine supplement therapy. Hypertension 33:1379–1384.

Muorohara, T., Asahara, T., Silver, M., Bauters, C., Masuda, H., Kalka, C., Kearney, M., Chen, D., Symes, J.F., Fishman, M.C., Huang, P.L., and Isner, J.M. 1998. Nitric oxide synthase modulates angiogenesis in response to tissue ischemia. J. Clin. Invest. 101:2567–2578.

Neufeld, G., Cohen, T., Gengrinovitch, S., and Poltorak, Z. 1999. Vascular endothelial growth factor (VEGF) and its receptors. FASEB. J. 13:9–22.

Noishiki, Y., Tomizawa, Y., Yamane, Y., and Matsumoto, A. 1996. Autocrine angiogenic vascular prosthesis with bone marrow transplantation. Nat. Med. 2:90–93.

Papapetropoulos, A., Garcia-Cardena, G., Madri, J.A., and Sessa, W.C. 1997. Nitric oxide production contributes to the angiogenic properties of vascular endothelial growth factor in human endothelial cells. J. Clin. Invest. 100:3131–3139.

Parenti, A., Morbidelli, L., Cui, X.L., and Douglas, J.G. 1998. Nitric oxide is an upstream signal of vascular endothelial growth factor-induced extracellular signal regulated kinase 1/2 activation in postcapillary endothelium. J. Biol. Chem. 273:4220–4226.

Rajanayagam, M.A., Shou, M., Thirumurti, V., Lazarous, D.F., Quyyumi, A.A., Goncalves, L., Stiber, J., Epstein, S.E., and Unger, E.F. 2000. Intracoronary basic fibroblast growth factor enhances myocardial collateral perfusion in dogs. J. Am. Coll. Cardiol. 35:519–526.

Ray, PS., Estrada-Hernandez, T., Sasaki, H., Zhu, L., and Maulik, N. 2000. Early effects of hypoxia/reoxygenation on VEGF, Ang-1, Ang-2 and their receptors in the rat myocardium: Implications for myocardial angiogenesis. Mol. Cell. Biochem. 213:145–153.

Sakakibara, Y., Nishimura, K., Tambara, K., Yamamoto, M., Lu, F., and Tabata, Y. 2002. Prevascularization with gelatin microspheres containing basic fibroblast growth factors enhances the benefits of cardiomyocyte transplantation. J. Thorac. Cardiovasc. Surg. 124:50–56.

Sasaki, H., Fukuda, S., Otani, H., Zhu, Li., Yamaura, G., Engelman, R.M., Das, D.K., and Maulik, N. 2002. Hypoxic preconditioning triggers myocardial angiogenesis: a novel approach to enhance contractile function reserve in rat with myocardial infarction. J. Mol. Cell. Cardiol. 34:335–348.

Sase, K., and Michel, T. 1997. Expression and regulation of endothelial nitric oxide synthase. Trends Cardiovasc. Med. 7:28–37.

Seghezzi, G., Patel, S., Ren, C.J., Gualandris, A., Pintucci, G. Robbins, E.S., Shapiro, R.L., Galoway, A.C., Rifkin, D.B., and Mignatti. P. 1998. Fibroblast growth factor-2 (FGF-2) induces vascular endothelial growth factor (VEGF) expression in the endothelial cells of forming capillaries: An autocrine mechanism contributing to angiogenesis. J. Cell Biol. 141:1659–1673.

Senger, D.R., Connolly, D.T., Van de Water, L., Feder, J., and Dvorak, H.F. 1990. Purification and NH_2-terminal amino acid sequence of guinea pig tumor-secreted vascular permeability factor. Cancer Res. 50:1774–1778.

Seymour, L.W., Shoaibi, M.A., Martin, A., Ahmed, A., Elvin, P., Kerr, D.J., and Wakelam, M.J.O. 1996. Vascular endothelial growth factor stimulates protein kinase C-dependent phospholipase D activity in endothelial cells. Lab. Invest. 75:427–437.

Shi, Q., Rafii, S., Wu, M.H., Wijelath, E.S., Yu, C., Ishida, A., Fujita, Y., Kothari, S., Mohle, R., Sauvage, L.R., Moore, M.A., Storb, R.F., and Hammond, W.P. 1998. Evidence for circulating bone marrow-derived endothelial cells. Blood 92:362–367.

Shima, D.T., Adamis, A.P., Ferrara, N., Yeo, K.T., Yeo, T.K., Allende, R., Folkman, J., and D'Amore, P.A. 1995. Hypoxic induction of endothelial cell growth factors in the retina: Identification and characterization of vascular endothelial growth factor (VEGF) as the mitogen. Mol. Med.1:182–193.

Shintani, S., Murohara, T., Ikeda, H., Ueno, T., Sasaki, K., Duan, J., and Imaizumi, T. 2001. Augmentation of postnatal neovascularization with autologous bone marrow transplantation. Circulation 103:897–903.

Shizukuda, Y., Iwamoto, T., Mallet, R.T., and Downey, H.F. 1993. Hypoxic preconditioning attenuates stunning caused by repeated coronary artery occlusion in dog heart. Cardiovas. Res. 27:559–564.

Shweiki, D., Itin, A., Soffer, D., and Keshet, E. 1992. Vascular endothelial growth factor induced by hypoxia may mediate hypoxia-initiated angiogenesis. Nature 359:843–845.

Silverstein, J., and Rifkin, D.B. 1987. Endothelial cell growth factors and the vessel wall. Semin. Thromb. Hemost. 13:504–513.

Simons, M., Annex, B.H., Laham, R.J., Kleiman, N., Henry, T., Dauerman, H., Udelson, J.E., Gervino, E.V., Pike, M., Whitehouse, M.J., Moon, M., and Chronos, N.A. 2002. Pharmacological treatment of coronary artery disease with recombinant fibroblast growth factor-2: double-blind, randomized, controlled clinical trial. Circulation 105:788–793.

Souroujon, M., and Mochly, R. 1998. Peptide modulators of protein-protein interactions in intracellular signaling. Nat. Biotechnol. 16:919–924.

Spyridopoulos, I., Brogi, E., Kearney, M., Sullivan, A.B., Cetrulo, C., Isner, J.M., and Losardo, D.W. 1997. Vascular endothelial growth factor inhibits endothelial cell apoptosis induced by tumor necrosis factor alpha balance between growth and death signals. J. Mol. Cell. Cardiol. 29:1321–1330.

Suzuki, K., Smolenski, R.T., and Jaykumar, J. 2000. Heat shock treatment enhances graft cell survival in skeletal myoblast transplantation to the heart. Circulation 102:III-216–III-221.

Suzuki, K., Murtuza, B., Smolenski, R., Sammut, I.A., Suzuki, N., Kaneda, Y., and Yacoub, M.H. 2001. Cell transplantation for the treatment of acute myocardial infarction using vascular endothelial growth factor-expression skeletal myoblasts. Circulation 104:I-207–I-212.

Svet-Moldavsky, G.J., and Chimishkyan, K.L. 1977. Tumour angiogenesis factor for revascularization in ischemia and myocardial infarction. Lancet 1:913.

Symes, J.F. Losordo, D.W., Vale, P.R., Lathi, K.G., Esakof, D.D., Mayskiy, M., and Isnr, J.M. 1999. Gene therapy with vascular endothelial growth factor for inoperable coronary artery disease. Ann. Thorac. Surg. 68:830–836.

Takahashi, T., Kalka, C., Masuda, H., Chen, D., Silver, M., Kearney, M., Magner, M., Isner, J.M., and Asahara, T. 1999. Ischemia- and cytokine-induced mobilization of bone marrow-derived endothelial progenitor cells for neovascularization. Nat. Med. 5:434–438.

Takeshita, S., Zheng, L.P., Brogi, E., Kearney, M., and Pu, L.Q. 1994. Therapeutic angiogenesis. A single intra arterial bolus of vascular endothelial growth factor augments revascularization in a rabbit ischemic hind limb model. J. Clin. Invest. 93:662–670.

Tateishi-Yuyama, E., Matsubara, H., Murohara, T., Ikeda, U., Shintani, S., Masaki, H., Amano, K., Kishimoto, Y., Yoshimoto, K., Akashi, H., Shimada, K., Iwasaka, T., and Imaizumi, T. 2002. Therapeutic Angiogenesis using Cell Transplantation (TACT) Study Investigators. Therapeutic angiogenesis for patients with limb ischaemia by autologous transplantation of bone-marrow cells: a pilot study and a randomised controlled trial. Lancet 360:427–435.

Tio, R.A., Tkebuchava, T., Scheuermann, T.H., Lebherz, C., Magner, M., Kearny, M., Esakof, D.D., Isner, J.M., and Symes, J.F. 1999. Intramyocardial gene therapy with naked DNA encoding vascular endothelial growth factor improves collateral flow to ischemic myocardium. Hum. Gene Ther. 10:2953–2960.

Tojo, N., Asakura, E., Koyama, M., Tanabe, T., and Nakamura, N. 1999. Effects of macrophage colony-stimulating factor (M-CSF) on protease production from monocyte, macrophage and foam cell *in vitro*: a possible mechanism for anti-atherosclerotic effect of M-CSF. Biochim. Biophys. Acta 1452:275–284.

Tomanek, R.J., Holifield, J.S., Reiter, R.S., Sandra, A., and Lin, J.J. 2002. Role of VEGF family members and receptors in coronary vessel formation. Dev. Dyn. 225:233–240.

Vale, P.R., Tkebuchava, T., Milliken, C.E., and Isner, J.M. 1999a. Percutaneous electromechanical mapping demonstrates efficacy of pVGI.1 (VEGF2) in an animal model of chronic myocardial ischemia. Circulation 100:I-512.

Vale, P.R., Losordo, D.W., Tkebuchava, T., Chen, D., Milliken, C.E., and Isner, J.M. 1999b. Catheter-based myocardial gene transfer utilizing nonfluoroscopic electromechanical left ventricular mapping. J. Am. Coll. Cardiol. 4:246–254.

Vale, P.R., Losordo, D.W., Milliken, C.E., Schatz, R.A., Fortuin, F.D., Esakof, D.D., Maysky, M., Symes, J.F., and Isner, J.M. 2000. Randomized, placebo-controlled clinical study of percutaneous catheter-based left ventricular endocardial gene transfer of VEGF-2 for myocardial angiogenesis in patients with chronic myocardial ischemia. Circulation102:II-563.

Van der Zee, R., Murohara, T., Luo, Z., Zollmann, F., Passeri, J., Lekutat, C., and Isner, J.M. 1997. Vascular endothelial growth factor/vascular permeability factor augments nitric oxide release from quiescent rabbit and human vascular endothelium. Circulation 95:1030–1037.

Waltenberger, J., Claesson-Welsh, L., Siegbahn, A., Shibuya, M., and Heldin, C.H. 1994. Different signal transduction properties of KDR and Flt-1, two receptors for vascular endothelial growth factor. J. Biol. Chem. 269: 26988–26995.

Wood, K.W., Sarnecki, C., Roberts, T.M., and Blenis, J. 1992. Ras mediates nerve growth factor receptor modulation of three signal transducing protein kinases: MAP kinases, Raf-1, and RSK. Cell 68:1041–1050.

Yanagisawa-Miwa, A., Uchida, Y., Nakamura, F., Tomaru, T., Kido, H., Kamijo, T., Sugimoto, T., Kaji, K., Utsuyama, M., and Kurashima, C. 1992. Salvage of infarcted myocardium by angiogenic action of basic fibroblast growth factor. Science 257:1401–1403.

Chapter 16
VEGF Signaling: A Therapeutic Target for Cardiovascular Disease

Dhanya Chandrasekharan Rajalakshmi, Aghila Rani Koippallil Gopalakrishnan, and Chandrasekharan Cheranellore Kartha

Abstract Vascular endothelial growth factor (VEGF) is one of the best-characterized peptide growth factors secreted by vascular endothelial cells. The cytokine regulates multiple biological functions in endothelial cells, including enhanced production of vasoactive mediators. The recognition that VEGF signaling pathways are critical in physiological angiogenesis has led to the concept that these pathways are suitable targets for therapeutic angiogenesis. This review focuses on VEGF signaling in the cardiovascular system, the role of VEGF in the pathogenesis of cardiovascular diseases such as atherosclerosis, ischemic heart disease, pulmonary hypertension, and vascular restenosis and discusses the therapeutic possibilities using angiogenesis modulators, which act at several sites in the VEGF signaling pathway. Current evidence from experimental studies suggests that many of the treatment strategies targeting VEGF signaling pathways have potential for clinical use in patients with cardiovascular diseases.

Introduction

Myocardial, cerebral, and peripheral limb ischemia resulting from diseases of blood vessels in the respective domains are among the major causes of morbidity and mortality in humans. A rational approach to mitigate the effects of ischemia is to stimulate growth of new collaterals. During the last decade, a variety of methods have been developed to induce angiogenesis. A breakthrough in therapeutic angiogenesis has been the discovery of vascular endothelial growth factor (VEGF) and the recognition that VEGF signaling represents a critical rate-limiting step in physiological angiogenesis (Folkman 1998; Ferrara et al. 2003).

This review focuses on recent evidence supporting the concept that VEGF signaling pathways are suitable targets for modulating angiogenesis linked to various cardiovascular diseases.

A. K. Srivastava, M. B. Anand-Srivastava (eds.) *Advances in Biochemistry in Health and Disease.* Signal Transduction in the Cardiovascular System in Health and Disease,

VEGF Biology

VEGF is one of the best-characterized peptide growth factors secreted by endothelial cells. A potent mitogen for micro- and macrovascular endothelial cells derived from arteries, veins, and lymphatics, VEGF stimulates capillary formation (Matsunaga et al. 2003). The cytokine also regulates multiple biological functions in endothelial cells, including enhanced production of vasoactive mediators, increased expression of components of the thrombolytic and coagulation pathways, and promotion of cell adhesion molecules such as VCAM-1 and ICAM-1 expression in endothelial cells.

The family of VEGF, a heparin-binding glycoprotein, encompasses seven members: VEGF-A, VEGF-B, VEGF-C, VEGF-D, VEGF-E, VEGF-F, and placental growth factor (PlGF) and three receptors for VEGF: VEGF receptor (VEGFR)-1 (Flt-1), VEGFR-2 (KDR/Flk-1), and VEGFR-3 (Flt-4) (Rosen 2005). All of these possess a common homology domain, which is composed of a cystein knot motif.

VEGF-A (or VEGF), the first discovered among them, is a 34- to 42-kDa dimeric disulfide-bound glycoprotein, which exists in seven homodimeric isoforms resulting from alternative splicing of eight exons of the VEGF gene (Hoeben et al. 2004). VEGF121 and VEGF165 are the most abundant isoforms. While VEGF121 diffuses freely in tissues, VEGF189 and half of VEGF165 bind to cell surface heparan sulfate proteoglycans in the extracellular matrix. Heparan sulfate proteoglycans are thus a major reservoir of VEGF, which can be mobilized by proteolysis (Tammela et al. 2005). VEGF165 is the principal effector of VEGF action (Rajagopalan et al. 2001).

The VEGF-B gene consists of seven exons, alternative splicing of which results in two isoforms, VEGF-B167 and VEGF-B186. The former (nonglycosylated) binds heparan sulfate proteoglycan and is sequestered in the extracellular matrix. VEGF-B186 (O-glycosylated) is freely diffusible.

Mature form of VEGF-C can induce mitogenesis, migration, and survival of endothelial cells of the lymphatic system. VEGF-D also promotes mitogenesis and lymphangiogenesis (Saharinen et al. 2004). VEGF-E is an Orf virus-encoded VEGF. PlGF is predominantly expressed in placenta, heart, and lungs (Persico et al. 1999).

Regulation of VEGF Gene Expression

Hypoxia or Oxygen Tension

Hypoxia plays a key role in the regulation of VEGF gene expression. Hypoxia inducible factor (HIF)-1 is the key mediator of hypoxia responses (Semenza 2002). Adenosine, which accumulates under hypoxic conditions, is involved in the induction of the VEGF gene which in turn increases VEGF mRNA levels through a protein kinase A-mediated pathway (Takagi et al. 1996).

Cytokines and Growth Factors

Several growth factors including EGF, TGF-α and β, IGF-1, FGF, and PDGF upregulate the expression of VEGF mRNA (Ferrara and Davis-Smyth 1997; Neufeld et al. 1999). Inflammatory cytokines, IL-1α and IL-6 also induce expression of VEGF in cell types such as synovial fibroblasts (Neufeld et al. 1999).

VEGF Signaling in Cardiovascular System

VEGF is responsible for inducing microvascular permeability, an important prerequisite for angiogenesis. Increased vascular permeability catalyzes plasma protein leakage and the deposition of fibrin in the extracellular matrix, which can serve as a scaffold for migrating endothelial cells. The migrated endothelial cells in turn express three VEGF receptors VEGFR-1, -2, and -3, which are transmembrane tyrosine kinases.

VEGF Receptors

VEGF signaling is initiated by the binding of the protein to its receptor on the cell surface. Three VEGFRs have been identified, namely, VEGFR-1 or fms-like tyrosine kinase (Flt-1), VEGFR-2 or kinase domain region (KDR/Flk-1), also referred to as fetal liver kinase, and VEGFR-3 or Flt-4. Each receptor has an extracellular domain, a single hydrophobic transmembrane region, and a consensus tyrosine kinase sequence in the cytoplasmic region within the catalytic domain (Hoeben et al. 2004). VEGFR-1 and -2 have seven extracellular immunoglobulin homology domains. But in VEGFR-3, the fifth immunoglobulin domain is cleaved on receptor processing into disulfide-linked subunits. VEGFR-1 and -2 mediate angiogenesis and VEGFR-3 is also involved in lymphangiogenesis. VEGFR-1 plays a regulatory role in monocyte/macrophage functions (Hiratsuka et al. 1998) whereby it mediates the VEGF-induced chemotactic responses. VEGFR1 also plays a significant part in the recruitment of hematopoietic stem cells (HSC) and endothelial progenitor cells (EPC) (Hattori et al. 2002).

VEGFR-1 binds VEGF-A, VEGF-B, and PlGF with high affinity. VEGFR-1 can heterodimerize with VEGFR-2 and this complex shows stronger signaling properties than VEGFR-1 or -2 homodimers (Huang et al. 2001).

VEGFR2 is produced in hematopoietic stem cells, megakaryocytes, platelets (Ziegler et al. 1999), and retinal progenitor cells (Yang and Cepko 1996). It binds VEGF-A, VEGF-C, VEGF-D, and VEGF-E. VEGF-E binds with high affinity to VEGFR-2. Binding affinity of VEGF to VEGFR-2 is lower than that to VEGFR-1, but VEGFR-2 is the primary receptor that transmits VEGF signals in endothelial cells (ECs) (Gille et al. 2001). VEGF-C and -D upregulate the expression of VEGFR-2 (Stacker et al. 2001).

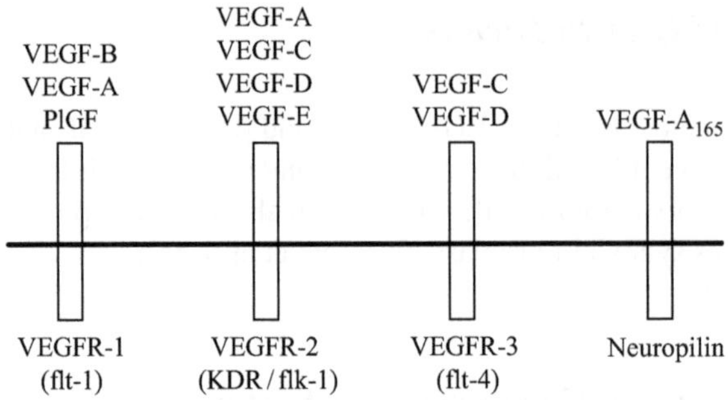

Fig. 16.1 The specificities of members of the VEGF family for their receptors.

During embryonic development, VEGFR-3 is mainly found in venous endothelium, but it becomes confined to lymphatic endothelial cells later. VEGFR-3 binds VEGF-C and D and does not bind A (Joukov et al. 1996).

There is another receptor called neuropilin-1 (NP-1), which is a receptor for VEGF165 and also recognizes VEGF-B and PlGF-2 (Soker et al. 1998; Makinen et al. 1999). NP-1 is expressed in the endocardium, coronary vessels, myocardial capillaries, and epicardial blood vessels. In endocardium and myocardial capillaries, it is co-expressed with VEGFR-1 and VEGFR-2. But in coronary vessels, it is co-expressed only with VEGFR-1 (Partanen et al. 1999). NP-1 regulates VEGF165 levels and is thought to be required for cardiovascular development (Kawasaki et al. 1999). The specificities of members of the VEGF family for their receptors are shown in Figure 16.1.

Downstream Signaling of VEGF and Its Biological Role

Binding of VEGF to its receptor results in receptor dimerization, followed by receptor transphosphorylation and autophosphorylation on the tyrosine residues. This activates the tyrosine kinase domain of the receptor and initiates various downstream signaling cascades that can reach the nucleus, thus modulating cellular proliferation and differentiation.

VEGF is a positive regulator of the cell cycle. VEGF stimulation induces the expression of genes for cell cycle regulators, such as cyclin and cyclin-dependent kinase (CDK). JNK (c-Jun N-terminal kinase) and molecules downstream from it affect VEGF-induced synthesis of cyclin D1, cdk4 activity, and G1/S cell cycling progression in ECs (Pedram et al. 1998). This is a result of upstream activation of SEK-1 (stress-activated protein/ERK kinase 1) and JNK through ERK stimulation. Studies in bovine aortic endothelial cells show that VEGF activates the STAT (STAT 1 and 6) pathway through VEGFR-2. STATs (tyrosine kinase-dependent transcription factors) are important modulators of cell growth responses in angiogenesis (Bartoli et al. 2000). In pathological conditions, activation of STATs in cardiac tissues has been observed (Negoro et al. 2000; Xuan et al. 2001). The activation of JAK-STAT signaling pathway is considered to activate cytoprotective genes and provide protection against ischemic stress (Negoro et al. 2001).

Angiogenic Mechanisms

Angiogenic mechanisms include ensuring EC survival, stimulating EC proliferation, and promoting EC migration.

Endothelial Cell Survival

VEGF activates endothelial cell survival mechanisms and antiapoptotic signaling, thereby maintaining endothelial integrity. VEGF inhibits apoptosis in human umbilical vein endothelial cells (HUVECs) by activating antiapoptotic kinase, Akt/PKB via PI3K-dependent pathway (Gerber et al. 1998b). Long-term effect is brought about by upregulating the antiapoptotic proteins Bcl-2 and A1 (Gerber et al. 1998a). This in turn upregulates two members of inhibitors of apoptosis (IAP), namely, survivin and X-chromosome-linked IAP (XIAP), thus inhibiting upstream caspases (Tran et al. 1999). The effect of VEGF on EC survival is mediated by increasing tyrosine phosphorylation of focal adhesion kinase (FAK) and focal adhesion association of FAK and paxillin, a FAK-associated protein. Studies indicate that Akt functions to promote cellular survival *in vivo*. The pathways involved in VEGF stimulation of cell survival mechanisms are depicted in Figure 16.2.

Endothelial Cell Proliferation

Binding of VEGF to VEGFR-2 mediates the DNA synthesis and proliferation in numerous cell types. Cell proliferation is mediated by VEGF through various signaling

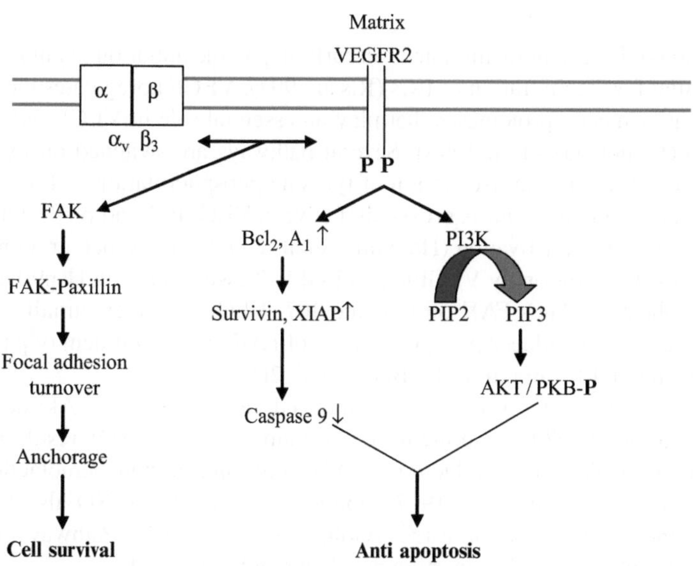

Fig. 16.2 Intracellular pathways involved in VEGF stimulation of cell survival mechanism.

pathways. One among them involves inducing the activity of extracellular signal-regulated kinases (ERKs) 1 and 2 that plays a central role in stimulating endothelial cell proliferation (Abedi and Zachary 1995).

VEGF stimulates PLCγ by phosphorylation of tyrosine residue. PLCγ converts phosphoinositol diphosphate (PIP2) to diacylglycerol (DAG) and inositol triphosphate (IP3), thereby activating protein kinase C (PKC). Activation of PKC is a major signaling event, which mediates VEGF-induced EC proliferation and angiogenesis. VEGF induces the translocation of calcium-dependent α and β_2 isoforms of PKC to plasma membrane (Xia et al. 1996). The mitogenic activity of VEGF is predominantly mediated by the PKC β_2 isoform (Harrington et al. 1997). Studies using PKC inhibitors indicate that PKC α and ξ isoforms are also involved in VEGF mitogen signaling (Xia et al. 1996; Higaki et al. 1999; Takahashi et al. 1999; Wellner et al. 1999). The isoform PKC δ retards cell cycle progression in microvascular endothelial cells. The reduced PKC δ activity contributes to VEGF-mediated cell proliferation. Mitogen activity of VEGF is also dependent on NO-mediated reduction in PKC δ activity (Shizukuda et al. 1999).

VEGFR-2 also associates with SH2 domain adapter proteins like ShC and Grb2, subsequently stimulating SOS, a guanine nucleotide exchange protein. This results in activation of Ras, followed by activation of the MAP kinase pathway leading to activation of c-Jun N-terminal protein kinase (Pedram et al. 1998). This signaling mediates cell proliferation. VEGFR-2 activates ERK cascade in a Ras-independent pathway also. This signaling is mediated by PKC (Takahashi et al. 1999; Gliki et al. 2001). Various signaling pathways involved in VEGF-induced EC proliferation are given in Figure 16.3.

Cell Migration

Degradation of basement membrane is an early step in the initiation of angiogenesis. It is essential for the migration of ECs (Risau 1997). VEGF upregulates the expression of matrix metalloproteinases that play an essential role in VEGF-induced EC migration (Lamoreaux et al. 1998). Several pathways are switched on by VEGF, resulting in cell migration. Activation of tyrosine phosphorylation of FAK at tyrosine 861 is one among them. A cross-talk between VEGFR-2 and $\alpha v\beta 3$ integrin is also involved in FAK activation (Hutchings et al. 2003). Heat shock protein Hsp90 facilitates FAK activation by VEGFR-2. VEGFR-2 associates with Hsp90 and subsequently phosphorylates FAK on tyrosine 407. VEGF activates initially GTPase RhoA, which in turn induces phosphorylation of FAK and recruitment of paxillin to FAK, resulting in EC migration (Le Boeuf et al. 2004).

Another pathway involved in VEGF-induced cell migration is Akt-dependent phosphorylation of eNOS at serine residue (Dimmeler et al. 2000), resulting in increased nitric oxide (NO) production. NO is implicated in nonchemotactic scalar movement of ECs (Noiri et al. 1997; Goligorsky et al. 1999). NO also regulates FAK tyrosine phosphorylation in ECs (Goligorsky et al. 1999). Pathways involved in VEGF stimulation of EC migration are shown in Figure 16.4.

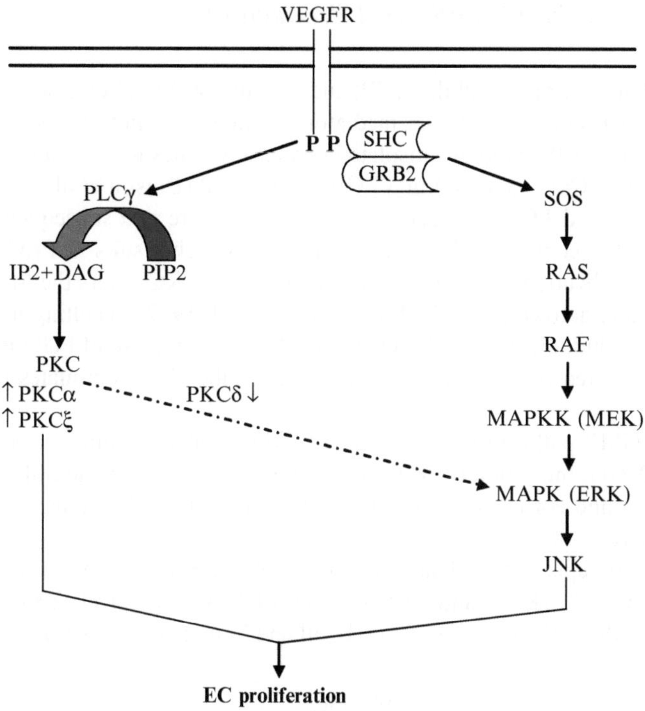

Fig. 16.3 Signaling pathways involved in VEGF-induced EC proliferation.

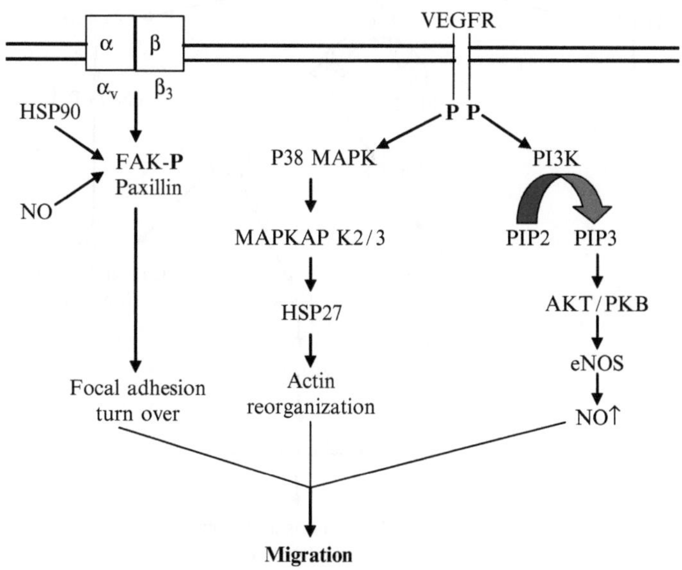

Fig. 16.4 Signaling pathways involved in VEGF-induced EC migration.

Role of VEGF in NO and PGI2 Production

VEGF mediates vasopermeability (VP) and vascular proliferation also, by stimulating the production of intercellular mediators, namely, NO and prostacyclin (PGI2) (Zachary et al. 2000). Studies in isolated coronary venules have confirmed the role of PLCγ, PKC, Ca^{2+}, and NO in VEGF-induced VP (Breslin et al. 2003). VEGF stimulates NO and PGI2 production by activating c-Src and subsequently PLCγ. Activated PLCγ converts PIP2 to DAG and IP3, which results in PKC activation and Ca^{2+} mobilization, respectively (He et al. 1999). PKC enters the MAPK pathway that finally activates PLA2 (Wheeler-Jones et al. 1997), resulting in the mobilization of arachidonic acid, and synthesis of PGI2. The release of PGI2 is mediated via Ca^{2+}-triggered pathway. An increase in cytosolic Ca^{2+} stimulates the release of PGI2.

eNOS–cGMP pathway regulates the VEGF-mediated endothelial permeability (Lal et al. 2001). Inhibition of NOS is found to block the activation of ERK1/2 by VEGF in porcine coronary venular ECs (Parenti et al. 1998), indicating that NO activates ERK.

NO also disrupts cytoskeletal protein complex formation and arrangement of actin fibers in ECs (Kroll and Waltenberger 1997; Lackey et al. 2000), resulting in dilation of the cells' tight junctions (Kroll and Waltenberger 1997).

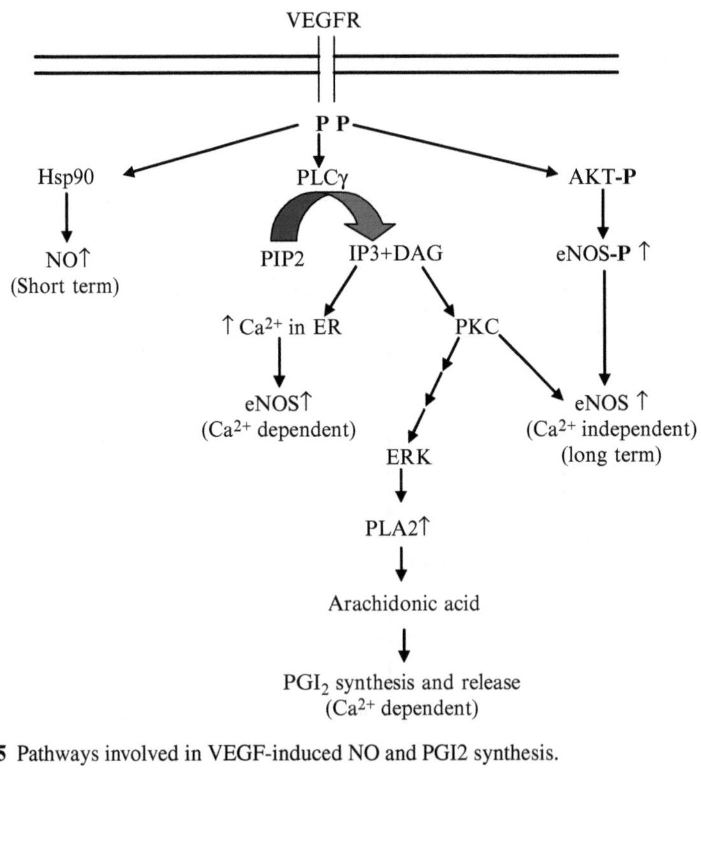

Fig. 16.5 Pathways involved in VEGF-induced NO and PGI2 synthesis.

VEGF induces both long- and short-term NO production. Short-term NO production is induced by the activation of constitutive eNOS isoforms through Ca^{2+} mobilization. NOS is also stimulated through the activation of Hsp90 or Hsp90-associated protein (Garcia-Cardena et al. 1998).

Long-term NO production is mediated by VEGF by the upregulation of eNOS mRNA and protein expression, which involves PKC (Dimmeler et al. 1999; Fulton et al. 1999; Shen et al. 1999). Pathways involved in VEGF-induced NO and PGI2 syntheses are shown in Figure 16.5.

Vascular Protective Actions of VEGF Mediated by NO and PGI2

NO and PGI2 produced by VEGF exhibit several other vascular protective effects. They inhibit vascular smooth muscle cell (VSMC) proliferation. The antiproliferative effect of NO and PGI2 on SMCs is mediated via the production of the intracellular messengers cGMP and cAMP, respectively (Asada et al. 1994; von der Leyen et al. 1995).

NO also inhibits leukocyte adhesion to the endothelium (von der Leyen et al. 1995). NO and PGI2 have antiplatelet actions as well. Thus, via NO synthesis, VEGF can provide protection against proatherogenic factors (Zachary 2001). The overall vascular protective action of VEGF mediated by NO and PGI2 is illustrated in Figure 16.6.

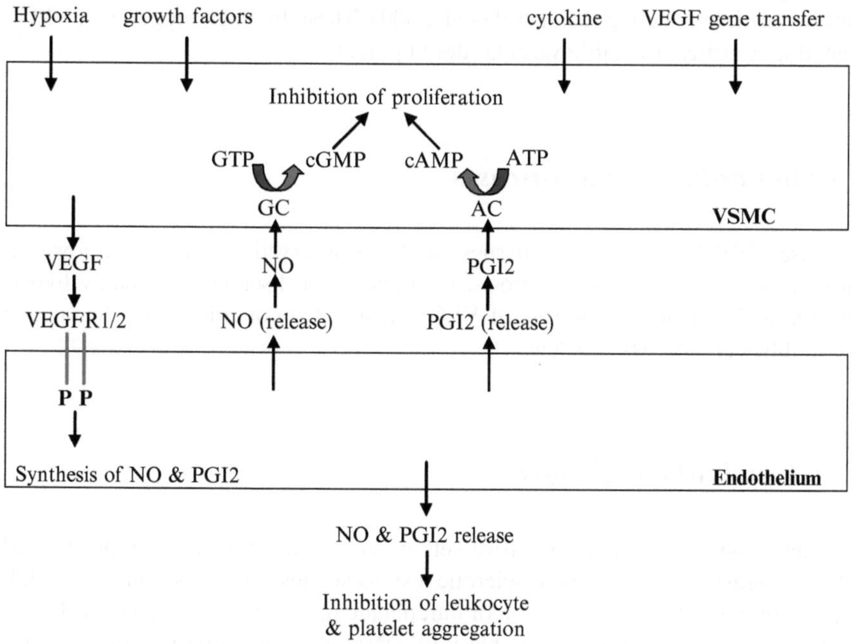

Fig. 16.6 Vascular protective action of VEGF mediated by NO and PGI2.

VEGF and the Cardiovascular System

Development of Heart and Vessels

VEGF plays a significant role in the development and replacement of the vascular system during embryonic growth. VEGF expression is strong at sites of active vasculogenesis and angiogenesis in embryos of mouse (Weinstein 1999). The expression of VEGF and its high-affinity binding receptors (Flt-1 and Flk-1) are critical for the development of embryonic vasculature. Flt-1 signaling pathway may regulate normal endothelial cell–cell or cell–matrix interactions during vascular development (Fong et al. 1995).

The levels of VEGF during development appear to be very critical and studies have shown that experimental mice lacking certain VEGF isoforms, die soon after birth and those that survive are prone to ischemic cardiomyopathy and multi-organ failure (Carmeliet et al. 1999a; Post and Waltenberger 2005). Mice expressing VEGF188 (homologue of human VEGF189) show impaired arteriolar development; a half of them die at birth (Stalmans et al. 2002). VEGF-B gene knockout mice deficient in VEGF-B have smaller heart and recover poorly from experimental myocardial infarction.

VEGF overexpression also produces similar results. The mutant embryos display several developmental anomalies consisting of an attenuated compact layer of myocardium, overproduction of trabeculae, defective ventricular septation, aberrant coronary development due to oversized epicardial vessels, and also abnormalities in outflow track remodeling (Miquerol et al. 2000). These findings suggest that VEGF signaling is critical for cardiovascular development.

Role in Cardiovascular Diseases

Increased VEGF expression in tissues has been observed in several cardiovascular diseases including atherosclerosis, ischemic heart diseases, pulmonary hypertension, and restenosis. The role of VEGF in selected conditions and therapeutic possibilities are discussed below.

VEGF and Atherosclerosis

The detection of intraplaque microvessels in advanced atherosclerotic plaque and VEGF expression in early atherosclerotic lesions suggests the importance of VEGF expression preceding the formation of intraplaque microvessels (Inoue et al. 1998).

Transfection of animals with plasmid DNA encoding murine soluble Flt-1 (sFlt-1) prevents the early vascular inflammation and late arteriosclerosis. Since sFlt-1 acts by blocking VEGF activity, the study indicates that VEGF mediates monocyte recruitment, necessary for the development of atherosclerosis.

Prolonged treatment with angiogenesis inhibitors such as angiostatin, endostatin, and fumagillin analogue TNP-40 reduces plaque growth and intimal neovascularization (Moulton et al. 1999). A reduced abundance in VEGF as a result of angiostatin administration accounts for the suppressed angiogenic activity (Moulton et al. 2003).

Celletti et al. (2001) have evaluated the impact of recombinant human VEGF (rhVEGF) administration on the progression and destabilization of atherosclerotic plaques. An increase in cross-sectional plaque area and circumferential plaque extension accompanied by increase in thickness was found in rhVEGF-treated apo E-deficient cholesterol-fed mice when compared to control mice, treated with albumin. This effect of rhVEGF is not species specific and has been observed in rabbits and in dogs as well.

VEGF and Myocardial Infarction

Animal studies reveal that ischemia induces the expression of VEGF mRNA in heart. VEGF-mediated angiogenesis is a compensatory response to ischemia. Studies by Kranz et al. (2000) revealed that 10 days after an acute myocardial infarction (AMI), VEGF-A levels in the serum reach a concentration capable of stimulating proliferation of ECs. Shinohara has studied the transcription and distribution of VEGF mRNA in normal human and myocardial infarcted hearts. In AMI, VEGF mRNA was detected in VSMC of arterioles as well as in macrophages that infiltrated the granulation tissues. Interestingly, in the normal heart, VEGF mRNA signals are only seen in the cytoplasm of cardiomyocytes and not in blood vessels (Shinohara 1994).

Within hours after ischemic injury, changes occur in the vasculature, which include alterations in vascular permeability and endothelial survival. VEGF binds to its receptor (Flk) and mediates Src kinase activity in the blood vessels. This regulates EC barrier function following AMI. Quiescent blood vessels contain a complex of Flk, VE-cadherin, and β-catenin (Carmeliet et al. 1999b; Shay-Salit et al. 2002; Zanetti et al. 2002). Upon VEGF binding to its receptor, Src is recruited to the receptor (Chou et al. 2002). Src kinase can phosphorylate E-cadherin, resulting in the dissociation of ECs from one another (Behrens et al. 1993). There is a transient disruption of the Flk–cadherin–catenin complex. This leads to dissociation of the junctional complex, thereby disrupting EC barrier function (Weis et al. 2004).

In the consecutive repair phase after AMI, VEGF directs angiogenesis and arteriogenesis resulting in the formation of collateral circulation (Garcia-Dorado and Oliveras 1993). Cells and molecules involved in ventricular remodeling after infarction influence VEGF expression. Angiotensin II expressed by myofibroblasts upregulates the expression of VEGF and Flt-1 (Chintalgattu et al. 2003). Endogenous monocyte chemoattractant protein (MCP-1) contributes to VEGF-mediated acute increase in VP in a dose-dependent manner (Yamada et al. 2003). *In vitro* studies show that anti-MCP-1 antibody significantly blocks VEGF-induced tubule formation.

Therapeutic effect of VEGF-induced angiogenesis in animal models of coronary ischemia has been extensively studied (Banai et al. 1994a; Harada et al. 1996). An increased collateral vessel development and increased blood flow to myocardium are seen after administration of VEGF or rhVEGF. Arteriogenesis can also be induced through signaling pathway mediated by VEGFR-1. Studies on experimental models of AMI revealed that PlGF induces the formation of vascular collaterals through VEGF signaling (Luttun et al. 2002; Autiero et al. 2003). When PlGF binds to VEGFR-1, it recruits monocytes and acts directly on ECs and SMCs and induces their growth. Studies on knockout mice of VEGF or receptors reveal the critical role played by VEGF in the development and formation of blood vessel networks (Fong et al. 1995; Shalaby et al. 1995).

In patients with ischemic syndromes, intra-arterial administration of rhVEGF significantly increases the development of collateral vessels (Isner and Asahara 1999), indicating its applicability in therapeutic angiogenesis. In patients who are not suitable for PTCA or CABG, administration of VEGF can be a novel treatment option.

An interesting finding is that Src blockade eliminates the VEGF-induced gaps and the physiological cascade which contributes to the severity of ischemic lesion in AMI (Weis et al. 2004). Blockade of Src kinase affects one of the several signaling pathways downstream of VEGF activation. It affects VEGF-induced Src and FAK phosphorylation without disrupting Flk or MAPK activation (Weis et al. 2004). This indicates that Src regulates VEGF- induced VP without affecting neovascularization (Eliceiri et al. 1999; Weis et al. 2004). A transient suppression of Src activity within several days after MI can reduce ischemia-induced heart injury without affecting VEGF-mediated revascularization thus preventing long-term myocardial damage. Therefore, pharmacological inhibition of Src and Yes (a member of SFK family) is possibly a therapeutic approach to provide protection to cardiac vessels and myocytes (Weis et al. 2004). An Src inhibitor, 4-amino-5-(4-chlorophenyl)-7-(t-butyl)pyrazolo [3,4-D] pyrimidine, inhibits Src-mediated migration and antiapoptosis (Abu-Ghazaleh et al. 2001).

VEGF in Ischemic Preconditioning

Studies by Kawata et al. (2001) showed that ischemic preconditioning (IP) enhances the expression of VEGF mRNA in ischemic cardiac myocytes and vascular ECs. They observed increase in capillary density in the infarcted area and a reduction in infarct size, 3 days after the onset of AMI preceded by IP. Several recent studies also provide evidence for an increased expression of VEGF triggered by IP (Fukuda et al. 2004). When rats with chronic MI were subjected to IP, an induction of survival factors including VEGF, Bcl-2, and survivin was observed (Fukuda et al. 2004).

Researchers suggest that co-administration or sequential gene therapy with VEGF, Bcl-2, and survivin can enhance myocardial collateral blood vessel function.

VEGF and Restenosis

The role of VEGF in restenosis is controversial. Studies by Ohtani et al. suggest that VEGF promotes restenosis (Ohtani et al. 2004). There are also reports that both exogenous and endogenous VEGF attenuate restenosis. The explanation for these paradoxical findings is that VEGF may have different effects depending on the extent of intimal injury (Hutter et al. 2004; Shiojima and Walsh 2004).

The role of VEGF in restenosis is also apparent in experimental studies after vascular intraluminal injury. Expression of VEGF and its receptors is upregulated with the development of neointimal formation in hypercholesterolemic mice after experimentally induced arterial injury (Zhao et al. 2004). sFlt-1 gene transfer attenuates neointimal formation after intraluminal injury in rabbits, rats, and mice. sFlt-1 gene transfer also inhibits the increased expression of VEGF (Ohtani et al. 2004), Flt-1, Flk-1, and inflammatory factors such as MCP-1. Mice with Flt-1 kinase deficiency exhibit a reduced neointimal formation, indicating the significant role played by VEGF and Flt-1 signals in the pathogenesis of neointimal formation (Zhao et al. 2004).

VEGF and Pulmonary Hypertension

VEGF plays a pivotal role in lung vascular growth in the embryo. Chronic intrauterine pulmonary hypertension markedly decreases lung VEGF expression (Grover et al. 2003). It has been reported that changes with selective inhibition of VEGF165 mimic the structural and physiological changes of experimental PPHN (persistent pulmonary hypertension of the newborn), suggesting a possibility that hypertension downregulates VEGF expression in the developing lung and that impaired VEGF signaling may contribute to the pathogenesis of PPHN. Inhibition of VEGF receptors in newborn rats decreases vascular growth and alveolarization and results in pulmonary hypertension (Tang et al. 2004).

Therapeutic Options

Discovery of the role of VEGF in normal cardiovascular development and in disease conditions has opened avenues to target the molecule for treatment of several cardiovascular diseases either to promote angiogenesis or to modulate vascular development. The strategies attempted target the VEGF signaling pathway or regulate VEGF expression in tissues.

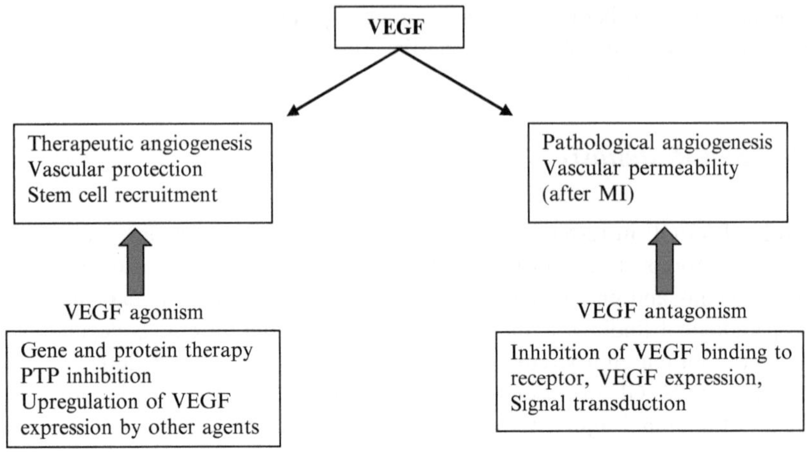

Fig. 16.7 Therapeutic strategies with the use of VEGF in cardiovascular diseases.

VEGF signaling can be targeted for therapy depending on the desired outcome. For angiogenesis and for promoting beneficial effects, VEGF agonists will be useful. To reduce vascular permeability and arrest the progression of atherosclerosis, VEGF antagonists can be applied. Several strategies have been developed for intervening the VEGF-mediated signal transduction pathways. These are shown in Figure 16.7.

VEGF Agonism

Several studies in animals have provided evidence for the beneficial effects of VEGF for therapeutic angiogenesis. Mould et al. observed an enhanced vascular growth in aortic explants from VEGF-B transgenic mice and in mice in which VEGF-B is overexpressed in ECs (Mould et al. 2005). Direct administration of adenoviral vector containing cDNA of VEGF165 induces coronary collateral vessel formation and improves regional myocardial perfusion and function in pigs (Zhang et al. 2002).

Several clinical trials for therapeutic angiogenesis have been conducted in patients with myocardial ischemia. Although an improvement in exercise time, angina class, and quality of life was observed in placebo and VEGF-delivered groups, there was no significant difference between treated patients and controls.

Gene therapy with VEGF is another strategy that is being used for clinical trials in humans. The VEGF gene can be transferred using adenoviral vector or as naked plasmid DNA, both of which produce only transient transfection (Hoeben et al. 2004). The administration of naked DNA encoding the VEGF-A165 gene comprises the largest phase I trial; an improvement in exercise time and time to angina was maintained 1 year after administration.

Another approach for inducing angiogenesis is the administration of a gene or a factor that is by itself not angiogenic, but is capable of stimulating the expression of angiogenic genes. Delivery of HIF-1α/VP16 (transactivated domain from herpes simplex virus VP16 protein) is found to promote improvement in perfusion of ischemic limb in rabbits (Vincent et al. 2000). Intramyocardial injection of naked DNA encoding HIF-1α/VP16 hybrid was found to reduce infarct size and enhance neovascularization in AMI model in rats (Shyu et al. 2002).

PTP Inhibitors

The first step in signal transduction of VEGF is the activation of RTK (receptors of VEGF). Studies reveal that RTK activation is downregulated by protein tyrosine phosphatases (PTPs). Numerous receptor type PTPs (HPTP-β and HPTP-γ, PTP-ε) are expressed in ECs. Retroviral overexpression of PTP-ε in HUVECs negatively affects survival and proliferation. Inhibitors of PTP-1B enhance VEGFR-2 mediated angiogenesis in Matrigel plug assay (Soeda et al. 2002). Therefore, an alternative approach to enhance collateral vessel development by activation of RTK is by blocking the action of PTPs (Ostman and Bohmer 2001; Zhang 2001). A group of scientists in Ohio found that a nonselective PTP inhibitor, bis (maltolato) oxovanadium IV (BMOV), enhanced the activation of VEGFR-2 and Tie2 *in vitro* in cultured ECs and enhanced VEGF- and angiopoietin-1-mediated cell survival. BMOV was also found to increase vessel sprouting in rat aortic ring explants, indicating that PTP inhibition can enhance angiogenesis in a multicellular system (Carr et al. 2004). Another interesting finding was that PTP inhibitor showed its efficacy *in vivo* even in the absence of exogenous growth factors. This effect may be mediated through inherent signals stimulated by changes in shear stress (Helisch and Schaper 2003; Wahlberg 2003). PTP inhibition and shear stress have been found to activate eNOS in a Ca^{2+}-dependent manner (Fleming et al. 1998), thereby promoting vascular remodeling and collateral blood flow. Therefore, PTPs that negatively regulate eRTK-mediated signaling must be targeted to enhance collateral development.

The use of PTP inhibitors as therapeutic agents is advantageous when compared to exogenous administration of VEGF. VEGF at high dose can produce side effects like vascular leakage, edema, inflammation, and angioma (Thurston et al. 2000; Masaki et al. 2002). The disadvantage of PTP inhibitors is that, even though low dose of BMOV promotes angiogenesis, high doses are found to inhibit vessel formation in aortic ring assays (Carr et al. 2004).

There are other agents that can regenerate the endothelium and thus can be employed for therapeutic angiogenesis. One of them is simvastatin, an HMG CoA reductase inhibitor. Recent studies in hamsters showed that treatment with simvastatin restored the carotid arterial blood flow after endothelial injury. Anti-VEGF antibody or anti-Flt-1 antibody reduced the simvastatin-induced endothelial healing, indicating that simvastatin promotes endothelial regeneration after vascular injury by overreleasing VEGF (Matsuno et al. 2004).

Luttun et al. (2002) studied the role of PlGF in stimulating angiogenesis and development of collateral vessels in the ischemic heart. When PlGF was systemically administered, an enhanced angiogenesis in ischemic myocardium and growth of collateral side branches in ischemic limb were observed. PlGF can heterodimerize with VEGF. It activates VEGFR-1 inducing phosphorylation of distinct tyrosine residue in tyrosine kinase domain of the receptor (Autiero et al. 2003). Thus, combined administration of PlGF and VEGF enhances the VEGF-driven angiogenesis (Park et al. 1994).

VEGF Antagonism

Inhibition of angiogenesis is a promising strategy for the treatment of pathological conditions associated with new blood vessel formation (Folkman 1995). Several approaches have been developed to antagonize the action of VEGF (Waltenberger 1997). Inhibiting the binding of the growth factor to its receptor is one among them. This can be achieved by using neutralizing antibodies against the growth factor. VEGF antagonism by the administration of mFlt(1-3)-IgG reduces edema and tissue damage after ischemic reperfusion injury in the mouse brain (van Bruggen et al. 1999).

Antisense oligonucleotides directed against mRNA of VEGF receptor (Flk-1 and Flt-1) inhibit VEGF-mediated angiogenesis (Marchand et al. 2002). Soluble ephrin EphA2-Fc receptor is also found to inhibit VEGF-induced EC survival, migration, and sprouting (Cheng et al. 2002).

As mentioned earlier, MCP-1 has a key role in mediating VEGF-induced vascular permeability and angiogenesis. Mutation studies by Yamada et al. indicate that the AP-1 binding site in the MCP-1 promoter region participate in VEGF-induced MCP-1 promoter activity and that AP-1 plays a key role in VEGF-induced MCP-1 expression and the subsequent monocyte and macrophage infiltration into the ECs of arterial walls (Yamada et al. 2003). AP-1 has been proposed to be a useful therapeutic target for VEGF-related cardiovascular diseases (Yamada et al. 2003).

Suramin, an antitrypanosomal agent, exerts an antiangiogenic action by inhibiting the action of VEGF. VEGF-induced mitogenic activity and chemotaxis of EC is inhibited by suramin in a dose-dependent manner. The drug inhibits VEGF-induced tyrosine phosphorylation of KDR *in vitro* by interacting with KDR and the drug thus represents an important treatment option (Waltenberger et al. 1996).

Jia et al. (2001) have identified a 12-amino-acid synthetic peptide derived from exon 6 (based on sequence of VEGF189) that inhibits VEGF binding to HUVECs and porcine aortic ECs. It also inhibits VEGF-stimulated ERK activation and prostacyclin production. The peptide blocks the interaction of VEGF with KDR and neuropilin-1.

A glutathione-S-transferase fusion protein corresponding to VEGF165 exon 7-encoded domain has been identified which inhibits the binding of VEGF165 to KDR/Flk-1 in HUVECs, thus antagonizing VEGF-mediated angiogenesis (Soker et al. 1997). Another inhibitor, Neovastat, a naturally occurring multifunctional antiangiogenic drug extracted from cartilage, is capable of blocking VEGF-dependent microvessel sprouting from Matrigel-embedded rat aortic rings and VEGF-induced

EC tubulogenesis *in vitro*. It acts by inhibiting the binding of VEGF to KDR and tyrosine phosphorylation of VEGFR-2 (Beliveau et al. 2002). HGF/NKF4 (hepatocyte growth factor containing N-terminal hairpin domain and four subsequent kringle domains of HGF) is another agent found to inhibit VEGF-induced angiogenesis in cultured ECs and *in vivo* rabbit models. Its action is by inhibiting phosphorylation of ERK and ets-1 expression.

PTK787/ZK222584 (1-[4-pyridyl methyl] phthalazine succinate) is a potent inhibitor of VEGF. It inhibits VEGF-induced autophosphorylation of KDR (Wood et al. 2000). Carboxymethyl dextran benzylamine (CMD B7), a heparinlike molecule, prevents VEGF165-induced KDR autophosphorylation, thus inhibiting the downstream signals. It interferes with the heparin binding to VEGF165 (Hamma-Kourbali et al. 2001).

For angiogenesis to occur, cell-to-cell contact between ECs in the parent vessel must first be lost to enhance the dynamic migration of ECs (Ausprunk and Folkman 1977). VE-cadherin, an EC-specific adhesion molecule, interacts with anchoring molecules like β-catenin in the cytoskeleton rendering intracellular signals (Aberle et al. 1996). VEGF stimulates the increased tyrosine phosphorylation of VE-cadherin, and this mechanism is involved in the EC migration and tubular formation. Modulating the phosphorylation of VE-cadherin can also therefore regulate angiogenesis. Targeting VE-cadherin with monoclonal antibodies (Corada et al. 2002) is found to impair remodeling and maturation of vascular network. However, the high cost of production, manufacturing difficulties, and immunogenic potential make the approach less advantageous (Lin et al. 2003).

Resveratrol

Resveratrol inhibits VEGF-induced angiogenesis by blocking Src-dependent tyrosine phosphorylation of VE-cadherin (Lin et al. 2003). Src enriches cell-to-cell contacts and is also capable of regulating the phosphorylation of cadherin and catenins (Papkoff 1997). Blockage of Src activity by resveratrol impairs tyrosine phosphorylation of VE-cadherin thereby impairing EC migration and tube formation (Lin et al. 2003). Studies in HUVECs show that VEGF-mediated Src activation and phosphorylation of VE-cadherin is dependent on reactive oxygen species (ROS). Resveratrol decreases the VEGF-induced generation of ROS (Lin et al. 2003). These findings have led to the suggestion that resveratrol usage may be a therapeutic modality to inhibit VEGF-induced angiogenesis.

Statins

Angiostatin (AS) and endostatin (ES) are antiangiogenic agents with low toxicity that inhibit the VEGF-mediated migration of primary human microvascular ECs.

AS and ES inhibit chemotaxis without affecting the signals which regulate the EC migration, proliferation, and survival (Eriksson et al. 2003). ES acts by transiently blocking the association of β-catenin and paxillin and binding to integrin, thereby disrupting the extracellular matrix interaction (Eriksson et al. 2003).

Statins are found to interfere with RhoA activation in ECs, important for VEGF-induced EC migration (van Nieuw Amerongen et al. 2000). Rho-kinase is upregulated in the arteriosclerotic coronary lesions and is possibly involved in constrictive remodeling and vasospastic activity of the arteriosclerotic coronary artery. Both of these activities of Rho-kinase can be reversed by long-term inhibition of the molecule. Rho-kinase for these reasons is regarded as a novel therapeutic target for arteriosclerotic vascular disease (Shimokawa et al. 2001).

There are several anticancer drugs that target VEGF. Antitumour agent topotecan (10-hydroxy-9-dimethylaminomethyl-(*S*)-camptothecin) is a water-soluble camptothecin analogue that inhibits angiogenesis. It acts by downregulating the PI3K-Akt signaling pathway (Nakashio et al. 2002). Antitumor agent apigenin (4′,5,7,-trihydroxyflavone), a common dietary flavonoid, inhibits HIF-1 and VEGF expression (Fang et al. 2005). Further studies are needed to know whether the drugs can also decrease vascular permeability and retard the progression of atherosclerosis.

VEGF and Endothelial Progenitor Cells

Studies by Asahara et al. (1997) and others (Shi et al. 1998) have demonstrated that bone marrow-derived endothelial progenitor cells (EPCs) in the peripheral blood may home to sites of neovascularization and differentiate into mature ECs *in situ*. EPCs thus are involved in endogenous repair mechanism as well as in ischemia-induced neovascularization. Mobilization of EPCs into the peripheral blood is regulated by several chemokines/cytokines that are released during endothelial injury or tissue ischemia. VEGF is one of the growth factors modulating EPC kinetics and differentiation (Asahara et al. 1999). Recently it has been reported that VEGF-induced EPC migration *in vitro* is mediated by both VEGFR-1 and VEGFR-2 receptors in contrast to VEGF-induced EC migration, wherein only VEGFR-1 is involved. Both VEGF receptors are necessary for the formation of functional vessels derived from exogenously administered human *ex vivo* expanded EPCs (Li et al. 2006).

A recent study by Yi et al. (2006) reported that EPCs combined with VEGF gene therapy represent a lucrative therapeutic option for endothelial regeneration and postnatal vasculogenesis.

Although the role of VEGF in enhancing EPC mobilization and differentiation is well established, the involved signaling pathways and mechanisms still remain elusive. Llevadot et al. (2001) in their studies on statins and EPCs have reported that the PI3K/Akt/eNOS signaling pathway mediates the proangiogenic effects of VEGF on EPCs *in vitro*. Statins and stromal cell-derived factor-1α (SDF-1α) are also reported to enhance the *in vivo* ischemia-induced vasculogenesis and angiogenesis by EPCs through the VEGF/eNOS-related pathway (Llevadot et al. 2001; Hiasa et al. 2004). A further elucidation of the more specific VEGF-induced signaling mechanisms may provide valuable therapeutic strategies to improve EPC function especially in patients with CAD.

Conclusion

The role of VEGF and its signaling pathways in the development and physiological processes of the cardiovascular system are well elucidated. Influence of the growth factor in the pathogenesis and/or progression of cardiovascular diseases is also recognized. Treatment strategies targeting these signaling pathways are, however, still in experimental stages. Be that as it may, several of them have potential for clinical use. Intensive studies are needed to translate all of the therapeutic possibilities for eventual use in patients.

References

Abedi, H., and Zachary, I. 1995. Signaling mechanisms in the regulation of vascular cell migration. *Cardiovasc. Res.* 30:544–556.

Aberle, H., Schwartz, H., and Kemler, R. 1996. Cadherin-catenin complex: protein interactions and their implications for cadherin function. *J. Cell. Biochem.* 61:514–523.

Abu-Ghazaleh, R., Kabir, J., Jia, H., Lobo, M., and Zachary, I. 2001. Src mediates stimulation by vascular endothelial growth factor of the phosphorylation of focal adhesion kinase at tyrosine 861, and migration and anti-apoptosis in endothelial cells. *Biochem. J.* 360:255–264.

Asada, Y., Kisanuki, A., Hatakeyama, K., Takahama, S., Koyama, T., et al. 1994. Inhibitory effects of prostacyclin analogue, TFC-132, on aortic neointimal thickening *in vivo* and smooth muscle cell proliferation *in vitro*. *Prostaglandins Leukotrienes Essent. Fatty Acids* 51:245–248.

Asahara, T., Murohara, T., Sullivan, A., Silver, M., van der Zee, R., et al. 1997. Isolation of putative progenitor endothelial cells for angiogenesis. *Science* 275:964–967.

Asahara, T., Takahashi, T., Masuda, H., Kalka, C., Chen, D., et al. 1999. VEGF contributes to postnatal neovascularization by mobilizing bone marrow-derived endothelial progenitor cells. *EMBO J.* 18:3964–3972.

Ausprunk, D. H., and Folkman, J. 1977. Migration and proliferation of endothelial cells in preformed and newly formed blood vessels during tumor angiogenesis. *Microvasc. Res.* 14:53–65.

Autiero, M., Waltenberger, J., Communi, D., Kranz, A., Moons L., et al. 2003. Role of PlGF in the intra- and intermolecular cross talk between the VEGF receptors Flt1 and Flk1. *Nat. Med.* 9:936–943.

Banai, S., Shweiki, D., Pinson, A., Chandra, M., Lazarovici, G., and Keshet, E. 1994a. Upregulation of vascular endothelial growth factor expression induced by myocardial ischaemia: implications for coronary angiogenesis. *Cardiovasc. Res.* 28:1176–1179.

Bartoli, M., Gu, X., Tsai, N. T., Venema, R. C., Brooks, S. E., et al. 2000. Vascular endothelial growth factor activates STAT proteins in aortic endothelial cells. *J. Biol. Chem.* 275:33189–33192.

Behrens, J., Vakaet, L., Friis, R., Winterhager, E., Van Roy, F., et al. 1993. Loss of epithelial differentiation and gain of invasiveness correlates with tyrosine phosphorylation of the E-cadherin/beta-catenin complex in cells transformed with a temperature-sensitive v-SRC gene. *J. Cell Biol.* 120:757–766.

Beliveau, R., Gingras, D., Kruger, E. A., Lamy, S., Sirois, P., et al. 2002. The antiangiogenic agent neovastat (AE-941) inhibits vascular endothelial growth factor-mediated biological effects. *Clin. Cancer Res.* 8:1242–1250.

Breslin, J. W., Pappas, P. J., Cerveira, J. J., Hobson, R. W., 2nd, and Duran, W. N. 2003. VEGF increases endothelial permeability by separate signaling pathways involving ERK-1/2 and nitric oxide. *Am. J. Physiol. Heart Circ. Physiol.* 284:H92–H100.

Carmeliet, P., Ng, Y.S., Nuyens, D., Theilmeier, G., Brusselmans, K., et al. 1999a. Impaired myocardial angiogenesis and ischemic cardiomyopathy in mice lacking the vascular endothelial growth factor isoforms VEGF164 and VEGF188. *Nat. Med.* 5:495–502.

Carmeliet, P., Lampugnani, M. G., Moons, L., Breviario, F., Compernolle, V., et al. 1999b. Targeted deficiency or cytosolic truncation of the VE-cadherin gene in mice impairs VEGF-mediated endothelial survival and angiogenesis. *Cell* 98:147–157.

Carr, A. N., Davis, M. G., Eby-Wilkens, E., Howard, B. W., Towne, B. A., et al. 2004. Tyrosine phosphatase inhibition augments collateral blood flow in a rat model of peripheral vascular disease. *Am. J. Physiol. Heart Circ. Physiol.* 287:H268–H276.

Celletti, F. L., Waugh, J. M., Amabile, P. G., Brendolan, A., Hilfiker, P. R., and Dake, M. D. 2001. Vascular endothelial growth factor enhances atherosclerotic plaque progression. *Nat. Med.* 7:425–429.

Cheng, N., Brantley, D. M., Liu, H., Lin, Q., Enriquez, M., et al. 2002. Blockade of EphA receptor tyrosine kinase activation inhibits vascular endothelial cell growth factor-induced angiogenesis. *Mol. Cancer Res.* 1:2–11.

Chintalgattu, V., Nair, D. M., and Katwa, L. C. 2003. Cardiac myofibroblasts: a novel source of vascular endothelial growth factor (VEGF) and its receptors Flt-1 and KDR. *J. Mol. Cell. Cardiol.* 35:277–286.

Chou, M. T., Wang, J., and Fujita, D. J. 2002. Src kinase becomes preferentially associated with the VEGFR, KDR/Flk-1, following VEGF stimulation of vascular endothelial cells. *BMC Biochem.* 3:32.

Corada, M., Zanetta, L., Orsenigo, F., Breviario, F., Lampugnani, M. G., et al. 2002. A monoclonal antibody to vascular endothelial-cadherin inhibits tumor angiogenesis without side effects on endothelial permeability. *Blood* 100:905–911.

Dimmeler, S., Fleming, I., Fisslthaler, B., Hermann, C., Busse, R., and Zeiher, A. M. 1999. Activation of nitric oxide synthase in endothelial cells by Akt-dependent phosphorylation. *Nature* 399:601–605.

Dimmeler, S., Dernbach, E., and Zeiher, A. M. 2000. Phosphorylation of the endothelial nitric oxide synthase at ser-1177 is required for VEGF-induced endothelial cell migration. *FEBS Lett.* 477:258–262.

Eliceiri, B. P., Paul, R., Schwartzberg, P. L., Hood, J. D., Leng, J., and Cheresh, D. A. 1999. Selective requirement for Src kinases during VEGF-induced angiogenesis and vascular permeability. *Mol. Cell.* 4:915–924.

Eriksson, K., Magnusson, P., Dixelius, J., Claesson-Welsh, L., and Cross, M. J. 2003. Angiostatin and endostatin inhibit endothelial cell migration in response to FGF and VEGF without interfering with specific intracellular signal transduction pathways. *FEBS Lett.* 536:19–24.

Fang, J., Xia, C., Cao, Z., Zheng, J. Z., Reed, E., and Jiang, B. H. 2005. Apigenin inhibits VEGF and HIF-1 expression via PI3K/AKT/p70S6K1 and HDM2/p53 pathways. *FASEB J.* 19:342–353.

Ferrara, N., and Davis-Smyth, T. 1997. The biology of vascular endothelial growth factor. *Endocr. Rev.* 18:4–25.

Ferrara, N., Gerber, H. P., and LeCouter, J. 2003. The biology of VEGF and its receptors. *Nat. Med.* 9:669–676.

Fleming, I., Bauersachs, J., Fisslthaler, B., and Busse, R. 1998. Ca2+-independent activation of the endothelial nitric oxide synthase in response to tyrosine phosphatase inhibitors and fluid shear stress. *Circ. Res.* 82:686–695.

Folkman, J. 1995. Seminars in Medicine of the Beth Israel Hospital, Boston. Clinical applications of research on angiogenesis. *N. Engl. J. Med.* 333:1757–1763.

Folkman, J. 1998. Angiogenic therapy of the human heart. *Circulation* 97:628–629.

Fong, G. H., Rossant, J., Gertsenstein, M., and Breitman, M. L. 1995. Role of the Flt-1 receptor tyrosine kinase in regulating the assembly of vascular endothelium. *Nature* 376:66–70.

Fukuda, S., Kaga, S., Sasaki, H., Zhan, L., Zhu, L., et al. 2004. Angiogenic signal triggered by ischemic stress induces myocardial repair in rat during chronic infarction. *J. Mol. Cell. Cardiol.* 36:547–559.

Fulton, D., Gratton, J. P., McCabe, T. J., Fontana, J., Fujio, Y., et al. 1999. Regulation of endothelium-derived nitric oxide production by the protein kinase Akt. *Nature* 399:597–601.

Garcia-Cardena, G., Fan, R., Shah, V., Sorrentino, R., Cirino, G., et al. 1998. Dynamic activation of endothelial nitric oxide synthase by Hsp90. *Nature* 392:821–824.

Garcia-Dorado, D., and Oliveras, J. 1993. Myocardial oedema: a preventable cause of reperfusion injury? *Cardiovasc. Res.* 27:1555–1563.

Gerber, H. P., Dixit, V., and Ferrara, N. 1998a. Vascular endothelial growth factor induces expression of the antiapoptotic proteins Bcl-2 and A1 in vascular endothelial cells. *J. Biol. Chem.* 273:13313–13316.

Gerber, H. P., McMurtrey, A., Kowalski, J., Yan, M., Keyt, B. A., et al. 1998b. Vascular endothelial growth factor regulates endothelial cell survival through the phosphatidylinositol 3′-kinase/Akt signal transduction pathway. Requirement for Flk-1/KDR activation. *J. Biol. Chem.* 273:30336–30343.

Gille, H., Kowalski, J., Li, B., LeCouter, J., Moffat, B., et al. 2001. Analysis of biological effects and signaling properties of Flt-1 (VEGFR-1) and KDR (VEGFR-2). A reassessment using novel receptor-specific vascular endothelial growth factor mutants. *J. Biol. Chem.* 276:3222–3230.

Gliki, G., Abu-Ghazaleh, R., Jezequel, S., Wheeler-Jones, C., and Zachary I. 2001. Vascular endothelial growth factor-induced prostacyclin production is mediated by a protein kinase C (PKC)-dependent activation of extracellular signal-regulated protein kinases 1 and 2 involving PKC-delta and by mobilization of intracellular Ca^{2+}. *Biochem. J.* 353:503–512.

Goligorsky, M. S., Abedi, H., Noiri, E., Takhtajan, A., Lense, S., et al. 1999. Nitric oxide modulation of focal adhesions in endothelial cells. *Am. J. Physiol.* 276:C1271–C1281.

Grover, T. R., Parker, T. A., Zengem, J. P., Markham, N. E., Kinsella, J. P., and Abman, S. H. 2003. Intrauterine hypertension decreases lung VEGF expression and VEGF inhibition causes pulmonary hypertension in the ovine fetus. *Am. J. Physiol. Lung Cell Mol. Physiol.* 284: L508–L517.

Hamma-Kourbali, Y., Vassy, R., Starzec, A., Le Meuth-Metzinger, V., Oudar, O., et al. 2001. Vascular endothelial growth factor 165 (VEGF(165)) activities are inhibited by carboxymethyl benzylamide dextran that competes for heparin binding to VEGF(165) and VEGF(165). KDR complexes. *J. Biol. Chem.* 276:39748–39754.

Harada, K., Friedman, M., Lopez, J. J., Wang, S. Y., Li, J., et al. 1996. Vascular endothelial growth factor administration in chronic myocardial ischemia. *Am. J. Physiol.* 270:H1791–H1802.

Harrington, E. O., Loffler, J., Nelson, P. R., Kent, K. C., Simons, M., and Ware, J. A. 1997. Enhancement of migration by protein kinase Cδ and inhibition of proliferation and cell cycle progression by protein kinase Cδ in capillary endothelial cells. *J. Biol. Chem.* 272:7390–7397.

Hattori, K., Heissig, B., Wu, Y., Dias, S., Tejada, R., et al. 2002. Placental growth factor reconstitutes hematopoiesis by recruiting VEGFR1(+) stem cells from bone-marrow microenvironment. *Nat. Med.* 8:841–849.

He, H., Venema, V. J., Gu, X., Venema, R. C., Marrero, M. B., and Caldwell, R. B. 1999. Vascular endothelial growth factor signals endothelial cell production of nitric oxide and prostacyclin through flk-1/KDR activation of c-Src. *J. Biol. Chem.* 274:25130–25135.

Helisch, A., and Schaper, W. 2003. Arteriogenesis: the development and growth of collateral arteries. *Microcirculation* 10:83–97.

Hiasa, K., Ishibashi, M., Ohtani, K., Inoue, S., Zhao, Q., et al. 2004. Gene transfer of stromal cell-derived factor-1alpha enhances ischemic vasculogenesis and angiogenesis via vascular endothelial growth factor/endothelial nitric oxide synthase-related pathway: next-generation chemokine therapy for therapeutic neovascularization. *Circulation* 109:2454–2461.

Higaki, T., Sawada, S., Kono, Y., Imamura, H., Tada, Y., et al. 1999. A role of protein kinase C in the regulation of cytosolic phospholipase A(2) in bradykinin-induced PGI(2) synthesis by human vascular endothelial cells. *Microvasc. Res.* 58:144–155.

Hiratsuka, S., Minowa, O., Kuno, J., Noda, T., and Shibuya, M. 1998. Flt-1 lacking the tyrosine kinase domain is sufficient for normal development and angiogenesis in mice. *Proc. Natl. Acad. Sci. USA* 95:9349–9354.

Hoeben, A., Landuyt, B., Highley, M. S., Wildiers, H., Van Oosterom, A. T., and De Bruijn, E.A. 2004. Vascular endothelial growth factor and angiogenesis. *Pharmacol. Rev.* 56:549–580.

Huang, K., Andersson, C., Roomans, G. M., Ito, N., and Claesson-Welsh, L. 2001. Signaling properties of VEGF receptor-1 and -2 homo- and heterodimers. *Int. J. Biochem. Cell Biol.* 33:315–324.

Hutchings, H., Ortega, N., and Plouet, J. 2003. Extracellular matrix-bound vascular endothelial growth factor promotes endothelial cell adhesion, migration, and survival through integrin ligation. *FASEB J.* 17:1520–1522.

Hutter, R., Carrick, F. E., Valdiviezo, C., Wolinsky, C., Rudge, J. S., et al. 2004. Vascular endothelial growth factor regulates reendothelialization and neointima formation in a mouse model of arterial injury. *Circulation* 110:2430–2435.

Inoue, M., Itoh, H., Ueda, M., Naruko, T., Kojima, A., et al. 1998. Vascular endothelial growth factor (VEGF) expression in human coronary atherosclerotic lesions: possible pathophysiological significance of VEGF in progression of atherosclerosis. *Circulation* 98:2108–2116.

Isner, J. M., and Asahara, T. 1999. Angiogenesis and vasculogenesis as therapeutic strategies for postnatal neovascularization. *J. Clin. Invest.* 103:1231–1236.

Jia, H., Jezequel, S., Lohr, M., Shaikh, S., Davis, D., et al. 2001. Peptides encoded by exon 6 of VEGF inhibit endothelial cell biological responses and angiogenesis induced by VEGF. *Biochem. Biophys. Res. Commun.* 283:164–173.

Joukov, V., Pajusola, K., Kaipainen, A., Chilov, D., Lahtinen, I., et al. 1996. A novel vascular endothelial growth factor, VEGF-C, is a ligand for the Flt4 (VEGFR-3) and KDR (VEGFR-2) receptor tyrosine kinases. *EMBO J.* 15:290–298.

Joukov, V., Sorsa, T., Kumar, V., Jeltsch, M., Claesson-Welsh, L., et al. 1997. Proteolytic processing regulates receptor specificity and activity of VEGF-C. *EMBO J.* 16:3898–3911.

Kawasaki, T., Kitsukawa, T., Bekku, Y., Matsuda, Y., Sanbo, M., et al. 1999. A requirement for neuropilin-1 in embryonic vessel formation. *Development* 126:4895–4902.

Kawata, H., Yoshida, K., Kawamoto, A., Kurioka, H., Takase, E., et al. 2001. Ischemic preconditioning upregulates vascular endothelial growth factor mRNA expression and neovascularization via nuclear translocation of protein kinase C epsilon in the rat ischemic myocardium. *Circ. Res.* 88:696–704.

Kranz, A., Rau, C., Kochs, M., and Waltenberger, J. 2000. Elevation of vascular endothelial growth factor-A serum levels following acute myocardial infarction. Evidence for its origin and functional significance. *J. Mol. Cell. Cardiol.* 32:65–72.

Kroll, J., and Waltenberger, J. 1997. The vascular endothelial growth factor receptor KDR activates multiple signal transduction pathways in porcine aortic endothelial cells. *J. Biol. Chem.* 272:32521–32527.

Lackey, K., Cory, M., Davis, R., Frye, S. V., Harris, P. A., et al. 2000. The discovery of potent cRaf1 kinase inhibitors. *Bioorg. Med. Chem. Lett.* 10:223–226.

Lal, B. K., Varma, S., Pappas, P. J., Hobson, R. W., and Duran, W. N. 2001. VEGF increases permeability of the endothelial cell monolayer by activation of PKB/akt, endothelial nitric-oxide synthase, and MAP kinase pathways. *Microvasc. Res.* 62:252–262.

Lamoreaux, W. J., Fitzgerald, M. E., Reiner, A., Hasty, K. A., and Charles, S. T. 1998. Vascular endothelial growth factor increases release of gelatinase A and decreases release of tissue inhibitor of metalloproteinases by microvascular endothelial cells *in vitro*. *Microvasc. Res.* 55:29–42.

Le Boeuf, F., Houle, F., and Huot, J. 2004. Regulation of vascular endothelial growth factor receptor 2-mediated phosphorylation of focal adhesion kinase by heat shock protein 90 and Src kinase activities. *J. Biol. Chem.* 279:39175–39185.

Li, B., Sharpe, E. E., Maupin, A. B., Teleron, A. A., Pyle, A. L., et al. 2006. VEGF and PlGF promote adult vasculogenesis by enhancing EPC recruitment and vessel formation at the site of tumor neovascularization. *FASEB J.* 20:1495–1497.

Lin, M. T., Yen, M. L., Lin, C. Y., and Kuo, M. L. 2003. Inhibition of vascular endothelial growth factor-induced angiogenesis by resveratrol through interruption of Src-dependent vascular endothelial cadherin tyrosine phosphorylation. *Mol. Pharmacol.* 64:1029–1036.

Llevadot, J., Murasawa, S., Kureishi, Y., Uchida, S., Masuda, H., et al. 2001. HMG-CoA reductase inhibitor mobilizes bone marrow-derived endothelial progenitor cells. *J. Clin. Invest.* 108:399–405.

Luttun, A., Tjwa, M., Moons, L., Wu, Y., Angelillo-Scherrer, A., et al. 2002. Revascularization of ischemic tissues by PlGF treatment, and inhibition of tumor angiogenesis, arthritis and atherosclerosis by anti-Flt1. *Nat. Med.* 8:831–840.

Makinen, T., Olofsson, B., Karpanen, T., Hellman, U., Soker, S., et al. 1999. Differential binding of vascular endothelial growth factor B splice and proteolytic isoforms to neuropilin-1. *J. Biol. Chem.* 274:21217–21222.

Marchand, G. S., Noiseux, N., Tanguay, J. F., and Sirois, M. G. 2002. Blockade of *in vivo* VEGF-mediated angiogenesis by antisense gene therapy: role of Flk-1 and Flt-1 receptors. *Am. J. Physiol. Heart Circ. Physiol.* 282:H194–H204.

Masaki, I., Yonemitsu, Y., Yamashita, A., Sata, S., Tanii, M., et al. 2002. Angiogenic gene therapy for experimental critical limb ischemia: acceleration of limb loss by overexpression of vascular endothelial growth factor 165 but not of fibroblast growth factor-2. *Circ. Res.* 90:966–973.

Matsunaga, T., Warltier, D. C., Tessmer, J., Weihrauch, D., Simons, M., and Chilian, W. M. 2003. Expression of VEGF and angiopoietins-1 and -2 during ischemia-induced coronary angiogenesis. *Am. J. Physiol. Heart Circ. Physiol.* 285:H352–H358.

Matsuno, H., Takei, M., Hayashi, H., Nakajima, K., Ishisaki, A., and Kozawa, O. 2004. Simvastatin enhances the regeneration of endothelial cells via VEGF secretion in injured arteries. *J. Cardiovasc. Pharmacol.* 43:333–340.

Miquerol, L., Langille, B. L., and Nagy, A. 2000. Embryonic development is disrupted by modest increases in vascular endothelial growth factor gene expression. *Development* 127:3941–3946.

Mould, A. W., Greco, S. A., Cahill, M. M., Tonks, I. D., Bellomo, D., et al. 2005. Transgenic over expression of vascular endothelial growth factor-B isoforms by endothelial cells potentiates postnatal vessel growth *in vivo* and *in vitro*. *Circ. Res.* 97:e60–e70.

Moulton, K. S., Heller, E., Konerding, M. A., Flynn, E., Palinski, W., and Folkman J. 1999. Angiogenesis inhibitors endostatin or TNP-470 reduce intimal neovascularization and plaque growth in apolipoprotein E-deficient mice. *Circulation* 99:1726–1732.

Moulton, K.S., Vakili, K., Zurakowski, D., Soliman, M., Butterfield, C., et al. 2003. Inhibition of plaque neovascularization reduces macrophage accumulation and progression of advanced atherosclerosis. *Proc. Natl. Acad. Sci. USA* 100:4736–4741.

Nakashio, A., Fujita, N., and Tsuruo, T. 2002. Topotecan inhibits VEGF- and bFGF-induced vascular endothelial cell migration via downregulation of the PI3K-Akt signaling pathway. *Int. J. Cancer* 98:36–41.

Negoro, S., Kunisada, K., Tone, E., Funamoto, M., Oh, H., et al. 2000. Activation of JAK/STAT pathway transduces cytoprotective signal in rat acute myocardial infarction. *Cardiovasc. Res.* 47:797–805.

Negoro, S., Kunisada, K., Fujio, Y., Funamoto, M., Darville, M. I., et al. 2001. Activation of signal transducer and activator of transcription 3 protects cardiomyocytes from hypoxia/reoxygenation-induced oxidative stress through the upregulation of manganese superoxide dismutase. *Circulation* 104:979–981.

Neufeld, G., Cohen, T., Gengrinovitch, S., and Poltorak, Z. 1999. Vascular endothelial growth factor (VEGF) and its receptors. *FASEB J.* 13:9–22.

Noiri, E., Hu, Y., Bahou, W. F., Keese, C. R., Giaever, I., and Goligorsky, M. S. 1997. Permissive role of nitric oxide in endothelin-induced migration of endothelial cells. *J. Biol. Chem.* 272:1747–1752.

Ohtani, K., Egashira, K., Hiasa, K., Zhao, Q., Kitamoto, S., et al. 2004. Blockade of vascular endothelial growth factor suppresses experimental restenosis after intraluminal injury by inhibiting recruitment of monocyte lineage cells. *Circulation* 110:2444–2452.

Ostman, A., and Bohmer, F. D. 2001. Regulation of receptor tyrosine kinase signaling by protein tyrosine phosphatases. *Trends Cell Biol.* 11:258–266.

Papkoff, J. 1997. Regulation of complexed and free catenin pools by distinct mechanisms. Differential effects of Wnt-1 and v-Src. *J. Biol. Chem.* 272:4536–4543.

Parenti, A., Morbidelli, L., Cui, X. L., Douglas, J. G., Hood, J. D., et al. 1998. Nitric oxide is an upstream signal of vascular endothelial growth factor-induced extracellular signal-regulated kinase1/2 activation in post capillary endothelium. *J. Biol. Chem.* 273:4220–4226.

Park, J. E., Chen, H. H., Winer, J., Houck, K. A., and Ferrara, N. 1994. Placenta growth factor. Potentiation of vascular endothelial growth factor bioactivity, *in vitro* and *in vivo*, and high affinity binding to Flt-1 but not to Flk-1/KDR. *J. Biol. Chem.* 269:25646–25654.

Partanen, T. A., Makinen, T., Arola, J., Suda, T., Weich, H. A., and Alitalo, K. 1999. Endothelial growth factor receptors in human fetal heart. *Circulation* 100:583–586.

Pedram, A., Razandi, M., and Levin, E. R. 1998. Extracellular signal-regulated protein kinase/Jun kinase cross-talk underlies vascular endothelial cell growth factor-induced endothelial cell proliferation. *J. Biol. Chem.* 273:26722–26728.

Persico, M. G., Vincenti, V., and DiPalma, T. 1999. Structure, expression and receptor-binding properties of placenta growth factor (PlGF). *Curr. Top. Microbiol. Immunol.* 237:31–40.

Post, M., and Waltenberger, J. 2005. Modulation of growth factor action in the cardiovascular system. *Cardiovasc. Res.* 65:547–549.

Rajagopalan, S., Shah, M., Luciano, A., Crystal, R., and Nabel, E. G. 2001. Adenovirus-mediated gene transfer of VEGF(121) improves lower-extremity endothelial function and flow reserve. *Circulation* 104:753–755.

Risau, W. 1997. Mechanisms of angiogenesis. *Nature* 386:671–674.

Rosen, L. S. 2005. VEGF-targeted therapy: therapeutic potential and recent advances. *Oncologist* 10:382–391.

Saharinen, P., Tammela, T., Karkkainen, M. J., and Alitalo, K. 2004. Lymphatic vasculature: development, molecular regulation and role in tumor metastasis and inflammation. *Trends Immunol.* 25:387–395.

Semenza, G. 2002. Signal transduction to hypoxia-inducible factor 1. *Biochem. Pharmacol.* 64:993–998.

Shalaby, F., Rossant, J., Yamaguchi, T. P., Gertsenstein, M., Wu, X. F., et al. 1995. Failure of blood-island formation and vasculogenesis in Flk-1-deficient mice. *Nature* 376:62–66.

Shay-Salit, A., Shushy, M., Wolfovitz, E., Yahav, H., Breviario, F., et al. 2002. VEGF receptor 2 and the adherens junction as a mechanical transducer in vascular endothelial cells. *Proc. Natl. Acad. Sci. USA* 99:9462–9467.

Shen, B. Q., Lee, D. Y., and Zioncheck, T. F. 1999. Vascular endothelial growth factor governs endothelial nitric-oxide synthase expression via a KDR/Flk-1 receptor and a protein kinase C signaling pathway. *J. Biol. Chem.* 274:33057–33063.

Shi, Q., Rafii, S., Wu, M. H., Wijelath, E. S., Yu, C., et al. 1998. Evidence for circulating bone marrow-derived endothelial cells. *Blood* 92:362–367.

Shimokawa, H., Morishige, K., Miyata, K., Kandabashi, T., Eto, Y., et al. 2001. Long-term inhibition of Rho-kinase induces a regression of arteriosclerotic coronary lesions in a porcine model *in vivo*. *Cardiovasc. Res.* 51:169–177.

Shinohara, K. 1994. The expression and the role of vascular endothelial growth factor (VEGF) in human normal and myocardial infarcted heart. *Hokkaido Igaku Zasshi* 69:978–993.

Shiojima, I., and Walsh, K. 2004. The role of vascular endothelial growth factor in restenosis: the controversy continues. *Circulation* 110:2283–2286.

Shizukuda, Y., Tang, S., Yokota, R., and Ware, J. A. 1999. Vascular endothelial growth factor-induced endothelial cell migration and proliferation depend on a nitric oxide-mediated decrease in protein kinase C delta activity. *Circ. Res.* 85:247–256.

Shyu, K. G., Wang, M. T., Wang, B. W., Chang, C. C., Leu, J. G., et al. 2002. Intramyocardial injection of naked DNA encoding HIF-1alpha/VP16 hybrid to enhance angiogenesis in an acute myocardial infarction model in the rat. *Cardiovasc. Res.* 54:576–583.

Soeda, S., Shimada, T., Koyanagi, S., Yokomatsu, T., Murano, T., et al. 2002. An attempt to promote neo-vascularization by employing a newly synthesized inhibitor of protein tyrosine phosphatase. *FEBS Lett.* 524:54–58.

Soker, S., Gollamudi-Payne, S., Fidder, H., Charmahelli, H., and Klagsbrun, M. 1997. Inhibition of vascular endothelial growth factor (VEGF)-induced endothelial cell proliferation by a peptide corresponding to the exon 7-encoded domain of VEGF165. *J. Biol. Chem.* 272:31582–31588.

Soker, S., Takashima, S., Miao, H. Q., Neufeld, G., and Klagsbrun, M. 1998. Neuropilin-1 is expressed by endothelial and tumor cells as an isoform-specific receptor for vascular endothelial growth factor. *Cell* 92:735–745.

Stacker, S. A., Caesar, C., Baldwin, M. E., Thornton, G. E., Williams, R. A., et al. 2001. VEGF-D promotes the metastatic spread of tumor cells via the lymphatics. *Nat. Med.* 7:186–191.

Stalmans, I., Ng, Y. S., Rohan, R., Fruttiger, M., Bouche, A., et al. 2002. Arteriolar and venular patterning in retinas of mice selectively expressing VEGF isoforms. *J. Clin. Invest.* 109:327–336.

Takagi, H., King, G. L., Robinson, G. S., Ferrara, N., and Aiello, L. P. 1996. Adenosine mediates hypoxic induction of vascular endothelial growth factor in retinal pericytes and endothelial cells. *Invest. Ophthalmol. Vis. Sci.* 37:2165–2176.

Takahashi, T., Ueno, H., and Shibuya, M. 1999. VEGF activates protein kinase C-dependent, but Ras-independent Raf-MEK-MAP kinase pathway for DNA synthesis in primary endothelial cells. *Oncogene* 18:2221–2230.

Tammela, T., Enholm, B., Alitalo, K., and Paavonen, K. 2005. The biology of vascular endothelial growth factors. *Cardiovasc. Res.* 65:550–563.

Tang, J. R., Markham, N. E., Lin, Y. J., McMurtry, I. F., Maxey, A., et al. 2004. Inhaled nitric oxide attenuates pulmonary hypertension and improves lung growth in infant rats after neonatal treatment with a VEGF receptor inhibitor. *Am. J. Physiol. Lung Cell Mol. Physiol.* 287:L344–L351.

Thurston, G., Rudge, J. S., Ioffe, E., Zhou, H., Ross, L., et al. 2000. Angiopoietin-1 protects the adult vasculature against plasma leakage. *Nat. Med.* 6:460–463.

Tran, J., Rak, J., Sheehan, C., Saibil, S. D., LaCasse, E., et al. 1999. Marked induction of the IAP family antiapoptotic proteins survivin and XIAP by VEGF in vascular endothelial cells. *Biochem. Biophys. Res. Commun.* 264:781–788.

van Bruggen, N., Thibodeaux, H., Palmer, J. T., Lee, W. P., Fu, L., et al. 1999. VEGF antagonism reduces edema formation and tissue damage after ischemia/reperfusion injury in the mouse brain. *J. Clin. Invest.* 104:1613–1620.

van Nieuw Amerongen, G. P., Vermeer, M. A., Negre-Aminou, P., Lankelma, J., Emeis, J. J., and van Hinsbergh, V. W. 2000. Simvastatin improves disturbed endothelial barrier function. *Circulation* 102:2803–2809.

Vincent, K. A., Shyu, K. G., Luo, Y., Magner, M., Tio, R. A., et al. 2000. Angiogenesis is induced in a rabbit model of hindlimb ischemia by naked DNA encoding an HIF-1alpha/VP16 hybrid transcription factor. *Circulation* 102:2255–2261.

von der Leyen, H. E., Gibbons, G. H., Morishita, R., Lewis, N. P, Zhang, L., et al. 1995. Gene therapy inhibiting neointimal vascular lesion: *in vivo* transfer of endothelial cell nitric oxide synthase gene. *Proc. Natl. Acad. Sci. USA* 92:1137–1141.

Wahlberg, E. 2003. Angiogenesis and arteriogenesis in limb ischemia. *J. Vasc. Surg.* 38:198–203.

Waltenberger, J. 1997. Modulation of growth factor action: implications for the treatment of cardiovascular diseases. *Circulation* 96:4083–4094.

Waltenberger, J., Mayr, U., Frank, H., and Hombach, V. 1996. Suramin is a potent inhibitor of vascular endothelial growth factor. A contribution to the molecular basis of its antiangiogenic action. *J. Mol. Cell. Cardiol.* 28:1523–1529.

Weinstein, B. M. 1999. What guides early embryonic blood vessel formation? *Dev. Dyn.* 215:2–11.

Weis, S., Shintani, S., Weber, A., Kirchmair, R., Wood, M., et al. 2004. Src blockade stabilizes a Flk/cadherin complex, reducing edema and tissue injury following myocardial infarction. *J. Clin. Invest.* 113:885–894.

Wellner, M., Maasch, C., Kupprion, C., Lindschau, C., Luft, F. C., et al. 1999. The proliferative effect of vascular endothelial growth factor requires protein kinase C-alpha and protein kinase C-zeta. *Arterioscler. Thromb. Vasc. Biol.* 19:178–185.

Wheeler-Jones, C., Abu-Ghazaleh, R., Cospedal, R., Houliston, R. A., Martin, J., and Zachary, I. 1997. Vascular endothelial growth factor stimulates prostacyclin production and activation of cytosolic phospholipase A2 in endothelial cells via p42/p44 mitogen-activated protein kinase. *FEBS Lett.* 420:28–32.

Wood, J. M., Bold, G., Buchdunger, E., Cozens, R., Ferrari, S., et al. 2000. PTK787/ZK 222584, a novel and potent inhibitor of vascular endothelial growth factor receptor tyrosine kinases, impairs vascular endothelial growth factor-induced responses and tumor growth after oral administration. *Cancer Res.* 60:2178–2189.

Xia, P., Aiello, L. P., Ishii, H., Jiang, Z. Y., Park, D. J., et al. 1996. Characterization of vascular endothelial growth factor's effect on the activation of protein kinase C, its isoforms, and endothelial cell growth. *J. Cell Sci.* 98:2018–2026.

Xuan, Y. T., Guo, Y., Han, H., Zhu, Y., and Bolli, R. 2001. An essential role of the JAK-STAT pathway in ischemic preconditioning. *Proc. Natl. Acad. Sci. USA* 98:9050–9055.

Yamada, M., Kim, S., Egashira, K., Takeya, M., Ikeda, T., et al. 2003. Molecular mechanism and role of endothelial monocyte chemoattractant protein-1 induction by vascular endothelial growth factor. *Arterioscler. Thromb. Vasc. Bioavail.* 23:1996–2001.

Yang, K., and Cepko, C. L. 1996. Flk-1, a receptor for vascular endothelial growth factor (VEGF), is expressed by retinal progenitor cells. *J. Neurosci.* 16:6089–6099.

Yi, C., Xia, W., Zheng, Y., Zhang, L., Shu, M., et al. 2006. Transplantation of endothelial progenitor cells transferred by vascular endothelial growth factor gene for vascular regeneration of ischemic flaps. *J. Surg. Res.* 135:100–106.

Zachary, I. 2001. Signaling mechanisms mediating vascular protective actions of vascular endothelial growth factor. *Am. J. Physiol. Cell Physiol.* 280:C1375–C1386.

Zachary, I., Mathur, A., Yla-Herttuala, S., and Martin, J. 2000. Vascular protection: A novel nonangiogenic cardiovascular role for vascular endothelial growth factor. *Arterioscler. Thromb. Vasc. Biol.* 20:1512–1520.

Zanetti, A., Lampugnani, M. G., Balconim, G., Breviariom, F., Corada, M., et al. 2002. Vascular endothelial growth factor induces SHC association with vascular endothelial cadherin: a potential feedback mechanism to control vascular endothelial growth factor receptor-2 signaling. *Arterioscler. Thromb. Vasc. Biol.* 22:617–622.

Zhang, D., Gai, L., Fan, R., Dong, W., and Wen, Y. 2002. Efficacy and safety of therapeutic angiogenesis from direct myocardial administration of an adenoviral vector expressing vascular endothelial growth factor 165. *Chin. Med. J. (Engl.)* 115:643–648.

Zhang, Z. Y. 2001. Protein tyrosine phosphatases: prospects for therapeutics. *Curr. Opin. Chem. Biol.* 5:416–423.

Zhao, Q., Egashira, K., Hiasa, K., Ishibashi, M., Inoue, S., et al. 2004. Essential role of vascular endothelial growth factor and Flt-1 signals in neointimal formation after periadventitial injury. *Arterioscler. Thromb. Vasc. Biol.* 24:2284–2289.

Ziegler, B. L., Valtieri, M., Porada, G. A., De Maria, R., Muller, R., et al. 1999. KDR receptor: a key marker defining hematopoietic stem cells. *Science* 285:1553–1558.

Chapter 17
Myristoyl-CoA:Protein *N*-Myristoyltransferase and Myristoyl-CoA Binding Protein from Bovine Cardiac Muscle

Ponniah Selvakumar and Rajendra K. Sharma

Abstract Protein myristoylation refers to the cotranslational addition of a myristoyl group to an amino-terminal glycine residue of a protein by the enzyme *N*-myristoyltransferase (NMT). The myristoylation reaction depends on the availability of the cellular pools of coenzymc A and myristate and their subsequent formation of myristoyl-CoA, the substrate of NMT. In this review, we discuss NMT and myristoyl-CoA binding protein from bovine cardiac muscle which was carried out in our laboratory.

Introduction

Myristoyl-CoA:protein *N*-myristoyltransferase (NMT; EC 2.1.3.97) catalyzes the cotranslational transfer of the C14:0 fatty acid from myristoyl-CoA to the N-terminal glycine of a variety of cellular proteins (Selvakumar et al. 2006, 2002; Farazi et al. 2002; Rajala et al. 2000; Resh 1999; Boutin 1997). The incorporation of myristate into the N-terminus of a number of oncogenic, viral, and eukaryotic gene products is extremely important for the function of these proteins (Selvakumar et al. 2006, 2002; Farazi et al. 2002; Rajala et al. 2000; Resh 1999; Boutin 1997). Biochemical evidence has indicated the presence of several distinct NMTs *in vivo*, often varying in either apparent molecular weight and/or subcellular distribution. King and Sharma (1992) provided the first evidence for the existence of multiple forms of bovine brain NMT. In addition, McIlhinney et al. (1993) identified two forms of NMT in bovine brain cortex. Subsequently, Glover and Felsted (1995) showed that bovine brain exists as a heterogeneous mixture of NMT subunits. Two types of NMT, NMT1 and NMT2, were identified and cloned (Rundle et al. 2002; Giang and Cravatt 1998). Comparisons between the NMT1 and NMT2 proteins revealed reduced levels of sequence identity (76–77%), indicating that NMT1 and NMT2 comprise two distinct families of NMTs (Giang and Cravatt 1998). These studies demonstrate a heretofore-unappreciated level of genetic complexity underlying the

A. K. Srivastava, M. B. Anand-Srivastava (eds.) *Advances in Biochemistry in Health and Disease.* Signal Transduction in the Cardiovascular System in Health and Disease,

enzymology of N-terminal myristoylation. It also suggests that the specific inhibition or regulation of either NMT *in vivo* may in turn allow for the selective control of particular myristoylation-dependent cellular functions.

The purpose of this review is to summarize the investigations of NMT and its binding protein from bovine cardiac muscle and its possible involvement in various signaling which have been carried out in our laboratory.

Cloning and Characterization of NMT from Bovine Cardiac Muscle

Previously, we reported that there is a low level of NMT activity in bovine cardiac muscle (Raju and Sharma 1997). Therefore, we are interested in determining whether the low activity in cardiac muscle could be due to the low NMT gene expression or whether it could be due to the presence of inhibitor(s) of the enzyme activity of the expressed proteins. We cloned and characterized the NMT gene from bovine cardiac muscle (Raju et al. 1998). The NMT gene contains an open reading frame of 1248 bp and codes for 416 amino acids with apparent molecular mass of 46,686 Da. At the nucleotide level, cardiac NMT has 99.9 and 92.2% homology to spleen NMT (Raju et al. 1997) and human NMT (Duronio et al. 1992), respectively. Although the N-termini are diverged in mammalian and yeast NMTs, these sequences are not required for catalysis (Raju et al. 1997; Ntwasa et al. 1997; Zhang et al. 1996; Duronio et al. 1992). It has been shown that the N-terminal domain is involved in targeting NMT to the ribosomal subcellular fraction (Glover et al. 1997).

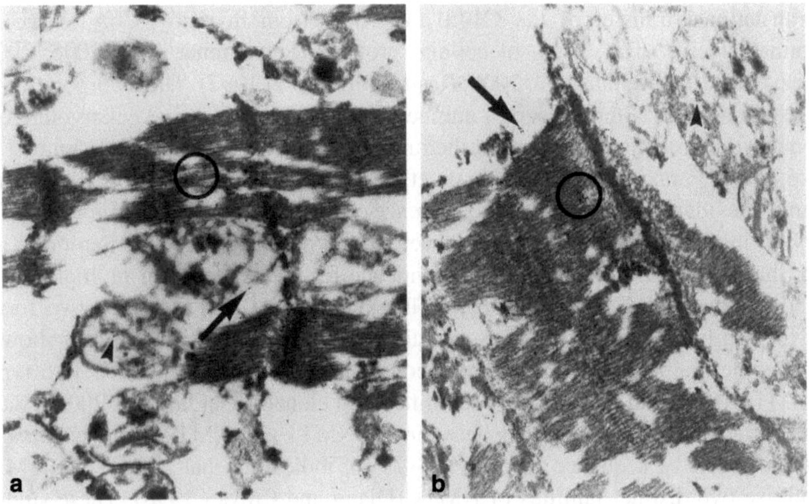

Fig. 17.1 Ultrastructural localization of NMT in human cardiac muscle by electron microscopy. For details see Raju et al. (1998).

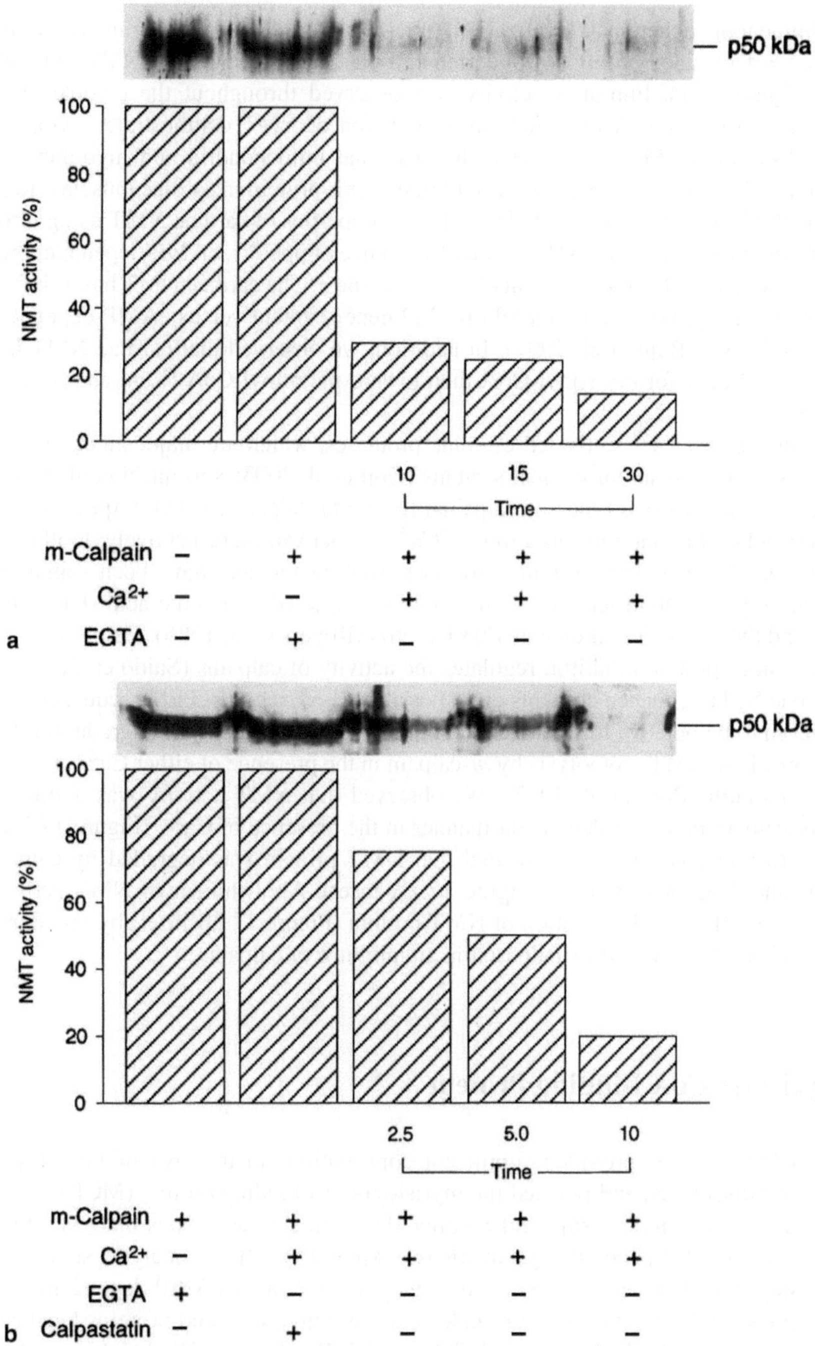

Fig. 17.2 (A) Effect of *m*-calpain on the activity and protein degradation of cardiac NMT. (B) Effect of calpastatin on cardiac NMT. For details see Raju et al. (1998).

Immunohistochemical localization of NMT from autopsy specimens of human heart revealed strong to moderate cytoplasmic staining with NMT antibodies (Figure 17.1). Immunoreactivity was observed throughout the cytoplasm of the myocardial cells. Ultrastructural localization of NMT demonstrated cytoplasmic distribution (Figure 17.1A) with occasional mitochondria and myofilaments (Figure 17.1B) suggesting an important role of this protein in cardiac muscles (Raju et al. 1997). We also investigated the kinetic properties of cardiac NMT using three peptides derived from the NH_2-terminal sequence of $pp60^{src}$, cAMP-dependent protein kinase, and M2 gene segment of reovirus type 3. The cardiac NMT has a sixfold lower K_m for $pp60^{src}$ and threefold for M2 gene segment versus cAMP-dependent protein kinase (Raju et al. 1997). In addition, we observed that cardiac NMT has a higher affinity for myristoyl-CoA than toward palmitoyl-CoA (Raju and Sharma 1997).

Calpains are Ca^{2+}-activated cysteine proteases, which are major mediators for Ca^{2+} signals in many biological systems (Goll et al. 2003; Sorimachi et al. 1997). There are at least two types of calpains, μ- and m-calpains, which require micromolar and millimolar concentrations of Ca^{2+} for activation, respectively (Goll et al. 2003). Ca^{2+}-dependent cysteine proteases, such as calpain, have been suggested to cleave polypeptide sequences enriched in proline (P), glutamic acid (E), serine (S), and threonine (T), known as PEST motifs (Rogers et al. 1986). Calpastatin, an endogenous protein inhibitor, regulates the activity of calpains (Saido et al. 1994). Cardiac NMT sequence contains eight possible PEST regions. PEST sequences are believed to be putative intramolecular signals for rapid proteolytic degradation. We examined the NMT proteolysis by m-calpain in the presence of either Ca^{2+}, EGTA, or calpastatin (Raju et al. 1997). We observed that NMT activity was abolished by m-calpain in a time-dependent manner in the presence of Ca^{2+} (Figure 17.2A). Western blot analysis also indicated that NMT protein was degraded by calpain in a time-dependent manner (Figure 17.2A inset). Abolishment of NMT activity (Figure 17.2B) and degradation of NMT protein (Figure 17.2B inset) by m-calpain was efficiently inhibited by calpastatin, an inhibitor of calpains.

Myristoyl-CoA Binding Protein

Our laboratory has provided significant contributions in the area of myristoylation. We discovered and purified the myristoyl-CoA binding protein (MCBP) from bovine cardiac muscle (Raju and Sharma 1997). In cardiac tissues there is a high level of cAMP-dependent protein kinase expression whose catalytic subunit is myristoylated (Carr et al. 1982). The catalytic subunit of cAMP-dependent protein kinase and the beta subunit of calcineurin are myristoylated proteins localized in the cytoplasm (Selvakumar et al. 2006, 2002; Rajala et al. 2000; Johnson et al. 1994; Carr et al. 1982; Aitken et al. 1982). Recently it has been shown that dephosphorylation of the catalytic subunit of myristoylated and nonmyristoylated cAMP-dependent protein kinase at Thr-197 by cellular protein phosphatase and protein

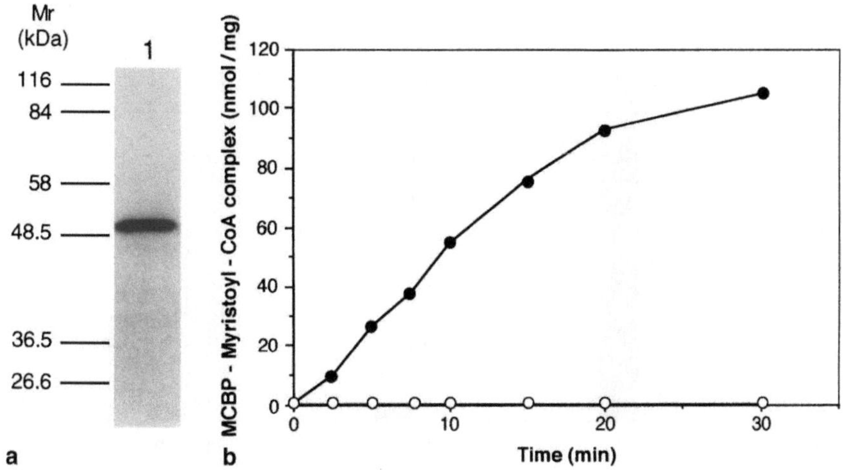

Fig. 17.3 (A) Analysis of purified MCBP by SDS-polyacrylamide gel electrophoresis. (B) Time course of acylation of protein-myristoyl-CoA complex. For details see Raju et al. (1997).

phosphatase-2A (PP-2A) indicated that myristoylated C subunit was more resistant to dephosphorylation than the nonmyristoylated enzyme (Liauw and Steinberg 1996). The substrate of NMT, myristoyl-CoA, is generated from the cellular pools of coenzyme A and myristate by the action of the enzyme acyl-CoA synthetase (Johnson et al. 1994). The myristoylation reaction is controlled by the availability of myristoyl-CoA in the cellular pools. There could be some mechanism by which the myristoyl-CoA pools can be regulated such as the presence of MCBPs. However, such a system has not been explored so far. Previously our laboratory identified, purified, and characterized a membrane-bound NMT inhibitor protein (NIP71) from bovine brain and suggested that this inhibitor protein could be a regulator of several mammalian NMT activities (King and Sharma 1993). Recently we identified the NIP71 has a homologue of heat shock cognate protein 70 (HSC70) (Selvakumar et al. 2006, 2005, 2004).

The purification of MCBP is simple, rapid, and applicable to large-scale preparations (Figure 17.3A) (Raju and Sharma 1997). Except for the cytosol, NMT activity in the various fractions obtained during purification was the same in the presence and absence of peptide substrates, suggesting that an endogenous protein in these fractions is myristoylated. Purified protein was incubated with radiolabeled myristoyl-CoA or myristic acid and activity assayed in the absence of peptide substrates. Figure 17.3B shows that there is a linear, time-dependent incorporation of radiolabeled myristoyl-CoA into the protein suggesting the formation of an acyl-protein complex. Substitution of [^{3}H]palmitoyl-CoA as the acyl donor resulted in less than 7% of the total radioactivity incorporated into the protein as compared to myristoyl-CoA (Raju and Sharma 1997). A higher concentration of CoA (2500-fold) is required for the complete inhibition of myristoyl-CoA binding to the protein. However, myristic acid did not serve as an acyl donor for the protein (Raju and Sharma 1997). However, the purified homogeneous 50-kDa protein

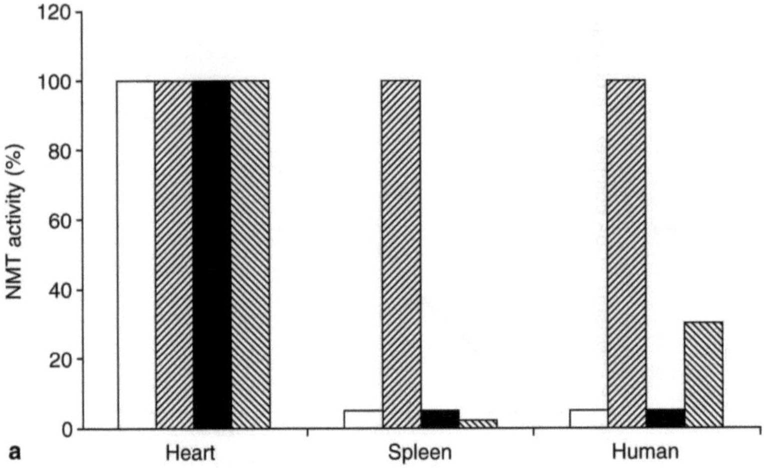

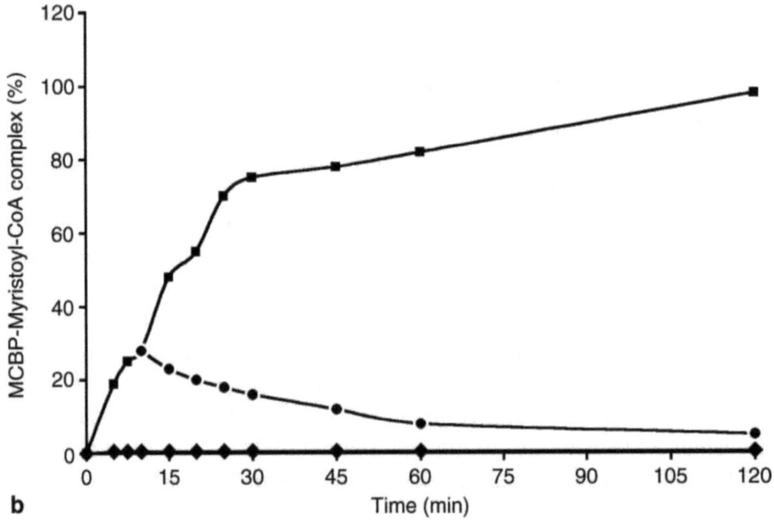

Fig. 17.4 (A) Inhibition of MCBP by NIP71. (B) Deacylation of protein-myristoyl-CoA complex. For details see Raju et al. (1997).

incorporated radiolabeled myristoyl-CoA in the absence of the peptide substrate in a time-dependent manner, suggesting the formation of a high-affinity protein–myristoyl-CoA complex (Raju and Sharma 1997). This complex formation was not observed in the cytosolic fraction. The association of [l-^{14}C]myristoyl-CoA with protein was sensitive to hydroxylamine, pH 7.0, consistent with acyl-protein linkage via an ester bond (Raju and Sharma 1997). These studies suggested that there exists an MCBP in cardiac muscle.

We examined whether the protein–myristoyl-CoA complex could be inhibited by NIP71. The results indicated that NIP71 did not inhibit the formation of complexation between MCBP and myristoyl-CoA (Figure 17.4A). On the other hand,

NMTs from bovine spleen and human were found to be completely inhibited (Figure 17.4A). The lack of effect of NIP71 on bovine cardiac muscle protein suggested that the protein–myristoyl CoA complex formation cannot be inhibited by NIP71. Furthermore, these results suggested that NIP71 could be competing for the peptide binding site or causing the inhibition of myristoyl-CoA–NMT–peptide ternary complex formation. Previous studies also indicated that NIP71 is not complexed with myristoyl-CoA (King and Sharma 1993). Our study shows that the binding protein is directly complexed with the myristoyl-CoA which was insensitive to NIP71 (Selvakumar et al. 2006; Raju and Sharma 1997).

Purified MCBP was incubated with [l-^{14}C]myristoyl-CoA in separate tubes and aliquots were removed at selected times. At time 10 min, cytosolic fraction was added to one tube and buffer was added to the other. Aliquots were removed at selected times up to 120 min. The vial which did not contain added cytosol exhibited the formation of a protein–myristoyl-CoA complex, while the tube which contained added cytosol exhibited deacylation (Figure 17.4B). These results suggest that the cytosolic fraction may contain thioesterases/proteinases which could modulate the acylation reaction *in vivo* (Raju and Sharma 1997). The absence of acyl-complex formation in the cytosol could be due to the presence of either esterases or proteinases. It has been reported that porcine phospholipase A contained thioesterase and deacylase activities (Nocito et al. 1996).

The physiological consequences of acylation and deacylation of proteins have been shown to be involved in regulating the activity and cellular localization of many proteins (Nocito et al. 1996). The deacylation process has been involved in the fusion between viral and cellular membranes (Schmidt 1982) and in the release of virus particles from infected cells (Schmidt 1983) showing a decrease in both activities when proteins are deacylated by hydroxylamine (Lambrecht and Schmidt 1986). Our studies indicated that the myristoylation reaction in cardiac muscle could be regulated by the relative distribution of NMT, myristoyl-CoA, coenzyme A, MCBP, acyl-synthetases(s), inhibitors, activators, etc. (Raju and Sharma 1997). The regulation of acyl-protein complexes in signal transduction systems about the specific deacylases has yet to be explored. Purification and characterization of deacylases will greatly facilitate our understanding of the regulation of protein myristoylation, especially the regulation of acyl-protein complexes.

Conclusion

The number of myristoylated proteins identified in viruses is constantly increasing. Nevertheless, substantial additional work will be required to shed light on the multiple functions that could be executed by the respective proteins as well as their involvement in the viral life cycle and elucidation of the role of the lipid anchor. Inhibition of NMT or replacing myristic acid by analogs in infected cells is known to affect numerous viruses of medical importance, for example, HIV (Bryant et al. 1989), hepatitis B virus (Parang et al. 1997), herpesviruses (Harper et al. 1993), and arenaviruses (Cordo et al. 1999). This approach might be generally suitable for influencing infections by viruses that take advantage of the eukaryotic myristoylation machinery. Further works are progressing in our laboratory.

Acknowledgement This work is supported by the Heart and Stroke Foundation of Saskatchewan, Canada. P.S. was a recipient of a Postdoctoral Fellowship from the Saskatchewan Health Research Foundation, Saskatoon, Canada.

References

Aitken, A., Cohen, P., Santikarn, S., Williams, D. H., Calder, A. G., Smith, A., and Klee, C. B. 1982. Identification of the NH_2-terminal blocking group of calcineurin B as myristic acid. FEBS Lett. 150:314–318.

Boutin, J. A. 1997. Myristoylation. Cell. Signal. 9:15–35.

Bryant, M. L., Heuckeroth, R. O., Kimata, J. T., Ratner, L., and Gordon, J. I. 1989. Replication of human immunodeficiency virus 1 and Moloney murine leukemia virus is inhibited by different heteroatom-containing analogs of myristic acid. Proc. Natl. Acad. Sci. USA 86:8655–8659.

Carr, S. A., Biemann, K., Shoji, S., Parmelee, D. C., and Titani, K. 1982. n-Tetradecanoyl is the NH2-terminal blocking group of the catalytic subunit of cyclic AMP-dependent protein kinase from bovine cardiac muscle. Proc. Natl. Acad. Sci. USA 79:6128–6131.

Cordo, S. M., Candurra, N. A., and Damonte, E. B. 1999. Myristic acid analogs are inhibitors of Junin virus replication. Microbes Infect. 1:609–614.

Duronio, R. J., Reed, S. I., and Gordon, J. I. 1992. Mutations of human myristoyl-CoA:protein N-myristoyltransferase cause temperature-sensitive myristic acid auxotrophy in *Saccharomyces cerevisiae*. Proc. Natl. Acad. Sci. USA 89:4129–4133.

Farazi, T. A., Waksman, G., and Gordon, J. I. 2002. The biology and enzymology of protein N-myristoylation. J. Biol. Chem. 277:39501–39504.

Giang, D. K., and Cravatt, B. F. 1998. A second mammalian N-myristoyltransferase. J. Biol. Chem. 273:6595–6598.

Glover, C. J., and Felsted, R. L. 1995. Identification and characterization of multiple forms of bovine brain N-myristoyltransferase. J. Biol. Chem. 270:23226–23233.

Glover, C. J., Hartman, K. D., and Felsted, R. L. 1997. Human N-myristoyltransferase amino-terminal domain involved in targeting the enzyme to the ribosomal subcellular fraction. J. Biol. Chem. 272:28680–28689.

Goll, D. E., Thompson, V. F., Li, H., Wei, W., and Cong, J. 2003. The calpain system. Physiol. Rev. 83:731–801.

Harper, D. R., Gilbert, R. L., Blunt, C., and McIlhinney, R. A. 1993. Inhibition of varicella-zoster virus replication by an inhibitor of protein myristoylation. J. Gen. Virol. 74:1181–1184.

Johnson, D. R., Bhatnagar, R. S., Knoll, L. J., and Gordon, J. I. 1994. Genetic and biochemical studies of protein N-myristoylation. Annu. Rev. Biochem. 63:869–914.

King, M. J., and Sharma, R. K. 1992. Demonstration of multiple forms of bovine brain myristoyl CoA:protein N-myristoyl transferase. Mol. Cell. Biochem. 113:77–81.

King, M. J., and Sharma, R. K. 1993. Identification, purification and characterization of a membrane-associated N-myristoyltransferase inhibitor protein from bovine brain. Biochem. J. 291:635–639.

Lambrecht, B., and Schmidt, M. F. 1986. Membrane fusion induced by influenza virus hemagglutinin requires protein bound fatty acids. FEBS Lett. 202:127–132.

Liauw, S., and Steinberg, R. A. 1996. Dephosphorylation of catalytic subunit of cAMP-dependent protein kinase at Thr-197 by a cellular protein phosphatase and by purified protein phosphatase-2A. J. Biol. Chem. 271:258–263.

McIlhinney, R. A., McGlone, K., and Willis, A. C. 1993. Purification and partial sequencing of myristoyl-CoA:protein N-myristoyltransferase from bovine brain. Biochem. J. 290:405–410.

Nocito, M., Roy, G., Villar, L. M., Palacios, C., Serrano, A., Alvarez-Cermeno, J. C., and Gonzalez-Porque, P. 1996. Thioesterase and protein deacylase activities of porcine pancreatic phospholipase A2. Biochim. Biophys. Acta 1299:17–22.

Ntwasa, M., Egerton, M., and Gay, N. J. 1997. Sequence and expression of *Drosophila* myristoyl-CoA: protein N-myristoyltransferase: evidence for proteolytic processing and membrane localisation. J. Cell Sci. 110:149–156.

Parang, K., Wiebe, L. I., Knaus, E. E., Huang, J. S., Tyrrell, D. L., and Csizmadia, F. 1997. *In vitro* antiviral activities of myristic acid analogs against human immunodeficiency and hepatitis B viruses. Antiviral Res. 34:75–90.

Rajala, R. V., Datla, R. S., Moyana, T. N., Kakkar, R., Carlsen, S.A., and Sharma, R.K. 2000. N-myristoyltransferase. Mol. Cell. Biochem. 204:135–155.

Raju, R. V., and Sharma, R. K. 1997. Demonstration and purification of a myristoyl-CoA binding protein from bovine cardiac muscle. Life Sci. 60:2145–2153.

Raju, R. V., Anderson, J. W., Datla, R. S., and Sharma, R. K. 1997. Molecular cloning and biochemical characterization of bovine spleen myristoyl CoA:protein N-myristoyltransferase. Arch. Biochem. Biophys. 348:134–142.

Raju, R. V., Kakkar, R., Datla, R. S., Radhi, J., and Sharma, R. K. 1998. Myristoyl-coA:protein N-myristoyltransferase from bovine cardiac muscle: molecular cloning, kinetic analysis, and *in vitro* proteolytic cleavage by *m*-calpain. Exp. Cell Res. 241:23–35.

Resh, M. D. 1999. Fatty acylation of proteins: new insights into membrane targeting of myristoylated and palmitoylated proteins. Biochim. Biophys. Acta 1451:1–16.

Rogers, S., Wells, R., and Rechsteiner, M. 1986. Amino acid sequences common to rapidly degraded proteins: the PEST hypothesis. Science 234:364–368.

Rundle, D. R., Rajala, R. V., and Anderson, R. E. 2002. Characterization of Type I and Type II myristoyl-CoA:protein N-myristoyltransferases with the Acyl-CoAs found on heterogeneously acylated retinal proteins. Exp. Eye Res. 75:87–97.

Saido, T. C., Sorimachi, H., and Suzuki, K. 1994. Calpain: new perspectives in molecular diversity and physiological-pathological involvement. FASEB J. 8:814–822.

Schmidt, M. F. 1982. Acylation of proteins—a new type of modification of membrane glycoproteins. Trends Biochem. Sci. 7:322–324.

Schmidt, M. F. 1983. Fatty acid binding: a new kind of posttranslational modification of membrane proteins. Curr. Top. Microbiol. Immunol. 102:101–129.

Selvakumar, P., Pasha, M. K., Ashakumary, L., Dimmock, J. R., and Sharma, R. K. 2002. Myristoyl-CoA:protein N-myristoyltransferase: a novel molecular approach for cancer therapy. Int. J. Mol. Med. 10:493–500.

Selvakumar, P., Lakshmikuttyamma, A., Pasha, M. K., King, M. J., Olson, D. J., Mori, S., Ross, A. R., Hayashi, K., Dimmock, J. R., and Sharma, R. K. 2004. N-myristoyltransferase inhibitor protein is homologous to heat shock cognate protein 70. J. Cell. Biochem. 92:573–578.

Selvakumar, P., Ashakumary, L., Shrivastav, A., and Sharma, R. K. 2005. Myristoylation and cell signaling: Involvement of heat shock protein 70 family. Recent Res. Dev. Life Sci. 3:75–82.

Selvakumar, P., Lakshmikuttyamma, A., Shrivastav, A., Das, S. B., Dimmock, J. R., and Sharma, R. K. 2007. Potential role of N-myristoyltransferase in cancer. Prog. Lipid Res. 46:1–36.

Sorimachi, H., Ishiura, S., and Suzuki, K. 1997. Structure and physiological function of calpains. Biochem. J. 328:721–732.

Zhang, L., Jackson-Machelski, E., and Gordon, J. I. 1996. Biochemical studies of *Saccharomyces cerevisiae* myristoyl-coenzyme A:protein N-myristoyltransferase mutants. J. Biol. Chem. 271:33131–33140.

Part IV
Calcium Signaling

Chapter 18
Ca^{2+}/Calmodulin-Dependent Protein Kinase II Signaling in Vascular Smooth Muscle

Suzanne J. House, Melissa M. Zachar, Roman G. Ginnan, Dee Van Riper, and Harold A. Singer

Abstract Ca^{2+} signaling pathways regulate diverse basic and differentiated cell functions including gene transcription, proliferation, and contraction of vascular smooth muscle. A key mediator of Ca^{2+} signals is the multifunctional serine/threonine protein kinase Ca^{2+}/calmodulin-dependent protein kinase II (CaMKII). CaMKII is structurally complex and has unusual autoregulatory properties. Because of this, there is a general lack of expertise and specific tools for studying its specific functions and relevant protein substrates have been difficult to identify. Biochemical and molecular studies have resulted in identification of multiple isoforms of the kinase expressed in differentiated vascular smooth muscle, where it has been implicated in the regulation of contractile function. Potential targets are diverse and the function of different isoforms is not well defined. In cultured vascular smooth muscle, different isoforms are expressed and associated with regulation of proliferation and migration. With recent accumulation of knowledge regarding the specific isoforms of CaMKII expressed in arterial smooth muscle and consequent development of isoform-specific molecular approaches for modifying expression and activity, progress on identifying CaMKII-dependent functions is foreseeable. Studies using these approaches indicate that dynamic regulation of CaMKII isozyme composition is an important determinant of, and contributor to, vascular smooth muscle phenotype modulation that is a component of fibroproliferative vascular disease.

Introduction

Intracellular free Ca^{2+} is a ubiquitous and conserved second messenger involved in the regulation of differentiated cellular functions including neurotransmission, exocytosis, and muscle contraction, as well as basic cellular functions such as gene transcription, migration, and proliferation. Many of the diverse actions of Ca^{2+} signals are known or suspected to be mediated through Ca^{2+}/calmodulin activated serine/threonine protein kinases. These protein kinases range from those with very specific substrate specificity, such as myosin light chain kinase (MLCK) or

A. K. Srivastava, M. B. Anand-Srivastava (eds.) *Advances in Biochemistry in Health and Disease.* Signal Transduction in the Cardiovascular System in Health and Disease,

phosphorylase kinase, to multifunctional protein kinases such as Ca^{2+}/calmodulin-dependent protein kinase II (CaMKII) that are capable of phosphorylating diverse protein substrates and integrating cellular function (Soderling and Stull 2001). Much of our current understanding of CaMKII structure and function has resulted from studies of the kinase as it occurs in mammalian brain, where it is abundant and is involved in regulating neurotransmission and memory (Braun and Schulman 1995). CaMKII, however, is ubiquitous and there is growing interest in its function in peripheral tissues, notably heart where it is involved in regulating Ca^{2+} dynamics and has been implicated in the progression of cardiac hypertrophy and heart failure (Maier et al. 2007). The present chapter summarizes our current knowledge of CaMKII structure, signaling, and function in vascular smooth muscle (VSM) where it has been implicated in the control of contractility in differentiated muscle, as well as gene transcription, proliferation, and migration in cultured VSM cells. We discuss recent evidence that dynamic regulation of CaMKII isozyme composition is an important determinant of VSM phenotypic modulation and cell proliferation and the implications of this with respect to vascular disease.

CaMKII Structure

Central to the understanding of CaMKII function is an appreciation of its complex structural properties and how these are determinants of enzymatic activity, autoregulation, and subcellular localization. In mammals, CaMKII is encoded by four genes (denoted α, β, δ, and γ) which were originally cloned from brain cDNA libraries. Based on Northern analysis of subunit mRNAs and immunological analysis of subunit proteins, α- and β-subunits have been found to have a distribution which is largely restricted to neuronal tissue while δ- and γ-subunits (Figure 18.1) are more widely and variably expressed in both brain and peripheral tissues, including heart and vascular smooth muscle (Tobimatsu and Fujisawa 1989).

Comparison of predicted primary sequences from the four CaMKII genes indicates that the first 315 amino acids (AA) are highly conserved between subunits (>90% identity) and have homology with other protein kinase catalytic domains (Figure 18.1). This region is referred to as the "catalytic/regulatory" domain since it has identifiable ATP binding, catalytic, calmodulin binding, and autoinhibitory sequences (Braun and Schulman 1995) and is fully active as a 32- to 35-kDa monomeric Ca^{2+}/CaM-dependent kinase when expressed *in vitro* (Yamagata et al. 1991). A variable region is followed by another conserved sequence of approximately 140 AA which has been denoted the "association" domain because it interacts resulting in formation of multimeric holoenzymes. A unique 21-AA carboxyl-terminal sequence is also variably expressed in some δ-subunits (Mayer et al. 1993). It is remarkable that the primary sequence of CaMKII homologs from *Drosophila* (Ohsako et al. 1993) and *C. elegans* (Rosenberg et al. 2005) are over 70% identical to the rat CaMKII α-subunit with

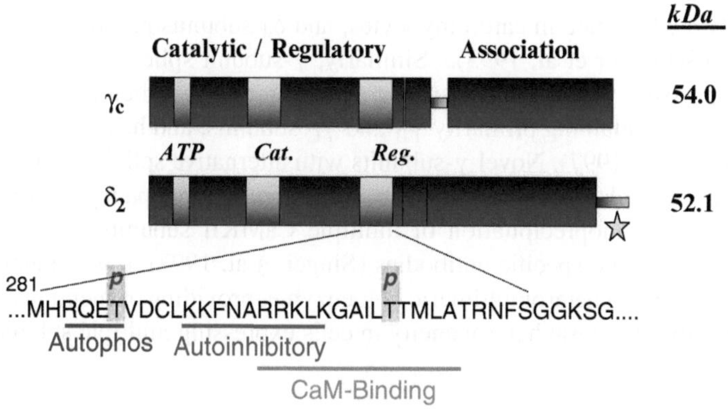

Fig. 18.1 Two structures representative of variants from four separate genes (α, β, δ, γ) are depicted. Conserved regions across subunits are indicated by large rectangles. Variable regions of alternative splicing are indicated by small rectangles. Conserved ATP binding, catalytic (Cat.), and regulatory (Reg.) domains are indicated. The regulatory domain encompasses an autoinhibitory sequence, calmodulin (CaM)-binding domain, and several autophosphorylation sites. A conserved association domain is responsible for formation of multimeric holoenzymes. An antipeptide polyclonal antibody directed to the alternatively spliced C-terminus of δ-isoforms (indicated by the star) was produced and used in studies depicted in Figure 18.2.

homology extending across both the catalytic/regulatory and association domains. This indicates strong evolutionary conservation of both subunit and holoenzyme structure and likewise suggests conserved essential Ca^{2+}-dependent functions mediated by the kinase.

CaMKII purified from brain (Bennett et al. 1983) and a number of peripheral tissues, including heart (Jett et al. 1987) and vascular smooth muscle (Singer et al. 1996), is generally described as a large 400- to 600-kDa holoenzyme. Electron microscopic analyses of purified brain CaMKII molecules indicate that holoenzymes composed of α- and β-subunits are dodecamers, consistent with the association of 12 "dumbbell" – shaped kinase subunits in a hub and spoke arrangement where the association domains form the hub and catalytic/regulatory domains the spokes (Kanaseki et al. 1991). Although a complete holoenzyme has not yet been crystallized, diffraction analysis of association domain crystals indicates a structure of two stacked rings of 6 subunits each forming the hub (Hoelz et al. 2003). Similarly, recombinant CaMKII multimers composed of δ- or γ-subunits expressed in mammalian cells are also extracted as 600-kDa holoenzymes, consistent with this structure (Singer et al. 1997).

More detailed RT-PCR analyses of mRNA from a variety of tissues have revealed additional structural complexity and identification of a large number of alternatively spliced subunit transcripts that differ in variable domain nucleic acid sequence (Tombes et al. 2003). These subunit variants are expressed in a tissue-specific manner. For example, the δ_1 subunit mRNA is expressed predominantly in brain, whereas δ_2 and δ_3 variants are expressed in rat aortic smooth muscle,

δ_3 subunits predominate in cardiomyocytes, and δ_4 subunits predominate in skeletal muscle (Schworer et al. 1993).[1] Similarly, γ-subunit splice variants are differentially expressed, with brain containing primarily the γ_E subunit, carotid arterial smooth muscle containing primarily γ_B and γ_C subunits, and heart γ_B and γ_E subunits (Singer et al. 1997). Novel γ-subunits with alternative splicing in the association domain have also been identified in ferret aorta (Gangopadhyay et al. 2003). Based on co-immunoprecipitation of multiple CaMKII subunits from tissue homogenates using type-specific antibodies (Singer et al. 1997), it is evident that the kinase can occur as a heteromultimeric holoenzyme, providing tremendous potential for holoenzyme structural heterogeneity in cells expressing multiple subunits.

Functional Consequences of Subunit Heterogeneity

Autoregulation

Given the high degree of CaMKII structural heterogeneity and cell-specific patterns of subunit expression, obvious questions arise regarding the functional influence of both the CaMKII subunit and holoenzyme structures. One major consequence of the multimeric structure is that upon binding of Ca^{2+}/CaM there is a cooperative intersubunit, intraholoenzyme autophosphorylation of subunits on Thr^{287} (Thr^{286} in the α-subunit). This phosphorylation event causes a several thousand-fold decrease in Ca^{2+}/CaM dissociation rate and the generation of an "autonomous" form of the kinase that has partial activity in the absence of bound Ca^{2+}/CaM (Hanson et al. 1994). Dissociation of Ca^{2+}/CaM from a subunit phosphorylated on Thr^{287} exposes threonines 306 and 307 in the calmodulin binding domain ($Thr^{305/306}$ in the α-subunit) and these residues can also be shown *in vitro* to be transphosphorylated in the holoenzyme (Hanson and Schulman 1992). Kinase subunits phosphorylated on Thr^{306} or Thr^{307} cannot rebind Ca^{2+}/CaM or regain full activity. Thus, phosphorylation of Thr^{287} provides CaMKII with the capacity to sustain activity for a period of time following transient increases in $[Ca^{2+}]_i$ and to accumulate activity in a frequency-dependent manner, while $Thr^{306/307}$ phosphorylation provides a means for limiting total Ca^{2+}/CaM-stimulated activity following a Ca^{2+} transient. Autophosphorylation of CaMKII has also been reported to affect its subcellular localization in postsynaptic densities (McNeill and Colbran 1995). The functional importance of both the Thr^{286} and $Thr^{306/307}$ autophosphorylated forms of CaMKIIα has been established in brain, where they contribute to Ca^{2+} transient frequency sensing, postsynaptic density remodeling, and memory (Giese et al. 1998; Elgersma et al. 2002). However, the function of these unique autophosphorylation events in peripheral tissues such as VSM or heart, and their regulation by specific protein phosphatases has not been explored.

[1] According to alternative nomenclatures (Schworer et al. 1993; Edman and Schulman 1994): CaMKIIδ_1 = CaMKIIδ_A; $\delta_2 = \delta_C$, $\delta_3 = \delta_B$, $\delta_4 = \delta_D$.

Table 18.1 Substrate (autocamtide-2) and calmodulin activation kinetics in recombinant CaMKII holoenzymes purified from Baculovirus-infected insect cells

	Autocamtide-2		Calmodulin	
Isozyme	K_M(μM)	V_{MAX}(μmol/min/mg)	K_M(nM)	V_{MAX}(μmol/min/mg)
α	2.02 ± 0.04	5.35 ± 0.14	24.96 ± 0.95	4.15 ± 0.31
δ_1	1.37 ± 0.13	6.39 ± 0.30	2.43 ± 1.23	4.71 ± 1.22
δ_2	3.68 ± 0.63	8.59 ± 1.27	10.83 ± 3.01	6.94 ± 0.73
δ_3	5.17 ± 0.67	11.77 ± 2.09	9.36 ± 3.21	9.78 ± 0.13
γ_B	3.49 ± 0.44	7.64 ± 0.88	3.73 ± 0.78	4.12 ± 0.80
γ_C	2.21 ± 0.04	5.30 ± 0.19	6.76 ± 2.20	4.07 ± 0.14

Kinetic constants were determined from concentration/activity curves fit to a hyperbolic plot. Values shown are means ± S.E.M. from two or three independent experiments.

Kinetics

In studies using highly purified recombinant CaMKII holoenzymes from Baculovirus-infected insect cells (produced in collaboration with Dr. R. Colbran, Vanderbilt University), some differences in the kinetic properties of holoenzymes composed of various α-, δ-, and γ-subunit splice variants can be demonstrated (Table 18.1). Under conditions of saturating [Ca^{2+}/CaM], the range of K_M and V_{max} for the peptide substrate autocamtide-2 across isoforms is reasonably consistent varying only 2- to 3-fold, suggesting that subunit structure has little effect on intrinsic substrate specificity. However, activation constants for calmodulin vary 10-fold across isoforms ranging from a K_a of 2.4 nM for the δ_1 isoform to a $K_a \sim 25$ nM for the α isoform. These δ and γ holoenzyme calmodulin activation constants are somewhat lower than might be expected from studies of purified brain kinase (reported K_a ranging between 25 and 100 nM calmodulin) which is composed mainly of α- and β-subunits (Bennett et al. 1983), but are consistent with that of CaMKII purified from porcine carotid arterial smooth muscle ($K_a \sim 7$ nM calmodulin; Singer et al. unpublished), which is a mixture of holoenzymes composed of δ- and γ-subunit variants (Singer et al. 1997). For comparison, the affinity of calmodulin for smooth muscle myosin light chain kinase under conditions of saturating Ca^{2+} is approximately 1 nM (Soderling and Stull 2001). Therefore, even small differences in subunit structure may be reflected in holoenzyme kinetics, raising the possibility that endogenous protein substrate affinities may also differ significantly between subunits, especially under conditions of submaximal activation where [Ca^{2+}/CaM] is limiting.

Subcellular Localization

Discrete subcellular localization and protein interaction are other mechanisms by which this otherwise multifunctional protein kinase could achieve some degree

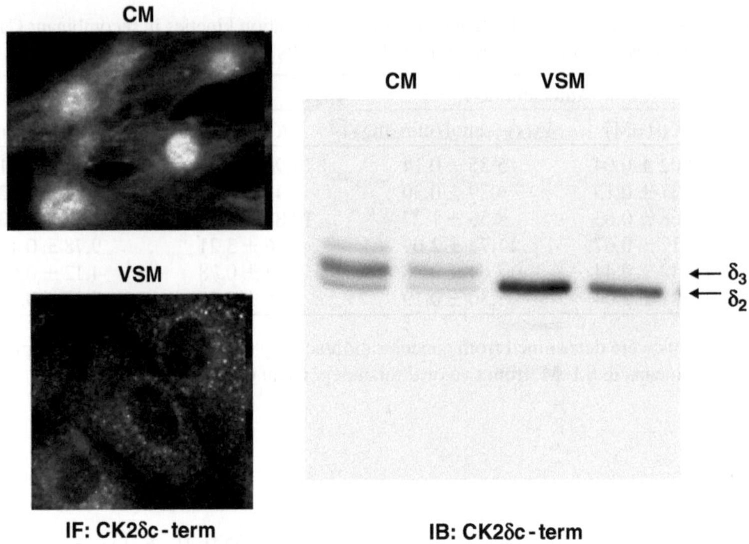

Fig. 18.2 CaMKII-δ isoform expression and distribution in neonatal rat cardiomyocytes (CM) and cultured rat aortic vascular smooth muscle (VSM) cells. The photomicrographs depict d-isozyme distribution assessed by indirect immunofluorescence (IF) using an antibody specific for the alternatively spliced δ-subunit C-terminus (CK2δc-term; see Figure 18.1). The immunoblot (IB) with the same antibody δ-subunit variants expressed in the cells. Two protein loadings were used for each cell lysate. The major subunits were identified by RT-PCR in previously published studies (Schworer et al. 1993).

of specificity with regard to substrate phosphorylation and function (Bayer and Schulman 2001). Alternatively spliced transcripts from the CaMKIIα gene have been shown to result in the formation of a protein (αKAP) that lacks kinase activity, but contains the association domain as well as a C-terminal membrane interaction domain. Co-assembly of αKAP into CaMKII holoenzymes targets the kinase to intracellular membranes (Bayer et al. 1998). Another clear example of subunit and holoenzyme structure affecting subcellular localization is a conserved 11-AA variable domain sequence (K-R-K-S-S-S-S-V-Q/H-L/M-M) expressed in four of the known CaMKII subunits (α_B, $\delta_{3,7}$, and γ_A) which results in nuclear targeting of CaMKII holoenzymes (Srinivasan et al. 1994). In transfected cells, the amount of overexpressed CaMKII localized to the nucleus varies according to the relative proportion of holoenzyme subunits containing the nuclear localization sequence (Srinivasan et al. 1994). Evidence for this in untransfected cardiovascular cells is shown in Figure 18.2. Immunolocalization of CaMKII in cultured neonatal rat cardiomyocytes (Figure 18.2B), using an antibody specific for the alternatively spliced C-terminus of δ-subunits, indicates a concentration of kinase in the nucleus, consistent with expression of δ_3-subunits containing the nuclear localization signal (Schworer et al. 1993). In contrast, cultured rat aortic vascular smooth muscle cells express predominantly the δ_2-subunit (Schworer et al. 1993), which lacks the nuclear localization signal, and this subunit is distributed primarily in the cytosolic compartment with a perinuclear concentration. Considering evidence that free [Ca^{2+}] may increase significantly in the nucleus following stimulation of cells with Ca^{2+} mobilizing agonists (Santella and Carafoli 1997), differential expression of these specific CaMKII subunits could provide a direct pathway for coupling Ca^{2+} signals to the phosphorylation of nuclear substrates involved in regulating gene transcription or the cell cycle.

Protein Interactions

Although the effects of CaMKII variable domain structure on protein/protein interactions are not fully explored, specific domains in some β-subunits (Urquidi and Ashcroft 1995) and γ-subunits (Gangopadhyay et al. 2003) have been implicated in SH3 domain-binding interactions. CaMKIIδ isoforms containing the 21-AA C-terminus also contain a short proline-rich region which conforms to a class II SH3-binding domain (Mayer and Eck 1995). It is interesting to speculate that such interactions could link Ca^{2+} signals to intracellular signaling pathways involving nonreceptor tyrosine kinases and functional consequences including ERK1/2 activation (Ginnan and Singer 2002) and/or regulation of cell proliferation (House et al. 2007) and migration (Pfleiderer et al. 2004) responses. Alternatively spliced domains in certain β-subunits have been implicated in actin filament interactions (Shen et al. 1998) and the holoenzyme structure of CaMKIIβ has been proposed to result in actin filament bundling in neuronal synapses (Okamoto et al. 2007). CaMKII isoforms have also been reported to differentially associate with other postsynaptic density proteins, including the *N*-methyl-D aspartate (NMDA) receptor NRB2 subunit, densin180, α-actinin-2, and PSD-95 (Robison et al. 2005). Several of these interactions are dependent on CaMKII $Thr^{286/287}$ autophosphorylation and the picture emerges of a CaMKII holoenzyme serving as a Ca^{2+}/calmodulin regulated protein scaffold. Outside of the central nervous system, this aspect of CaMKII structure and function has not been studied to any significant extent although our recent results implicating CaMKIIδ in regulating iNOS trafficking (Jones et al. 2007) and VSM cell migration (Pfleiderer et al. 2004) may be partially explained by scaffolding functions of this nature.

Activation of CaMKII in VSM Cells

The activation properties of CaMKII have been well documented *in vitro* and are understood in considerable molecular detail. In contrast, much less is known about CaMKII activation kinetics in intact cells that display complex and localized Ca^{2+} transients as well as phosphatase activities that reverse autophosphorylation events. Using autonomous activity as a quantitative index of CaMKII activation, we have defined CaMKII activation kinetics in cultured (Abraham et al. 1996) and intact (Rokolya and Singer 2000) arterial smooth muscle. Remarkably high levels of autonomous activity (as much as 44% of total Ca^{2+}/CaM-stimulated activity) are generated in cultured rat aortic VSM cells in response to physiological Ca^{2+}-mobilizing agonists including angiotensin II, vasopressin, and platelet-derived growth factor (Abraham et al. 1996). In intact carotid arterial smooth muscle, histamine transiently increased CaMKII autonomous activity to nearly 80% of total, while KCl depolarization stimulated slow increases in autonomous activity reaching about 30% of total (Rokolya and Singer 2000). Following addition of the Ca^{2+} ionophore, ionomycin, generation of autonomous activity in cultured cells can reach as much as 60% of the total activity (Abraham et al. 1996). Considering that the maximal level of autonomous activity reached under optimal activation conditions *in vitro* is about 75% of the total activity (Abraham et al. 1996), the levels of autonomous activity reached in the intact VSM cells reflect nearly complete activation and autophosphorylation of the kinase pool *in situ*. Although the functional importance of free calmodulin availability has been raised in the context of smooth muscle contraction (Geguchadze et al. 2004), the studies discussed above indicate that in intact arterial smooth muscle cells, free calmodulin levels do not appear to be limiting with respect to CaMKII activation.

High-efficiency CaMKII activation and generation of autonomous activity in intact VSM reflects, at least in part, a close functional coupling between the kinase and intracellular sources of activator Ca^{2+}. By correlating changes in $[Ca^{2+}]_i$, measured by Fura-2 fluorescence, with the changes in Ca^{2+}/CaM-stimulated (total) and -independent (autonomous) CaMKII activity in detergent lysates of rat aortic VSM, an apparent EC_{50} for Ca^{2+} activation of 304 nM was determined in cultured aortic VSM (Abraham et al. 1996). However, a twofold decrease in apparent Ca^{2+} sensitivity ($Ca^{2+}EC_{50} = 616\,nM$) was determined following depletion of intracellular Ca^{2+} stores with thapsigargin. Our interpretation of these data is that the higher "apparent" sensitivity of the kinase to Ca^{2+} in intact cells reflects the effect of high local concentrations of intracellular Ca^{2+} coupled with a cooperative mechanism of autophosphorylation (Abraham et al. 1996) and CaMKII holoenzymes concentrated in the proximity of the intracellular Ca^{2+} pools. Direct support for this was provided from studies of CaMKII localization and dynamics in cultured VSM cells stimulated by ionomycin which demonstrated CaMKIIδ_2 colocalization with endoplasmic reticulum markers (Van Riper and Singer 2000). CaMKIIδ_2 has also been reported to colocalize with endoplasmic reticulum in cultured astrocytes (Takeuchi et al. 2000). Thus, localization of CaMKII in those intracellular domains that experience large Ca^{2+} transients could provide a mechanism by which the kinase (1) compensates for a relatively low intrinsic sensitivity to activator Ca^{2+}/CaM, (2) optimizes cooperative autophosphorylation and generation of autonomous activity, and (3) achieves functional specificity. The subcellular distribution of CaMKII in differentiated VSM has not been studied extensively, although a large fraction of kinase activity can be purified from myofibril preparations (Singer et al. 1996) and novel isoforms of the γ-subunit identified in ferret aorta appear to localize with cytoskeletal elements including vimentin filaments and α-actinin-containing cytosolic dense bodies (Marganski et al. 2005).

CaMKII Function in Vascular Smooth Muscle

Because CaMKII can phosphorylate numerous substrates, is ubiquitous, and is controlled by a universal second messenger (Ca^{2+}), it has been implicated in numerous cellular functions through the regulation of a large number of protein substrates. CaMKII function has been inferred from studies of its cellular and subcellular distributions, comparisons of *in vitro* and *in vivo* substrate specificities, and through the use of both peptide and chemical inhibitors which selectively interfere with calmodulin binding to the kinase. More recently, molecular approaches using overexpression of CaMKII mutants or introduction of antisense and small interfering RNAs to suppress endogenous kinase expression have led to new insights into CaMKII function (Pfleiderer et al. 2004; Marganski et al. 2005; House et al. 2007). The following discussion is limited to a consideration of possible CaMKII regulated functions that are supported by multiple lines of evidence and are of particular importance in cardiovascular control. The major mechanisms and interactions are summarized in a model depicted in Figure 18.3.

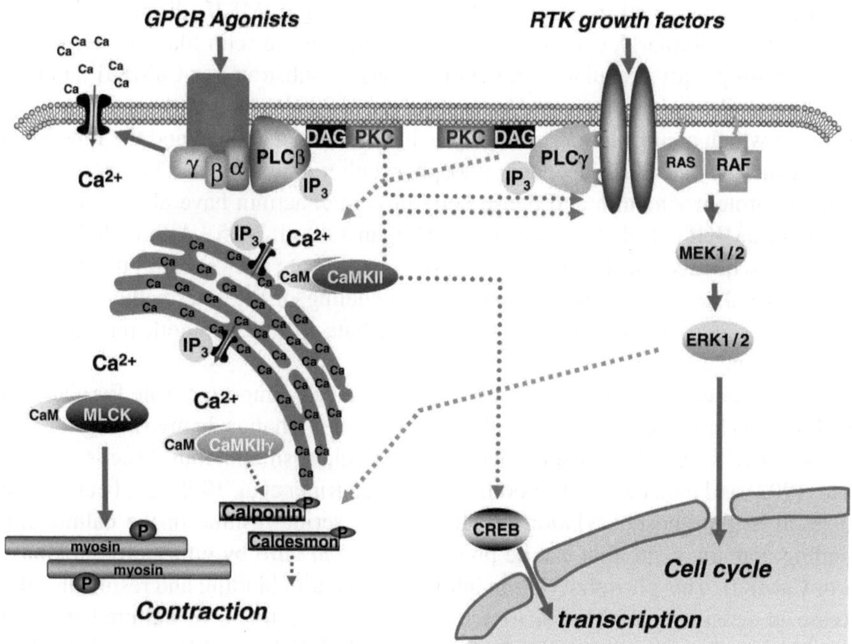

Fig. 18.3 Model for CaMKII-dependent regulation of vascular smooth muscle contraction, cell cycle, and gene transcription. CaMKIIγ isoforms are proposed to regulate contractile function stimulated by G-protein-coupled receptor (GPCR) agonists with potential targets including myosin light chain kinase (MLCK) and thin filament associated regulatory proteins, calponin and caldesmon. CaMKIIδ isoforms are proposed to mediate transcriptional responses, promote cell proliferation and (not shown) cell migration. Potential interactions of GPCR-stimulated CaMKIIδ and protein kinase C (PKC)-dependent pathways with receptor tyrosine kinase (RTK)-dependent pathways and ERK1/2 signaling are indicated by dotted arrows. Potential reciprocal interactions with pathways regulating contraction are also indicated.

Differentiated Contractile Function

A number of contractile proteins and associated regulatory proteins are substrates for CaMKII *in vitro* (Singer et al. 1996). In smooth muscle, a CaM kinase activity co-purifying with the actin filament-associated regulatory protein caldesmon has been identified as CaMKII (Scott-Woo et al. 1990). Caldesmon inhibits smooth muscle actin-activated myosin ATPase activity *in vitro* and phosphorylation of caldesmon *in vitro* by either CaMKII or protein kinase C reverses its inhibition of actomyosin ATPase activity. However, the functional significance of this putative regulatory pathway *in vivo* is still in question because the sites phosphorylated in caldesmon extracted from contracting arterial smooth muscle are not the same as those phosphorylated by CaMKII *in vitro*. In fact, the sites in caldesmon which are phosphorylated *in situ* are consistent with the actions of a proline-directed protein kinase such as ERK1/2 (reviewed in Morgan and Gangopadhyay 2001). This does not necessarily exclude a role for CaMKII as a physiologically relevant regulator of caldesmon since several studies in both cultured (Ginnan and Singer 2002) and intact (Marganski et al. 2005) VSM indicate that Ca^{2+}-dependent activation of ERK1/2 is mediated in part by pathways involving CaMKII (discussed in more detail below). Similarly, calponin, another smooth muscle actin filament-associated protein with putative regulatory functions, is also a substrate for CaMKII and protein kinase C *in vitro* (reviewed in Isotani et al. 2004). In this case, the sites in calponin which are phosphorylated *in vivo* have not been defined and the functional significance of CaMKII-dependent phosphorylation is unknown. The intermediate filament protein vimentin and dense body protein α-actinin have also been implicated as CaMKII cytoskeletal substrates (Marganski et al. 2005). Although the functional consequences of this in arterial smooth muscle are not yet clear, it is interesting to speculate that CaMKII-dependent remodeling of the intermediate filament and/or actin cytoskeletal network could contribute to tonic smooth muscle force maintenance.

A stronger case has been made for a physiologically important role for CaMKII in phosphorylating and modulating the activity of smooth muscle myosin light chain kinase (MLCK). Ca^{2+}-dependent activation of tracheal smooth muscle cells (Tansey et al. 1992) and intact arterial smooth muscle (Van Riper et al. 1995) has been shown to result in the phosphorylation of MLCK on a serine residue in the calmodulin binding domain, a site that can be phosphorylated *in vitro* by either protein kinase A or CaMKII. This phosphorylation inhibits calmodulin binding and results in a decrease in sensitivity of MLCK to activator Ca^{2+}. Treatment of cultured tracheal smooth muscle cells with inhibitors of CaMKII (KN-62, $CaMK_{281-309}$) inhibits this phosphorylation event and subsequent desensitization of MLCK in agonist-stimulated cells, resulting in potentiation of myosin regulatory light chain phosphorylation (Tansey et al. 1994). Inhibitory phosphorylation of MLCK in these cells was also found to be dependent on free intracellular [Ca^{2+}] with a Ca^{2+} EC_{50} of about 500 nM, which is consistent with the sensitivity of CaMKII to activator Ca^{2+} in VSM cells. Thus, CaMKII negatively modulates MLCK activity, resulting in the attenuation of myosin light chain phosphorylation in cultured tracheal smooth muscle.

Based on this, one would predict that treatment of intact smooth muscle preparations with an inhibitor of CaMKII should potentiate myosin light chain phosphorylation and the rapid phase of force development. In contrast, we have found that in carotid arterial smooth muscle, KN-93 inhibition of CaMKII activation in response to physiological contractile stimuli correlates with a marked inhibition of tonic force responses (Rokolya and Singer 2000), suggesting an alternative dominant action of CaMKII on the smooth muscle contractile apparatus.

Another general mode of smooth muscle contractile regulation by CaMKII is through effects of the kinase on Ca^{2+} regulatory proteins, including membrane ion channels and ATPase pumps. A substantial amount of attention has been focused on the endogenous cardiac sarcoplasmic reticulum (SR) CaMKII and its potential role in regulating SR Ca^{2+} homeostasis (reviewed in Grueter et al. 2007). Briefly summarized, *in vitro* studies suggest that this could occur by multiple mechanisms including: (1) the phosphorylation and activation of Ca^{2+}-release ryanodine receptors, (2) activation of the SERCA2a Ca^{2+}/ATPase by direct phosphorylation, and (3) phosphorylation and inactivation of the Ca^{2+}/ATPase inhibitory protein, phospholambam. Other potential CaMKII targets include membrane channels such as voltage-gated Ca^{2+} channels (Grueter et al. 2006) and Na^{+} channels (Wagner et al. 2006). Recent genetic studies evaluating conditional overexpression of a CaMKII peptide inhibitor have strongly indicated the physiological significance of the kinase in regulating cardiac homeostasis (Wu et al. 2006). Although only limited studies along these lines have been carried out in vascular smooth muscle (McCarron et al. 1992), it is reasonable to expect similar CaMKII-dependent regulation of Ca^{2+} homeostasis in these cells. An important future area of research will be to determine the function of CaMKII and the relative importance of each of these mechanisms in the regulation of free intracellular $[Ca^{2+}]$ in differentiated vascular smooth muscle.

Synthetic Phenotype Function

VSM cells in the medial wall of blood vessels are quiescent and differentiated for contraction and regulation of blood vessel diameter and tone. VSM myocytes, however, can undergo marked alterations in phenotype leading to development of atherosclerotic plaques or neointimal hyperplasia following balloon angioplasty and vascular grafting (Owens et al. 2004). Transition of differentiated VSM cells to a "synthetic" phenotype is characterized by downregulation of smooth muscle-specific contractile proteins and upregulation of proteins involved in matrix remodeling, DNA synthesis, cell migration, and proliferation (Owens 1995; Turley 2001; Owens et al. 2004). A similar modulation of phenotype is observed *in vitro* following primary culture of freshly isolated VSM cells (Owens 1995; Owens et al. 2004).

Stimulation of cultured VSM cell migration by PDGF (Pauly et al. 1995) or by scrape wounding a monolayer of VSM cells grown in serum (Zhang et al. 2003) is associated with activation of CaMKII. Simple adherence of cultured VSM cells to extracellular matrix-coated substrates or to poly-l-lysine also results in significant CaMKII activation and downstream signaling (Lu et al. 2005). Thus, as a multifunctional serine/threonine kinase, CaMKII appears to be well positioned to coordinate complex cellular events involved in cell migration. Consistent with this view, pharmacological inhibition of CaMKII inhibits PDGF-stimulated VSM cell migration, suggesting a net positive role for the kinase in regulating this process (Pauly et al. 1995; Lu et al. 2005). However, this interpretation is complicated by conflicting data resulting from application of molecular approaches to modify CaMKII activity. Whereas stable overexpression of a constitutively active mutant of the CaMKIIα isoform (an isoform not endogenously expressed in VSM) is associated with enhanced migration (Pauly et al. 1995), transient overexpression of a constitutively active mutant of CaMKIIδ_2 (the endogenous isoform) was reported to *inhibit* VSM cell migration and overexpression of a kinase-negative δ_2 mutant enhanced migration (Pfleiderer et al. 2004). Thus, although available evidence suggests a role(s) for CaMKII in regulating VSM cell migration, the precise mechanisms and substrates are not yet known and interpretation of results is complicated by potential nonspecific effects of chemical inhibitors and/or CaMKII isoform-specific effects.

The CaMKII substrates involved in cell migration have not been identified, but results from work in differentiated muscle suggest cytoskeletal proteins are likely targets. It is possible, for example, that CaMKII could act to modulate myosin light chain kinase activity, or calponin function, both of which have been implicated in the control of smooth muscle cell motility (Klemke et al. 1996; Jiang et al. 1997). Another general mechanism by which CaMKII could exert diverse cellular effects during a complex process like cell migration is through interaction with other intracellular signaling pathways. There is now substantial evidence that CaMKII can mediate activation of MAP kinase (ERK1/2) signaling by Ca^{2+}-dependent stimuli (Abraham et al. 1997; Ginnan and Singer 2002; Muthalif et al. 1998). This conclusion is supported by both pharmacological and molecular approaches, including recent studies using siRNA to selectively suppress CaMKIIδ_2 in VSM cells (Lu et al. 2005). At least one of the mechanisms coupling CaMKII activation to ERK1/2 signaling involves the action of nonreceptor (src-family) tyrosine kinases and trans-activation activation of EGF receptor tyrosine kinases (Ginnan and Singer 2002). CaMKII-dependent activation of tyrosine kinases and ERK1/2 signaling provides a general mechanism for the Ca^{2+}-dependent regulation of diverse cellular responses including gene transcription, cell growth/proliferation, and cell migration.

In addition to cell migration, another characteristic function of the VSM synthetic phenotype is proliferation. CaMKII has been implicated in cell cycle control in a number of systems, but investigations in this area have yield mixed results. Most studies to date have relied on pharmacological inhibitors of CaMKII such as KN-62 or KN-93 (Tombes et al. 1995) and point to a positive role for CaMKII in mediating the cell cycle. Conversely, overexpression of a constitutively active mutant of CaMKIIα suggested a negative role for kinase (Beauman et al. 2003). Interpretation of the latter studies is complicated by potential nonspecific effects of overexpressing a persistently active multifunctional protein kinase and the fact that CaMKIIα was not an endogenous isoform in the cell types studied. Selective suppression of CaMKIIγ using siRNA has also been reported to disrupt regulation of mitotic spindles in several human cell lines, inhibiting mitosis (Holmfeldt et al. 2005).

Until recently, the function of CaMKII in regulating VSM cell proliferation had not been reported, although heparin, a selective inhibitor of VSM proliferation, was reported to inhibit serum-stimulated activation and autophosphorylation of CaMKII in VSM (Mishra-Gorur et al. 2002). Using molecular approaches to inhibit endogenous CaMKIIδ_2 activity (overexpression of kinase-negative CaMKIIδ_2 mutant) or suppress expression (siRNA targeting CaMKIIδ), we have demonstrated positive regulation of cultured VSM cell proliferation through effects at the G2/M transition and/or cytokinesis (House et al. 2007). One potential substrate for CaMKII at the G2/M transition is the Cdc25C phosphatase that can be activated by CaMKII, resulting in dephosphorylation and activation of the cyclin B/CDK1 complex required for transition through G2/M (Patel et al. 1999). Potential substrates involved in cytokinesis include many of the aforementioned cytoskeletal proteins or associated regulatory proteins.

CaMKII in Vascular Disease

Molecular mechanisms governing phenotypic modulation of VSM are still incompletely understood. Most studies have focused on identification of environmental factors that induce changes in phenotype (e.g., growth factors and cytokines, oxidative stress, mechanical stress, endothelial factors) or on transcriptional regulation of phenotype-specific genes (Owens 1995). A few studies have raised the possibility that changes in intracellular signaling components may also contribute to VSM phenotype modulation. For example, cGMP-dependent protein kinase has been shown to be downregulated in synthetic VSM cells and reexpression promotes characteristics of the differentiated contractile phenotype (Lincoln et al. 2006). Subcellular localization of phosphodiesterase 1A, which selectively hydrolyzes cGMP, was recently shown to change upon VSM phenotype modulation *in vitro*, with a nuclear localization promoting the synthetic phenotype and a cytosolic localization promoting a differentiated contractile phenotype (Nagel et al. 2006). Changes in Ca^{2+} signaling dynamics have been linked to transition of cerebral VSM to a proliferative phenotype in explant cultures (Wilkerson et al. 2006). One mechanism may involve differential activation of CREB and NFAT transcription factors in response to specific sources or dynamics of the Ca^{2+} signal (Dolmetsch et al. 2001; Barlow et al. 2006). Other alterations in ion channel expression (Landsberg and Yuan 2004; Wilkerson et al. 2006) that may affect Ca^{2+} signaling have also been demonstrated following VSM cell culture and conversion to a proliferative phenotype. These studies support the concept that altered patterns of Ca^{2+} signaling may be important components of VSM phenotype modulation and contribute to regulation of phenotype-specific

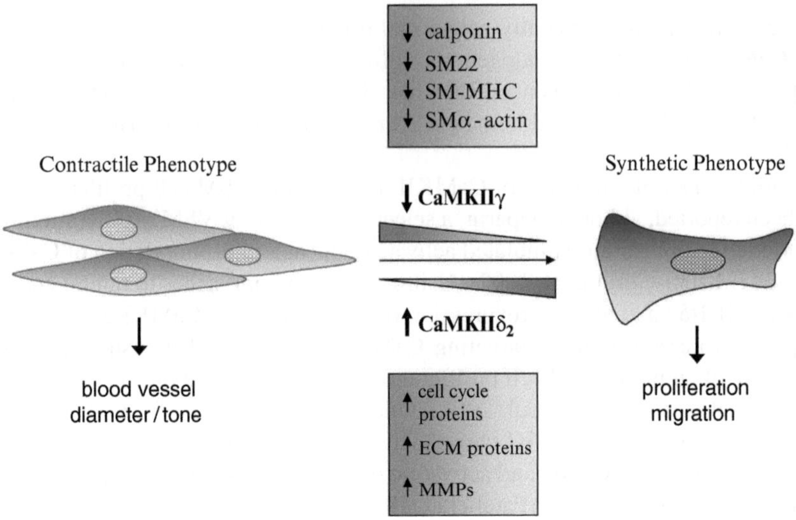

Fig. 18.4 Model depicting changes in CaMKII γ and δ isoform expression as a function of vascular smooth muscle phenotype modulation.

cell functions. By extrapolation, altered patterns of expression or activity of a major Ca^{2+} signaling mediator, such as CaMKII, could be integral to VSM phenotype modulation.

Upregulation of CaMKIIδ isoforms has been described in models of cardiac hypertrophy and heart failure, and mechanisms involving CaMKII-dependent regulation of histone deacetylase have been proposed to account for alterations in gene expression associated with these pathologies (Grueter et al. 2007). Similarly, the CaMKIIδ_2 isoform is upregulated in regenerating skeletal muscle following injury (Abraham and Shaw 2006). While it is tempting to speculate that similar mechanisms might contribute to vascular pathologies involving VSM cell hypertrophy or proliferation, this has not yet been critically examined. In support of this concept, our recent *in vitro* studies have for the first time clearly demonstrated a rapid shift in CaMKII isoform expression upon acute transition of VSM cells from a differentiated to synthetic phenotype in primary culture (House et al. 2007). Upregulation of the CaMKIIδ_2 isoform and reciprocal downregulation of CaMKIIγ isoforms over the course of hours was found to coincide with, or precede, the onset of VSM cell proliferation. Furthermore, in keeping with a positive role for CaMKIIδ_2 in regulating VSM cell proliferation, treatment with siRNA to prevent upregulation of the δ_2 isoform attenuated subsequent VSM proliferation. In light of the rapid changes observed in CaMKII isoform expression in response to cell culturing, CaMKII isoform modulation may be a proximal event contributing to transition of VSM from a contractile to synthetic phenotype through regulation of gene expression patterns promoting VSM cell proliferation and migration-functions characteristic of synthetic VSM cells (Figure 18.4). Since atherosclerosis, vein graft failure, and transplant vasculopathy are dependent on pathologic VSM cell proliferation, we predict a role for CaMKII isoform modulation in each of these processes. In addition, increases in VSM cell ploidy, similar to those observed following suppression of endogenous CaMKIIδ_2 in cultured aortic VSM cells, have been linked to vascular remodeling associated with hypertension (Hixon and Gualberto 2003), and in this context, it may be interesting to consider a role of CaMKII isoform modulation as a factor contributing to vascular remodeling in hypertensive disease.

Summary and Future Directions

As a comparatively abundant multifunctional serine/threonine protein kinase regulated by a universal second messenger, CaMKII is almost certainly a physiologically important mediator of Ca^{2+} signals in vascular smooth muscle. Yet because of its multifunctional nature, structural complexity, and to this point, a general lack of specific tools for modulating expression and activity, a reliable description of specific functions and relevant protein substrates has been difficult to construct. With the recent accumulation of knowledge regarding the specific isoforms of CaMKII expressed in arterial smooth muscle and consequent development of isoform-specific molecular approaches for modifying expression and activity, it is now possible to critically evaluate the function(s) of this kinase in differentiated VSM cells as well as its participation in VSM phenotype modulation and vascular disease.

References

Abraham, S. T., and Shaw, C. 2006. J. Cell Biochem. 97:621–632.
Abraham, S. T., Benscoter, H., Schworer, C. M., and Singer, H. A. 1996. J. Biol. Chem. 271:2506–2513.
Abraham, S. T., Benscoter, H. A., Schworer, C. M., and Singer, H. A. 1997. Circ. Res. 81:575–584.
Barlow, C. A., Rose, P., Pulver-Kaste, R. A., and Lounsbury, K. M. 2006. J. Physiol. (London) 570:59–64.
Bayer, K. U., and Schulman, H. 2001. Biochem. Biophys. Res. Commun. 289:917–923.
Bayer, K. U., Harbers, K., and Schulman, H. 1998. EMBO J. 17:5598–5605.
Beauman, S. R., Campos, B., Kaetzel, M. A., and Dedman, J. R. 2003. Cell Signal. 15:1049–1057.
Bennett, M. K., Erondu, N. E., and Kennedy, M. B. 1983. J. Biol. Chem. 258:12735–12744.
Braun, A. P., and Schulman, H. (1995). Annu. Rev. Physiol. 57:417–445.
Dolmetsch, R. E., Pajvani, U., Fife, K., Spotts, J. M., and Greenberg, M. E. 2001. Science 294:333–339.
Edman, C. F., and Schulman, H. 1994. Biochim. Biophys. Acta 1221:89–101.
Elgersma, Y., Fedorov, N. B., Ikonen, S., Choi, E. S., Elgersma, M., Carvalho, O. M., Giese, K. P., and Silva, A. J. 2002. Neuron 36:493–505.
Gangopadhyay, S. S., Barber, A. L., Gallant, C., Grabarek, Z., Smith, J. L., and Morgan, K. G. 2003. Biochem. J. 372:347–357.
Geguchadze, R., Zhi, G., Lau, K. S., Isotani, E., Persechini, A., Kamm, K. E., and Stull, J. T. 2004. FEBS Lett. 557:121–124.
Giese, K. P., Fedorov, N. B., Filipkowski, R. K., and Silva, A. J. 1998. Science 279:870–873.
Ginnan, R., and Singer, H. A. 2002. Am. J. Physiol. (Cell Physiol.) 282:C754–C761.

Grueter, C. E., Abiria, S. A., Dzhura, I., Wu, Y., Ham, A. J., Mohler, P. J., Anderson, M. E., and Colbran, R. J. 2006. Mol. Cell 23:641–650.
Grueter, C. E., Colbran, R. J., and Anderson, M. E. 2007. J. Mol. Med. 85:5–14.
Hanson, P. I., and Schulman, H. 1992. J Biol. Chem. 267:17216–17224.
Hanson, P. I., Meyer, T., Stryer, L., and Schulman, H. 1994. Neuron 12:943–956.
Hixon, M. L., and Gualberto, A. 2003. Cell Cycle 2:105–110.
Hoelz, A., Nairn, A. C., and Kuriyan, J. 2003. Mol. Cell 11:1241–1251.
Holmfeldt, P., Zhang, X., Stenmark, S., Walczak, C. E., and Gullberg, M. 2005. EMBO J. 24:1256–1266.
House, S. J., Ginnan, R., Armstrong, S., and Singer, H. A. 2007. Am. J. Physiol. (Cell Physiol.) 292(6):C2276-87. Epub 2007 Jan 31.
Isotani, E., Zhi, G., Lau, K. S., Huang, J., Mizuno, Y., Persechini, A., Geguchadze, R., Kamm, K. E., and Stull, J. T. 2004. Proc. Natl. Acad. Sci. USA 101:6279–6284.
Jett, M.-T., Schworer, C. M., Bass, M., and Soderling, T. R. 1987. Arch. Biochem. Biophys. 255:354–360.
Jiang, Z., Grange, R. W., Walsh, M. P., and Kamm, K. E. 1997. FEBS 413:441–445.
Jones, R. J., Jourd'heuil, D., Salerno, J. C., Smith, S. M., and Singer, H. A. 2007. Am. J. Physiol. Heart Circ. Physiol. 292(6):H2634-42. Epub 2007 Feb 9.
Kanaseki, T., Ikeuchi, Y., Sugiura, H., and Yamauchi, T. 1991. J. Cell Biol. 115:1049–1060.
Klemke, R. L., Cai, S., Giannini, A. L., Gallagher, P. L., De Lanerolle, P., and Cheresh, D. A. 1996. J. Cell Biol. 137:481–492.
Landsberg, J. W., and Yuan, J. X. 2004. News Physiol. Sci. 19:44–50.
Lincoln, T. M., Wu, X., Sellak, H., Dey, N., and Choi, C. S. 2006. Front. Biosci. 11:356–367.
Lu, K. K., Armstrong, S. E., Ginnan, R., and Singer, H. A. 2005. Am. J. Physiol. (Cell Physiol.) 289:C1343–C1350.
Maier, L. S., Bers, D. M., and Brown, J. H. 2007. Cardiovasc. Res. 73:629–630.
Marganski, W. A., Gangopadhyay, S. S., Je, H. D., Gallant, C., and Morgan, K. G. 2005. Circ. Res. 97:541–549.
Mayer, B. J., and Eck, M. J. 1995. Curr. Biol. 5:364–367.
Mayer, P., Mohlig, M., Schatz, H., and Pfeiffer, A. 1993. FEBS Lett. 333:315–318.
McCarron, J. G., McGeown, J. G., Reardon, S., Ikebe, M., Fay, F. S., and Walsh, J. V., Jr. 1992. Nature 357:74–77.
McNeill, R. B., and Colbran, R. J. 1995. J. Biol. Chem. 270:10043–10049.
Mishra-Gorur, K., Singer, H. A., and Castellot, J. J., Jr. 2002. Am. J. Pathol. 161:1893–1901.
Morgan, K. G., and Gangopadhyay, S. S. (2001). J. Appl. Physiol. 91:953–962.
Muthalif, M. M., Benter, I. F., Karzoun, N., Fatima, S., Harper, J., Uddin, M. R., and Malik, K. U. 1998. Proc. Natl. Acad. Sci. USA 95:12701–12706.
Nagel, D. J., Aizawa, T., Jeon, K. I., Liu, W., Mohan, A., Wei, H., Miano, J. M., Florio, V. A., Gao, P., Korshunov, V. A., Berk, B. C., and Yan, C. 2006. Circ. Res. 98:777–784.
Ohsako, S., Nishida, Y., Ryo, H., and Yamauchi, T. 1993. J. Biol. Chem. 268:2052–2062.
Okamoto, K., Narayanan, R., Lee, S. H., Murata, K., and Hayashi, Y. 2007. Proc. Natl. Acad. Sci. USA 104:6418–6423.
Owens, G. K. 1995. Physiol. Rev. 75:487–517.
Owens, G. K., Kumar, M. S., and Wamhoff, B. R. 2004. Physiol. Rev. 84:767–801.
Patel, R., Holt, M., Philipova, R., Moss, S., Schulman, H., Hidaka, H., and Whitaker, M. 1999. J. Biol. Chem. 274:7958–7968.
Pauly, P. R., Bilato, C., Sollott, S. J., Monticone, R., Kelly, P. T., Lakatta, E. G., and Crow, M. T. 1995. Circ Res 91:1107–1115.
Pfleiderer, P. J., Lu, K. K., Crow, M. T., and Singer, H. A. 2004. Am. J. Physiol. (Cell Physiol.) 286(6):C1238-45. Epub 2004 Feb 4.
Robison, A. J., Bass, M. A., Jiao, Y., MacMillan, L. B., Carmody, L. C., Bartlett, R. K., and Colbran, R. J. 2005. J. Biol. Chem. 280:35329–35336.
Rokolya, A., and Singer, H. A. 2000. Am. J. Physiol. (Cell Physiol.) 278:C537–C545.
Rosenberg, O. S., Deindl, S., Sung, R. J., Nairn, A. C., and Kuriyan, J. 2005. Cell 123:849–860.

Santella, L., and Carafoli, E. 1997. FASEB J. 11:1091–1109.
Schworer, C. M., Rothblum, L. I., Thekkumkara, T. J., and Singer, H. A. 1993. J. Biol. Chem. 268:1443–1449.
Scott-Woo, G. C., Sutherland, C., and Walsh, M. P. 1990. Biochem. J. 268:367–370.
Shen, K., Teruel, M. N., Subramanian, K., and Meyer, T. 1998. Neuron 21:593–606.
Singer, H. A., Abraham, S. T., and Schworer, C. M. 1996. In: *Biochemistry of Smooth Muscle Contraction*, M. Barany (ed.). Academic Press, San Diego,143–153.
Singer, H. A., Benscoter, H. A., and Schworer, C. M. 1997. J. Biol. Chem. 272:9393–9400.
Soderling, T. R., and Stull, J. T. 2001. Chem. Rev. 101:2341–2352.
Srinivasan, M., Edman, C. F., and Schulman, H. 1994. J. Cell Biol. 126:839–852.
Takeuchi, Y., Yamamoto, H., Fukunaga, K., Miyakawa, T., and Miyamoto, E. 2000. J. Neurochem. 74:2557–2567.
Tansey, M. G., Word, R. A., Hidaka, H., Singer, H. A., Schworer, C. M., Kamm, K. E., and Stull, J. T. 1992. J. Biol. Chem. 267:12511–12516.
Tansey, M. G., Luby-Phelps, K., Kamm, K. E., and Stull, J. T. 1994. J. Biol. Chem. 269:9912–9920.
Tobimatsu, T., and Fujisawa, H. 1989. J. Biol. Chem. 264:17907–17912.
Tombes, R. M., Grant, S., Westin, E. H., and Krystal, G. 1995. Cell Growth Differ. 6:1063–1070.
Tombes, R. M., Faison, M. O., and Turbeville, J. M. 2003. Gene 322:17–31.
Turley, E. A. 2001. Circ. Res. 88:2–4.
Urquidi, V., and Ashcroft, S. J. H. 1995. FEBS Lett 358:23–26.
Van Riper, D. A, and Singer, H. A. 2000. Mol. Cell Biochem. 213:83–92.
Van Riper, D. A., Weaver, B. A., Stull, J. T., and Rembold, C. M. (1995). Am. J. Physiol. 268:H2466–H2475.
Wagner, S., Dybkova, N., Rasenack, E. C., Jacobshagen, C., Fabritz, L., Kirchhof, P., Maier, S. K., Zhang, T., Hasenfuss, G., Brown, J. H., Bers, D. M., and Maier, L. S. 2006. J. Clin. Invest. 116:3127–3138.
Wilkerson, M. K., Heppner, T. J., Bonev, A. D., and Nelson, M. T. 2006. Am. J. Physiol. Heart Circ. Physiol. 290:H240–H247.
Wu, Y., Shintani, A., Grueter, C., Zhang, R., Hou, Y., Yang, J., Kranias, E. G., Colbran, R. J., and Anderson, M. E. 2006. J. Mol. Cell. Cardiol. 40:213–223.
Yamagata, Y., Czernik, A. J., and Greengard, P. 1991. J. Biol. Chem. 266:15391–15397.
Zhang, S., Yang, Y., Kone, B. C., Allen, J. C., and Kahn, A. M. 2003. Circulation 107:1539–1544.

Chapter 19
The KCa3.1 Channel in Endothelial Cells as New Target for an EDHF-Based Control of Vascular Tone: From Structure to Regulation and Pharmacological Properties

Umberto Banderali, Line Garneau, Manuel Simoes, Hélène Klein, and Rémy Sauvé

Abstract The vasodilator process known as endothelium-derived hyperpolarizing factor (EDHF) has been documented to critically depend on the presence of the KCa2.3 (SK3) and KCa3.1 calcium-activated potassium channels in endothelial cells. Studies on the regulatory mechanisms and structure of the KCa3.1 channel were further prompted by the observation that KCa3.1 inhibition prevented restenosis after angioplasty. In fact, both KCa2.3 and KCa3.1 are now considered as new targets for a pharmacological control of vascular tone. The KCa3.1 channel belongs to the six-transmembrane-domain channel family with a Ca^{2+} sensitivity conferred by the Ca^{2+}-binding protein calmodulin that is constitutively bound to the membrane-proximal region of the channel intracellular C-terminus. Recent data have shown that the activation of KCa3.1 by Ca^{2+} also requires the presence of phosphatidylinositol 3-phosphate (PI(3)P) and is inhibited by MTMR6, a PI(3)P phosphatase bound to KCa3.1 in C-terminus. Furthermore, channel activity was reported to be regulated by ATP through the phosphorylation by the nucleoside diphosphate kinase NDPK-B of a histidine residue located within the KCa3.1 domain responsible for the Ca^{2+}-dependent binding of calmodulin to the channel in C-terminus. Despite these important advances in our understanding of the channel cellular regulation, structural and functional data concerning the KCa3.1 channel pore and gating regions remain scarce. With the solved 3D structures of the KcsA, MthK, and Kv1.2 potassium channels, we combined predictions from computer-based homology modeling and substituted cysteine accessibility measurements to obtain structural information on the KCa3.1 pore region and gate location. Residues lining the channel pore were for the first time identified providing the structural bases to characterize the molecular interactions between KCa3.1 and channel inhibitors such as TRAM-34 and clotrimazole. Similarly, the conformational changes involved in channel opening could be investigated at the molecular level enabling a detailed description of how channel openers and intracellular factors such as free radicals affect the channel open probability. This information is essential to improve drug design targeted at KCa3.1 control and EDHF regulation.

A. K. Srivastava, M. B. Anand-Srivastava (eds.) *Advances in Biochemistry in Health and Disease.* Signal Transduction in the Cardiovascular System in Health and Disease,

Introduction

Endothelial cells are recognized to play a prominent role in the regulation of vascular tone by releasing vasoactive compounds such as NO and prostacyclin in response to hormonal or physical stimuli. Numerous studies have in addition demonstrated that there is an important endothelial pathway, distinct from cyclooxygenase and NO-synthase, also involved in the relaxation of vascular smooth muscle cells (Busse et al. 2002; Feletou and Vanhoutte 2004). This mechanism globally termed EDHF (endothelium-dependent hyperpolarizing factor) depends to a large extent on the presence of K^+-selective channels that regulate the endothelial cell membrane potential. Increasing evidence indicates that the calcium-activated KCa2.3 and KCa3.1 channels constitute key determinants to the EDHF response in various blood vessels from different species including human (Feletou and Vanhoutte 2004). In this chapter we present an overview of the KCa3.1 channel cellular regulation and pharmacological, properties in relation to emerging structural data obtained from cysteine-scanning mutagenesis experiments and computer-based molecular modeling.

Ca^{2+}-Activated K^+ Channels in Endothelial Cells

Three main classes of Ca^{2+}-activated K^+ channels have been to date identified on the basis of their permeation properties and pharmacological profiles. These include the charybdotoxin- and iberiotoxin-sensitive KCa1.1 channels (MaxiK) of large conductance (150–220 pS) (Miller et al. 1985), the apamin-sensitive and -insensitive SK channels of small conductance (KCa2.1, KCa2.2, KCa2.3) (Kohler et al. 1996), and the intermediate-conductance (20–50 pS) KCa3.1 (IKCa) channel inhibited by CLT (clotrimazole) (Rittenhouse et al. 1997) and TRAM-34 (Wulff et al. 2000). It is now generally recognized that the variations in membrane potential in response to an increase in internal Ca^{2+} in endothelial cells are governed by KCa2.3 and KCa3.1 channels, whereas KCa1.1 would play a determinant role in controlling the voltage-dependent Ca^{2+} influx in nonproliferating VSMC (vascular smooth muscle cells).

Few electrophysiological studies on endothelial cells have provided direct evidence for KCa2.x channels of small conductance. A voltage-independent apamin-sensitive channel of 6–7 pS was identified in the endothelium of intact rat aorta (Marchenko and Sage 1996) and single-channel measurements performed on pig coronary artery endothelial cells have revealed an apamin-sensitive channel of 6.8 pS as well. The latter observations were corroborated by immunofluorescent labeling and reverse transcriptase-polymerase chain reaction (RT-PCR) measurements supporting the presence of KCa2.3 channels in these cells (Burnham et al. 2002). Despite the limited number of single-channel studies providing direct evidence for KCa2.x channels in endothelial cells, the large number of studies where an apamin-dependent inhibition of EDHF was measured (Dong et al. 2000; Andersson et al. 2000) clearly argues for a broad expression of KCa2.3 in various endothelial preparations.

The presence of KCa3.1 channels was similarly confirmed in several single-channel studies on vascular endothelial cells (Sauvé et al. 1988; Vaca et al. 1992; Colden-Stanfield et al. 1992; Kohler et al. 2000; Bychkov et al. 2002). These observations were subsequently complemented by experiments performed in human mesenteric artery and rat carotid artery endothelial cells using single–cell RT-PCR analysis and patch clamping (Kohler et al. 2000, 2001). Again, most of the evidence supporting a broad distribution of KCa3.1 in endothelial cells comes from the observation that apamin in combination with TRAM-34 or CLT, two rather specific inhibitors of KCa3.1, impairs EDHF in a large number of blood vessels. In fact, numerous studies have revealed that KCa3.1 in endothelial cells are rapidly activated in response to humoral and/or mechanical stimuli leading to a hyperpolarization of the cell membrane. It has often been proposed that a KCa3.1-induced hyperpolarization of the endothelium indirectly modulates vascular tone by increasing the electrodriving force for Ca^{2+} entry and thus promoting the Ca^{2+}-dependent synthesis and release of vasoactive agents such as NO and prostacyclin. Studies in intact peripheral arteries have shown, however, that the agonist-mediated Ca^{2+} influx in endothelial cells responds poorly to KCa3.1 + KCa2.3 blocking agents (such as CLT or TRAM-34 plus apamin) suggesting that the main contribution of KCa3.1 to the control of vascular tone is downstream of the endothelial Ca^{2+} response (Ghisdal and Morel 2001; Marrelli et al. 2003). As the theoretical electrochemical potential for Ca^{2+} ions at a membrane voltage of −30 mV approximates −260 mV, a 20- to 30-mV hyperpolarization is indeed expected to contribute little extra to the electrodriving force acting on a passive Ca^{2+} entry. The mechanism by which KCa3.1 activation translates into VSMC hyperpolarization is not likely therefore to involve an increase in NO and prostacyclin release. Evidence has been presented, however, supporting an electrical coupling between the endothelium and the adjacent VSMC via myoendothelial gap junctions (Yamamoto et al. 1999; Chaytor et al. 2001) (see Figure 19.1). Other studies suggest in addition a stimulation of the smooth muscle Kir channels and/or the Na-K-ATPase by a modest increase of the extracellular K^+ concentration (Edwards et al. 1998; Busse et al. 2002). Despite the diversity in functional mechanisms proposed to date to account for the VSMC hyperpolarization following KCa3.1 activation, it is generally recognized that the stimulation of KCa3.1 constitutes an obligatory step to the EDHF vasodilator process. In support of this proposal is the recent observation that KCa3.1-knockout mice showed a significant increase in arterial blood pressure and a mild left ventricular hypertrophy (Si et al. 2006). These results confirmed that the endothelial KCa3.1 is a fundamental determinant of endothelial hyperpolarization and EDHF signaling.

The role of KCa3.1 channels in endothelial cells is not strictly limited to the hyperpolarization of the cell potential. Of special interest is the observation that KCa3.1 channels play a pivotal role in cell proliferation. For instance, the expression of KCa3.1 has been reported to increase in mesenteric artery endothelial cells of patients with colonic adenocarcinoma, resulting in an enhanced bradykinin-induced endothelial hyperpolarization compared with controls (Kohler et al. 2000, 2001). These findings support previous results obtained with fibroblasts where it was shown that the increase in KCa3.1 expression observed following mitogen stimulation was related to the activation of the Raf/Ras pathways (Pena and Rane 1997). Evidence

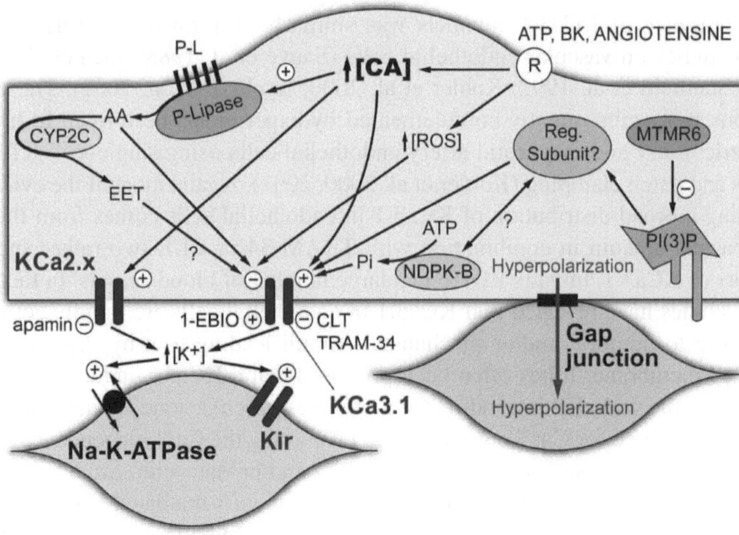

Fig. 19.1 Regulation of KCa3.1 in endothelial cells. In the proposed scheme, stimulation of receptors linked to the phosphoinositide cascade causes an increase in internal Ca^{2+} concentration which in turn results in KCa2.x and KCa3.1 channel activation via a CaM-dependent process and in a hyperpolarization of the endothelial cells. In one possible mechanism underlying EDHF, the resulting voltage change is transmitted to adjacent smooth muscle cells directly via gap junctions leading to vascular relaxation. Alternatively, KCa2.x and KCa3.1 channel activation may induce an elevation of the local K^+ concentration in the interstitial space causing a hyperpolarization of the smooth muscle cells through the activation of the Na-K-ATPase and/or Kir channels. Endothelial cell stimulation by Ca^{2+}-mobilizing agonists may also be accompanied by an increase in the internal arachidonic acid (AA) concentration resulting in an inhibition of KCa3.1 through a direct effect of AA on KCa3.1. AA may in turn be converted into epoxyeicosatrienoic acid (EETs) via the CYP2C epoxygenase pathway and exert a stimulatory effect on KCa1.1 (Maxi KCa) in smooth muscle cells. Evidence has also been provided indicating that the activation of KCa3.1 by Ca^{2+} critically depends on PI(3)P and is inhibited by MTMR6, a PI(3)P phosphatase bound to KCa3.1 in C-terminus. PI(3)P did not appear to interact directly with KCa3.1, but more probably exerts its effect via a yet-to-be-identified accessory protein. In addition, KCa3.1 can be activated through the phosphorylation by NDPK-B of a histidine residue at position 358 in C-terminus but the nature of the interactions between NDPK-B and PI(3)P is not completely understood. A PKA and a PKC related modulation of KCa3.1 activity was similarly reported but these effects appeared tissue dependent. Finally, KCa3.1 activity was found to be ROS sensitive with an overproduction of hydroxyl radicals OH^- leading to channel inhibition.

was also provided for an impaired expression of KCa3.1 channels in regenerated endothelium from rat carotid arteries after angioplasty (Kohler et al. 2001). It was concluded on the basis of these observations that the regenerated endothelium has a reduced capacity to hyperpolarize and dilate due to a decrease of the KCa3.1 expression. The functional link between KCa3.1 channel activity and DNA synthesis remains, however, to be established. An altered KCa3.1 expression has also been reported in rat coronary artery VSMC following balloon catheter injury (Kohler et al. 2003). In this case, neointimal formation was associated with a switch in expression from exclusively KCa1.1 in mature contractile VSMC to downregulated KCa1.1 and upregulated KCa3.1 expression during VSMC proliferation. More importantly, the use of the KCa3.1 inhibitor TRAM-34 was found to prevent restenosis, thus suggesting a novel therapeutic strategy for the prevention of restenosis after angioplasty (Kohler et al. 2003). An upregulation of KCa3.1 expression was similarly observed in human endothelium under long-term laminar shear stress conditions (Brakemeier et al. 2003). Shear stress-induced KCa3.1 expression might therefore represent an important mechanism in long-term vessel adaptation in response to an increase in hemodynamic forces as present in hypertension. Taken together, these observations argue for KCa3.1 as being a key determinant to the short- and long-term endothelium regulation of vascular tone.

KCa3.1 Channel Regulation

Ca^{2+}-Dependent Regulation

An important contribution to our understanding of the KCa3.1 channel molecular identity came from the cloning of the human KCa3.1 channel (Joiner et al. 1997). KCa3.1 channels are tetrameric proteins with each subunit comprising 427 amino acids organized in six transmembrane segments S1–S6 with a pore motif between segments 5 (S5) and 6 (S6) (Figure 19.2). The KCa3.1 Ca^{2+} sensitivity is conferred

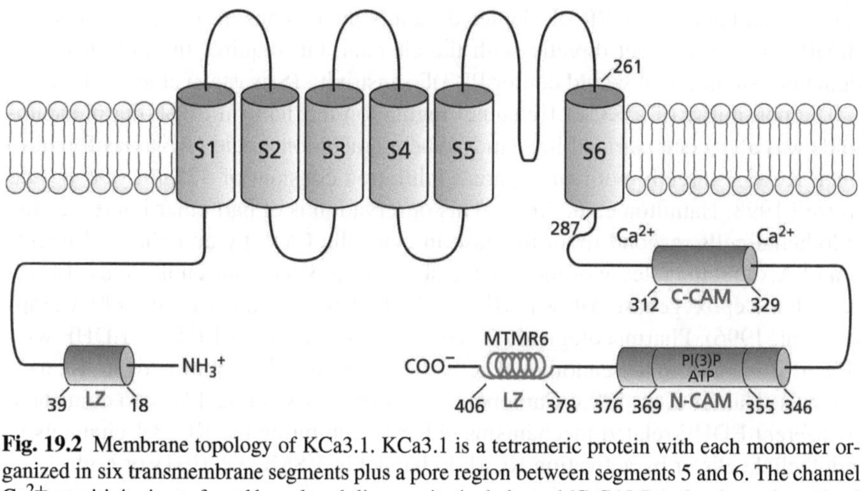

Fig. 19.2 Membrane topology of KCa3.1. KCa3.1 is a tetrameric protein with each monomer organized in six transmembrane segments plus a pore region between segments 5 and 6. The channel Ca^{2+} sensitivity is conferred by calmodulin constitutively bound (C-CAM) to the channel proximal region in C-terminus [aa 312–329]. A stretch of 14 amino acids (aa 355–369) has been identified in C-terminus as being essential for channel regulation by ATP and PI(3)P while the binding of the MTMR6 phosphatase to the channel is thought to involve the coil-coil segment of MTMR6 and the coil-coil domain of KCa3.1 in C-terminus (aa 383–406). Finally, KCa3.1contains two leucine zipper motifs (LZ) in N[aa 18–39] and C[aa 378–406] terminus, respectively, critical for channel assembly and trafficking.

by the Ca^{2+}-binding protein calmodulin (CaM) the C-lobe of which is constitutively bound to a domain in the membrane-proximal region of the intracellular C-terminus of the channel (Maylie et al. 2004). This is at variance with the KCa1.1 channel where Ca^{2+} sensitivity is voltage dependent and arises from direct interactions of Ca^{2+} with different segments of the channel C-terminus (Zeng et al. 2005). CaM also regulates KCa3.1 channel assembly and trafficking and as such represents an essential accessory protein for proper KCa3.1 functioning (Joiner et al. 2001). With the exception of CaM, no auxiliary protein has been to date documented for KCa3.1. The KCa3.1 amino acid sequence also contains two leucine zipper motifs (LZ) respectively located at the channel N- and C-terminal ends (Figure 19.2). The presence of these motifs is critical for channel assembly and trafficking (Syme et al. 2003). KCa3.1 presents in addition a putative N-linked glycosylation site adjacent to the pore, the importance of which remains to be established.

Phospholipids

The activation of KCa3.1 by Ca^{2+} was recently reported to specifically require PI(3)P and to be inhibited by MTMR6 (phosphatase myotubularin-related protein 6), a PI(3)P phosphatase bound to the coil coil segment of KCa3.1 in C-terminus (Srivastava et al. 2006a) (Figures 19.1 and 19.2). A PI(3)P-dependent channel activation in the presence of Ca^{2+} was not observed with the KCa2.2 channel, indicating that PI(3)P represents a unique signaling pathway for KCa3.1 regulation. Because PI(3)P added internally in inside-out patch clamp experiments failed to restore channel activity in PI(3)P-depleted membrane patches, it was concluded that PI(3)P does not interact directly with the channel, but requires the presence of a regulatory subunit that would confer PI(3)P sensitivity (Srivastava et al. 2006a).

The possibility of a KCa3.1 channel regulation mechanism involving phospholipids was also demonstrated in a study where arachidonic acid (AA) was found to block KCa3.1 directly with an apparent inhibition constant of 425 nM (Devor and Frizzell 1998; Hamilton et al. 2003). This observation is of particular interest as the endothelial cells respond to an increase in cytosolic Ca^{2+} by an enhanced liberation of AA and the release of the AA metabolites epoxyeicosatrienoic acids (EETs) through the epoxygenase pathway (Figure 19.1) (Channon and Leslie 1990; Campbell et al. 1996). Pharmacological evidence supports a role of EETs in EDHF with EETs diffusing from the endothelial cells to the VSMC where they activate KCa1.1 channels (Harder et al. 1995; Campbell et al. 1996). AA is thus likely to contribute to different EDHF-related mechanisms either by inhibiting the KCa3.1 channels in endothelial cells or by activating KCa1.1 channel in VSMC via EETs or both.

ATP-Dependent Regulation

There is also evidence from inside-out patch-clamp experiments that the human and canine KCa3.1 channels respectively expressed in *Xenopus leavis* oocytes and

HEK-293 cells are activated by ATP in the presence of Ca^{2+}. Similar results were obtained with the KCa3.1 channels present in submandibular gland. The ATP-dependent stimulation of KCa3.1 could be reversed by both alkaline and acid phosphatases, suggesting that ATP activates KCa3.1 via a membrane-delimited kinase. Such a behavior contrasts with thc absence of ATP-dependent control of the KCa2.1 and KCa2.2 channel activity (Gerlach et al. 2001). The contribution of PKA and PKC to KCa3.1 regulation remains, however, tissue dependent as the stimulatory action of ATP on KCa3.1 could be prevented using the PKA inhibitor PKI_{5-24} in T84 cells, but not in HEK-293 cells expressing KCa3.1 (Gerlach et al. 2000). Furthermore, PKA was reported to inhibit, not activate, the rat homologue KCa3.1 channel expressed in *X. laevis* oocytes through direct phosphorylation of a modulatory locus in the cytoplasmic region of the channel (Neylon et al. 2004). These observations argue for an ATP-dependent regulation of the KCa3.1 channel activity that involves multiple intracellular sites and several ATP-sensitive auxiliary proteins. An analysis of the KCa3.1 sequence shows a single PKA phosphorylation consensus site at position 334 in C-terminus, and four potential sites for phosphorylation by PKC (T101, S178, T329, S388). Mutating the PKA-consensus site S334 did not, however, affect the basal and/or ATP-activated currents, arguing for KCa3.1 being not itself the target for phosphorylation at this site (Gerlach et al. 2001). An extensive study in which both truncated KCa3.1 and KCa3.1-KCa2.2 chimeras were functionally analyzed, led to the identification of a region in C-terminus of KCa3.1 critical for ATP-dependent activation (Gerlach et al. 2000, 2001). Indeed, the segment R355-A413 was reported to be necessary and sufficient for an ATP-dependent regulation of KCa3.1 activity, with the first 14 amino acids playing a critical role as they impart partial ATP dependence when transferred to the KCa2.2 channel (Figure 19.2). This 14-residue segment is located within the Ca^{2+}-dependent CaM binding domain of KCa3.1 in C-terminus (Figure 19.2) suggesting an ATP-related modulation of the Ca^{2+}-dependent binding of CaM to KCa3.1 (Srivastava et al. 2006a). Notably, the same 14-amino-acid span was reported to confer PI(3)P sensitivity to KCa3.1 and to critically contribute to the recruiting and binding of the nucleoside diphosphate kinase NDPK-B to the channel (Figures 19.1 and 19.2). Once bound, NDPK-B is thought to directly phosphorylate a His at position 358, leading to channel activation (Srivastava et al. 2006b). The interplay between PI(3)P, NDPK-B, and KCa3.1 still remains to be clarified, but the evidence gathered to date does not support a model involving the binding of NDPK-B to PI(3)P or a control of the NDPK-B kinase activity by PI(3)P (Srivastava et al. 2006b). The possibility of a yet-to-be-identified regulatory protein mediating the stimulatory action of PI(3)P on KCa3.1 cannot currently be ruled out (Figure 19.1).

ROS-Related Regulation

Work from our laboratory has provided single-channel evidence for a redox control of the KCa3.1 channel in bovine aortic endothelial cells (Cai and Sauvé 1997). In these cells, exposure to the reactive oxygen species (ROS) generating system

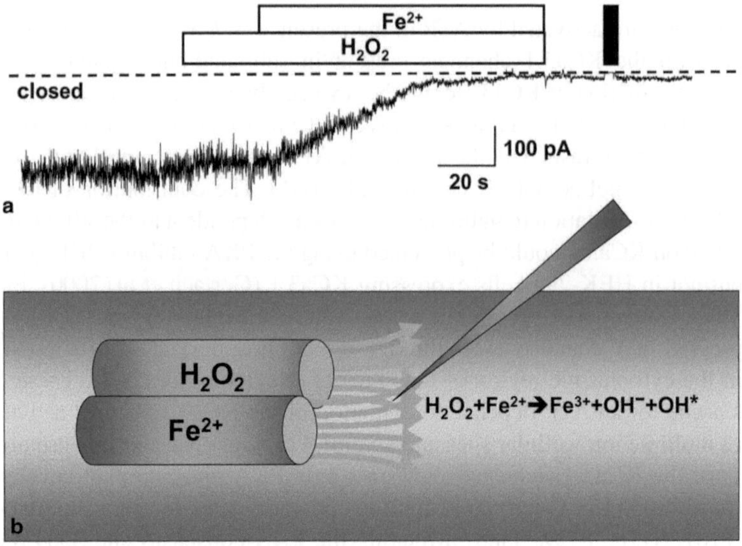

Fig. 19.3 Inhibition of KCa3.1 by ROS. **(A)** Inside-out patch-clamp experiment illustrating KCa3.1 inhibition by hydroxyl radicals $OH^•$. The channel appeared insensitive to H_2O_2 but a clear decrease in channel activity was observed following the mixture of the H_2O_2 and Fe^{2+} solutions. Current recording performed at a pipette potential of 60 mV. Perfusion with a Ca^{2+}-free solution is represented by the black rectangle. **(B)** $OH^•$ were generated according to the Fenton reaction by mixing two 200 mM K_2SO_4 plus 25 μM free Ca^{2+} solutions respectively containing 10 mM H_2O_2 and 10 μM Fe^{2+} directly at the tip of the patch pipette.

H_2O_2 caused a channel inhibition that was partly reversible by dithiothreitol (DTT). Patch excised experiments performed on KCa3.1 channels expressed in *X. laevis* oocytes confirmed that KCa3.1 can be irreversibly inhibited by H_2O_2 in the presence of Fe^{2+}, suggesting that hydroxyl radicals ($OH^•$) specifically react with KCa3.1 residues (cysteine, methionine, histidine, lysine) leading to a nonfunctional channel (Figure 19.3). This proposal is further supported by the observation that KCa3.1 remained insensitive to the $OH^•$ generating system $H_2O_2 + Fe^{2+}$ in the presence of vitamin C, a $OH^•$ scavenger (data not shown). As the production of ROS has been associated with dysfunctions of the endothelial cells in pathologies such as hypertension and diabetes, an inhibitory action of ROS on KCa3.1 would further exacerbate the impaired vasorelaxation function of the endothelium in ROS conditions.

Molecular Basis of KCa3.1 Regulation and Pharmacological Properties

Structural Features of the KCa3.1 Channel Pore

Although the primary structure of KCa3.1 has been established in a variety of cell preparations, little is actually known on the 3D structure of KCa3.1. In the absence of X-ray structural data, our laboratory has implemented an approach combining substituted cysteine accessibility measurements and computer-based homology modeling to generate the first molecular description of the open/closed KCa3.1 pore region (Banderali et al. 2004). To model the KCa3.1 channel in the open conformation we used as template the bacterial Ca^{2+}-activated potassium channel MthK. The resulting structure predicts that the V275, T278, and V282 residues should be lining the channel pore with V275 and T278 contributing to the formation of a central inner cavity approximately 10 Å wide and with pore dimensions exceeding 18 Å in diameter for the region extending from the channel central cavity to the internal medium (Figure 19.4A). Results from an analysis of the KCa3.1 channel voltage-dependent block by the quaternary ammonium cation tributylammonium (TBA) (Banderali et al. 2004) have indicated in this regard that the pore diameter of the open KCa3.1 should be wide enough to accommodate a molecule with a diameter greater than 5 Å with more than 81% of the applied transmembrane potential restricted to the selectivity filter region. It was concluded that the open channel pore structure is compatible with the structure reported for the bacterial MthK channel, with a pore inner vestibule wide enough for K^+ ions to have an unrestricted access to the channel cavity up to the selectivity filter (Figure 19.4A).

A combined SCAM (substituted cysteine accessibility method) and molecular modeling approach was also used to address the question of the KCa3.1 channel pore structure in the closed configuration. We had proposed in a previous work a 3D homology-based model of the pore-forming S6 transmembrane segment for the closed KCa3.1 configuration using the bacterial KcsA channel structure as template (Simoes et al. 2002). This model predicts that the diameter of the KCa3.1 conducting pathway should vary along the channel central axis of diffusion with a minimum van der Waals diameter of 2.0 Å at the level of the V282 residue (Figure 19.4B). The pore structure for the closed KCa3.1 channel based on a KcsA template would thus be characterized by a bundle-crossing region extending from V282 to A286 with the presence of a tight hydrophobic seal at the level of the V282 residue (Figure 19.4B). Data supporting this model would strongly argue for a KCa3.1 activation gate located at the C-terminal end of the transmembrane S6 segments capable of controlling K^+ ion flow. The SCAM results we obtained showed, however, that the bundle-crossing region of closed KCa3.1 is permeable to hydrophilic molecules with a diameter as large as 4.8 Å, indicating (1) that the bundle-crossing region of KCa3.1 in the closed state cannot be adequately represented by a KcsA-like structure and (2) that the Ca^{2+}-dependent gating of KCa3.1 involves different sections of the S6 segment with the C-terminal end of S6 not constituting a hydrophobic seal capable of controlling K^+ ion flow (Klein et al. 2007).

A Molecular View of the KCa3.1 Ca^{2+}-Dependent Activation

The exact molecular mechanism underlying KCa3.1 opening in response to Ca^{2+} binding to the CaM–KCa3.1 complex in C-terminus remains to be elucidated.

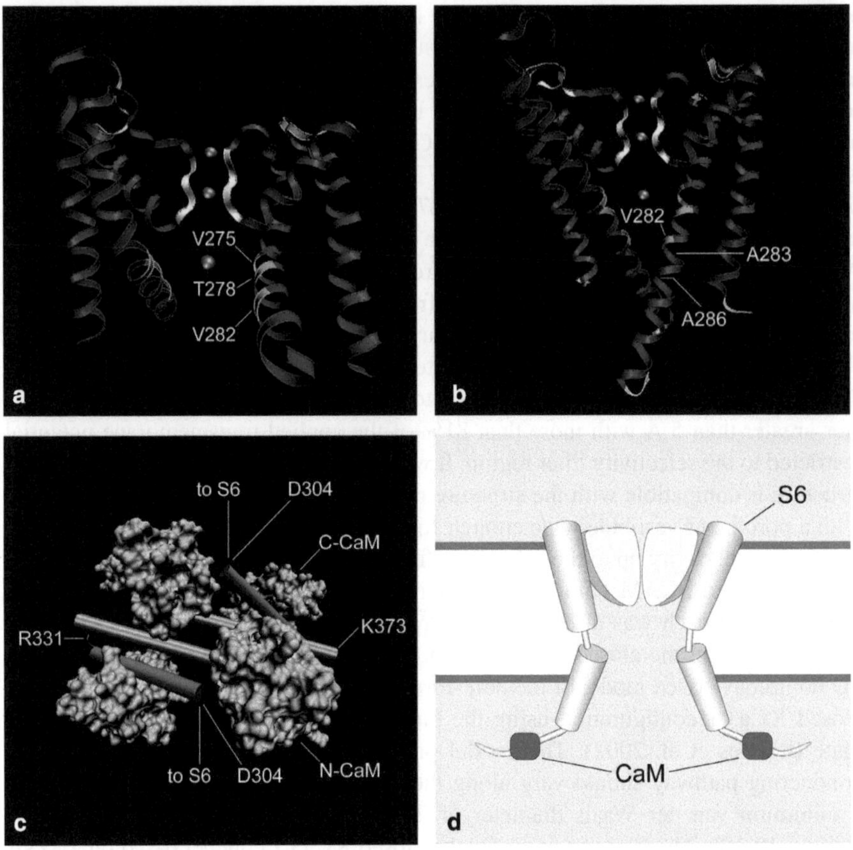

Fig. 19.4 Model structures of the S5–pore–S6 segment of KCa3.1 channel generated by homology modeling. (**A**) Model of the S5–pore–S6 region for the open KCa3.1 conformation computed using the MthK structure as template. In green, residues predicted to be lining the channel pore and to form the channel central cavity (V275, T278, V282). Also illustrated in white is the selectivity filter region with three K^+ ions (in gold). (**B**) Model of the KCa3.1 S5–pore–S6 segment for the closed IKCa3.1 generated using the KcsA potassium channel as template. According to this representation, the bundle-crossing region (V282 in green, A283 and A286 in blue) should form a constriction in the channel pore tight enough to generate an impermeable barrier to K^+ ion flow. (**C**) Model representation of two CaM molecules (in blue) bridging two KCa3.1 C-terminal segments extending from D304 to K373 (orange cylinders) leading to the formation of a dimeric structure between two adjacent KCa3.1 monomers. This bridging process has been proposed to induce a rotation/translation of the S6 transmembrane segments resulting in channel opening. The model was generated by homology modeling using the crystallized KCa2.2–CaM complex (PDB: 1G4Y) as template. Molecular representations using VMD (University of Illinois). (**D**) Schematic representation of the model proposed for the KCa3.1 closed configuration based on SCAM measurements of the KCa3.1 pore structure performed in zero Ca^{2+} conditions. This model predicts that C-terminal end of the S6 segments cannot form a barrier to the diffusion of K^+ ions, but that the control of K^+ ion flow by Ca^{2+} involves a structural change at the level of the selectivity filter or a region close to it. (See color plate.)

Structural information pertinent to channel gating has, however, been obtained through the crystallization of CaM bound to the rat KCa2.2–CaM binding domain in the presence of Ca^{2+}. These experiments confirmed that CaM binds endogenously to KCa2.2 in C-terminus close to the S6 transmembrane domain. Because the amino acid sequence of the CaM binding domains in KCa2.2 and KCa3.1 are highly homologous, a computer-based homology modeling approach was implemented to obtain structural information on the CaM binding domain of KCa3.1. The resulting model predicts that the CaM binding domain in KCa3.1 consists of a helical segment of 91 amino acids starting at the residue D304, with a hairpin folding 5 amino acids long centered at R331 (Figure 19.4C). The C-lobe of CaM would be constitutively bound to the KCa3.1 segment extending from K312 to T329 with a stoichiometry of 1 CaM molecule per channel subunit. Upon Ca^{2+} binding, it has been proposed that a large-scale conformational rearrangement would take place in which the N-lobe of CaM would interact with the C-terminus of an adjacent channel subunit at L356–M376, resulting in a subunit–subunit dimerization process. This rearrangement would in turn lead to a rotation/translation of the associated S6 transmembrane domains and to the opening of the ion-conducting pore (Schumacher et al. 2001, 2004; Wissmann et al. 2001; Maylie et al. 2004). Our SCAM results confirmed that Ca^{2+} binding to the CaM–KCa3.1 complex in C-terminus initiates a widening of the pore at the C-terminal end of the S6 segments (Klein et al. 2007). This conformational change would not, however, be responsible for the Ca^{2+}-dependent channel opening as cysteine residues engineered inside the channel cavity were found to be accessible to reagents as large as 4.8 Å with KCa3.1 in the closed state. This observation argues for a closed KCa3.1 structure (not conducting) where K^+ ions can freely diffuse in and out of the channel cavity. Our results would thus support a model whereby the S6 segments act as transducers conveying the structural changes occurring at the C-terminal end of the S6 segments to the selectivity filter. Changes in the selectivity filter structure or in a region close to it could in turn affect the passage of K^+ ion inside the pore and thus be responsible for KCa3.1 activation (Figure 19.4D).

KCa3.1 Structure and Pharmacological Properties

The channel inhibitors CLT (Rittenhouse et al. 1997) and TRAM-34 (Wulff et al. 2000) have been extensively used to confirm the contribution of KCa3.1 channels to the endothelial hyperpolarization underlying the EDHF-induced vascular smooth muscle relaxation (Edwards et al. 1998, 1999, 2000; Yamamoto et al. 1999; Dong et al. 2000; Coleman et al. 2001). More importantly, TRAM-34 was shown to prevent restenosis after angioplasty (Kohler et al. 2003) and abolish the endothelial cell proliferation induced by either the basic fibroblast growth factor or vascular endothelial growth factor *in vitro*, suggesting that KCa3.1 blockade represents a new strategy to prevent tumor angiogenesis (Grgic et al. 2005). There is therefore an urgent need for the rational design of selective KCa3.1 blockers as therapeutic agents. The design of a specific KCa3.1 blocking agent will largely depend, however, on the information available concerning the KCa3.1 pore structure. The model of the KCa3.1 central cavity derived on the basis of our SCAM results (Simoes et al. 2002; Klein et al. 2007) presents the V275, T278, and V282 residues as lining the channel pore. According to our current model, both T250 and V275 should be projecting inside the channel cavity with T250 located at the inner entrance of the selectivity filter (Figure 19.5B). These residues are thus likely to form a coordination site for several different inhibitors, such as TBA (Dunn 1998; Hamilton et al. 2003; Banderali et al. 2004; Lenaeus et al. 2005), CLT (Rittenhouse et al. 1997), and TRAM-34 (Wulff et al. 2000) (see Figure 19.5A). In support of this proposal is the observation that mutating the Val at position 275 substantially modifies the potency of TBA and TRAM-34 to inhibit KCa3.1. For instance, substituting the Val at 275 by a Cys increased by 10-fold the affinity of KCa3.1 to TBA (Banderali et al. 2004) whereas the mutations V275A and T250S decreased by more than 1000-fold the sensitivity of KCa3.1 to TRAM-34 (Wulff et al. 2001). Both T250 and V275 are also likely to account for the direct inhibitory effect of AA on KCa3.1 as the double mutation T250S–V275C was found to eliminate the AA sensitivity of KCa3.1 (Hamilton et al. 2003). Taken together these results argue for a rational design of specific KCa3.1 inhibitors based on the molecular structure proposed for the KCa3.1 channel cavity. Docking calculations have already proven in this regard that it is possible to qualitatively account for the interactions between TRAM-34 and KCa3.1. Figures 19.5B and 5C illustrate a schematic representation of one of the docking configurations obtained using the Affinity program (Accelrys) where both the ligand and the binding pocket are allowed to be flexible during the calculations. In the proposed configuration, the pyrazole group of TRAM-34 is seen to strongly interact with the T250 residue of monomer C, while two of the phenyl rings are stabilized by strong hydrophobic interactions at the level of the T250 and V275 residues on monomer A and T278 on the adjacent monomer D. These predictions are in accordance with the mutagenesis results reported on the binding of TRAM-34 to KCa3.1 (Wulff et al. 2001).

KCa3.1 is also sensitive to toxins applied externally. For instance, charybdotoxin (ChTX) was reported to inhibit KCa3.1 from the external side. In accordance with the ChTX blocking mechanism described for the Kv1.3 and KCa1.1 channels, the blocking action of ChTX on KCa3.1 most likely involves the formation of a hydrogen bond between the residue K27 of the toxin and the Tyr at position 253 deep in the channel selectivity filter. The toxin would be further stabilized by additional hydrogen bonds in the channel external vestibule with the interaction N30 (ChTX)–D239 (KCa3.1) playing the major role (Gao and Garcia 2003; Visan et al. 2004). These modeling results tend to support the overall topology of the KCa3.1 external vestibule derived from computer-based homology modeling using KcsA as template.

Finally, several molecules such as 1-EBIO (1-ethyl-2-benzimidazolinone) or 7,8-benzoquinoline have been identified as KCa3.1 channel openers. These molecules do not substitute for Ca^{2+} as their stimulatory effect is Ca^{2+} dependent. The

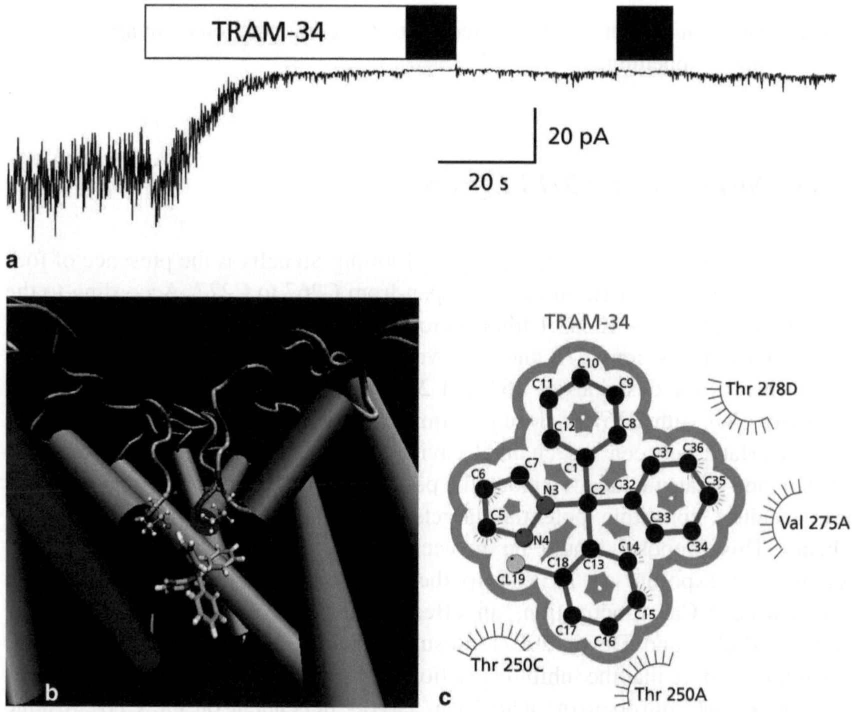

Fig. 19.5 Molecular determinants of KCa3.1 block by TRAM-34. **(A)** Inside-out patch clamp recording illustrating the inhibitory effect of TRAM-34 applied internally. Experiment performed in symmetrical 200 mM K_2SO_4 solutions plus 25 μM free internal Ca^{2+} at an applied pipette potential of 60 mV. Black boxes refer to perfusion with a Ca^{2+}-free solution. **(B)** TRAM-34 molecule trapped in the channel central cavity. The S6 segments of three monomers are presented as cylinders whereas TRAM-34 and T250 residues are shown in a CPK atomic presentation. Molecular representation by VMD (University of Illinois). Docking calculations were performed using the Affinity module of InsightII (Accelrys). No hydrogen bonding or tethered subsets were defined. The ligand was confined to stay within 5 Å of its starting position at the center of the cavity. The TRAM-34 molecule plus all of the KCa3.1 channel residues within a 7.5-Å radius from the center of the cavity were flexible during the calculations. A Monte Carlo search was performed leading to 100 structures with no acceptance filter. We used the nonbonded method: Quartic VDW with coulombic interactions off and VDW scaled to 0.1, flexible ligand with an energy range of 200 and tolerance to 10^6 and 100 steps minimization. **(C)** Schematic representation of the docking results using the LIGPLOT software. This analysis illustrates hydrophobic interactions between the pyrazole moiety of TRAM-34 and the T250 residue of monomer C, while predicting that the stabilization of two of the phenyl rings by strong hydrophobic interactions involves the T250 and V275 residues on monomer A and T278 on the adjacent monomer D. (See color plate.)

molecular mechanism underlying the action of KCa3.1 openers remains ill defined, but an analysis of the action of EBIO on chimeric KCa3.1/KCa2.2 channels suggests a modification of the interaction between the CaM N-lobe and the CaM binding domain of KCa3.1 resulting in an apparent increase in Ca^{2+} sensitivity (Pedarzani et al. 2001). A better description of how KCa3.1 openers affect channel gating will largely depend on our future understanding of the conformational changes responsible for channel opening.

KCa3.1 Structure and ROS Effects

One of the key features of the KCa3.1 pore-forming S6 helix is the presence of four cysteine residues over a 10-amino-acid span from C267 to C277. According to the model proposed for the KCa3.1 filter region and central cavity (Figures 19.4A, B), the cysteines at positions 276 and 277 would be oriented opposite to the central cavity whereas the cysteines at 267 and 269 would be located at the level of the selectivity filter with C267 in close proximity (less than 3.5 Å) of the Y253 residue at the interface between the channel cavity and the selectivity filter region. Because of their strategic location along the pore-forming S6 transmembrane segment, these residues constitute potential targets for a regulation of KCa3.1 by free radicals. This proposal would be in accordance with a recent observation indicating that exposure of KCa3.1 to the oxidative agent thimerosal results in an enhanced KCa3.1 activation, an effect attributable to the C276 and C277 residues (Bailey and Devor 2007). Results obtained from our laboratory have shown furthermore that the inhibitory action of the cysteine-specific reagent DTNB [5, 5′-dithio-bis(2-nitrobenzoic acid)] was largely dependent on the C267 residue with 50% of the DTNB-induced current inhibition being lost with the C267S mutant. Notably, a DTNB-insensitive channel could only be observed following the mutation of the 9 cysteine residues located either in the channel transmembrane segments or intracellular linkers, suggesting a synergic action of several cysteines to the KCa3.1 channel activation and integrity. Our results thus support a model where C267 constitutes a potential target for the redox control of KCa3.1 in endothelial cells owing to its critical location near the selectivity filter region.

Concluding Remarks

KCa3.1 is now recognized as a major constituent of the Ca^{2+} signaling pathway underlying the control of vascular tone by the endothelium. Accumulating evidence indicates furthermore that KCa3.1 is critically involved in the proliferation process of endothelial and vascular smooth muscle cells. It is therefore not surprising that KCa3.1 has become an important pharmacological target for the treatment of vascular diseases related to the endothelium (Jensen et al. 2002). Key residues involved in drug binding inside the channel pore have already been identified, and SCAM results have provided the first mechanistic model of the conformation changes underlying channel activation by Ca^{2+}. Additional structural information is still needed, however, to define new potential sites for drug binding and develop alternative strategies for the pharmacological control of the endothelium function.

References

Andersson, D. A., Zygmunt, P. M., Movahed, P., Andersson, T. L., and Hogestatt, E. D. 2000. Effects of inhibitors of small- and intermediate-conductance calcium-activated potassium channels, inwardly-rectifying potassium channels and Na(+)/K(+) ATPase on EDHF relaxations in the rat hepatic artery. Br. J. Pharmacol. 129:1490–1496.

Bailey, M. A., and Devor, D. C. 2007. Modification of a conserved cysteine residue within S6 alters the calcium activation of hIK1 and rSK3. Biophysical Society 95a.

Banderali, U., Klein, H., Garneau, L., Simoes, M., Parent, L., and Sauvé, R. 2004. New insights on the voltage dependence of the KCa3.1 channel block by internal TBA. J. Gen. Physiol. 124:333–348.

Brakemeier, S., Kersten, A., Eichler, I., Grgic, I., Zakrzewicz, A., Hopp, H., Kohler, R., and Hoyer, J. 2003. Shear stress-induced up-regulation of the intermediate-conductance Ca^{2+}-activated K^+ channel in human endothelium. Cardiovasc. Res. 60:488–496.

Burnham, M. P., Bychkov, R., Feletou, M., Richards, G. R., Vanhoutte, P. M., Weston, A. H., and Edwards, G. 2002. Characterization of an apamin-sensitive small-conductance Ca^{2+}-activated K^+ channel in porcine coronary artery endothelium: relevance to EDHF. Br. J. Pharmacol. 135:1133–1143.

Busse, R., Edwards, G., Feletou, M., Fleming, I., Vanhoutte, P. M., and Weston, A. H. 2002. EDHF: bringing the concepts together. Trends Pharmacol. Sci. 23:374–380.

Bychkov, R., Burnham, M. P., Richards, G. R., Edwards, G., Weston, A. H., Feletou, M., and Vanhoutte, P. M. 2002. Characterization of a charybdotoxin-sensitive intermediate conductance Ca^{2+}-activated K^+ channel in porcine coronary endothelium: relevance to EDHF. Br. J. Pharmacol. 137:1346–1354.

Cai, S., and Sauvé, R. 1997. Effects of thiol-modifying agents on a K(Ca^{2+}) channel of intermediate conductance in bovine aortic endothelial cells. J. Membr. Biol. 158:147–158.

Campbell, W. B., Gebremedhin, D., Pratt, P. F., and Harder, D. R. 1996. Identification of epoxyeicosatrienoic acids as endothelium-derived hyperpolarizing factors. Circ. Res. 78:415–423.

Channon, J. Y., and Leslie, C. C. 1990. A calcium-dependent mechanism for associating a soluble arachidonoyl-hydrolyzing phospholipase A2 with membrane in the macrophage cell line RAW 264.7. J. Biol Chem. 265:5409–5413.

Chaytor, A. T., Martin, P. E., Edwards, D. H., and Griffith, T. M. 2001. Gap junctional communication underpins EDHF-type relaxations evoked by ACh in the rat hepatic artery. Am. J. Physiol. Heart Circ. Physiol. 280:H2441–H2450.

Colden-Stanfield, M., Cramer, E. B., and Gallin, E. K. 1992. Comparison of apical and basal surfaces of confluent endothelial cells: patch clamp and viral studies. Am. J. Physiol. 263:C573–C583.

Coleman, H. A., Tare, M., and Parkington, H. C. 2001. EDHF is not K^+ but may be due to spread of current from the endothelium in guinea pig arterioles. Am. J. Physiol. Heart Circ. Physiol. 280:H2478–H2483.

Devor, D. C., and Frizzell, R. A. 1998. Modulation of K^+ channels by arachidonic acid in T84 cells. I. Inhibition of the Ca^{2+}-dependent K^+ channel. Am. J. Physiol. 274:C138–C148.

Dong, H., Jiang, Y., Cole, W. C., and Triggle, C. R. 2000. Comparison of the pharmacological properties of EDHF-mediated vasorelaxation in guinea-pig cerebral and mesenteric resistance vessels. Br. J. Pharmacol. 130:1983–1991.

Dunn, P. M. 1998. The action of blocking agents applied to the inner face of Ca^{2+}-activated K^+ channels from human erythrocytes. J. Membr. Biol. 165:133–143.

Edwards, G., Dora, K. A., Gardener, M. J., Garland, C. J., and Weston, A. H. 1998. K^+ is an endothelium-derived hyperpolarizing factor in rat arteries. Nature 396:269–272.

Edwards, G., Gardener, M. J., Feletou, M., Brady, G., Vanhoutte, P. M., and Weston, A. H. 1999. Further investigation of endothelium-derived hyperpolarizing factor (EDHF) in rat hepatic artery: studies using 1-EBIO and ouabain. Br. J. Pharmacol. 128:1064–1070.

Edwards, G., Thollon, C., Gardener, M. J., Feletou, M., Vilaine, J., Vanhoutte, P. M., and Weston, A. H. 2000. Role of gap junctions and EETs in endothelium-dependent hyperpolarization of porcine coronary artery. Br. J. Pharmacol. 129:1145–1154.

Feletou, M., and Vanhoutte, P. M. 2004. EDHF: new therapeutic targets? Pharmacol. Res. 49:565–580.

Gao, Y. D., and Garcia, M. L. 2003. Interaction of agitoxin2, charybdotoxin, and iberiotoxin with potassium channels: selectivity between voltage-gated and Maxi-K channels. Proteins 52:146–154.

Gerlach, A. C., Gangopadhyay, N. N., and Devor, D. C. 2000. Kinase-dependent regulation of the intermediate conductance, calcium-dependent potassium channel, hIK1. J. Biol. Chem. 275:585–598.

Gerlach, A. C., Syme, C. A., Giltinan, L., Adelman, J. P., and Devors, D. C. 2001. ATP-dependent activation of the intermediate conductance, Ca^{2+}-activated K^{+} channel, hIK1, is conferred by a C-terminal domain. J. Biol. Chem. 276:10963–10970.

Ghisdal, P., and Morel, N. 2001. Cellular target of voltage and calcium-dependent K^{+} channel blockers involved in EDHF-mediated responses in rat superior mesenteric artery. Br. J. Pharmacol. 134:1021–1028.

Grgic, I., Eichler, I., Heinau, P., Si, H., Brakemeier, S., Hoyer, J., and Kohler, R. 2005. Selective blockade of the intermediate-conductance Ca^{2+}-activated K^{+} channel suppresses proliferation of microvascular and macrovascular endothelial cells and angiogenesis *in vivo*. Arterioscler. Thromb. Vasc. Biol. 25:704–709.

Hamilton, K. L., Syme, C. A., and Devor, D. C. 2003. Molecular localization of the inhibitory arachidonic acid binding site to the pore of hIK1. J. Biol. Chem. 278:16690–16697.

Harder, D. R., Campell, W. B., and Roman, R. J. 1995. Role of cytochrome P-450 enzymes and metabolites of arachidonic acid in the control of vascular tone. J. Vasc. Res. 32:79–92.

Jensen, B. S., Hertz, M., Christophersen, P., and Madsen, L. S. 2002. The Ca^{2+}-activated K^{+} channel of intermediate conductance: a possible target for immune suppression. Expert Opin. Ther. Targets. 6:623–636.

Joiner, W. J., Wang, L.-Y., Tang, M. D., and Kaczmarek, L. K. 1997. hSK4, a member of a novel subfamily of calcium-activated potassium channel. Proc. Natl. Acad. Sci. USA 94:11013–11018.

Joiner, W. J., Khanna, R., Schlichter, L. C., and Kaczmarek, L. K. 2001. Calmodulin regulates assembly and trafficking of SK4/IK1 Ca^{2+}-activated K^{+} channels. J. Biol. Chem. 276:37980–37985.

Klein, H., Garneau, L., Banderali, U., Simoes, M., Parent, L., and Sauvé, R. 2007. Structural determinants of the closed KCa3.1 channel pore in relation to channel gating: Results from a substituted cysteine accessibility analysis. J. Gen. Physiol. 129:299–315.

Kohler, M., Hirschberg, B., Bond, C. T., Kinzie, J. M., Marrion, N. V., Maylie, J., and Adelman, J. P. 1996. Small-conductance,calcium-activated potassium channels from mammalian brain. Science 273:1709–1714.

Kohler, R., Degenhardt, C., Kuhn, M., Runkel, N., Paul, M., and Hoyer, J. 2000. Expression and function of endothelial Ca^{2+}-activated K^{+} channels in human mesenteric artery: A single-cell reverse transcriptase-polymerase chain reaction and electrophysiological study in situ. Circ. Res. 87:496–503.

Kohler, R., Brakemeier, S., Kuhn, M., Behrens, C., Real, R., Degenhardt, C., Orzechowski, H. D., Pries, A. R., Paul, M., and Hoyer, J. 2001. Impaired hyperpolarization in regenerated endothelium after balloon catheter injury. Circ. Res. 89:174–179.

Kohler, R., Wulff, H., Eichler, I., Kneifel, M., Neumann, D., Knorr, A., Grgic, I., Kampfe, D., Si, H., Wibawa, J., Real, R., Borner, K., Brakemeier, S., Orzechowski, H. D., Reusch, H. P., Paul, M., Chandy, K. G., and Hoyer, J. 2003. Blockade of the intermediate-conductance calcium-activated potassium channel as a new therapeutic strategy for restenosis. Circulation 108:1119–1125.

Lenaeus, M. J., Vamvouka, M., Focia, P. J., and Gross, A. 2005. Structural basis of TEA blockade in a model potassium channel. Nat. Struct. Mol. Biol. 12:454–459.

Marchenko, S. M., and Sage, S. O. 1996. Calcium-activated potassium channels in the endothelium of intact rat aorta. J. Physiol. (London) 492:53–60.

Marrelli, S. P., Eckmann, M. S., and Hunte, M. S. 2003. Role of endothelial intermediate conductance KCa channels in cerebral EDHF-mediated dilations. Am. J. Physiol. Heart Circ. Physiol. 285:H1590–H1599.

Maylie, J., Bond, C. T., Herson, P. S., Lee, W. S., and Adelman, J. P. 2004. Small conductance Ca^{2+}-activated K^+ channels and calmodulin. J. Physiol. (London) 554:255–261.

Miller, C., Moczydlowski, E., Latorre, R., and Phillips, M. 1985. Charybdotoxin, a protein inhibitor of single Ca^{2+}-activated K^+ channels from mammalian skeletal muscle. Nature 313:316–318.

Neylon, C. B., D'Souza, T., and Reinhart, P. H. 2004. Protein kinase A inhibits intermediate conductance Ca^{2+}-activated K^+ channels expressed in Xenopus oocytes. Pflugers Arch. 448:613–620

Pedarzani, P., Mosbacher, J., Rivard, A., Cingolani, L. A., Oliver, D., Stocker, M., Adelman, J. P., and Fakler, B. 2001. Control of electrical activity in central neurons by modulating the gating of small conductance Ca^{2+}-activated K^+ channels. J. Biol. Chem. 276:9762–9769.

Pena, T. L., and Rane, S. G. 1997. The small conductance calcium-activated potassium channel regulates ion channel expression in C3H10T1/2 cell ectopically expressing the muscle regulatory factor MRF4. J. Cell. Biochem. 272:21909–21916.

Rittenhouse, A. R., Vandorpe, D. H., Brugnara, C., and Alper, S. L. 1997. The antifungal imidazole clotrimazole and its major *in vivo* metabolite are potent blockers of the calcium-activated channel in murine erythroleukemia cells. J. Membr. Biol. 157:177–191.

Sauvé, R., Parent, L., Simoneau, C., and Roy, G. 1988. External ATP triggers a biphasic activation process of a calcium-dependent K^+ channel in cultured bovine aortic endothelial cells. Pflügers Arch. 41:469–481.

Schumacher, M. A., Rivard, A. F., Bachinger, H. P., and Adelman, J. P. 2001. Structure of the gating domain of a Ca^{2+}-activated K^+ channel complexed with Ca^{2+}/calmodulin. Nature 410:1120–1124.

Schumacher, M. A., Crum, M., and Miller, M. C. 2004. Crystal structures of apocalmodulin and an apocalmodulin/SK potassium channel gating domain complex. Structure (Cambridge) 12:849–860.

Si, H., Heyken, W. T., Wolfle, S. E., Tysiac, M., Schubert, R., Grgic, I., Vilianovich, L., Giebing, G., Maier, T., Gross, V., Bader, M., de, W. C., Hoyer, J., and Kohler, R. 2006. Impaired endothelium-derived hyperpolarizing factor-mediated dilations and increased blood pressure in mice deficient of the intermediate-conductance Ca^{2+}-activated K^+ channel. Circ. Res. 99:537–544.

Simoes, M., Garneau, L., Klein, H., Banderali, U., Hobeila, F., Roux, B., Parent, L., and Sauve, R. 2002. Cysteine mutagenesis and computer modeling of the S6 region of an intermediate conductance IKCa channel. J. Gen. Physiol. 120:99–116.

Srivastava, S., Choudhury, P., Li, Z., Liu, G., Nadkarni, V., Ko, K., Coetzee, W. A., and Skolnik, E. Y. 2006a. Phosphatidylinositol 3-phosphate indirectly activates KCa3.1 via 14 amino acids in the carboxy terminus of KCa3.1. Mol. Biol. Cell 17:146–154.

Srivastava, S., Li, Z., Ko, K., Choudhury, P., Albaqumi, M., Johnson, A. K., Yan, Y., Backer, J. M., Unutmaz, D., Coetzee, W. A., and Skolnik, E. Y. 2006b. Histidine phosphorylation of the potassium channel KCa3.1 by nucleoside diphosphate kinase B is required for activation of KCa3.1 and CD4 T cells. Mol. Cell 24:665–675.

Syme, C. A., Hamilton, K. L., Jones, H. M., Gerlach, A. C., Giltinan, L., Papworth, G. D., Watkins, S. C., Bradbury, N. A., and Devor, D. C. 2003. Trafficking of the Ca^{2+}-activated K^+ channel, hIK1, is dependent upon a C-terminal leucine zipper. J. Biol. Chem. 278:8476–8486.

Vaca, L., Schilling, W. P., and Kunze, D. L. 1992. G-protein-mediated regulation of a Ca^{2+}-dependent K^+ channel in cultured vascular endothelial cells. Pflügers Arch. 422:66–74.

Visan, V., Fajloun, Z., Sabatier, J. M., and Grissmer, S. 2004. Mapping of maurotoxin binding sites on hKv1.2, hKv1.3, and hIKCa1 channels. Mol. Pharmacol. 66:1103–1112.

Wissmann, R., Bildl, W., Neumann, H., Rivard, A. F., Klocker, N., Weitz, D., Schulte, U., Adelman, J. P., Bentrop, D., and Fakler, B. 2002. A helical region in the C-terminus of small-conductance

Ca^{2+}-activated K^+ channels controls assembly with apo-calmodulin. J. Biol. Chem. 277:4558–4564.

Wulff, H., Miller, M. J., Hansel, W., Grissmer, S., Cahalan, M. D., and Chandy, K. G. 2000. Design of a potent and selective inhibitor of the intermediate-conductance Ca^{2+}-activated K^+ channel, IKCa1: a potential immunosuppressant. Proc. Natl. Acad. Sci. USA 97:8151–8156.

Wulff, H., Gutman, G. A., Cahalan, M. D., and Chandy, K. G. 2001. Delineation of the clotrimazole/TRAM-34 binding site on the intermediate conductance calcium-activated potassium channel, IKCa1. J. Biol. Chem. 276:32040–32045.

Yamamoto, Y., Imaeda, K., and Suzuki, H. 1999. Endothelium-dependent hyperpolarization and intercellular electrical coupling in guinea-pig mesenteric arterioles. J. Physiol. (London) 514(Pt 2):505–513.

Zeng, X. H., Xia, X. M., and Lingle, C. J. 2005. Divalent cation sensitivity of BK channel activation supports the existence of three distinct binding sites. J. Gen. Physiol. 125:273–286.

Part V
Adipokine Signaling

Chapter 20
Leptin Signaling in the Cardiovascular System

Morris Karmazyn, Daniel M. Purdham, Venkatesh Rajapurohitam, and Asad Zeidan

Abstract Since its initial discovery as a satiety product of the obesity gene, leptin has received extensive attention especially in terms of its potential role in appetite suppression and regulation of energy expenditure. The primary sources of leptin production are adipocytes and the peptide exerts its principal effects on the hypothalamus by acting on its receptors, termed OBR. Plasma leptin levels are greatly elevated in obese individuals and this finding has been closely related to the degree of adipocity. Although once considered to be solely derived from adipose tissue, it is now apparent that leptin can be produced by various tissues including those comprising the cardiovascular system. Moreover, identification of OBR expression has been demonstrated in numerous cardiovascular tissues as well as bloodborne factors such as platelets suggesting that leptin exerts biological effects beyond those initially identified and related to appetite suppression. In terms of the cardiovascular system, OBR have been identified in both vascular and cardiac tissues. The increased cardiovascular risk associated with obesity is well known and many of the effects of leptin appear to be compatible with its potential role as a contributing factor to increased cardiovascular morbidity associated with obesity. In both myocardial and vascular tissues, leptin exerts its effects via multifaceted cell signaling mechanisms. In terms of leptin's ability to produce vascular or cardiomyocyte hypertrophy or hyperplasia, the effects appear to MAPK-dependent. Recent evidence suggests that p38 activation is of particular importance although how this occurs is uncertain. Leptin also activates the RhoA/ROCK pathway resulting in altered actin dynamics which in turn may be important to p38 activation resulting in hypertrophy. Understanding the cell signaling mechanisms underlying the effects of leptin is of major importance in terms of developing therapeutic intervention targeting the leptin system as a novel approach for treating cardiovascular disorders, particularly those associated with hyperleptinemia.

Introduction

The discovery and cloning of the obesity gene (*ob*) in 1994 (Zhang et al. 1994) has led to an explosion of studies aimed at unraveling the molecular and cellular basis

A. K. Srivastava, M. B. Anand-Srivastava (eds.) *Advances in Biochemistry in Health and Disease.* Signal Transduction in the Cardiovascular System in Health and Disease,

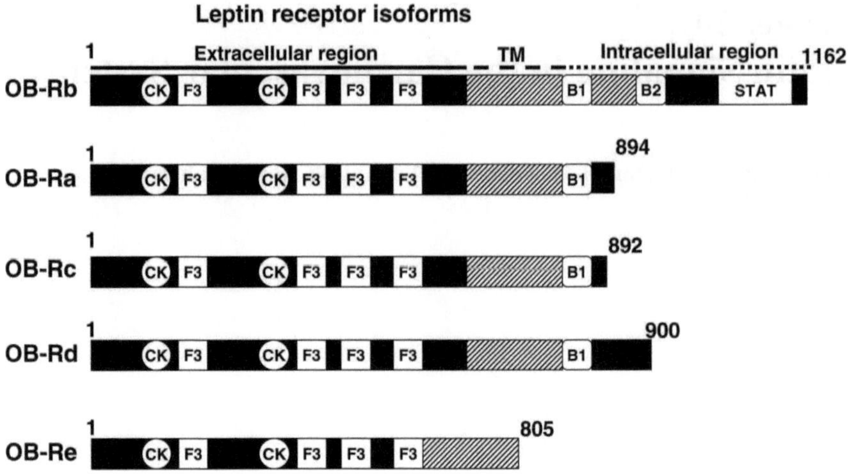

Fig. 20.1 Leptin receptor isoforms. Leptin receptor isoforms (Ob-Ra to Ob-Re) are the result of alternative gene splicing. Ob-Rb is considered to be the active isoform for carrying out downstream signaling. The extracellular region of the receptor is comprised of two CK (cytokine receptor domain) and three F3 (fibronectin domain) domains. The extracellular domain is followed by the transmembrane domain. The intracellular region consists of B1 (box 1) and B2 (box 2) for JAK2 binding and STAT (signal transducers and activators of transcription) binding domain.

of obesity and its accompanying disorders. Although originally thought to represent a disease reflecting an imbalance between food intake and energy expenditure, the identification of *ob* and the demonstration of its overexpression in obesity lent credence to the notion of a biochemical and molecular basis for obesity. The finding that *ob* encodes a circulating satiety peptide, subsequently named leptin for the Greek word *leptos*,meaning thin, was particularly exciting in view of the peptide's potential for treating obesity.

Leptin exerts its effects via membrane-bound receptors, termed OBR, which include various short forms possessing short intracellular domains (OBRa, OBRc, OBRd, OBRf) and a long form (OBRb) which is highly homologous to the type I cytokine receptor family. Their structures are illustrated in Figure 20.1. OBRb has a long intracellular domain and its activation is coupled to the activation of the Janus kinase (JAK)-signal transducers and activators of transcription (STAT) pathway. As will be discussed later, cardiovascular tissues possess a multiplicity of OBR linked to various components of cell signaling processes.

Leptin Synthesis and Structure

Leptin is located on chromosome 7q31.1, 4, and 6 in human, rat, and mouse, respectively. The gene spans ∼20 kb consisting of three exons and two introns. The promoter region is ~3 kb having TATA box, multiple C/EBP (CCAAT/enhancer binding protein) sites, glucocorticoid response element (GRE), and many cAMP response element-binding protein (CREB) sites.

Leptin is a 16-kDa, 167-amino-acid polypeptide with a 21-amino-acid signal sequence at the amino-terminus which is cleaved following the translocation of leptin into microsomes and then secreted into the bloodstream. Thus, circulating leptin in humans is actually a polypeptide of 146 amino acids having a molecular mass of 14–16 kDa. Mouse leptin is 84 and 83% homologous to human and rat, respectively. The peptide contains a single intramolecular disulfide bond that is conserved in mouse, rat, and human, resulting in a loop at the C-terminal region and a N-terminal region. It has been proposed that the N-terminal region and not the C-terminal region or disulfide bond is essential for leptin's biological activity and receptor binding activity (Imagawa et al. 1998), although it has also been suggested that the disulfide bond is of importance for the secretion, stability, and solubility of the peptide (Montague et al. 1997).

Leptin Resistance

The concept of leptin resistance is important for understanding the biological effects of leptin and also in terms of explaining the failure of leptin as a potential treatment for obesity. The latter reflects leptin resistance at the central level in obese individuals with concomitant chronic hyperleptinemia although it should also be noted that resistance to leptin may also occur at peripheral tissues. Humans develop resistance to leptin-induced effects resulting in decreased central nervous system signaling and no reduction in body weight (Schwartz et al. 1996). The precise mechanisms for resistance are not known with certainty although it has been suggested that leptin resistance may be due to defective transport of leptin through the blood–brain barrier, a concept supported by the observation that obese individuals have disproportionately low cerebral spinal fluid concentrations of leptin compared with plasma levels (Banks et al. 1996; Caro et al. 1996; Schwartz et al. 1996). It has recently been postulated that chronic elevations in leptin result in activation of STAT-3 which produces elevations of SOCS-3, a natural inhibitor of leptin signaling. This would result in attenuated or abolished leptin-induced OBR activation and downstream signaling (Munzberg and Myers 2005). A recent study also suggested that excess NO production in the hypothalamic regions contributes to central leptin resistance (Jang et al. 2007).

There are various examples of development of peripheral resistance to leptin. For instance, leptin loses its restrictive effects on insulin secretion from pancreatic β-cells in obese individuals (Seufert 2004). As will be alluded to below, cardiac myocytes from spontaneously hypertensive and hyperleptinemic rats do not respond to leptin while normotensive controls show a decrease in contraction (Wold et al. 2002). The mechanism of peripheral resistance is unknown but may involve effects on receptor/cell signaling functions (Clement et al. 1998; Vaisse et al. 1996), although this does not appear to involve downregulation of the OBRb leptin receptor (Wold et al. 2002). Recently, it has been proposed that leptin binding to C reactive protein contributes to peripheral leptin resistance (Chen et al. 2006).

Is Leptin a Link Between Obesity and Increased Cardiovascular Risk?

The relationship between obesity and increased risk for the development of cardiovascular disease is well known (Okerberg and Hamilton 2003; Sowers 2003; Sundell 2005) and obesity produces distinct changes in myocardial biochemistry, structure, and function (Ricci et al. 2006; Wong and Marwick 2007). Although a clear mechanistic basis for increased cardiovascular risk in obese individuals is uncertain (Hall et al. 2002), leptin has received some attention as a potential causative, or at least a contributing, factor to this phenomenon particularly as it pertains to hypertension (Aneja et al. 2004; El-Atat et al. 2003) but also to a host of other cardiovascular conditions. One of the manifestations of obesity is increased sympathetic tone which results in catecholamine-related cardiovascular dysfunction and as such the ability of leptin to stimulate sympathetic activity could suggest its involvement in this phenomenon (Grassi 2004; Hall et al. 2003; Rahmouni et al. 2005). Leptin has also been shown to directly stimulate catecholamine synthesis in cultured bovine adrenal medullary cells through a mechanism involving tyrosine hydroxylase and MAPK activation (Shibuya et al. 2002). This would suggest that leptin-induced catecholamine elevation may occur via two mechanisms, activation of the sympathetic nervous system and direct stimulation of catecholamine synthesis in adrenal medulla.

A significant correlation between plasma leptin concentrations and systolic blood pressure has been reported by various investigators although a cause-and-effect relationship has not been established (El-Gharbawy et al. 2002; Guagnano et al. 2003; Henriksen et al. 2000; Hu et al. 2001; Livshits et al. 2005). Such a relationship was also observed in obese children who also demonstrated a strong relationship between plasma leptin and insulin levels suggesting that leptin could be a marker or contributor to insulin resistance in obese subjects (Nishina et al. 2003). Plasma leptin has also been associated with a potential for increased thromboembolic risk in obesity via two potential mechanisms, first through impaired fibrinolysis (Skurk et al. 2002) and second through increased levels of fibrinogen (Gomez-Ambrosi et al. 2002). Dietary-mediated reduction in body weight reduces blood pressure as well as leptin concentrations in obese hypertensive individuals (Mori et al. 2004). The link between leptin and blood pressure appears to be rather convincing although it should be noted that the effects of leptin on vascular reactivity are multifaceted since the peptide likely directly produces vasorelaxation via a nitric oxide

(NO)-dependent process (Fruhbeck 1999). In this regard, NO production following leptin administration has been shown to be markedly depressed in obese animals (Beltowski et al. 2003). Interestingly, dietary-induced obesity in hypertensive rats results in slower recovery from stress-induced elevations in blood pressure and heart rate which was associated with myocardial hypertrophy and hyperleptinemia (Sedova et al. 2004).

Expression of Leptin Receptors in Cardiovascular Tissues

The first demonstration of the presence of OBR gene expression in cardiac tissue was reported in 1996 upon the discovery of the gene encoding the *db/db* mutation (Lee et al. 1996). Further characterization of OBR isoforms indicated that cardiac tissue expressed OBRa, OBRb, and OBRe (Lollmann et al. 1997; Wold et al. 2002). Recent work from the authors' laboratory suggests that OBR gene expression in the heart differs in terms of regional distribution and is also affected by gender (Purdham et al. 2004). Semiquantitative real-time polymerase chain reaction revealed that in both males and females all three isoforms investigated were expressed in both atria, left and right ventricular walls as the interventricular septum although the greatest gene abundance was found in the atria. In terms of gender differences, OBR expression was generally higher in tissues from female rats especially in the right atria (Purdham et al. 2004).

The functions of each of the OBR isoforms in the heart, the relevance of regional distribution expression patterns, or the influence of gender are currently unclear although some potential functions for leptin signaling in the heart will be discussed later in this review. The identification of OBRe in cardiac tissue was of particular interest since this soluble receptor represents the primary binding protein for leptin in plasma and may thus dictate leptin availability to tissues. It is possible that the presence of OBRe in cardiac tissues is a consequence of proteolytic cleavage of the extracellular domains of one of the other isoforms (Ge et al. 2002; Maamra et al. 2001). Although the function of OBRe in the heart is currently unknown, it is interesting to speculate that its local tissue production serves to "fine tune" leptin concentrations in that specific tissue which would be in keeping with its role as a clearance receptor, although evidence for this hypothesis needs to be obtained with further studies.

In addition to cardiac tissue, leptin receptors have also been identified in both cerebral and coronary vessels (Bjorbaek et al. 1997; Knudson et al. 2005). With respect to the latter it was proposed that OBR-mediated leptin-induced vasodilation occurs through an NO-dependent process and which was abolished by hyperleptinemia. This finding emphasizes the potential dual role of leptin on vascular tissue, a direct NO-dependent vasodilation and vasoconstriction occurring secondarily to central stimulation of the sympathetic nervous system. These effects will be discussed below in greater detail.

Cardiovascular Actions of Leptin

Effect on Cardiomyocyte Function

Under *in vivo* conditions, the cardiovascular actions of leptin can be predicted based on the central sympathetic stimulatory effect of the peptide resulting in sympathetic nervous system-dependent effects such as elevations in blood pressure and positive inotropic and chronotropic effects. However, leptin can exert direct effects on both the heart and blood vessels through OBR-dependent cell signaling mechanisms. In isolated ventricular myocytes, leptin produces a negative inotropic effect via an NO-dependent pathway as the effect was abrogated by NO synthase inhibition with L-NAME and associated with increased NO synthase activity (Nickola et al. 2000). The negative inotropism is also associated with both JAK-STAT as well as MAP kinase p38 activation (Hintz et al. 2003; Wold et al. 2002). The negative inotropic effect of leptin can also be significantly augmented by ceramide. Leptin has also been shown to stimulate fatty acid oxidation in working perfused rat hearts in the absence of any effect on glucose oxidation while lowering cardiac triglyceride content (Atkinson et al. 2002).

Cardiomyocyte Hypertrophic Effects of Leptin

Evidence for leptin as a hypertrophic and progrowth factor stems primarily from studies examining the direct effect of the peptide on myocyte preparations. For example, our laboratory reported that leptin produces marked hypertrophy in cultured neonatal rat ventricular myocytes as manifested by increased cell size, elevated protein synthesis, and upregulation of a number of genetic hypertrophic markers (Rajapurohitam et al. 2003). Leptin-induced hypertrophy was associated with MAPK activation including both the p44/42 and p38 pathways whereas the hypertrophy was prevented only by p38 inhibition (Rajapurohitam et al. 2003). Although the latter suggests a p38-dependent pathway of leptin-induced hypertrophy, the mechanism may be multifaceted and involve other contributing factors. Thus, Xu et al. (2004) demonstrated that leptin-induced endothelin-1 release from neonatal rat ventricular myocytes results in activation of the endothelin-1 ETA receptor which then stimulates production of reactive oxygen species, the latter inducing cardiomyocyte hypertrophy (Xu et al. 2004). This study suggests that leptin does not induce hypertrophy directly *per se* but rather as a consequence of upregulation of other prohypertrophic factors. Accordingly, both ETA receptor blockade as well as catalase were effective in abrogating the hypertrophic response (Xu et al. 2004). In view of the fact that endothelin-1 and other hypertrophic factors such as angiotensin II are upregulated in obesity (reviewed by Barton et al. 2003), this study describes an important potential synergistic relationship between various neurohumoral factors in the overall hypertrophic process. This relationship between leptin is further highlighted by recent evidence from our laboratory that leptin mediates the hypertrophic effects of endothelin-1 and angiotensin II in cultured myocytes (Rajapurohitam et al. 2006). In this study, the hypertrophic effects of either endothelin-1 or angiotensin II were associated with increased OBR expression and release of leptin into the culture medium. Moreover, anti-OBR antibodies completely abrogated the hypertrophic responses to both endothelin-1 and angiotensin II (Rajapurohitam et al. 2006). These results need to be confirmed in other models but if validated they suggest that leptin plays a critical paracrine or autocrine obligatory role in mediating the hypertrophic responses to both endothelin-1 and angiotensin II and possibly other prohypertrophic factors.

Leptin has been shown to increase hyperplasia of the murine atrial HL-1 cell line as well as pediatric cardiomyocytes (Tajmir et al. 2004). Activation of ERK and phosphatidylinositol 3-kinase was demonstrated and implicated in the increase in cell number. It should be noted that leptin-induced hypertrophy has also been shown in human pediatric ventricular myocytes which was associated with increased ERK, p38, and JAK phosphorylation (Madani et al. 2006).

Vascular Effects of Leptin

Leptin appears to exert a multiplicity of effects on vascular tisssues which may increase the risk of pathology. One such effect involves decreasing vascular compliance possibly via stimulation of calcification. Indeed, leptin receptors have been identified on calcifying vascular cells (Parhami et al. 2001) and the peptide has been shown to induce vascular calcification by increasing alkaline phosphatase activity, a marker of osteogenic differentiation and calcification of osteoblastic cells (Parhami et al. 2001). Vascular vessel distensibility could also be reduced by smooth muscle cell proliferation and migration or the formation of neointima. In this regard, Oda and coworkers demonstrated that leptin stimulates vascular smooth muscle cell proliferation through a mechanism involving, at least in part, increased phosphatidylinositol 3-kinase activity (Oda et al. 2001). The activation of cell proliferation and migration can lead to acceleration of neointimal formation *in vivo* (Schafer et al. 2004; Stephenson et al. 2003) which would reduce arterial distensibility (Oda et al. 2001).

Leptin also exerts proangiogenic effects which likely occurs through activation of various factors including VEGF (Suganami et al. 2004), serine/threonine kinase Akt (Akt, protein kinase B, or Rac kinase) (Goetze et al. 2002), ERK1/2 (Bouloumie et al. 1998; Jin et al. 2003), and enhancement of metalloproteinase (MMP) expression (Park et al. 2001). How Akt mediates angiogenesis is not known although it has been shown to induce reorganization of actin cytoskeleton resulting in increased cell migration, an important part of the angiogenesis process (Morales-Ruiz et al. 2000). MMPs are members of the zinc-dependent proteinase family, which have an important function in cell migration by virtue of their ability to digest extracellular components (Park et al. 2001) as well as their ability to release growth factors leading to cell proliferation (reviewed by Stetler-Stevenson 1999). Other angiogenic factors which are increased by leptin are reactive oxygen species (Bouloumie et al. 1999; Maingrette and Renier 2003). Oxidative stress has the ability to induce angiogenesis indirectly by increasing the secretion of atherogenic factors such as lipoprotein lipase from macrophages (Maingrette and Renier 2003). In addition, smooth muscle cell proliferation and migration are important processes in angiogenesis (reviewed by D'Amore and Thompson 1987) and leptin has been shown to stimulate vascular smooth muscle proliferation and migration in cell cultures (Oda et al. 2001) as well as under *in vivo* conditions by acceleration of neointimal hyperplasia formation after injury (Schafer et al. 2004). Interestingly, the formation of neointima in *ob/ob* leptin-deficient mice was markedly attenuated compared to wild-type animals (Stephenson et al. 2003) suggesting the involvement of endogenous leptin in smooth muscle proliferation and migration.

Lastly, leptin also exerts potent hypertrophic effects on vascular tissue (Zeidan et al. 2005). Interestingly, the effects appeared to be dependent on both the angiotensin and endothelin systems as demonstrated by a marked upregulation of these pathways following leptin administration coupled with the ability of both angiotensin and endothelin receptor antagonists to mitigate leptin's hypertrophic influence (Zeidan et al. 2005).

Leptin as a Cardioprotective Agent

Although the hypertrophic/pro-remodeling effect of leptin is suggestive of an important role in cardiac pathology, salutary effects of the peptide have also been demonstrated particularly in terms of its effect in protecting cardiac tissue against hypoxia or ischemia. Indeed, leptin has been shown to exert protection against hypoxia/reoxygenation injury in cultured myocytes (Erkasap et al. 2006) or in isolated perfused mouse hearts subjected to ischemia and reperfusion in terms of infarct size reduction possibly acting via the PI3-Akt and ERK pathways (Smith et al. 2006). The rapid upregulation of expression of leptin and OBR following initiation of myocardial ischemia (Matsui et al. 2007; Purdham et al. 2004) suggests that stimulation of the leptin system represents a potential endogenous cardioprotective mechanism. It has even been suggested that mild obesity, and its attendant elevation in leptin levels, may offer cardiac protection following ischemia (Heusch 2006) which may explain reported improved outcomes observed in some clinical studies in obese patients following coronary events (Kennedy et al. 2005).

Postreceptor Leptin Signaling

In general, the complexity and diversity of leptin's effects is exemplified by its ability to activate several signal transduction pathways. It is beyond the scope of this

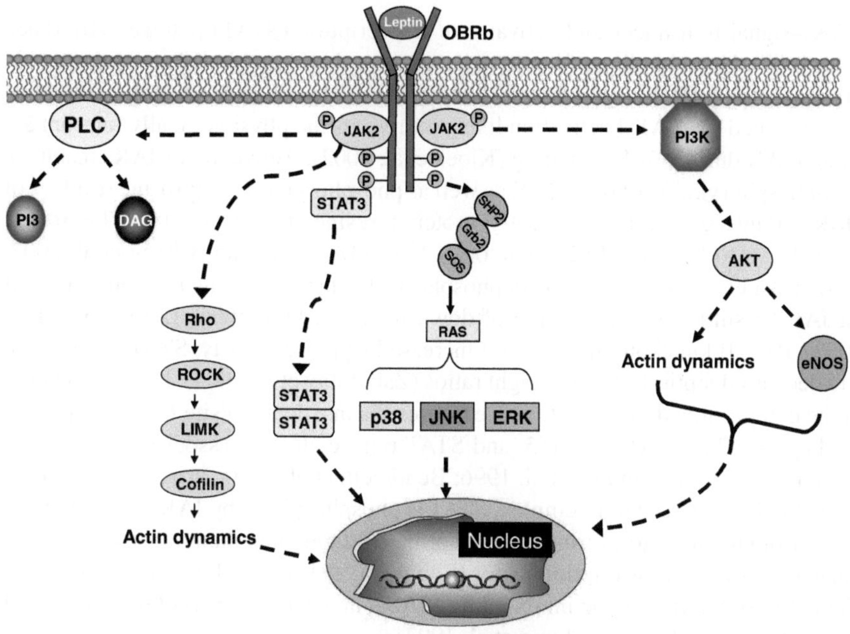

Fig. 20.2 A general schematic representation of leptin signaling. Stimulation of the long form of the leptin receptor (OBRb) can result in the activation of various cell signaling components many of which are dependent on the activation of the JAK/STAT. This results in a multiplicity of cell signaling responses many of which subsequently target the nucleus resulting in transcriptional changes. See text for further discussion.

review to comprehensively discuss leptin-mediated signaling in all tissues or organ systems. Instead, the discussion below focuses on signaling pathways which have been elucidated in cardiovascular tissue or which appears to be particularly relevant for understanding leptin-mediated cardiac signaling and its possible relationship to pathology particularly in view of harnessing these pathways for cardiac therapeutics. For a more general treatise of this subject, interested readers can consult various reviews appearing elsewhere (Ahima and Flier, 2000; Hegyi et al. 2004; Sweeney et al. 2001; Zabeau et al. 2003). A general summary of cell signaling pathways which have been proposed to mediate the actions of leptin, and which are discussed in the following sections, is illustrated in Figure 20.2.

The JAK-STAT Pathway

It is generally accepted that OBRb is the fully competent signal transduction isoform of the receptor, and that the short-form OBRs (a, c, d, f) while capable of signal transduction, do so to a lesser extent (Bjorbaek et al. 1997; Uotani et al. 1999). The major signaling pathway activated by leptin binding to OBRb is the Janus kinase

(JAK)–signal transducer and activator of transcription (STAT) pathway (Bjorbaek et al. 1997). Upon binding of leptin to its receptor, JAK1 and JAK2 are both capable of associating with the cytoplasmic domain of OBRb; however, recently it has been demonstrated that JAK2 activation likely represents the physiologically relevant activated JAK during OBR signaling (Kloek et al. 2002). Activation of JAKs results in transphosphorylation of other JAK as well as phosphorylation of tyrosine residues of OBRb (Banks et al. 2000). Recently, protein tyrosine phosphatase 1B (PTP1B) has been shown to be a negative regulator of JAK-STAT signaling (Cheng et al. 2002; Kaszubska et al. 2002). PTP1B dephosphorylates the consensus recognition motif on JAK2 resulting in inactivation of downstream STAT proteins (Zabolotny et al. 2002). PTP1B knockout mice exhibit increased leptin sensitivity, STAT3 activation, and decreased leptin-to-body-weight ratios (Zabolotny et al. 2002). Phosphorylation of the cytoplasmic domain of the receptor results in a docking site for STAT protein binding. STAT1, STAT3, STAT5, and STAT 6 have all been associated with leptin signaling *in vitro* (Baumann et al. 1996; Bendinelli et al. 2000; Briscoe et al. 2001). Upon binding the receptor complex, STAT is phosphorylated by JAK where it dissociates from the receptor, forms a homo- or heterodimer, and then translocates to the nucleus to act as a transcription factor (Baumann et al. 1996; Bendinelli et al. 2000; Heim 1996). STAT3 can be inhibited by PIAS3, an endogenous protein inhibitor of this transcriptional factor (Chung et al. 1997).

Evidence for JAK-STAT-dependent signaling in cardiac tissue in terms of physiological effects is at present limited but recent evidence suggests that it may be involved in the negative inotropic effect of leptin in cardiomyocytes based on the ability of the JAK2 inhibitor AG-490 to abrogate these effects (Wold et al. 2002). Interestingly, however, the effect of AG-490 was mimicked by the MAP kinase inhibitor SB203580 suggesting that leptin exerts its effects via multiple, and likely independent, cell signaling pathways (Wold et al. 2002).

MAP Kinases

Mitogen-activated protein kinase (MAPK) represents an additional target for leptin-mediated effects. In fact, OBRa has signal transduction capabilities through MAPK pathways both dependently and independently of JAK phosphorylation (Banks et al. 2000; Bjorbaek et al. 1997, 2001). JAK2 phosphorylation of OBR tyrosine residue 985 (Tyr^{985}) results in docking of an SH2-domain-containing protein tyrosine phosphatase (SHP-2), which associates with an adapter molecule, Grb-2, to activate extracellular regulated kinase (ERK) signaling (Banks et al. 2000). Although ERK activation is possible in the absence of Tyr^{985}, it still requires SHP-2 phosphatase activity (Banks et al. 2000). SHP-2 activation by leptin–OBR interaction leads to ERK activation, possibly through MEK1, but this has not as yet been confirmed (Hegyi et al. 2004). Activation of ERK results in alterations in gene expression patterns for several genes including c-fos (Bjorbaek et al. 2001). Another MAPK, p38, has not been studied as extensively as ERK, but has been shown to be activated by leptin in mononuclear cells (van den Brink et al. 2000). In contrast, leptin was shown to reduce insulin-induced p38 activation, while having no effect on p38 activation on its own (Sweeney et al. 2001). The role of leptin signaling through c-jun NH_2-terminal protein kinase (JNK) has not been well characterized. However, there are two reports of leptin activating JNK in endothelial cells (Bouloumie et al. 1999) and in prostate cancer cells (Onuma et al. 2003).

In the cardiovascular system, leptin has been demonstrated to activate components of the MAPK pathways. In cultured neonatal myocytes, ERK1/2 and p38, but not JNK, were activated by leptin; inhibiting ERK had no effect, while inhibition of p38 completely inhibited leptin-induced cardiomyocyte hypertrophy (Rajapurohitam et al. 2003). Leptin has also been shown to induce hyperplasia in the immortalized atrial HL-1 cell line via an ERK-dependent pathway (Tajmir et al. 2004). The results from studies using the HL-1 cell line are difficult to compare to primary culture of ventricular myocytes since the two models would likely respond to stimuli differently in view of the fact that the primary response of HL-1 cells is hyperplasia, not hypertrophy.

Stimulation of p38 may also mediate hypertrophic responses to leptin in vascular tissue. Recently, leptin has been shown to produce hypertrophy in aortic smooth muscle which was associated with p38 activation and which was blocked by p38 inhibition whereas neither ERK nor JNK inhibition exerted any effect (Shin et al. 2005). In cultured rat portal veins, leptin-induced hypertrophy is associated with p38, JNK, and ERK (unpublished data), suggesting that cell signaling mechanisms involved in leptin-induced hypertrophy may reflect the nature of the vascular tissues used to assess leptin-induced hypertrophy.

Pivotal Role for the RhoA/ROCK System in Mediating the Hypertrophic Effects of Leptin

It has become apparent that the Rho/ROCK pathway, a downstream target protein of small GTP-binding protein Rho important for regulation of cell morphology, is likely also an important contributor to hypertrophy although the mechanism leading to activation of Rho GTPases and subsequently to cardiac hypertrophy has not been well characterized (reviewed by Loirand et al. 2006; Noma et al. 2006). RhoA activates several protein kinases, including Rho kinases (ROCK). This leads to the activation of LIM kinase-2 (LIMK2) resulting in phosphorylation (inactivation) of the actin-binding protein cofilin, an important factor in the regulation of actin dynamics which in turn leads to depletion of globular actin (G-actin) pool and enhanced actin polymerization (F-actin). Work from our laboratory has recently shown that leptin is a potent activator of the RhoA/ROCK pathway leading to a decrease in the G/F-actin ratio (Zeidan et al. 2006). The precise mechanism of how activation of this pathway leads to cardiac hypertrophy is not known with certainty. Interestingly, however, activation of RhoA/ROCK by leptin results in a caveolae-dependent selective translocation of p38, but not other MAPK isoforms, to the nucleus

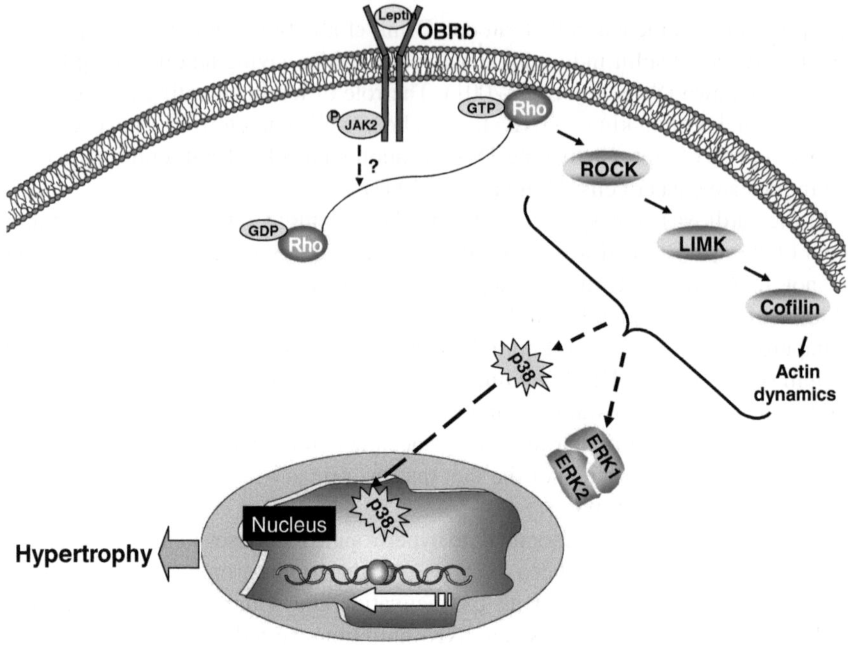

Fig. 20.3 Schematic showing a pivotal role of RhoA/ROCK activation as a mediator of leptin-induced hypertrophy and its interaction with p38. Activation of RhoA/ROCK results in increased cofilin phosphorylation and altered actin dynamics as demonstrated by a decreased G/F-actin ratio. Stimulation of this pathway results in an exclusive and selective translocation of p38 MAPK into the nucleus which results in an increased protein synthesis to an as yet to be identified mechanism.

(Zeidan et al. 2008) a finding in agreement with our initial observation that leptin-induced hypertrophy can be blocked by p38, but not by ERK inhibition (Rajapurohitam et al. 2003). A summary of the potential role of the RhoA/ROCK system in mediating the hypertrophic effect of leptin is shown in Figure 20.3.

AMP-Activated Protein Kinase (AMPK)

Although it has been known for several years that leptin can activate AMPK, recently this pathway has come into focus as it was demonstrated to be of utmost importance in regulating food intake in the hypothalamus (Minokoshi et al. 2002, 2004). AMPK plays a particularly important role in the regulation of fatty acid oxidation. Fatty acids which are not oxidized are stored in cytoplasm as triglycerides. Malonyl CoA is an important fatty acid in maintaining a balance between storage of fatty acids and transport into the mitochondria for oxidation. Elevated malonyl CoA results in impaired transport into the mitochondria and consequently prevents oxidative phosphorylation (McGarry et al. 1977). Leptin has been shown to cause phosphorylation and activation of AMPK resulting in phosphorylation of acetyl CoA carboxylase, the enzyme responsible for production of malonyl CoA (Minokoshi et al. 2002; Steinberg et al. 2003). Phosphorylation results in ACC inactivation thus decreasing the amount of malonyl CoA in the cell and stimulating oxidative phosphorylation (Winder and Hardie 1999). In the isolated working heart, leptin was shown to increase fatty acid oxidation independently of AMPK activation (Atkinson et al. 2002). Moreover, Gonzalez and colleagues demonstrated that AMPK activity in the heart was not altered by calorie restriction or fasting in mice despite a reduction in plasma leptin concentrations (Gonzalez et al. 2004).

Modulation of Leptin Receptor Expression in the Cardiovascular System

The ability of various factors, especially pathological factors, to modulate OBR expression is important in the understanding of the role of the leptin system in cardiovascular pathology. Compared to changes in leptin *per se*, modulation of its receptors has not been extensively studied. Isolated rat hearts subjected to 30 minutes of global complete ischemia resulted in substantial downregulation of OBRa, b, and e isoforms which was more pronounced in hearts from male versus female animals (Purdham et al. 2004). The relevance of this finding is presently uncertain but the results demonstrate the dynamic nature of these receptors and suggest at least that OBR expression can be altered by pathological changes. Whether this occurs *in vivo* or whether receptor expression is dependent on severity of ischemia are other important questions to address. Cardiac OBRb protein abundance has also been studied in spontaneously hypertensive rats. Ventricular myocytes from these animals exhibited resistance to the negative inotropic effect of leptin although this was not associated with any changes in OBRb protein levels and more likely reflective of defective postreceptor cell signaling function (Wold et al. 2002).

Summary and Conclusions

Since the first identification of leptin as a key satiety regulator, there has been a tremendous explosion of research into this intriguing peptide which is now known to exert a myriad of effects on a large number of tissues. The identification of leptin receptors in different tissues coupled with findings that diverse organs and tissues can produce leptin leads to the conclusion that leptin exerts effects which are substantially more extensive than initially thought. This clearly applies to the cardiovascular system where leptin and its receptors have been identified in many cell types. Indeed, leptin has diverse cardiovascular effects which are mediated by complex cell signaling mechanisms. Identification of leptin-induced cell signaling is important to fully appreciate the basis for the peptide's cardiovascular effects as well as for development of therapeutic targets involving the leptin system. Therefore, a potentially important outcome of such research will be the development of novel therapeutic strategies for the treatment of cardiovascular disease associated with hyperleptinemic conditions.

Acknowledgments The work described from the authors' laboratory is supported by the Canadian Institutes of Health Research. M.K. holds a Canada Research Chair in Experimental Cardiology. D.M.P. holds a scholarship from the Heart and Stroke Foundation of Canada. A.Z. is supported by the Heart and Stroke Foundation of Ontario Program in Heart Failure.

References

Ahima, R. S., and Osei, S. Y. 2004. Leptin signaling. Physiol. Behav. 81:223–241.

Aneja, A., El-Atat, F., McFarlane, S. I., and Sowers, J. R. 2004. Hypertension and obesity. Recent Prog. Horm. Res. 59:169–205.

Atkinson, L. L., Fischer, M. A., and Lopaschuk, G. D. 2002. Leptin activates cardiac fatty acid oxidation independent of changes in the AMP-activated protein kinase-acetyl-CoA carboxylase-malonyl-CoA axis. J. Biol. Chem. 277:29424–29430.

Banks, A. S., Davis, S. M., Bates, S. H., and Myers, M. G., Jr. 2000. Activation of downstream signals by the long form of the leptin receptor. J. Biol. Chem. 275:14563–14572.

Banks, W. A., Kastin, A. J., Huang, W., Jaspan, J. B., and Maness, L. M. 1996. Leptin enters the brain by a saturable system independent of insulin. Peptides 17:305–311.

Barton, M., Carmona, R., Ortmann, J., Krieger, J. E., and Traupe, T. 2003. Obesity-associated activation of angiotensin and endothelin in the cardiovascular system. Int. J. Biochem. Cell Biol. 35:826–837.

Baumann, H., Morella, K. K., White, D. W., Dembski, M., Bailon, P. S., Kim, H., Lai, C. F., and Tartaglia, L. A. 1996. The full-length leptin receptor has signaling capabilities of interleukin 6-type cytokine receptors. Proc. Natl. Acad. Sci. USA 93:8374–8378.

Beltowski, J., Wojcicka, G., and Jamroz, A. 2003. Stimulatory effect of leptin on nitric oxide production is impaired in dietary-induced obesity. Obes. Res. 11:1571–1580.

Bendinelli, P., Maroni, P., Pecori, G. F., and Piccoletti, R. 2000. Leptin activates Stat3, Stat1 and AP-1 in mouse adipose tissue. Mol. Cell. Endocrinol. 168:11–20.

Bjorbaek, C., Uotani, S., da Silva, B., and Flier, J. S. 1997. Divergent signaling capacities of the long and short isoforms of the leptin receptor. J. Biol. Chem. 272:32686–32695.

Bjorbaek, C., Buchholz, R. M., Davis, S. M., Bates, S. H., Pierroz, D. D., Gu, H., Neel, B. G., Myers, M. G., Jr., and Flier, J. S. 2001. Divergent roles of SHP-2 in ERK activation by leptin receptors. J. Biol. Chem. 276:4747–4755.

Bouloumie, A., Drexler, H. C., Lafontan, M., and Busse, R. 1998. Leptin, the product of Ob gene, promotes angiogenesis. Circ. Res. 83:1059–1066.

Bouloumie, A., Marumo, T., Lafontan, M., and Busse, R. 1999. Leptin induces oxidative stress in human endothelial cells. FASEB J. 13:1231–1238.

Briscoe, C. P., Hanif, S., Arch, J. R., and Tadayyon, M. 2001. Fatty acids inhibit leptin signalling in BRIN-BD11 insulinoma cells. J. Mol. Endocrinol. 26:145–154.

Caro, J. F., Kolaczynski, J. W., Nyce, M. R., Ohannesian, J. P., Opentanova, I., Goldman, W. H., Lynn, R. B., Zhang, P. L., Sinha, M. K., and Considine, R. V. 1996. Decreased cerebrospinal-fluid/serum leptin ratio in obesity: a possible mechanism for leptin resistance. Lancet 348:159–161.

Chen, K., Li, F., Li, J., Cai, H., Strom, S., Bisello, A., Kelley, D. E., Friedman-Einat, M., Skibinski, G. A., McCrory, M. A., Szalai, A. J., and Zhao, A. Z. 2006. Induction of leptin resistance through direct interaction of C-reactive protein with leptin. Nat. Med. 12:425–432.

Cheng, A., Uetani, N., Simoncic, P. D., Chaubey, V. P., Lee-Loy, A., McGlade, C. J., Kennedy, B. P., and Tremblay, M. L. 2002. Attenuation of leptin action and regulation of obesity by protein tyrosine phosphatase 1B. Dev. Cell 2:497–503.

Chung, C. D., Liao, J., Liu, B., Rao, X., Jay, P., Berta, P., and Shuai, K. 1997. Specific inhibition of Stat3 signal transduction by PIAS3. Science 278:1803–1805.

D'Amore, P. A., and Thompson, R. W. 1987. Mechanisms of angiogenesis. Annu. Rev. Physiol. 49:453–464.

El-Atat, F., Aneja, A., Mcfarlane, S., and Sowers, J. 2003. Obesity and hypertension. Endocrinol. Metab. Clin. North Am. 32:823–854.

El-Gharbawy, A. H., Kotchen, J. M., Grim, C. E., Kaldunski, M., Hoffmann, R. G., Pausova, Z., Hamet, P., and Kotchen, T. A. 2002. Gender-specific correlates of leptin with hypertension-related phenotypes in African Americans. Am. J. Hypertens. 15:989–993.

Erkasap, N., Ikizler, M., Shneyvays, V., Zinman, T., Mamedova, L. K., Uyar, R., and Shainberg, A. 2006. Leptin protects the cardiac myocyte cultures from hypoxic damage. Life Sci. 78:1098–1102.

Fruhbeck, G. 1999. Pivotal role of nitric oxide in the control of blood pressure after leptin administration. Diabetes 48:903–908.

Ge, H., Huang, L., Pourbahrami, T., and Li, C. 2002. Generation of soluble leptin receptor by ectodomain shedding of membrane-spanning receptors *in vitro* and *in vivo*. J. Biol. Chem. 277:45898–45903.

Goetze, S., Bungenstock, A., Czupalla, C., Eilers, F., Stawowy, P., Kintscher, U., Spencer-Hansch, C., Graf, K., Nurnberg, B., Law, R. E., Fleck, E., and Grafe, M. 2002. Leptin induces endothelial cell migration through Akt, which is inhibited by PPARgamma-ligands. Hypertension 40:748–754.

Gomez-Ambrosi, J., Salvador, J., Silva, C., Pastor, C., Rotellar, F., Gil, M. J., Cienfuegos, J. A., and Fruhbeck, G. 2006. Increased cardiovascular risk markers in obesity are associated with body adiposity: role of leptin. Thromb. Haemost. 95:991–996.

Gonzalez, A. A., Kumar, R., Mulligan, J. D., Davis, A. J., Weindruch, R., and Saupe, K. W. 2004. Metabolic adaptations to fasting and chronic caloric restriction in heart, muscle, and liver do not include changes in AMPK activity. Am. J. Physiol. Endocrinol. Metab. 287:E1032–E1037.

Grassi, G. 2004. Leptin, sympathetic nervous system, and baroreflex function. Curr. Hypertens. Rep. 6:236–240.

Guagnano, M. T., Manigrasso, M. R., Ballone, E., Della, V. R., Riccioni, G., Marinopiccoli, M., Nutini, M., Sensi, S., and Davi, G. 2003. Association between serum leptin levels and 24-hour blood pressure in obese women. Obes. Res. 11:549–555.

Hall, J. E., Crook, E. D., Jones, D. W., Wofford, M. R., and Dubbert, P. M. 2002. Mechanisms of obesity-associated cardiovascular and renal disease. Am. J. Med. Sci. 324:127–137.

Hall, J. E., Jones, D. W., Kuo, J. J., da Silva, A., Tallam, L. S., and Liu, J. 2003. Impact of the obesity epidemic on hypertension and renal disease. Curr. Hypertens. Rep. 5:386–392.

Hegyi, K., Fulop, K., Kovacs, K., Toth, S., and Falus, A. 2004. Leptin-induced signal transduction pathways. Cell Biol. Int. 28:159–169.

Heim, M. H. 1996. The Jak-STAT pathway: specific signal transduction from the cell membrane to the nucleus. Eur. J. Clin. Invest. 26:1–12.

Henriksen, J. H., Holst, J. J., Moller, S., Andersen, U. B., Bendtsen, F., and Jensen, G. 2000. Elevated circulating leptin levels in arterial hypertension: relationship to arteriovenous overflow and extraction of leptin. Clin. Sci. (London) 99:527–534.

Heusch, G. 2006. Obesity—a risk factor or a RISK factor for myocardial infarction? Br. J. Pharmacol. 149:1–3.

Hintz, K. K., Aberle, N. S., and Ren, J. 2003. Insulin resistance induces hyperleptinemia, cardiac contractile dysfunction but not cardiac leptin resistance in ventricular myocytes. Int. J. Obes. Relat. Metab. Disord. 27:1196–1203.

Hu, F. B., Chen, C., Wang, B., Stampfer, M. J., and Xu, X. 2001. Leptin concentrations in relation to overall adiposity, fat distribution, and blood pressure in a rural Chinese population. Int. J. Obes. Relat. Metab. Disord. 25:121–125.

Imagawa, K., Numata, Y., Katsuura, G., Sakaguchi, I., Morita, A., Kikuoka, S., Matumoto, Y., Tsuji, T., Tamaki, M., Sasakura, K., Teraoka, H., Hosoda, K., Ogawa, Y., and Nakao, K. 1998. Structure-function studies of human leptin. J. Biol. Chem. 273:35245–35249.

Jang, E. H., Park, C. S., Lee, S. K., Pie, J. E., and Kang, J. H. 2007. Excessive nitric oxide attenuates leptin-mediated signal transducer and activator of transcription 3 activation. Life Sci. 80:609–617.

Jin, X., Fukuda, N., Su, J., Takagi, H., Lai, Y., Lin, Z., Kanmatsuse, K., Wang, Z. W., and Unger, R. H. 2003. Effects of leptin on endothelial function with OB-Rb gene transfer in Zucker fatty rats. Atherosclerosis 169:225–233.

Kaszubska, W., Falls, H. D., Schaefer, V. G., Haasch, D., Frost, L., Hessler, P., Kroeger, P. E., White, D. W., Jirousek, M. R., and Trevillyan, J. M. 2002. Protein tyrosine phosphatase 1B negatively regulates leptin signaling in a hypothalamic cell line. Mol. Cell. Endocrinol. 195:109–118.

Kennedy, L. M., Dickstein, K., Anker, S. D., Kristianson, K., and Willenheimer, R. 2005. The prognostic importance of body mass index after complicated myocardial infarction. J. Am. Coll. Cardiol. 45:156–158.

Kloek, C., Haq, A. K., Dunn, S. L., Lavery, H. J., Banks, A. S., and Myers, M. G., Jr. 2002. Regulation of Jak kinases by intracellular leptin receptor sequences. J. Biol. Chem. 277:41547–41555.

Knudson, J. D., Dincer, U. D., Zhang, C., Swafford, A. N., Jr., Koshida, R., Picchi, A., Focardi, M., Dick, G. M., and Tune, J. D. 2005. Leptin receptors are expressed in coronary arteries, and hyperleptinemia causes significant coronary endothelial dysfunction. Am. J. Physiol. Heart Circ. Physiol. 289:H48–H56.

Lee, G. H., Proenca, R., Montez, J. M., Carroll, K. M., Darvishzadeh, J. G., Lee, J. I., and Friedman, J. M. 1996. Abnormal splicing of the leptin receptor in diabetic mice. Nature 379:632–635.

Livshits, G., Pantsulaia, I., and Gerber, L. M. 2005. Association of leptin levels with obesity and blood pressure: possible common genetic variation. Int. J. Obes. (London) 29:85–92.

Loirand, G., Guerin, P., and Pacaud, P. 2006. Rho kinases in cardiovascular physiology and pathophysiology. Circ. Res. 98:322–334.

Lollmann, B., Gruninger, S., Stricker-Krongrad, A., and Chiesi, M. 1997. Detection and quantification of the leptin receptor splice variants Ob-Ra, b, and, e in different mouse tissues. Biochem. Biophys. Res. Commun. 238:648–652.

Maamra, M., Bidlingmaier, M., Postel-Vinay, M. C., Wu, Z., Strasburger, C. J., and Ross, R. J. 2001. Generation of human soluble leptin receptor by proteolytic cleavage of membrane-anchored receptors. Endocrinology 142:4389–4393.

Madani, S., De Girolamo, S., Munoz, D. M., Li, R. K., and Sweeney, G. 2006. Direct effects of leptin on size and extracellular matrix components of human pediatric ventricular myocytes. Cardiovasc. Res. 69:716–725.

Maingrette, F., and Renier, G. 2003. Leptin increases lipoprotein lipase secretion by macrophages: involvement of oxidative stress and protein kinase C. Diabetes 52:2121–2128.

Matsui, H., Motooka, M., Koike, H., Inoue, M., Iwasaki, T., Suzuki, T., Kurabayashi, M., and Yokoyama, T. 2007. Ischemia/reperfusion in rat heart induces leptin and leptin receptor gene expression. Life Sci. 80:672–680.

McGarry, J. D., Mannaerts, G. P., and Foster, D. W. 1977. A possible role for malonyl-CoA in the regulation of hepatic fatty acid oxidation and ketogenesis. J. Clin. Invest. 60:265–270.

Minokoshi, Y., Kim, Y. B., Peroni, O. D., Fryer, L. G., Muller, C., Carling, D., and Kahn, B. B. 2002. Leptin stimulates fatty-acid oxidation by activating AMP-activated protein kinase. Nature 415:339–343.

Minokoshi, Y., Alquier, T., Furukawa, N., Kim, Y. B., Lee, A., Xue, B., Mu, J., Foufelle, F., Ferre, P., Birnbaum, M. J., Stuck, B. J., and Kahn, B. B. 2004. AMP-kinase regulates food intake by responding to hormonal and nutrient signals in the hypothalamus. Nature 428:569–574.

Montague, C. T., Farooqi, I. S., Whitehead, J. P., Soos, M. A., Rau, H., Wareham, N. J., Sewter, C. P., Digby, J. E., Mohammed, S. N., Hurst, J. A., Cheetham, C. H., Earley, A. R., Barnett, A. H., Prins, J. B., and O'Rahilly, S. 1997. Congenital leptin deficiency is associated with severe early-onset obesity in humans. Nature 387:903–908.

Morales-Ruiz, M., Fulton, D., Sowa, G., Languino, L. R., Fujio, Y., Walsh, K., and Sessa, W. C. 2000. Vascular endothelial growth factor-stimulated actin reorganization and migration of endothelial cells is regulated via the serine/threonine kinase Akt. Circ. Res. 86:892–896.

Mori, T. A., Burke, V., Puddey, I. B., Shaw, J. E., and Beilin, L. J. 2004. Effect of fish diets and weight loss on serum leptin concentration in overweight, treated-hypertensive subjects. J. Hypertens. 22:1983–1990.

Munzberg, H., and Myers, M. G., Jr. 2005. Molecular and anatomical determinants of central leptin resistance. Nat. Neurosci. 8:566–570.

Nickola, M. W., Wold, L. E., Colligan, P. B., Wang, G. J., Samson, W. K., and Ren, J. 2000. Leptin attenuates cardiac contraction in rat ventricular myocytes. Role of NO. Hypertension 36:501–505.

Nishina, M., Kikuchi, T., Yamazaki, H., Kameda, K., Hiura, M., and Uchiyama, M. 2003. Relationship among systolic blood pressure, serum insulin and leptin, and visceral fat accumulation in obese children. Hypertens. Res. 26:281–288.

Noma, K., Oyama, N., and Liao, J. K. 2006. Physiological role of ROCKs in the cardiovascular system. Am. J. Physiol. Cell. Physiol. 290:C661–C668.

Oda, A., Taniguchi, T., and Yokoyama, M. 2001. Leptin stimulates rat aortic smooth muscle cell proliferation and migration. Kobe J. Med. Sci. 47:141–150.

Okerberg, K., and Hamilton, D. A. 2003. Cardiovascular consequences of obesity. Curr. Womens Health Rep. 3:110–115.

Onuma, M., Bub, J. D., Rummel, T. L., and Iwamoto, Y. 2003. Prostate cancer cell-adipocyte interaction: leptin mediates androgen-independent prostate cancer cell proliferation through c-Jun NH2-terminal kinase. J. Biol. Chem. 278:42660–42667.

Parhami, F., Tintut, Y., Ballard, A., Fogelman, A. M., and Demer, L. L. 2001. Leptin enhances the calcification of vascular cells: artery wall as a target of leptin. Circ. Res. 88:954–960.

Park, H. Y., Kwon, H. M., Lim, H. J., Hong, B. K., Lee, J. Y., Park, B. E., Jang, Y., Cho, S. Y., and Kim, H. S. 2001. Potential role of leptin in angiogenesis: leptin induces endothelial cell proliferation and expression of matrix metalloproteinases *in vivo* and *in vitro*. Exp. Mol. Med. 33:95–102.

Purdham, D. M., Zou, M. X., Rajapurohitam, V., and Karmazyn, M. 2004. Rat heart is a site of leptin production and action. Am. J. Physiol. Heart Circ. Physiol. 287:H2877–H2884.

Rahmouni, K., Correia, M. L., Haynes, W. G., and Mark, A. L. 2005. Obesity-associated hypertension: new insights into mechanisms. Hypertension 45:9–14.

Rajapurohitam, V., Gan, X. T., Kirshenbaum, L. A., and Karmazyn, M. 2003. The obesity-associated peptide leptin induces hypertrophy in neonatal rat ventricular myocytes. Circ. Res. 93:277–279.

Rajapurohitam, V., Javadov, S., Purdham, D. M., Kirshenbaum, L. A., and Karmazyn, M. 2006. An autocrine role for leptin in mediating the cardiomyocyte hypertrophic effects of angiotensin II and endothelin-1. J. Mol. Cell. Cardiol. 41:265–274.

Ricci, E., Smallwood, S., Chouabe, C., Mertani, H. C., Raccurt, M., Morel, G., and Bonvallet, R. 2006. Electrophysiological characterization of left ventricular myocytes from obese Sprague-Dawley rat. Obesity (Silver Spring). 14:778–786.

Schafer, K., Halle, M., Goeschen, C., Dellas, C., Pynn, M., Loskutoff, D. J., and Konstantinides, S. 2004. Leptin promotes vascular remodeling and neointimal growth in mice. Arterioscler. Thromb. Vasc. Biol. 24:112–117.

Schwartz, M. W., Peskind, E., Raskind, M., Boyko, E. J., and Porte, D., Jr. 1996. Cerebrospinal fluid leptin levels: relationship to plasma levels and to adiposity in humans. Nat. Med. 2:589–593.

Sedova, L., Berube, J., Gaudet, D., Dumont, M., Tremblay, J., Hamet, P., and Pausova, Z. 2004. Diet-induced obesity delays cardiovascular recovery from stress in spontaneously hypertensive rats. Obes. Res. 12:1951–1958.

Seufert, J. 2004. Leptin effects on pancreatic beta-cell gene expression and function. Diabetes. 53 Suppl 1:S152–S158.

Shibuya, I., Utsunomiya, K., Toyohira, Y., Ueno, S., Tsutsui, M., Cheah, T. B., Ueta, Y., Izumi, F., and Yanagihara, N. 2002. Regulation of catecholamine synthesis by leptin. Ann. N.Y. Acad. Sci. 971:522–527.

Shin, H. J., Oh, J., Kang, S. M., Lee, J. H., Shin, M. J., Hwang, K. C., Jang, Y., and Chung, J. H. 2005. Leptin induces hypertrophy via p38 mitogen-activated protein kinase in rat vascular smooth muscle cells. Biochem. Biophys. Res. Commun. 329:18–24.

Skurk, T., van Harmelen, V., Lee, Y. M., Wirth, A., and Hauner, H. 2002. Relationship between IL-6, leptin and adiponectin and variables of fibrinolysis in overweight and obese hypertensive patients. Horm. Metab. Res. 34:659–663.

Smith, C. C., Mocanu, M. M., Davidson, S. M., Wynne, A. M., Simpkin, J. C., and Yellon, D. M. 2006. Leptin, the obesity-associated hormone, exhibits direct cardioprotective effects. Br. J. Pharmacol. 149:5–13.

Sowers, J. R. 2003. Obesity as a cardiovascular risk factor. Am. J. Med. 115 Suppl 8A:37S–41S.

Steinberg, G. R., Rush, J. W., and Dyck, D. J. 2003. AMPK expression and phosphorylation are increased in rodent muscle after chronic leptin treatment. Am. J. Physiol. Endocrinol. Metab. 284:E648–E654.

Stephenson, K., Tunstead, J., Tsai, A., Gordon, R., Henderson, S., and Dansky, H. M. 2003. Neointimal formation after endovascular arterial injury is markedly attenuated in db/db mice. Arterioscler. Thromb. Vasc. Biol. 23:2027–2033.

Stetler-Stevenson, W. G. 1999. Matrix metalloproteinases in angiogenesis: a moving target for therapeutic intervention. J. Clin. Invest. 103:1237–1241.

Suganami, E., Takagi, H., Ohashi, H., Suzuma, K., Suzuma, I., Oh, H., Watanabe, D., Ojima, T., Suganami, T., Fujio, Y., Nakao, K., Ogawa, Y., and Yoshimura, N. 2004. Leptin stimulates ischemia-induced retinal neovascularization: possible role of vascular endothelial growth factor expressed in retinal endothelial cells. Diabetes 53:2443–2448.

Sundell, J. 2005. Obesity and diabetes as risk factors for coronary artery disease: from the epidemiological aspect to the initial vascular mechanisms. Diabetes Obes. Metab. 7:9–20.

Sweeney, G., Keen, J., Somwar, R., Konrad, D., Garg, R., and Klip, A. 2001. High leptin levels acutely inhibit insulin-stimulated glucose uptake without affecting glucose transporter 4 translocation in l6 rat skeletal muscle cells. Endocrinology 142:4806–4812.

Tajmir, P., Ceddia, R. B., Li, R. K., Coe, I. R., and Sweeney, G. 2004. Leptin increases cardiomyocyte hyperplasia via extracellular signal-regulated kinase- and phosphatidylinositol 3-kinase-dependent signaling pathways. Endocrinology 145:1550–1555.

Uotani, S., Bjorbaek, C., Tornoe, J., and Flier, J. S. 1999. Functional properties of leptin receptor isoforms: internalization and degradation of leptin and ligand-induced receptor downregulation. Diabetes 48:279–286.

Vaisse, C., Halaas, J. L., Horvath, C. M., Darnell, J. E., Jr., Stoffel, M., and Friedman, J. M. 1996. Leptin activation of Stat3 in the hypothalamus of wild-type and ob/ob mice but not db/db mice. Nat. Genet. 14:95–97.

van den Brink, G. R., O'Toole, T., Hardwick, J. C., van den Boogaardt, D. E., Versteeg, H. H., van Deventer, S. J., and Peppelenbosch, M. P. 2000. Leptin signaling in human peripheral blood mononuclear cells, activation of p38 and p42/44 mitogen-activated protein (MAP) kinase and p70 S6 kinase. Mol. Cell Biol. Res. Commun. 4:144–150.

Winder, W. W., and Hardie, D. G. 1999. AMP-activated protein kinase, a metabolic master switch: possible roles in type 2 diabetes. Am. J. Physiol. 277:E1–E10.

Wold, L. E., Relling, D. P., Duan, J., Norby, F. L., and Ren, J. 2002. Abrogated leptin-induced cardiac contractile response in ventricular myocytes under spontaneous hypertension: role of Jak/STAT pathway. Hypertension 39:69–74.

Wong, C., and Marwick, T. H. 2007. Alterations in myocardial characteristics associated with obesity: detection, mechanisms, and implications. Trends Cardiovasc. Med. 17:1–5.

Xu, F. P., Chen, M. S., Wang, Y. Z., Yi, Q., Lin, S. B., Chen, A. F., and Luo, J. D. 2004. Leptin induces hypertrophy via endothelin-1-reactive oxygen species pathway in cultured neonatal rat cardiomyocytes. Circulation 110:1269–1275.

Zabeau, L., Lavens, D., Peelman, F., Eyckerman, S., Vandekerckhove, J., and Tavernier, J. 2003. The ins and outs of leptin receptor activation. FEBS Lett. 546:45–50.

Zabolotny, J. M., Bence-Hanulec, K. K., Stricker-Krongrad, A., Haj, F., Wang, Y., Minokoshi, Y., Kim, Y. B., Elmquist, J. K., Tartaglia, L. A., Kahn, B. B., and Neel, B. G. 2002. PTP1B regulates leptin signal transduction *in vivo*. Dev. Cell 2:489–495.

Zeidan, A., Purdham, D. M., Rajapurohitam, V., Javadov, S., Chakrabarti, S., and Karmazyn, M. 2005. Leptin induces vascular smooth muscle cell hypertrophy through angiotensin II- and endothelin-1-dependent mechanisms and mediates stretch-induced hypertrophy. J. Pharmacol. Exp. Ther. 315:1075–1084.

Zeidan, A., Javadov, S., and Karmazyn, M. 2006. Essential role of Rho/ROCK-dependent processes and actin dynamics in mediating leptin-induced hypertrophy in rat neonatal ventricular myocytes. Cardiovasc. Res. 72:101–111.

Zeidan, A., Javadov, S., Chakrabarti, S., and Karmazyn, M. 2008. Leptin-induced cardiomyocyte hypertrophy involves selective caveolae and RhoA/ROCK-dependent p38 MAPK translocation to nuclei. Cardiovasc. Res. 77:64–72.

Zhang, Y., Proenca, R., Maffei, M., Barone, M., Leopold, L., and Friedman, J. M. 1994. Positional cloning of the mouse obese gene and its human homologue. Nature 372:425–432.

Chapter 21
AMPK Regulation of Cardiac Metabolism in Heart Disease

Ding An, Min-Suk Kim, and Brian Rodrigues

Abstract AMP-activated protein kinase (AMPK) plays an important role in regulating cardiac metabolism, and once stimulated, AMPK turns on ATP-generating mechanisms believed to be essential in maintaining a normal heart function. AMPK is activated during various physiological or pathophysiological conditions. During cellular stresses, such as glucose deprivation, ischemia, hypoxia, and oxidative stress, ATP generation is compromised, leading to a rise in AMP/ATP ratio and activation of AMPK. During physiological conditions such as exercise, increased ATP consumption, rather than impaired ATP generation, changes the AMP/ATP ratio and stimulates AMPK. In cardiac muscle, AMPK promotes glucose uptake and glycolysis (through recruiting GLUT4 to the plasma membrane), fatty acid (FA) utilization (through its control of acetyl-CoA carboxylase), FA delivery (through its regulation of the FA transporter CD36 and lipoprotein lipase), and glycogen metabolism. Changes in metabolism have been closely linked to the development of heart disease. For example, in the heart, elevated FA use has been implicated in "lipotoxicity." In addition to lipotoxicity, excessive FA compromises glucose oxidation, and glycogen accumulates, with attendant effects on cardiac contractility. Understanding the regulation of this important stress kinase is expected to shed light on how changes in cardiac metabolism play a crucial role in the development of heart disease.

Introduction

As uninterrupted contraction is a unique feature of the heart, cardiac muscle has a high demand for provision of energy. Under normal physiological conditions, the heart can utilize multiple substrates, including fatty acid (FA), carbohydrate, amino acids, and ketones (Avogaro et al. 1990). Among these substrates, carbohydrate and FA are the major sources from which the heart derives most of its energy. In a normal heart, 70% of ATP generation is through FA oxidation, while glucose and lactate account for approximately 30% of energy provided to the cardiac muscle (Saddik and Lopaschuk 1991; Neely et al. 1972; Gertz et al. 1988). It should be noted that the

A. K. Srivastava, M. B. Anand-Srivastava (eds.) *Advances in Biochemistry in Health and Disease.* Signal Transduction in the Cardiovascular System in Health and Disease,

heart can rapidly switch its substrate selection to accommodate different physiological and pathphysiological conditions involving altered extracellular hormones, substrate availability, and work load (energy demand) (Atkinson et al. 2002; Rodrigues et al. 1995; Stanley et al. 1997; Schonekess 1997; King et al. 2005). Acutely or chronically, this regulation occurs through various mechanisms. Although substrate switching is essential to ensure continuous ATP generation to maintain heart function, it has also been associated with deleterious consequences (Barger and Kelly 2000; Young et al. 2002).

Recent studies have suggested that AMP-activated protein kinase (AMPK) plays an important role in regulating cardiac metabolism. During metabolic stresses associated with energy depletion like ischemia (when manufacture of ATP is hindered) or exercise (when ATP expenditure is augmented), changes in intracellular AMP/ATP levels promote threonine (Thr172) phosphorylation and activation of AMPK (Hardie and Carling 1997; Hong et al. 2003). Once stimulated, AMPK switches off energy-consuming processes like triglyceride (TG) and protein synthesis whereas ATP-generating mechanisms are turned on. These regulations are believed to be essential in maintaining a normal heart function. Increasing evidence suggests that AMPK plays an important role in regulating cardiac metabolism and is implicated in a number of myocardial diseases.

Structure of AMPK

Mammalian AMPK is a heterotrimer with α, β, and λ subunits (Hardie et al. 2003; Carling 2004). Each of these subunits has multiple isoforms. Up to now, two α isoforms (α1 and α2), two β isoforms (β1 and β2), and three λ isoforms (λ1, λ2, and λ3) have been identified. Therefore, a total of 12 different combinations of heterotrimer exist, and show tissue specificity. In the heart, α2 and β2, rather than α1 and β1, is highly expressed (Sambandam and Lopaschuk 2003). λ1 and λ2 are both expressed in the heart, while λ3 is only found in skeletal muscle (Cheung et al. 2000).

The C-terminus of the α subunit contains the binding sites for β and λ subunits, which forms a complex with the other two subunits (Crute et al. 1998). The N-terminus contains a serine/threonine protein kinase catalytic domain (Hanks et al. 1988). The α subunit also contains several residues that can be phosphorylated, such as Thr172, Thr258, and Ser485 (Woods et al. 2003b). Thr172 is located in the activation loop of the catalytic domain, and its phosphorylation leads to the activation of AMPK. The β subunit contains two conservative regions: KIS and ASC (Jiang and Carlson 1997). A recent study has indicated that the ASC domain functions as an anchor to connect α and λ subunits, whereas the KIS domain is a glycogen-binding domain (GBD) (Hudson et al. 2003). The exact function of GBD is still unclear. Interestingly, AMPK is localized with glycogen, and removal of GBD abolished this co-localization (Hudson et al. 2003). Thus, one hypothesis is that GBD is able to regulate AMPK activity by binding with glycogen. The λ subunit has four tandem repeats of cystothionine β synthase (CBS) domains (Hardie and Hawley 2001). Two CBS form a dimer as a functional unit. Recent studies have revealed that CBS dimers contain a binding site for AMP and ATP, and binding with AMP is essential for the activation of AMPK (Scott et al. 2004).

Regulation of AMPK

Cellular regulation of AMPK has been suggested through multiple mechanisms, and phosphorylation of Thr172 on AMPK is a key mechanism that controls AMPK activity. AMPK phosphorylation is regulated through both AMP-dependent and -independent mechanisms.

As constant ATP generation and supply is fundamental for a cell to maintain its function, a balance of ATP/ADP turnover is required (10:1 under normal aerobic conditions) (Hardie 2004). During energy stress, an increase in AMP/ATP ratio activates AMPK through three mechanisms. First, AMP allosterically activates AMPK by binding to the λ subunit (Sambandam and Lopaschuk 2003). This mechanism only mildly activates AMPK. The second mechanism by which AMP activates AMPK is through an indirect process. Binding with AMP also makes AMPK a worse substrate for protein phosphatases, which are known to dephosphorylate Thr172 and inactivate AMPK (Davies et al. 1995). Finally, binding with AMP also makes AMPK a better substrate to its upstream kinase, which was recently identified as LKB1 (Hawley et al. 2003; Woods et al. 2003a; Shaw et al. 2004). Following AMP binding, the Thr172 site of the α subunit is phosphorylated by LKB1, leading to 50–100 times increase in AMPK activity (Hawley et al. 1995). Indeed, Thr172 phosphorylation is essential for AMPK activity. Using an antibody that specifically recognizes AMPK with Thr172 phosphorylated, studies have shown that phosphorylation of Thr172 mirrors AMPK activity (Carling 2004). Mutation of this site abolishes activation of AMPK (Crute et al. 1998; Stein et al. 2000). Besides Thr172, several other sites in the α subunit are also phosphorylated by LKB1 (Woods et al. 2003a). However, phosphorylation of these sites does not change the activity of AMPK (Woods et al. 2003b). The roles of phosphorylation in these sites remain unknown. It should be noted that a recent study suggests that LKB1 is constitutive active and can phosphorylate AMPK independent of AMP. The authors suggest that AMP activates AMPK through two mechanisms: direct allosteric activation and by making AMPK a poor substrate for protein phosphatase 2C. Compared to AMP, ATP competitively inhibits AMPK. A high concentration of ATP has been shown to antagonize the AMP activation of AMPK (Hawley et al. 1996). As AMP activates AMPK while ATP antagonizes this activation, a rise of AMP/ATP ratio is a more accurate indicator of AMPK activation.

Additionally, AMPK can also be regulated through AMP/ATP-independent mechanisms. In response to increases in intracellular calcium concentration, calmodulin-dependent protein kinase kinases (CaMKK) can phosphorylate Thr172 and activate AMPK. Moreover, both leptin and adiponectin, hormones secreted from adipose tissue, activate AMPK in the skeletal muscle and liver, through ATP/AMP-independent mechanisms (Minokoshi et al. 2002; Tomas et al. 2002; Yamauchi et al. 2002). Moreover, insulin antagonizes AMPK activation in the heart during ischemia through an ATP-independent mechanism (Gamble and Lopaschuk 1997; Beauloye et al. 2001). Given that overexpression of active Akt in cardiomyocytes suppressed AMPK activity, it is likely that activation of Akt mediates the effect of insulin (Kovacic et al. 2003). Finally, a physiological concentration of fatty acid has been shown to increase AMPK phosphorylation in the heart with unclear mechanism. These studies suggest that AMPK may sense the substrate availability and regulate metabolism accordingly. Interestingly, high concentration of fatty acids has been shown to decrease AMPK phosphorylation in the heart. The mechanisms that mediate such an action are currently unclear. In addition, long-chain acyl-CoA esters prevent AMPK phosphorylation by the upstream kinase AMPKK (Kirchgessner et al. 1989). In another study using ob/ob mice or ZDF rats, decreased cardiac AMPK phosphorylation was found to be associated with augmented lipid oversupply and overexpression of protein phosphatase 2C, which is known to dephosphorylate and inactivate AMPK (Pillutla et al. 2005).

AMPK is activated during various physiological or pathophysiological conditions. During cellular stresses, such as glucose deprivation, ischemia, hypoxia, and

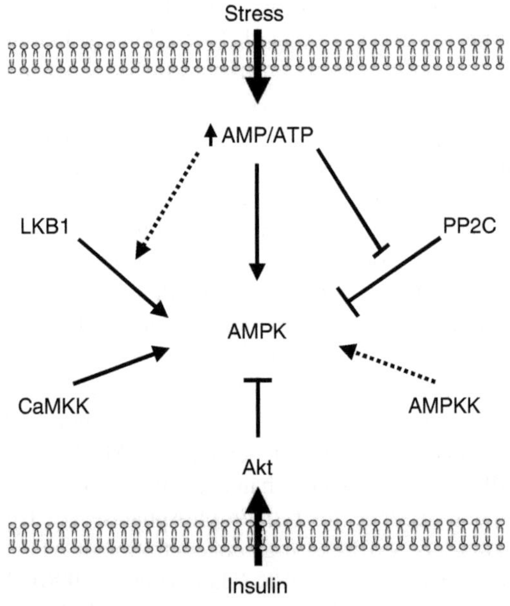

Fig. 21.1 Regulation of AMPK in cardiomyocyte. External metabolic stresses induce an increase in AMP/ATP ratio. Binding with AMP makes AMPK a better substrate to its upstream kinases, and a poor substrate for protein phosphatases 2C, which is known to dephosphorylate and inactivate AMPK. Moreover, insulin through activation of Akt also decreases AMPK activity in cardiomyocyte.

oxidative stress, ATP generation is compromised, leading to a rise in AMP/ATP ratio and activation of AMPK (Kahn et al. 2005). Moreover, during physiological conditions, such as exercise and muscle contraction, increased ATP consumption, rather than impaired ATP generation, changes the AMP/ATP ratio and stimulates AMPK (Kahn et al. 2005). Thus, following exercise, AMPK activation has been observed in heart, skeletal muscle, and liver (Coven et al. 2003; Vavvas et al. 1997; Ruderman et al. 2003). Additionally, recent studies have demonstrated that antidiabetic drugs also promote AMPK activity. Metformin, a drug commonly used in the management of Type 2 diabetes, activates AMPK through an unknown mechanism (Zhou et al. 2001; Fryer et al. 2002). Interestingly, a recent study has demonstrated that LKB1 knockout in liver abolished the effects of metformin to activate AMPK and lower blood glucose, suggesting LKB1 is required by metformin for AMPK activation (Shaw et al. 2004). Thiazolidinediones, PPAR-γ agonists, also activate AMPK, likely though inhibition of complex 1 in the mitochondrial respiratory chain, and changes in AMP/ATP ratio (Fryer et al. 2002; Brunmair et al. 2004) (Figure 21.1).

Downstream Targets of AMPK

AMPK acts as a cellular "fuel gauge." Through regulating a number of downstream targets, AMPK maintains cellular energy status by two main events: increasing ATP production and decreasing ATP-consuming processes. Thus, upon activation, AMPK rapidly modulates downstream targets acetyl-coA carboxylase (ACC) to promote FA oxidation (Hardie 2004; Kahn et al. 2005). Moreover, AMPK phosphorylates Akt substrate 160 (AS160) and phosphofructokinase 2 (PFK), leading to increased glucose transporter 4 translocation, glucose uptake, and glycolysis. All of these regulations promote energy generation. Conversely, through regulation of targets like ACC1 and FAS, AMPK reduces energy consumption (Hardie 2004). Besides these acute actions, through regulation of gene and protein expression, AMPK also has long-term effects to modulate energy metabolism. For example, through upregulating PGC1, AMPK promotes expression of numerous genes involved in mitochondria biogenesis (Terada et al. 2002). Moreover, through downregulating SREBP expression, AMPK decreases expression of genes involved in lipogenesis (Zhou et al. 2001). AMPK also influences protein synthesis, which accounts for 20% of energy turnover in growing cells. This effect is mediated through three different mechanisms. First, through phosphorylation and activation of elongation factor 2, AMPK inhibits protein synthesis (Horman et al. 2002). Another mechanism of AMPK modulating protein synthesis is through inhibition of target of rapamycin (TOR) pathway, which is a main regulator of protein synthesis (Bolster et al. 2002). Finally, AMPK also affects mRNA stability. Through reducing RNA-binding protein HuR, which is known to stabilize specific mRNAs, AMPK reduces mRNA levels, leading to a decrease in protein translation (Wang et al. 2002, 2003).

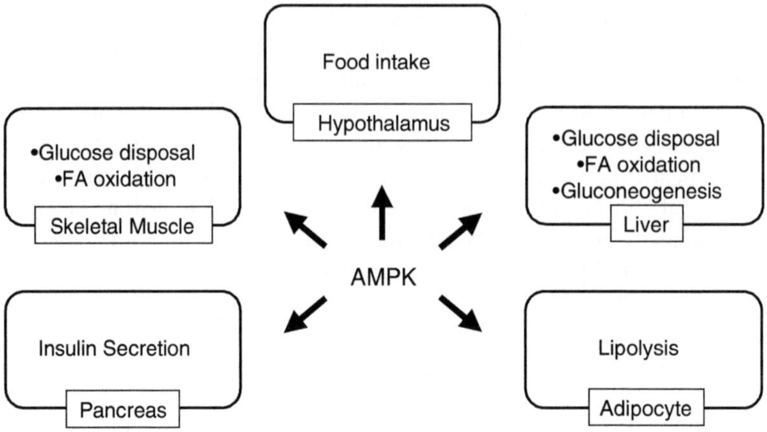

Fig. 21.2 Major effects of AMPK activation on numerous tissues. AMPK plays a key role in regulating whole body energy storage and expenditure. In hypothalamus, AMPK is involved in regulation of satiety and food intake. Activation of AMPK in the hypothalamus increases food intake, whereas inhibition decreases intake. In peripheral tissues such as skeletal muscle and liver, activation of AMPK increases energy expenditure by stimulating mitochondrial genesis and energy substrate utilization. AMPK also regulates lipolysis in adipose tissue and insulin secretion in pancreas.

Physiological Roles of AMPK

As expression of different isoforms of AMPK subunits is in a tissue-specific manner, it is likely that these isoforms may show different functions in the regulation of metabolism in different organs. Indeed, numerous studies demonstrate that AMPK regulates whole body metabolism through its effects on different organs. In the hypothalamus, fasting activates AMPK, leading to augmentation of food intake (Minokoshi et al. 2002). Inhibition of AMPK in the hypothalamus by high levels of glucose, leptin, or insulin suppresses food intake and weight gain (Minokoshi et al. 2002). Besides controlling food intake, AMPK also regulates energy expenditure in peripheral tissues. In skeletal muscle and liver, activation of AMPK increases glucose disposal and FA oxidation (Merrill et al. 1997). In the liver, AMPK also inhibits gluconeogenesis (Lochhead et al. 2000). In pancreatic β cells, activation of AMPK reduces insulin secretion (da Silva Xavier et al. 2003; Tsuboi et al. 2003). Taken together, AMPK, through its regulation of food intake and energy expenditure, is believed to play a key role in controlling whole body energy balance (Figure 21.2).

AMPK Regulates Cardiac Metabolism

Cardiac AMPK is activated during pathological conditions, such as ischemia or hypoxia, when the energy generation is hindered (Young et al. 2005). Moreover, during physiological conditions, like exercise, increased ATP expenditure

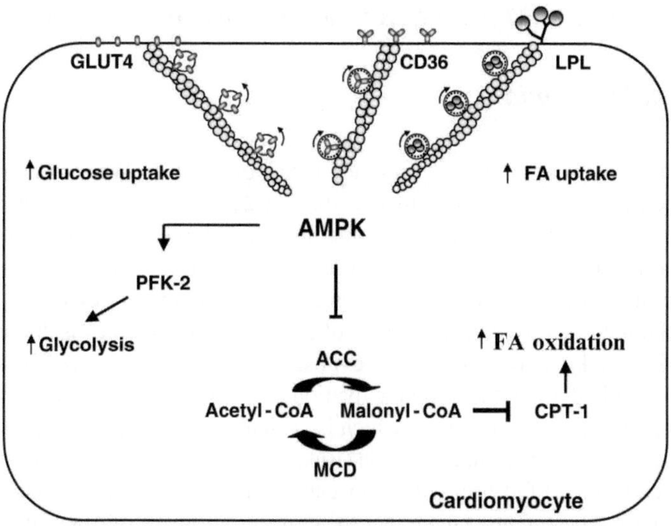

Fig. 21.3 Role of AMPK in regulating metabolism in cardiomyocyte. AMPK increases both glucose and fatty acid utilization in cardiomyocyte. Through recruiting translocation of GLUT4 to the plasma membrane and activation of phospho-fructokinase 2, AMPK promotes glucose uptake and glycolysis, respectively. Moreover, AMPK recruits lipoprotein lipase to the coronary lumen as well as increases translocation of the fatty acid transporter CD36 to the plasma membrane, leading to increased fatty acid uptake. Through inhibiting ACC, AMPK is able to decrease malonyl-CoA and minimize its inhibition of CPT-1, resulting in enhanced fatty acid oxidation.

also activated AMPK (Young et al. 2005). Once stimulated, AMPK switches off energy-consuming processes like protein synthesis, whereas ATP-generating mechanisms, such as FA oxidation and glycolysis, are turned on (Horman et al. 2002; Hardie 2003). In cardiac muscle, AMPK promotes glucose uptake and glycolysis through recruiting GLUT4 to the plasma membrane and activating PFK2. Moreover, AMPK also facilitates FA utilization through its control of ACC (Kudo et al. 1995, 1996). As ACC catalyzes the conversion of acetyl-CoA to malonyl-CoA, AMPK by inhibiting ACC is able to decrease malonyl-CoA and minimize its inhibition of CPT-1, the rate-limiting enzyme controlling FA oxidation. AMPK has also been implicated in FA delivery to cardiomyocytes through its regulation of the FA transporter, CD36 (Luiken et al. 2003). Additionally, a strong correlation between activation of whole heart AMPK and increases in coronary lumen LPL activity has been demonstrated (An et al. 2005). Finally, AMPK regulates glycogen metabolism. Mutations of gamma subunit of AMPK induce myocardial glycogen accumulation. In isolated perfused rat hearts, AICAR increased glycogen breakdown without affecting glycogen synthase or glycogen phosphorylase activity. The mechanism(s) that mediates AMPK in controlling glycogen synthesis and breakdown remain to be elucidated. Interestingly, recent studies using transgenic mice with a dominant negative form of AMPK have demonstrated that lack of AMPK activity does not affect cardiac metabolism under physiological conditions (Russell et al. 2004; Xing et al. 2003). At present, it is unclear as to what compensatory mechanisms are developed following knockout of AMPK (Figure 21.3).

Role of AMPK in Myocardial Disease: Ally or Adversary?

Myocardial ischemia

Myocardial ischemia is the most common cause of death in Western countries. In 1995, ischemic heart disease was responsible for 21% of deaths in Canada (Health Statistics Division 1997). During ischemia, reduced blood supply decreases oxidative metabolism of both glucose and fatty acid, leading to impaired ATP production, cell death, and cardiac dysfunction. Interestingly, myocardial AMPK is activated during ischemia. To investigate the role of AMPK in myocardial ischemia, hearts from AMPK-α2 knockout mice were exposed in ischemia *ex vivo* (Russell et al. 2004). This study demonstrated that activation of AMPK during ischemia promotes glycolysis and fatty acid oxidation, which is critical to lower energy depletion, limit cell death, and preserve cardiac function (Russell et al. 2004). It should be noted that a low concentration of fatty acid in the absence of lipoproteins was used in heart perfusion in these experiments. Given that the concentrations of plasma free fatty acid and lipoprotein are much higher *in vivo* (Lopaschuk et al. 1994), this study may not represent real *in vivo* situations. In the presence of high fatty acid, AMPK-stimulated fatty acid oxidation can suppress glucose oxidation, leading to dissociation of glycolysis and glucose oxidation (Folmes et al. 2006). As a consequence, activation of AMPK by ischemia may contribute to ischemic injuries. Moreover, even though depletion of AMPK makes the heart more vulnerable to ischemia/reperfusion, it is unknown if activation of myocardial AMPK through pharmacological manipulation can protect the heart against ischemic damages. Taken together, whether activation of AMPK during ischemia is beneficial or detrimental remains unclear. Future studies should investigate if activation of AMPK is able to protect the heart from ischemic injuries using *in vivo* models.

Myocardial hypertrophy

Myocardial hypertrophy is an adaptational response of the heart to increased work load. It can be developed in response to exercise so the heart can pump more efficiently. Myocardial hypertrophy may also be caused by a variety of pathological conditions, such as hypertension and valvular heart disease. Although the molecular mechanism(s) have not yet been resolved, recent studies suggest that the AMPK pathway is implicated in myocardial hypertrophy (Allard et al. 2007; Tian et al. 2001; Daniel and Carling 2002; Shibata et al. 2004). Activation of AMPK is observed during myocardial hypertrophy, associated with depletion of intracellular energy (Tian et al. 2001). A recent study also reported activation of AMPK in aortic-constriction-induced hypertrophy hearts, which is through an energy-independent mechanism (Allard et al. 2007). Interestingly, the activation of AMPK is associated with augmented glucose uptake, indicating it may be involved in the development of myocardial hypertrophy. However, another opinion is that AMPK negatively regulates cell growth and suppresses development of myocardial hypertrophy. In support of this hypothesis, activation of AMPK by adiponectin is reported to suppress hypertrophy (Shibata et al. 2004). Moreover, Akt1-induced hypertrophy is accompanied by inactivation of AMPK (Chan et al. 2004). Activation of AMPK through pharmacological manipulation (Chan et al. 2004) or expression of active AMPK kinase LKB1 inhibits protein synthesis associated with hypertrophy (Noga et al. 2007). To further elucidate the role of AMPK in myocardial hypertrophy, transgenic mice with decreased activity of AMPK were studied. These mice exhibited normal cardiac mass and function (Russell et al. 2004; Xing et al. 2003). Conversely, inactivation of AMPK through a mutation in the $\gamma 2$ subunit of AMPK decreases AMPK activity and contributed to hypertrophic growth (Sidhu et al. 2005).

Due to these discrepancies, it remains unclear if AMPK is beneficial or detrimental for the heart during hypertrophy. One hypothesis is that AMPK activation may play dual roles in the development of myocardial hypertrophy (Dyck and Lopaschuk 2006). In the early stage of myocardial hypertrophy, pharmacological activation of AMPK may play a role in inhibiting hypertrophic growth. However, the activation of AMPK during pathological hypertrophy may be an adaptive response to ensure the energy generation. Further studies are required to elucidate the role of AMPK in the development of myocardial hypertrophy.

Cardiomyocyte apoptosis

Cardiomyocyte apoptosis is accelerated during diabetes or heart disease and is one of the major factors causing heart failure. Numerous studies have shown that AMPK is implicated in regulating cardiomyocyte apoptosis. In an early study, incubation of cardiomyocytes with palmitic acid decreased AMPK activity and fatty acid oxidation, leading to augmentation of intracellular triglyceride, ceramide, and apoptosis (Hickson-Bick et al. 2002). This study suggests that impaired AMPK activity decreases fatty acid oxidative capacity in the cardiomyocyte, which initiates cell apoptosis. In agreement with this study, activation of AMPK by metformin has been shown to decrease palmitic acid-induced cardiomyocyte apoptosis (An et al. 2006). AMPK is also suggested to prevent ischemia/reperfusion-induced cell apoptosis through impeding energy depletion (Russell et al. 2004). Additionally, activation of AMPK by AICAR has been shown to protect cardiomyocytes against hypoxia through attenuation of endoplasmic reticulum stress (Terai et al. 2005). Similarly, cardiomyocytes treated with adiponectin were also resistant to hypoxia-induced apoptosis, an effect likely through activation of AMPK (Shibata et al. 2004). In the same study, adiponectin deficiency augmented myocardial apoptosis in response to ischemia and reperfusion. Activation of AMPK through adiponectin supplementation lowered apoptosis (Shibata et al. 2004). Taken together, all of these studies suggest that AMPK can provide protection to cardiomyocytes against apoptosis. However, activation of AMPK by high concentration of metformin induces cell death, likely through acidosis resulting from uncoupling of glycolysis and glucose oxidation (An et al. 2006). Moreover, a recent study suggests that AMPK induces translocation of Bax to mitochondria, an early apoptotic step in ischemia that occurs via AMPK activation of p38 MAPK (Capano and Crompton 2006).

Conclusion

It is generally accepted that AMPK plays an important role in regulating myocardial metabolism and apoptosis. Even though the mechanisms are not fully understood, increasing evidence suggests that AMPK is implicated in various heart diseases. It should be noted that a recent study demonstrated that chronic activation of AMPK in the absence of energy deficiency induces glycogen storage cardiomyopathy (Luptak et al. 2007). Thus, one should be cautious when considering AMPK as a treatment target. Future studies are necessary to improve our understanding of the role of AMPK in myocardial diseases.

References

Allard, M. F., Parsons, H. L., Saeedi, R., Wambolt, R. B., and Brownsey, R. 2007. AMPK and metabolic adaptation by the heart to pressure overload. Am J Physiol Heart Circ Physiol 292(1): H140–H148.

An, D., Pulinilkunnil, T., Qi, D., Ghosh, S., Abrahani, A., and Rodrigues, B. 2005. The metabolic "switch" AMPK regulates cardiac heparin-releasable lipoprotein lipase. Am J Physiol Endocrinol Metab 288(1): E246–E253.

An, D., Kewalramani, G., Chan, J. K., Qi, D., Ghosh, S., Pulinilkunnil, T., Abrahani, A., Innis, S. M., and Rodrigues, B. 2006. Metformin influences cardiomyocyte cell death by pathways that are dependent and independent of caspase-3. Diabetologia 49(9): 2174–2184.

Atkinson, L. L., Fischer, M. A., and Lopaschuk, G. D. 2002. Leptin activates cardiac fatty acid oxidation independent of changes in the AMP-activated protein kinase-acetyl-CoA carboxylase-malonyl-CoA axis. J Biol Chem 277(33): 29424–29430.

Avogaro, A., Nosadini, R., Doria, A., Fioretto, P., Velussi, M., Vigorito, C., Sacca, L., Toffolo, G., Cobelli, C., Trevisan, R., et al. 1990. Myocardial metabolism in insulin-deficient diabetic humans without coronary artery disease. Am J Physiol 258(4 Pt 1): E606–E618.

Barger, P. M., and Kelly, D. P. 2000. PPAR signaling in the control of cardiac energy metabolism. Trends Cardiovasc Med 10(6): 238–245.

Beauloye, C., Marsin, A. S., Bertrand, L., Krause, U., Hardie, D. G., Vanoverschelde, J. L., and Hue, L. 2001. Insulin antagonizes AMP-activated protein kinase activation by ischemia or anoxia in rat hearts, without affecting total adenine nucleotides. FEBS Lett 505(3): 348–352.

Bolster, D. R., Crozier, S. J., Kimball, S. R., and Jefferson, L. S. 2002. AMP-activated protein kinase suppresses protein synthesis in rat skeletal muscle through down-regulated mammalian target of rapamycin (mTOR) signaling. J Biol Chem 277: 23977–23980.

Brunmair, B., Staniek, K., Gras, F., Scharf, N., Althaym, A., Clara, R., Roden, M., Gnaiger, E., Nohl, H., Waldhausl, W., and Furnsinn, C. 2004. Thiazolidinediones, like metformin, inhibit respiratory complex I: a common mechanism contributing to their antidiabetic actions? Diabetes 53: 1052–1059.

Capano, M., and Crompton, M. 2006. Bax translocates to mitochondria of heart cells during simulated ischaemia: involvement of AMP-activated and p38 mitogen-activated protein kinases. Biochem J 39: 57–64.

Carling, D. 2004. The AMP-activated protein kinase cascade—a unifying system for energy control. Trends Biochem Sci 29: 18–24.

Chan, A. Y., Soltys, C. L., Young, M. E., Proud, C. G., and Dyck, J. R. 2004. Activation of AMP-activated protein kinase inhibits protein synthesis associated with hypertrophy in the cardiac myocyte. J Biol Chem 279: 32771–32779.

Cheung, P. C., Salt, I. P., Davies, S. P., Hardie, D. G., and Carling, D. 2000. Characterization of AMP-activated protein kinase gamma-subunit isoforms and their role in AMP binding. Biochem J 346 (Pt 3): 659–669.

Coven, D. L., Hu, X., Cong, L., Bergeron, R., Shulman, G. I., Hardie, D. G., and Young, L. H. 2003. Physiological role of AMP-activated protein kinase in the heart: graded activation during exercise. Am J Physiol Endocrinol Metab 285: E629–E636.

Crute, B. E., Seefeld, K., Gamble, J., Kemp, B. E., and Witters, L. A. 1998. Functional domains of the alpha1 catalytic subunit of the AMP-activated protein kinase. J Biol Chem 273: 35347–35354.

Daniel, T., and Carling, D. 2002. Functional analysis of mutations in the gamma 2 subunit of AMP-activated protein kinase associated with cardiac hypertrophy and Wolff-Parkinson-White syndrome. J Biol Chem 277: 51017–51024.

da Silva Xavier, G., Leclerc, I., Varadi, A., Tsuboi, T., Moule, S. K., and Rutter, G. A. 2003. Role for AMP-activated protein kinase in glucose-stimulated insulin secretion and preproinsulin gene expression. Biochem J 371(Pt 3): 761–774.

Davies, S. P., Helps, N. R., Cohen, P. T., and Hardie, D. G. 1995. 5′-AMP inhibits dephosphorylation, as well as promoting phosphorylation, of the AMP-activated protein kinase. Studies using bacterially expressed human protein phosphatase-2C alpha and native bovine protein phosphatase-2AC. FEBS Lett 377: 421–425.

Dyck, J. R., and Lopaschuk, G. D. 2006. AMPK alterations in cardiac physiology and pathology: enemy or ally? J Physiol 574(Pt 1): 95–112.

Folmes, C. D., Clanachan, A. S., and Lopaschuk, G. D. 2006. Fatty acids attenuate insulin regulation of 5′-AMP-activated protein kinase and insulin cardioprotection after ischemia. Circ Res 99: 61–68.

Fryer, L. G., Parbu-Patel, A., and Carling, D. 2002. The anti-diabetic drugs rosiglitazone and metformin stimulate AMP-activated protein kinase through distinct signaling pathways. J Biol Chem 277: 25226–25232.

Gamble, J., and Lopaschuk, G. D. 1997. Insulin inhibition of 5′ adenosine monophosphate-activated protein kinase in the heart results in activation of acetyl coenzyme A carboxylase and inhibition of fatty acid oxidation. Metabolism 46: 1270–1274.

Gertz, E. W., Wisneski, J. A., Stanley, W. C., and Neese, R. A. 1988. Myocardial substrate utilization during exercise in humans. Dual carbon-labeled carbohydrate isotope experiments. J Clin Invest 82: 2017–2025.

Hanks, S. K., Quinn, A. M., and Hunter, T. 1988. The protein kinase family: conserved features and deduced phylogeny of the catalytic domains. Science 241: 42–52.

Hardie, D. G. 2003. Minireview: the AMP-activated protein kinase cascade: the key sensor of cellular energy status. Endocrinology 144: 5179–5183.

Hardie, D. G. 2004. The AMP-activated protein kinase pathway–new players upstream and downstream. J Cell Sci 117(Pt 23): 5479–5487.

Hardie, D. G., and Carling, D. 1997. The AMP-activated protein kinase—fuel gauge of the mammalian cell? Eur J Biochem 246: 259–273.

Hardie, D. G., and Hawley, S. A. 2001. AMP-activated protein kinase: the energy charge hypothesis revisited. Bioessays 23: 1112–1119.

Hardie, D. G., Scott, J. W., Pan, D. A., and Hudson, E. R. 2003. Management of cellular energy by the AMP-activated protein kinase system. FEBS Lett 546: 113–120.

Hawley, S. A., Selbert, M. A., Goldstein, E. G., Edelman, A. M., Carling, D., and Hardie, D. G. 1995. 5′-AMP activates the AMP-activated protein kinase cascade, and Ca^{2+}/calmodulin activates the calmodulin-dependent protein kinase I cascade, via three independent mechanisms. J Biol Chem 270: 27186–27191.

Hawley, S. A., Davison, M., Woods, A., Davies, S. P., Beri, R. K., Carling, D., and Hardie, D. G. 1996. Characterization of the AMP-activated protein kinase kinase from rat liver and identification of threonine 172 as the major site at which it phosphorylates AMP-activated protein kinase. J Biol Chem 271: 27879–27887.

Hawley, S. A., Boudeau, J., Reid, J. L., Mustard, K. J., Udd, L., Makela, T. P., Alessi, D. R., and Hardie, D. G. 2003. Complexes between the LKB1 tumor suppressor, STRAD alpha/beta and MO25 alpha/beta are upstream kinases in the AMP-activated protein kinase cascade. J Biol 2: 28.

Health Statistics Division, S. C. 1997. Heart disease and stroke in Canada.

Hickson-Bick, D. L., Sparagna, G. C., Buja, L. M., and McMillin, J. B. 2002. Palmitate-induced apoptosis in neonatal cardiomyocytes is not dependent on the generation of ROS. Am J Physiol Heart Circ Physiol 282: H656–H664.

Hong, S. P., Leiper, F. C., Woods, A., Carling, D., and Carlson, M. 2003. Activation of yeast Snf1 and mammalian AMP-activated protein kinase by upstream kinases. Proc Natl Acad Sci USA 100: 8839–8843.

Horman, S., Browne, G., Krause, U., Patel, J., Vertommen, D., Bertrand, L., Lavoinne, A., Hue, L., Proud, C., and Rider, M. 2002. Activation of AMP-activated protein kinase leads to the phosphorylation of elongation factor 2 and an inhibition of protein synthesis. Curr Biol 12: 1419–1423.

Hudson, E. R., Pan, D. A., James, J., Lucocq, J. M., Hawley, S. A., Green, K. A., Baba, O., Terashima, T., and Hardie, D. G. 2003. A novel domain in AMP-activated protein kinase causes glycogen storage bodies similar to those seen in hereditary cardiac arrhythmias. Curr Biol 13: 861–866.

Jiang, R., and Carlson, M. 1997. The Snf1 protein kinase and its activating subunit, Snf4, interact with distinct domains of the Sip1/Sip2/Gal83 component in the kinase complex. Mol Cell Biol 17: 2099–2106.

Kahn, B. B., Alquier, T., Carling, D., and Hardie, D. G. 2005. AMP-activated protein kinase: ancient energy gauge provides clues to modern understanding of metabolism. Cell Metab 1: 15–25.

King, K. L., Okere, I. C., Sharma, N., Dyck, J. R., Reszko, A. E., McElfresh, T. A., Kerner, J., Chandler, M. P., Lopaschuk, G. D., and Stanley, W. C. 2005. Regulation of cardiac malonyl-CoA content and fatty acid oxidation during increased cardiac power. Am J Physiol Heart Circ Physiol 289: H1033–H1037.

Kirchgessner, T. G., LeBoeuf, R. C., Langner, C. A., Zollman, S., Chang, C. H., Taylor, B. A., Schotz, M. C., Gordon, J. I., and Lusis, A. J. 1989. Genetic and developmental regulation of the lipoprotein lipase gene: loci both distal and proximal to the lipoprotein lipase structural gene control enzyme expression. J Biol Chem 264: 1473–1482.

Kovacic, S., Soltys, C. L., Barr, A. J., Shiojima, I., Walsh, K., and Dyck, J. R. 2003. Akt activity negatively regulates phosphorylation of AMP-activated protein kinase in the heart. J Biol Chem 278: 39422–39427.

Kudo, N., Barr, A. J., Barr, R. L., Desai, S., and Lopaschuk, G. D. 1995. High rates of fatty acid oxidation during reperfusion of ischemic hearts are associated with a decrease in malonyl-CoA levels due to an increase in 5′-AMP-activated protein kinase inhibition of acetyl-CoA carboxylase. J Biol Chem 270: 17513–17520.

Kudo, N., Gillespie, J. G., Kung, L., Witters, L. A., Schulz, R., Clanachan, A. S., and Lopaschuk, G. D. 1996. Characterization of 5′ AMP-activated protein kinase activity in the heart and its role in inhibiting acetyl-CoA carboxylase during reperfusion following ischemia. Biochim Biophys Acta 1301: 67–75.

Lochhead, P. A., Salt, I. P., Walker, K. S., Hardie, D. G., and Sutherland, C. 2000. 5-aminoimidazole-4-carboxamide riboside mimics the effects of insulin on the expression of the 2 key gluconeogenic genes PEPCK and glucose-6-phosphatase. Diabetes 49: 896–903.

Lopaschuk, G. D., Collins-Nakai, R., Olley, P. M., Montague, T. J., McNeil, G., Gayle, M., Penkoske, P., and Finegan, B. A. 1994. Plasma fatty acid levels in infants and adults after myocardial ischemia. Am Heart J 128: 61–67.

Luiken, J. J., Coort, S. L., Willems, J., Coumans, W. A., Bonen, A., van der Vusse, G. J., and Glatz, J. F. 2003. Contraction-induced fatty acid translocase/CD36 translocation in rat cardiac myocytes is mediated through AMP-activated protein kinase signaling. Diabetes 52: 1627–1634.

Luptak, I., Shen, M., He, H., Hirshman, M. F., Musi, N., Goodyear, L. J., Yan, J., Wakimoto, H., Morita, H., Arad, M., Seidman, C. E., Seidman, J. G., Ingwall, J. S., Balschi, J. A., and Tian, R. 2007. Aberrant activation of AMP-activated protein kinase remodels metabolic network in favor of cardiac glycogen storage. J Clin Invest 117: 1432–1439.

Merrill, G. F., Kurth, E. J., Hardie, D. G., and Winder, W. W. 1997. AICA riboside increases AMP-activated protein kinase, fatty acid oxidation, and glucose uptake in rat muscle. Am J Physiol 273(6 Pt 1): E1107–E1112.

Minokoshi, Y., Kim, Y. B., Peroni, O. D., Fryer, L. G., Muller, C., Carling, D., and Kahn, B. B. 2002. Leptin stimulates fatty-acid oxidation by activating AMP-activated protein kinase. Nature 415: 339–343.

Neely, J. R., Rovetto, M. J., and Oram, J. F. 1972. Myocardial utilization of carbohydrate and lipids. Prog Cardiovasc Dis 15: 289–329.

Noga, A. A., Soltys, C. L., Barr, A. J., Kovacic, S., Lopaschuk, G. D., and Dyck, J. R. 2007. Expression of an active LKB1 complex in cardiac myocytes results in decreased protein synthesis associated with phenylephrine-induced hypertrophy. Am J Physiol Heart Circ Physiol 292: H1460–H1469.

Pillutla, P., Hwang, Y. C., Augustus, A., Yokoyama, M., Yagyu, H., Johnston, T. P., Kaneko, M., Ramasamy, R., and Goldberg, I. J. 2005. Perfusion of hearts with triglyceride-rich particles reproduces the metabolic abnormalities in lipotoxic cardiomyopathy. Am J Physiol Endocrinol Metab 288: E1229–E1235.

Rodrigues, B., Cam, M. C., and McNeill, J. H. 1995. Myocardial substrate metabolism: implications for diabetic cardiomyopathy. J Mol Cell Cardiol 27: 169–179.

Ruderman, N. B., Park, H., Kaushik, V. K., Dean, D., Constant, S., Prentki, M., and Saha, A. K. 2003. AMPK as a metabolic switch in rat muscle, liver and adipose tissue after exercise. Acta Physiol Scand 178: 435–442.

Russell, R. R., 3rd, Li, J., Coven, D. L., Pypaert, M., Zechner, C., Palmeri, M., Giordano, F. J., Mu, J., Birnbaum, M. J., and Young, L. H. 2004. AMP-activated protein kinase mediates ischemic glucose uptake and prevents postischemic cardiac dysfunction, apoptosis, and injury. J Clin Invest 114: 495–503.

Saddik, M., and Lopaschuk, G. D. 1991. Myocardial triglyceride turnover and contribution to energy substrate utilization in isolated working rat hearts. J Biol Chem 266: 8162–8170.

Sambandam, N., and Lopaschuk, G. D. 2003. AMP-activated protein kinase (AMPK) control of fatty acid and glucose metabolism in the ischemic heart. Prog Lipid Res 42: 238–256.

Schonekess, B. O. 1997. Competition between lactate and fatty acids as sources of ATP in the isolated working rat heart. J Mol Cell Cardiol 29: 2725–2733.

Scott, J. W., Hawley, S. A., Green, K. A., Anis, M., Stewart, G., Scullion, G. A., Norman, D. G., and Hardie, D. G. 2004. CBS domains form energy-sensing modules whose binding of adenosine ligands is disrupted by disease mutations. J Clin Invest 113: 274–284.

Shaw, R. J., Kosmatka, M., Bardeesy, N., Hurley, R. L., Witters, L. A., DePinho, R. A., and Cantley, L. C. 2004. The tumor suppressor LKB1 kinase directly activates AMP-activated kinase and regulates apoptosis in response to energy stress. Proc Natl Acad Sci USA 101: 3329–3335.

Shibata, R., Ouchi, N., Ito, M., Kihara, S., Shiojima, I., Pimentel, D. R., Kumada, M., Sato, K., Schiekofer, S., Ohashi, K., Funahashi, T., Colucci, W. S., and Walsh, K. 2004. Adiponectin-mediated modulation of hypertrophic signals in the heart. Nat Med 10: 1384–1389.

Sidhu, J. S., Rajawat, Y. S., Rami, T. G., Gollob, M. H., Wang, Z., Yuan, R., Marian, A. J., DeMayo, F. J., Weilbacher, D., Taffet, G. E., Davies, J. K., Carling, D., Khoury, D. S., and Roberts, R. 2005. Transgenic mouse model of ventricular preexcitation and atrioventricular reentrant tachycardia induced by an AMP-activated protein kinase loss-of-function mutation responsible for Wolff-Parkinson-White syndrome. Circulation 111: 21–29.

Stanley, W. C., Lopaschuk, G. D., and McCormack, J. G. 1997. Regulation of energy substrate metabolism in the diabetic heart. Cardiovasc Res 34: 25–33.

Stein, S. C., Woods, A., Jones, N. A., Davison, M. D., and Carling, D. 2000. The regulation of AMP-activated protein kinase by phosphorylation. Biochem J 345 Pt 3: 437–443.

Terada, S., Goto, M., Kato, M., Kawanaka, K., Shimokawa, T., and Tabata, I. 2002. Effects of low-intensity prolonged exercise on PGC-1 mRNA expression in rat epitrochlearis muscle. Biochem Biophys Res Commun 296: 350–354.

Terai, K., Hiramoto, Y., Masaki, M., Sugiyama, S., Kuroda, T., Hori, M., Kawase, I., and Hirota, H. 2005. AMP-activated protein kinase protects cardiomyocytes against hypoxic injury through attenuation of endoplasmic reticulum stress. Mol Cell Biol 25: 9554–9575.

Tian, R., Musi, N., D'Agostino, J., Hirshman, M. F., and Goodyear, L. J. 2001. Increased adenosine monophosphate-activated protein kinase activity in rat hearts with pressure-overload hypertrophy. Circulation 104: 1664–1669.

Tomas, E., Tsao, T. S., Saha, A. K., Murrey, H. E., Zhang, C. C., Itani, S. I., Lodish, H. F., and Ruderman, N. B. 2002. Enhanced muscle fat oxidation and glucose transport by ACRP30 globular domain: acetyl-CoA carboxylase inhibition and AMP-activated protein kinase activation. Proc Natl Acad Sci USA 99: 16309–16313.

Tsuboi, T., da Silva Xavier, G., Leclerc, I., and Rutter, G. A. 2003. 5′-AMP-activated protein kinase controls insulin-containing secretory vesicle dynamics. J Biol Chem 278: 52042–52051.

Vavvas, D., Apazidis, A., Saha, A. K., Gamble, J., Patel, A., Kemp, B. E., Witters, L.A., and Ruderman, N. B. 1997. Contraction-induced changes in acetyl-CoA carboxylase and 5′-AMP-activated kinase in skeletal muscle. J Biol Chem 272: 13255–13261.

Wang, W., Fan, J., Yang, X., Furer-Galban, S., Lopez de Silanes, I., von Kobbe, C., Guo, J., Georas, S. N., Foufelle, F., Hardie, D. G., Carling, D., and Gorospe, M. 2002. AMP-activated kinase regulates cytoplasmic HuR. Mol Cell Biol 22: 3425–3436.

Wang, W., Yang, X., Lopez de Silanes, I., Carling, D., and Gorospe, M. 2003. Increased AMP:ATP ratio and AMP-activated protein kinase activity during cellular senescence linked to reduced HuR function. J Biol Chem 278: 27016–27023.

Woods, A., Johnstone, S. R., Dickerson, K., Leiper, F. C., Fryer, L. G., Neumann, D.,Schlattner, U., Wallimann, T., Carlson, M., and Carling, D. 2003a. LKB1 is the upstream kinase in the AMP-activated protein kinase cascade. Curr Biol 13: 2004–2008.

Woods, A., Vertommen, D., Neumann, D., Turk, R., Bayliss, J., Schlattner, U., Wallimann, T., Carling, D., and Rider, M. H. 2003b. Identification of phosphorylation sites in AMP-activated protein kinase (AMPK) for upstream AMPK kinases and study of their roles by site-directed mutagenesis. J Biol Chem 278: 28434–28442.

Xing, Y., Musi, N., Fujii, N., Zou, L., Luptak, I., Hirshman, M. F., Goodyear, L. J., and Tian, R. 2003. Glucose metabolism and energy homeostasis in mouse hearts overexpressing dominant negative alpha2 subunit of AMP-activated protein kinase. J Biol Chem 278: 28372–28377.

Yamauchi, T., Kamon, J., Minokoshi, Y., Ito, Y., Waki, H., Uchida, S., Yamashita, S., Noda, M., Kita, S., Ueki, K., Eto, K., Akanuma, Y., Froguel, P., Foufelle, F., Ferre, P., Carling, D., Kimura, S., Nagai, R., Kahn, B. B., and Kadowaki, T. 2002. Adiponectin stimulates glucose utilization and fatty-acid oxidation by activating AMP-activated protein kinase. Nat Med 8: 1288–1295.

Young, L. H., Li, J., Baron, S. J., and Russell, R. R. 2005. AMP-activated protein kinase: a key stress signaling pathway in the heart. Trends Cardiovasc Med 15: 110–118.

Young, M. E., McNulty, P., and Taegtmeyer, H. 2002. Adaptation and maladaptation of the heart in diabetes: Part II: potential mechanisms. Circulation 105: 1861–1870.

Zhou, G., Myers, R., Li, Y., Chen, Y., Shen, X., Fenyk-Melody, J., Wu, M., Ventre, J., Doebber, T., Fujii, N., Musi, N., Hirshman, M. F., Goodyear, L. J., and Moller, D. E. 2001. Role of AMP-activated protein kinase in mechanism of metformin action. J Clin Invest 108: 1167–1174.

Signal Transduction in the Cardiovascular System in Health and Disease

Madhu B. Anand-Srivastava and Ashok K. Srivastava

This book has addressed the contributions of several key signal transduction pathways which are central to our understanding of cardiovascular physiology and pathophysiology. Aberrations in these signaling events have been suggested to be involved in a host of cardiovascular pathologies, such as cardiac arrhythmias, congenital heart failure, and hypertension.

All the chapters have been written by well-known leaders in the fields, which cover a wide range of intracellular events that regulate various aspects of cardiovascular functions. This book will be of interest to both basic as well as clinical scientists seeking to understand the molecular basis of cardiovascular diseases, and also to those interested in defining targets for cardiovascular pharmacotherapy.

About the Editors

Dr. Madhu B. Anand-Srivastava is a Professor, Department of Physiology, and a member of the Group de recherche sur le systéme nerveux autonome, University of Montreal. In 1990, she received a prestigious Scientist Award from the Medical Research Council of Canada, and in 2004, she was bestowed with the Vincenzo Panagia Distinguished Lecturer award from the Institute of Cardiovascular Sciences of the University of Manitoba, Winnipeg. She is an internationally recognized expert in the area of G-protein and natriuretic peptide receptor-C signaling, and has made significant contributions in defining the role of Giα proteins in the pathogenesis of hypertension. She has published more than 125 papers in refereed journals and edited 3 books in the area of G-proteins, hypertension, and NPR-C receptor and cell signaling, and has been invited to several national and international conferences and academic institutions to present her work. Dr. Anand-Srivastava has served or is currently on different committees of the Canadian Institutes of Health Research and the Heart and Stroke Foundation of Canada, and is also on the editorial boards of several scientific publications, including the *Journal of Molecular and Cellular Cardiology* and *Molecular and Cellular Biochemistry*. She has also organized several national and international conferences on cell signaling in the cardiovascular system.

Dr. Ashok K. Srivastava is a Professor, Department of Medicine, University of Montreal, and Head of the Laboratory of Cell Signaling at the Research Center of the Centre hospitalier de l'Université de Montréal (CHUM). He has over 100 publications in the fields of insulin, vasoactive peptide and redox signaling, and cardiovascular complications. He has served or is serving as a guest editor of many journals such as *Antioxidant and Redox Signaling*, *Canadian Journal of Physiology and Pharmacology*, *Cell Biochemistry and Biophysics*, and *Molecular and Cellular Biochemistry*, and is currently a member of the editorial boards of *Advances in Biochemistry in Health and Disease* and *Recent Patents on Endocrine, Metabolic and Immune Drug Discovery*. Dr. Srivastava is also on the grant review panels of the Canadian Institutes of Health Research, the Heart and Stroke Foundation of Canada, and the National Institutes of Health. He has also organized several international symposia and workshops on second messengers and cell signaling.

Index

X

Color Plates

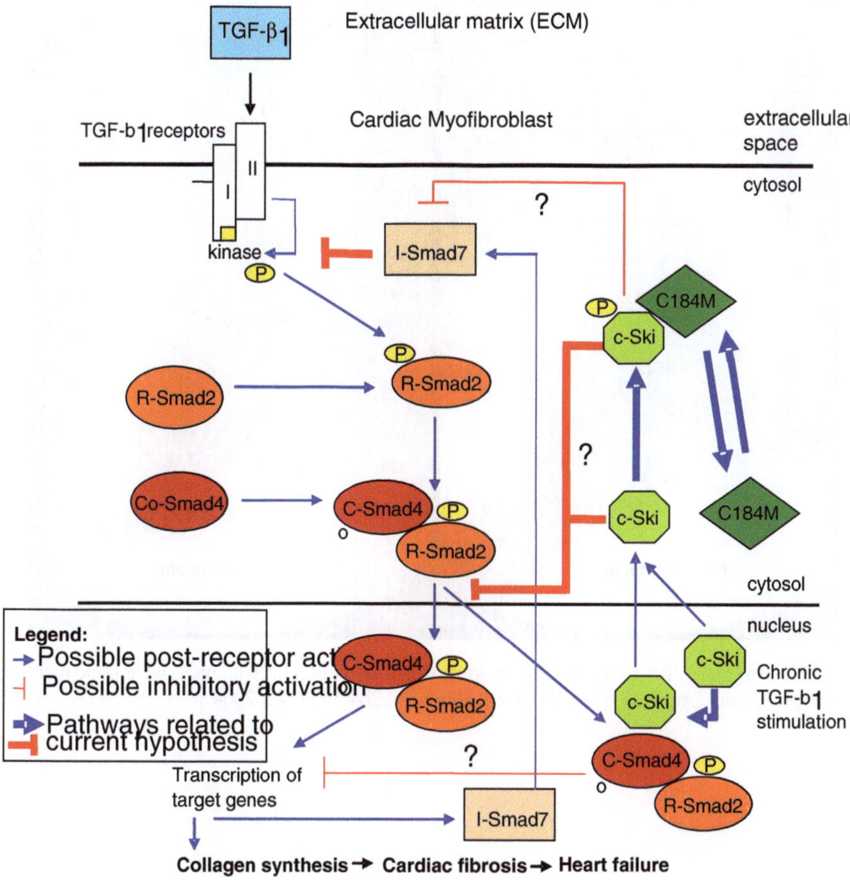

Fig. 13.2 R-Smad signaling and the putative roles of Smad7 and c-Ski. Schematic representation of integrated Smad signaling in cardiac myofibroblasts in response to TGF-β_1 stimulation. Receptor-activated Smad2 (R-Smad2) is phosphorylated by TGF-β_1; this potentiates enhanced synthesis of fibrillar collagens. Inhibition of R-Smads occurs via the expression and activation of Smad7. c-Ski is highly expressed in the post-MI heart, and we suggest that c-Ski, along with its novel binding partner C184M, may function like Smad7 and inhibit R-Smads in cardiac myofibroblasts.

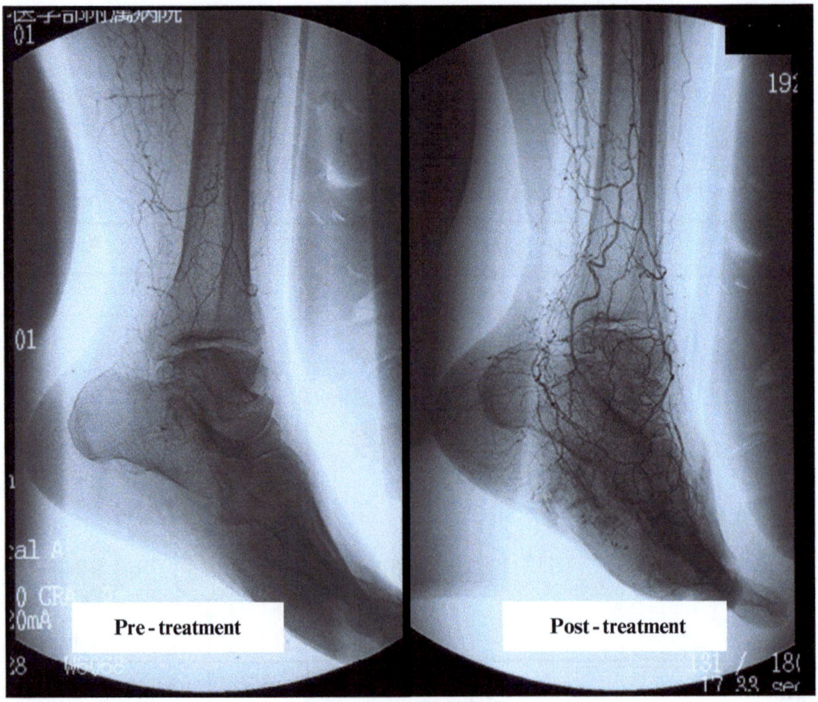

Fig. 15.1 Angiographies of a patient with arteriosclerosis obliterans who underwent bone marrow cell implantation. (Adapted from Mol. Cell. Biochem. 264:143–149, 2004.)

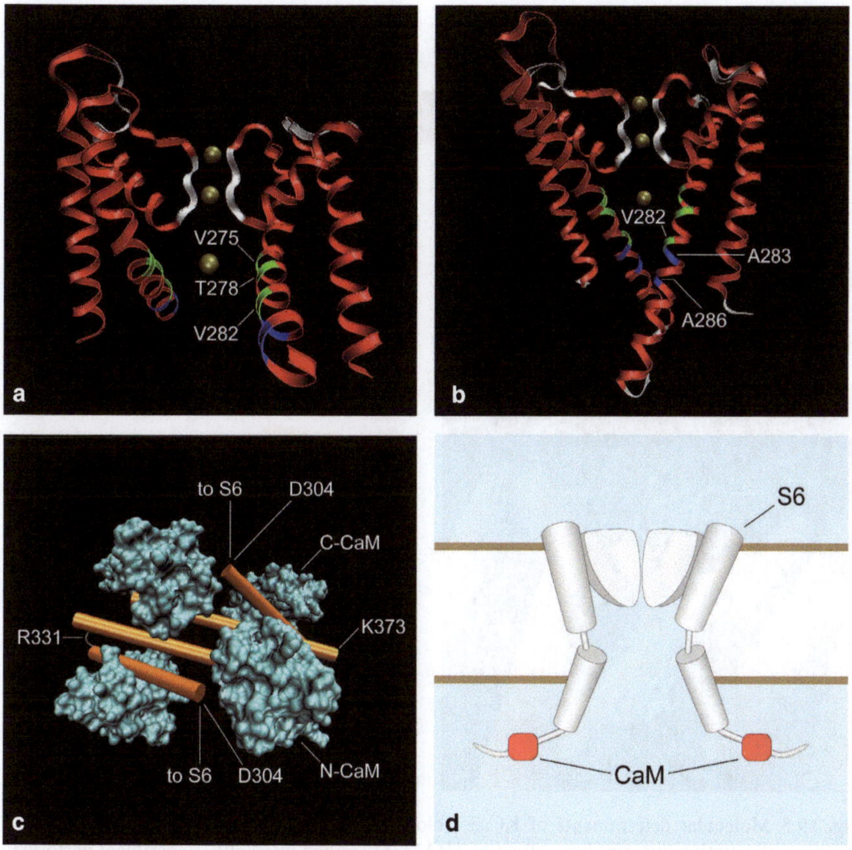

Fig. 19.4 Model structures of the S5–pore–S6 segment of KCa3.1 channel generated by homology modeling. (**A**) Model of the S5–pore–S6 region for the open KCa3.1 conformation computed using the MthK structure as template. In green, residues predicted to be lining the channel pore and to form the channel central cavity (V275, T278, V282). Also illustrated in white is the selectivity filter region with three K^+ ions (in gold). (**B**) Model of the KCa3.1 S5–pore–S6 segment for the closed IKCa3.1 generated using the KcsA potassium channel as template. According to this representation, the bundle-crossing region (V282 in green, A283 and A286 in blue) should form a constriction in the channel pore tight enough to generate an impermeable barrier to K^+ ion flow. (**C**) Model representation of two CaM molecules (in blue) bridging two KCa3.1 C-terminal segments extending from D304 to K373 (orange cylinders) leading to the formation of a dimeric structure between two adjacent KCa3.1 monomers. This bridging process has been proposed to induce a rotation/translation of the S6 transmembrane segments resulting in channel opening. The model was generated by homology modeling using the crystallized KCa2.2–CaM complex (PDB: 1G4Y) as template. Molecular representations using VMD (University of Illinois). (**D**) Schematic representation of the model proposed for the KCa3.1 closed configuration based on SCAM measurements of the KCa3.1 pore structure performed in zero Ca^{2+} conditions. This model predicts that C-terminal end of the S6 segments cannot form a barrier to the diffusion of K^+ ions, but that the control of K^+ ion flow by Ca^{2+} involves a structural change at the level of the selectivity filter or a region close to it.

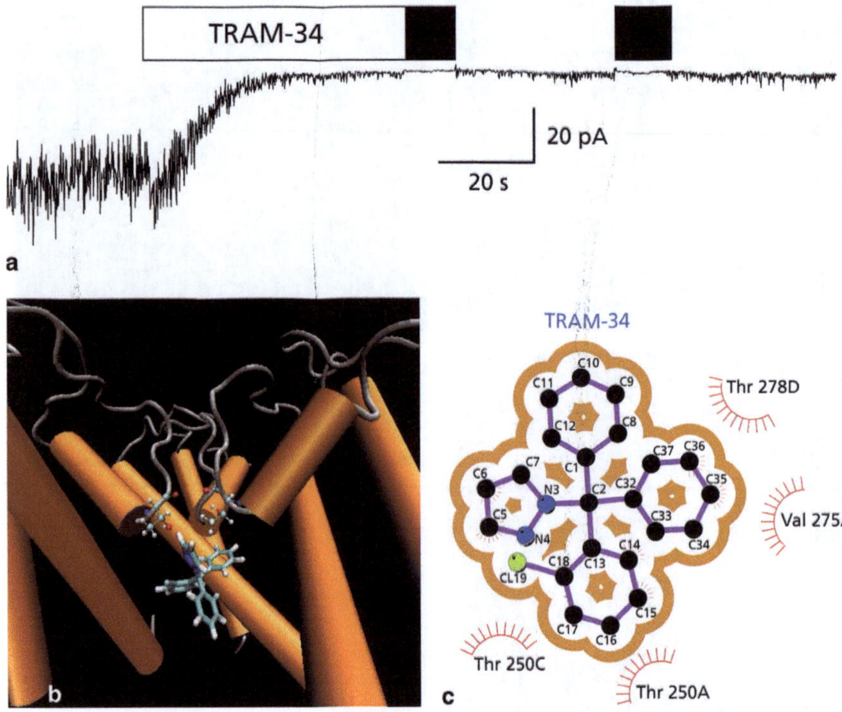

Fig. 19.5 Molecular determinants of KCa3.1 block by TRAM-34. **(A)** Inside-out patch clamp recording illustrating the inhibitory effect of TRAM-34 applied internally. Experiment performed in symmetrical 200 mM K_2SO_4 solutions plus 25 μM free internal Ca^{2+} at an applied pipette potential of 60 mV. Black boxes refer to perfusion with a Ca^{2+}-free solution. **(B)** TRAM-34 molecule trapped in the channel central cavity. The S6 segments of three monomers are presented as cylinders whereas TRAM-34 and T250 residues are shown in a CPK atomic presentation. Molecular representation by VMD (University of Illinois). Docking calculations were performed using the Affinity module of InsightII (Accelrys). No hydrogen bonding or tethered subsets were defined. The ligand was confined to stay within 5 Å of its starting position at the center of the cavity. The TRAM-34 molecule plus all of the KCa3.1 channel residues within a 7.5-Å radius from the center of the cavity were flexible during the calculations. A Monte Carlo search was performed leading to 100 structures with no acceptance filter. We used the nonbonded method: Quartic VDW with coulombic interactions off and VDW scaled to 0.1, flexible ligand with an energy range of 200 and tolerance to 10^6 and 100 steps minimization. **(C)** Schematic representation of the docking results using the LIGPLOT software. This analysis illustrates hydrophobic interactions between the pyrazole moiety of TRAM-34 and the T250 residue of monomer C, while predicting that the stabilization of two of the phenyl rings by strong hydrophobic interactions involves the T250 and V275 residues on monomer A and T278 on the adjacent monomer D.

MIX
Papier aus verantwortungsvollen Quellen
Paper from responsible sources
FSC® C105338

If you have any concerns about our products,
you can contact us on
ProductSafety@springernature.com

In case Publisher is established outside the EU,
the EU authorized representative is:
Springer Nature Customer Service Center GmbH
Europaplatz 3, 69115 Heidelberg, Germany

Printed by Libri Plureos GmbH
in Hamburg, Germany